Nobel Lectures

PHYSICS

2016 – 2020

Nobel Lectures

Including Presentation Speeches
and Laureates' Biographies

Physics

Chemistry

Physiology or Medicine

Economic Sciences

PREFACE

The Nobel Prizes are awarded to some of the most accomplished scientists in the world, and it is exciting to get to know their work and sources of inspiration through first-hand accounts. This volume covers the Physics Prize and the Laureates for the years from 2016 through to 2020.

Physics was the first field of science mentioned in the will of Alfred Nobel from 1895, to be awarded with a yearly Prize "to the one who in the field of physics has made the most important discovery or invention". The will stipulates to share the annual dividends of the Prize fund equally in five parts to "those who during the past year have conferred the greatest benefit to mankind", where the other fields beside physics spelled out by Alfred Nobel are chemistry, physiology or medicine, literature, and peace. The first set of Nobel Prizes was awarded in 1901. (A prize in economic sciences in memory of Alfred Nobel was initiated much later, by the Swedish Central Bank, in 1968.)

One of the most important aspects of the Nobel Prize is its international character, as the will states that the Prize should be awarded to the worthiest, irrespective of nationality. The Laureates of Physics and Chemistry Prizes are decided by the Royal Swedish Academy of Sciences, with the proposals prepared by its respective Nobel Committee. The will of Alfred Nobel is respected to the fullest degree possible, even if the requirement of "during the past year" has been changed in the bylaws to mean that the awarded work should have gained recent actuality, even if performed earlier. Although Alfred Nobel was a very productive inventor with more than 350 patents, the most well-known being that for dynamite in 1867, the best invention of all, in retrospect, seems to be the large, international Prize which bears his name and now is the most prestigious Prize in the world — the Nobel Prize.

There are several things that are not specified in the will, and therefore for the Nobel Prizes in Physics and Chemistry this has been addressed in the bylaws of the Royal Swedish Academy of Sciences. For the Physics Prize, the sentence about "most important discovery or invention" is of course the crucial one, and unchanged. However, the number of

Laureates each year has been limited to at most three. The Academy has so far chosen not to award the Physics Prize to an institution or collaboration, in contrast to the Peace Prize of the Norwegian Parliament where this has been done several times.

We now present a short overview of the theme of this volume, the Physics Laureates between 2016 and 2020.

The Nobel Prize in Physics of 2016 was awarded in one half to David Thouless and the other half shared between Duncan Haldane and Michael Kosterlitz, "for theoretical discoveries of topological phase transitions and topological phases of matter". Thus, this is a Prize rewarding work in theoretical condensed matter physics which employs the use of new physical methods, based on the mathematical field of topology. There are everyday examples of topology when one, for instance, considers the shape of a baked doughnut. It may be stretched and deformed but still a doughnut — the defining property is that there is a hole surrounded by dough. Similarly, a pretzel usually has three holes whereas an ordinary bun has zero holes. Thus, the number of holes defines classes of bakery, and of course also other items of general use — rings, footballs, mugs with one or more handles, etc. The number of holes in these objects is the simplest example of a topological invariant, which stays the same for continuous deformations.

In condensed matter physics, the concept of phases of matter has been developing with time, starting with gas, liquid and solid, to which later was added superconductors and superfluids. The three Laureates of this year have made remarkable discoveries which add phases of different topologies to this list.

As is sometimes common in physics, it takes time for the physics community to digest and fully understand new concepts such as in this case topology — where experimental verification is a crucial demand for scientific acceptance. In was thus already in the beginning of the 1970s that Kosterlitz and Thouless theoretically described a new type of phase transition in two-dimensional systems where the topology of defects was crucial. The theory was applicable to certain magnetic materials and to films of superconductors and superfluids. A decade later Thouless and Haldane found methods to describe phases of matter beyond their symmetry properties, employing topological invariants. In a subsequent paper, Thouless with collaborators explained the quantization of the Hall conductance (first discovered in its simplest form by Klaus von Klitzing, in 1980, Nobel Laureate in 1985) in two-dimensional electron gases using topology. Finally, Haldane derived a theory for chains of spin using topology in a clever way. He could show that, surprisingly, integer and half-integer spin chains behave differently, something that was later confirmed by experiments.

In recent years, the whole field has been flowering with a host of developments such as topological band theory, where theoretical insights have been found which opens for applications in electronics and quantum information science. One spectacular area, with later experimental discoveries, is that of topological insulators in two and three dimensions.

At the time of the Nobel award ceremonies in December 2016, David Thouless was, although present in Stockholm, not fit to give a traditional lecture, but Duncan Haldane and Michael Kosterlitz included his achievements in a very respectful manner in their lectures. Also, David Thouless with wife Margaret and daughter Helen provided a very interesting biography, included in this volume. This also contains a list of those papers which David Thouless himself judged to be his most important work.

Sometimes completely new methods of studying our universe are discovered. The Nobel Prize in Physics of 2017 was indeed awarded for such a spectacular discovery, with one half awarded to Rainer Weiss and the other half shared between Barry Barish and Kip Thorne "for decisive contributions to the LIGO detector and the observation of gravitational waves". This opened a new era of gravitational wave astrophysics, which has flourished since then, with also new detectors planned globally.

The field of gravitational waves and how to detect them directly had long looked like a dream, with estimates of the magnitude of the gravitational strain on Earth even by a dramatic event such as the gravitational merger of two multi-solar mass black holes realistically being as small as one part in 10^{21}. (The indirect detection of gravitational waves by accurate timing of the reduction of the orbital period of a binary pulsar, was on the other hand awarded a Nobel Prize in Physics in 1993 to Russell A. Hulse and Joseph H. Taylor Jr.)

The Laser Interferometer Gravitational-Wave Observatory, LIGO, developed from a medium scale physics experiment to one involving more than a thousand scientists by the time gravitational waves were finally observed in September 2015. One of the three most important LIGO scientists, as judged by the Nobel Committee for Physics, was Barry Barish, who had experience from particle physics to lead large projects, and therefore could take LIGO through various hurdles to success. The main early developments of the ingenious laser technology needed for the impressive interferometric accuracy were done by Rainer Weiss, who besides technological skills also had a solid background in general relativity applied to cosmology. Major theoretical developments were further elegantly performed by Kip Thorne, who among other contributions used advanced computer tools to solve the complicated non-linear equations of general relativity numerically.

General relativity (GR), developed by Albert Einstein around 1915, is a fundamental theory which connects matter, space, and time. Several of the main predictions of GR had been verified during the first decades of the 1900s, such as the perihelion precession of Mercury, and the bending of light rays near the Sun. However, as space-time is dynamical in this theory, small fluctuations of relative distances should appear, caused by violent processes deforming the local space-time such as near the merger of two stars, or black holes. This was a both theoretically and experimentally challenging effect to search for — Einstein himself was uncertain about the reality of gravitational waves emanating from such events. In the period 1970 to 1990 the necessary theoretical developments were completed, and it was shown that indeed there should be physical effects of gravitational waves, in the stretching and contraction of distances between masses fixed to the Earth.

LIGO was developed as an extremely accurate interferometer, where minute changes in distances can be measured through interference of laser waves traveling in orthogonal directions. This is achieved through the sophisticated use of lasers, with a mirror system like a detector used already by A. A. Michelson (Nobel Prize in Physics, 1907), and with seismically extremely well-shielded mirrors and detectors. To exclude local disturbances, causing the instrument to vibrate, the observatory consists of two identical detectors in the American continent, separated by about 3000 km. In the morning on September 14, 2015, both detectors registered almost identical signals from the passing wave, with less than 7 milliseconds delay. This was consistent with a source on the Southern sky — the first direct detection of gravitational waves, caused by the merger of two multi-solar mass black holes.

The laser, "light amplification through stimulated emission of radiation", is an extremely useful tool in today's physics research. This was witnessed, e.g., in the work which led to the Physics Nobel Prize of 2017, where the detection of extremely tiny signals from gravitational waves in the LIGO detector were possible thanks to the use of interference of stable and accurate laser beams. Of course, in everyday life we also encounter various applications of lasers — in medicine, communications, precision measurements, bar code readers, laser pointers, etc.

The first Nobel Prize in Physics, which included the laser, was awarded in 1964 to Charles H. Townes, Nicolay G. Basov, and Aleksandr M. Prokhorov. In the decades following this basic discovery, lasers have been the tools for several Nobel Prizes in Physics, e.g., Bloembergen and Schawlow in 1981; Steven Chu, Claude Cohen-Tannoudji, and William D. Phillips in 1997; Roy J. Glauber, John L. Hall, and Theodor W. Hänsch in 2005; and Serge Haroche and David J. Wineland in 2012.

Continuing this remarkable success of investigating basic processes in Nature, using lasers, the Nobel Prize of 2018 was awarded to Arthur Ashkin, Gérard Mourou, and Donna Strickland for important new developments and groundbreaking inventions in the field of laser physics. Arthur Ashkin is the father of the "optical tweezers", where intense laser light is used in biological systems to give radiation pressure large enough to mechanically move microscopic particles in given directions. By using a lens for focusing the light, he could in fact lift particles out of biological tissues, as if they were grabbed by tweezers, but without touching them. He was also able to study various types of activity within cells such as the tiny molecular "motors" that are important components of biological cells.

An important property of laser light is given by the time structure of laser pulses, from continuous to extremely short, where the short pulses awarded "snapshots" of processes which cannot otherwise be studied. Gérard Mourou and Donna Strickland are being rewarded for an invention in the field of intensive, ultra-short laser pulses — chirped pulse amplification, CPA, by which one may create extremely intense, short laser pulses. The road to create more intense laser pulses had been investigated since the 1960s when the first laser was built. However, this turned out to be difficult, as the high intensity of the laser light was detrimental to the laser itself. By using Chirped Pulse Amplification, Mourou and Strickland stretched out the laser pulse in time, whereby its intensity was reduced such that it could be amplified. Then, compressing the length of the pulse again to its original value, the intensity was correspondingly increased. This method was transformative to the field of high intensity lasers and could be used in even small laboratories around the world. The production of shorter pulses has moved towards the attosecond level (one attosecond equals one billionth of a billionth of a second). This paves the way for studying the movements of electrons in atoms and molecules.

The Nobel Prize in Physics for 2019 went to three scientists, with the common motivation "for contributions to our understanding of the evolution of the universe and Earth's place in the cosmos" with a prize share of one half to James Peebles "for theoretical discoveries in physical cosmology", and Michel Mayor and Didier Queloz sharing the other half "for the discovery of an exoplanet orbiting a solar-type star".

The Physics Prize was awarded to James Peebles for deep theoretical discoveries about the structure and evolution of the universe, and to Michel Mayor and Didier Queloz for the first observation of an extra-solar planet, an exoplanet, orbiting another star of solar type, some 50 lightyears from our solar system. Through this Prize, two essential new pieces of information about our universe, in the past and the present, were awarded.

James Peebles succeeded in developing the physical basis for cosmology through studies of the cosmic microwave background radiation (detected by A. Penzias and R. Wilson in 1964, Nobel Prize in Physics, 1978). He worked out the effects of the dynamical interaction of matter and radiation in the hot big bang cosmology. He found that the baryons would oscillate and create "baryonic acoustic oscillations" (BAO) which were subsequently detected in the distribution of galaxies. He also, with graduate student Jer Yu, showed how these baryonic oscillations would produce a series of roughly equally spaced bumps, at the wavelengths favored by the decoupling of matter and radiation. This resulted in the peaks in the angular correlation function which were detected by the COBE satellite (J. Mather and G. Smoot, Nobel Prize in Physics, 2006). Still, there were many pieces of the puzzle of the Universe missing, and Peebles did important early work relating to the present "Standard Model" of the universe, the so-called ΛCDM model. For example, he introduced in 1982 a model which contains other slowly moving matter, "cold" nonbaryonic and non-visible dark matter, which is only interacting gravitationally. This Cold Dark Matter (CDM) component seems in fact to dominate the energy density of matter in the universe. The other component, Λ, is a bit more mysterious. It may just be, as Peebles assumed in 1984, Einstein's cosmological constant, which seems to give an even larger contribution at the present epoch than CDM to the energy density of the Universe. Peebles shows in his Nobel Lecture how the ΛCDM model was put together piece by piece. And he remains surprised and impressed at how well ΛCDM passes ever more demanding tests but continues to hope that a still better and more complete theory will be found.

The other half of the 2019 Physics Prize also concerns astrophysics, but at much smaller scale — it brings us to the second part of the common motivation to the Prize, "the Earth's place in the cosmos", namely the quest for planets orbiting other stars Sun, exoplanets.

The first observation of an exoplanet, orbiting the star 51 Pegasi, was published in 1995 by Michel Mayor and Didier Queloz, using the spectrograph ELODIE which had started in scientific operation in 1994 using a telescope at the French Observatoire de Haute Provence. The discovery was made using very precise radial velocity measurements caused by the "wiggling" of the position of the star, as the star and the planet are orbiting around their common center of mass. Since the star is much heavier than the planet, the amplitude of the radial velocity of the star is quite small, even for planets of Jupiter size or bigger. For 51 Pegasus, Mayor and Queloz measured with the help of ingenious methods of cross-correlation of many different spectral lines an amplitude of around 60 m/s and a period of only 4.2 days. The estimated mass of the planet, now named 51 Pegasi b, was around half that of Jupiter, meaning that this was an almost Jupiter-sized mass planet orbiting very close to the star. This was indeed unexpected in the planet formation theories of the time, leading to the development of new scenarios where planets migrate from their formation distance to the star to even much closer distances. The discovery of Mayor and Queloz was soon followed by many findings of exoplanets of similar type, and since then the field of exoplanets has really exploded with currently many thousands discovered. The Nobel Lectures by Mayor and Queloz contain, besides a detailed account of the discovery of the first exoplanet, elegant descriptions of the development of the field up to 2019.

The Nobel Prize in Physics for 2020 was awarded with one half to Roger Penrose "for the discovery that black hole formation is a robust prediction of the general theory of relativity" and with Reinhard Genzel and Andrea Ghez sharing the other half "for the discovery of a supermassive compact object at the center of our galaxy".

The Prize was thus again in the field of astrophysics, but now dealing with some of the most interesting and peculiar objects of the universe: black holes. Quoting Academician Ulf Danielsson in his presentation speech "Black holes are bodies with gravity so strong that not even light can escape. To create a black hole, the earth would have to be squeezed down to the size of a pea and the sun compressed into a sphere comparable in size to the central parts of Stockholm."

Black holes were predicted in simple, symmetric solutions of Albert Einstein's general theory of relativity, such as the solution found by Karl Schwarzschild already in 1916, only two months after Einstein published his theory. The spherically symmetric black hole solution of Schwarzschild was for a very simple object with its mass in a point and appeared to have a singularity at the origin and a black hole "horizon" bounding the region from which no light ray could escape. It was unclear whether realistic astrophysical matter would behave in the same way — having the same kind of space-time singularity and a black hole horizon. Roger Penrose proved, using clever mathematical methods to a large part invented by himself, that this is indeed the case. No matter how one perturbs the solution away from the spherical symmetry assumed by Schwarzschild, a space-time singularity would persist. Penrose also showed using his concept of a "trapped surface" that gravitational collapse cannot be stopped after a trapped surface of arbitrary symmetry is formed. Penrose's result marked a breakthrough, as it was the first post-Einsteinian result, showing that the existence of black holes is inevitable in general relativity. It may be remarked that Penrose also has made many other important contributions to mathematics, some relevant to physics and chemistry (see, e.g., the Penrose tiling of the plane with five-fold symmetry, shown in Dan Schechtman's Nobel Lecture in Chemistry, 2011).

The existence of black holes in astrophysics had been suspected since the discovery in the early 1960s of extragalactic "quasars" containing super-massive objects at the center of very massive galaxies. Also, X-ray studies of binaries in the Milky Way (Riccardo Giacconi et al. 1962, see Giacconi's Nobel Lecture, 2002), implied masses higher than those expected for neutron stars, and are thus black hole candidates. So, evidence has been rather convincing, but indirect, that black holes are plentiful in the Universe. In 1971, Lynden-Bell and Martin Rees argued for the existence of a supermassive black hole in the Galactic center and proposed key observations to explore the nature of the compact object.

The Nobel Laureates of 2020 Reinhold Genzel and Andrea Ghez have provided the so-far most compelling arguments that a several million solar-mass black hole indeed exists also at the center of our own, Milky Way, galaxy. They were leading two independent observational teams, one by Genzel at the Max Planck Institute for Extraterrestrial Physics and the other by Ghez at the University of California, Los Angeles. Both have in impressive detail monitored the motions of stars orbiting very close to the Galactic center for dozens of years. Genzel's group used telescopes in Chile operated by the European Southern Observatory, while Ghez and her colleagues used the Keck Observatory in Hawaii. Excellent spatial resolution has allowed following the motion of individual stars, using near-infrared wavelengths. At these long wavelengths the mean-free path of photons is large, allowing observations of high quality. In the contribution of Genzel to this volume, a detailed history of the field — more than 40 years old — can be found, as well as a description of how the results point uniquely to the existence of a 4 million solar-mass black hole at the Milky Way center. (Ghez has not provided any written material.)

Lars Bergström

CONTENTS

Physics 2016

one half to

David J. Thouless

and the other half to

F. Duncan M. Haldane and J. Michael Kosterlitz

"for theoretical discoveries of topological phase transitions and topological phases of matter"

The Nobel Prize in Physics, 2016

Presentation speech by Professor Thors Hans Hansson, Member of the Royal Swedish Academy of Sciences; Member of the Nobel Committee for Physics, 10 December 2016.

Your Majesties, Your Royal Highnesses, Esteemed Nobel Laureates, Dear Colleagues in the International Community of Scholars, Ladies and Gentlemen,

Physics is a human activity, and like all such activities it evolves in the context of historical, geographic, and linguistic settings. Nonetheless, the central results of physics – natural laws – transcend these barriers. They are the same today as they were at inception, and they will also be valid in the future. They are the same here on earth as on the exotic planets that we now know are orbiting distant stars. The laws of physics are universal, and they are formulated in the language of mathematics.

This extraordinary universality is closely connected to fundamental properties of space, time and matter, which are reflected in the form of symmetries in the laws of physics. A branch of mathematics used to describe symmetries is called group theory and was substantially developed during the 19th century. Only later did this mathematics turn out to be an indispensable tool in the new quantum theory that revolutionised physics during the 20th century.

The role of group theory in quantum mechanics is but one example of how major scientific breakthroughs often require a new language. And since the language of physics is mathematical, breakthroughs in physics often require new mathematics. Newton developed the theory of derivatives and integrals at the same time as his mechanics, and Einstein formulated his general theory of relativity in terms of the then contemporary mathematics of curved spaces, which was unfamiliar to physicists.

Mathematical language is abstract, but it has two major virtues. A mathematically formulated theory is quantitative and can therefore often be used to in detail, and with high precision, explain and predict the results of experiments. Mathematics is deductive, so mathematical arguments can be used to predict the results of future experiments – the discovery of the Higgs particle is a striking example. Deduction is indeed an important tool in all the sciences, but it excels in theoretical physics.

Many philosophers, physicists and mathematicians have marvelled at the astonishing ability of mathematics to describe nature. In a famous essay, Eugene Wigner called this "the unreasonable effectiveness of mathematics".

This year's Nobel Prize in Physics rewards contributions to theoretical physics that are examples of precisely this unreasonable effectiveness. By using new concepts taken from the branch of mathematics called topology, the 2016 Laureates did two things. First, they explained striking observations and second they made theoretical discoveries, that is, they predicted new phenomena that were observed later. Topology describes the robust properties of objects. Topologically, an egg and a football belong to the same category – the category of three-dimensional objects without holes. But a doughnut, along with a wedding ring, belongs to the category of objects with one hole. The number of holes, which is always an integer, is an example of a topological invariant.

Quantum mechanics was born as a theory of microscopic objects: individual atoms and molecules. But it also soon came to be used in order to understand the usual phases of matter: gases, liquids and solids. Superfluids and superconductors were also added to these. Here, too, symmetry concepts proved extremely important. By studying symmetry properties, Lev Landau was able to classify possible phases of matter and describe the mechanisms of the transitions between them.

This year's Laureates showed that Landau's classification of the phases of matter was incomplete – there are additional so-called "topological phases" with characteristic values for different topological invariants. With the aid of topology they also showed how one can explain observed transitions between different phases of matter that should not be possible according to earlier theory.

It often takes time before new concepts, and new ways of thinking, have an impact. But during the past decade, research in the "topological phases of matter" has virtually exploded. Today's young physicists work with topological concepts as self-confidently and skilfully as earlier generations have done with derivatives, symmetries and geometry.

Not only have this year's Laureates made important special discoveries but in addition – and perhaps most importantly – they have set the stage for a new way of describing matter.

They have given us a rich mathematical language with deep and beautiful abstract concepts. But in the natural sciences, beauty is not enough – mathematics is full of beautiful results. Truth is also necessary. This requires experiments and measurements. The theory of topological matter has passed this test with distinction. It combines truth with beauty. This is theoretical physics at its best.

Professor Thouless, Professor Haldane, Professor Kosterlitz:

You have been awarded the Nobel Prize in Physics for "theoretical discoveries of topological phase transitions and topological phases of matter". On behalf of the Royal Swedish Academy of Sciences it is my honour and my pleasure to convey to you the warmest congratulations. I now ask you to step forward to receive your Nobel Prizes from the hands of His Majesty the King.

David J. Thouless. © Nobel Prize Outreach AB. Photo: A. Mahmoud

David J. Thouless*

Biography

THE NAME

The name Thouless is very rare. Fewer than 150 people with the name live in Britain, almost all of whom are connected to Norwich. This is because it is a relatively new spelling of an old name spelt variously Thules, Thewless, Thewlis etc. Five generations before David, his ancestor John was born Thules. When John's son, James, was born, both parents were illiterate and the clergyman filling out the baptism certificate wrote what he thought he heard— 'Thouless.' It seems that almost everyone with this particular spelling of the name is descended from James.

MOTHER'S FAMILY BACKGROUND

David's mother was born Ella Grafton Gorton in 1898. She changed her name to Priscilla when she was studying in Italy, where she found that having a name that translated to "She" in Italian was inconvenient. The family name originates in Gorton, a suburb of Manchester. Her father's family had been Church of England clergymen back into the 1600s and continued this tradition through her generation. Her grandfather, father, three of her brothers and two of her brothers-in-law were all clergymen. The most prominent of these was her brother Neville Gorton, the Bishop of Coventry who was deeply involved in building the new cathedral after the war.

Priscilla was the sixth of seven children. She was taught by a governess who did not like mathematics and influenced her pupil to feel the same way. After her father died, she went to Altrincham High School and then got a scholarship to

* by David, Margaret and Helen Thouless

Manchester University. She gained a BA and MA in literature and taught English at Manchester until David's older sister Susan was born in 1925.

David's generation, which included 14 first cousins on his mother's side, were not involved with the church or science. The most notable of his cousins was Assheton Gorton, who was an artistic director of a number of well known films such as *Blow Up* and *The French Lieutenant's Woman*.

FATHER'S FAMILY BACKGROUND AND SCIENTIFIC INTERESTS

David's father Robert Henry Thouless was born in Norwich in 1894. Robert's father Henry James Thouless married Maud Harper from Devon who was studying at a Teacher Training College in Norwich. Henry James was a company secretary at Barnards, a Norwich engineering firm. However, his passion in life was natural history, with a particular interest in insects, specifically moths. He served a term as the President of the Norfolk and Norwich Natural History Society and had a bungalow on the edge of a marsh in Wroxham, which was ideal for finding insects. He bequeathed a collection of insects he had collected, which included two named after himself, to the museum in Norwich Castle.

Robert had two sisters, Sybil and Margaret. Sybil became a nun and taught school in the order of Notre Dame. The younger sister, Margaret, also became a teacher. She studied at Oxford University before they allowed women to take degrees. Once women were allowed to take degrees, she returned to study Latin for a year, as this was a requirement for graduation. Margaret had wanted to study science, but she was considered too frail to do the lab work required. Instead, she taught English literature and foreign languages at St. Mary's Calne, a girls' private boarding school.

David's father Robert attended King Edward VI School in Norwich. In 1912 he went as a scholar to Corpus Christi College, Cambridge. In 1914 he was awarded a bachelor's degree in natural sciences. He joined the Royal Engineers as a signaller. After a couple of years, in 1917 he went to the Salonika Front, from which by his account he was lucky to have come home alive. He became a lifelong pacifist but it did not stop him joining the home guard to defend his own town, Cambridge, during the Second World War.

After the First World War, Robert returned to Cambridge and did a PhD in psychology. He then became a lecturer at Manchester University before moving to Scotland to start the Psychology Department at the University of Glasgow. While at Glasgow, he did his most important work on how an object is perceived, introducing the term "phenomenal regression" in 1931. In the 1930s this was a very unfashionable line of research, and it did not enter mainstream psychology until the 1950s. Robert was offered the resources to study this phenomenon in Australia after he had retired, but he replied that he did not have the strength and brain power he had had when he wanted to study the topic 40 years earlier. David has followed his father's originality of thought, which sometimes came before the rest of the world is ready to engage in a topic.

Robert Thouless was also known for his radio programmes on how to critically analyse flaws in reasoning and arguments, which he later turned into the book *Straight and Crooked Thinking*. This is known as *How to Think Straight* in the US. It has been a required textbook for many generations of students of rhetoric. His grandson Christopher Thouless has revised the last two editions, so it has been in print for over 85 years.

Later in life, Robert concentrated his research efforts on studies of the paranormal. He was elected President of the Society for Psychical Research in 1942. Although a frequent result of his painstaking investigations was the detection of cheating in apparent cases of psychic powers, he continued to believe in the possible existence of such abilities.

DAVID'S EDUCATION WITH EMPHASIS ON MATHEMATICS

Even as a four year old, David was precocious in mathematics. One day, after discussing with his father how far counting goes, David decided to take the experimental approach. His family was bored by the time he reached 500 and even more bored by the time he reached the second thousand.

Just before his fifth birthday David and his sister were evacuated to his grandmother's house in Devon at the outbreak of the Second World War. While there, David taught himself to read and write, with the help of his grandmother's housekeeper.

As soon as it appeared that a German invasion would not happen immediately, David and Susan returned to Cambridge and David started school. At this point David stopped asking how to spell words and started thinking about arithmetic. With the aid of a simple abacus he worked out problems for himself. He worked out the 2 times table, working out what 2 times 27 was before he got bored with the project. At age 7 he set himself the task of working out how many seconds there are in a year. From age seven through eight, he spent two years as the only boy in the top class of the school, with the rest of the class being large 9- to 11-year-old girls, which was not a situation he enjoyed at that age. Much to his relief, his next school St. Faith's was a boys' prep school.

David's father had a big influence on his intellectual development. "When I was 5 my father taught me to play chess, at which I slowly acquired competence

Figure 1. David Thouless, three years old.

but not brilliance. I think I was a teenager before I had a good chance of beating him at chess, but I was generally much better than my friends." In fact David continued to play chess until he was in graduate school, when he felt the mental effort was too similar to physics. He and his friend played chess in their heads on long Territorial Army marches.

> My formal education in science was close to non-existent until I was nearly fourteen. I can remember one young man trying to teach chemistry. Half the boys knew what he was talking about, but I had no idea why a chemical should go in one particular direction, rather than to any other end-product that had the same number of each atomic species. Fortunately my father was always willing and able to fill such gaps in my understanding.
>
> I cannot enumerate all the things I learned from my father. He certainly told me about Wegener's theory of continental drift, which was very unfashionable at that time. His enthusiasm for probabilistic reasoning was something he shared with me quite early; he was an early follower of Cyril Burt in stressing the importance of careful statistical analysis of psychological tests. He showed me

> how base 2 arithmetic could be used to win the game of Nim. I saw, but never really absorbed, the Boolean notation he used to solve problems in logic.

David met interesting visitors his father invited to the house. "A frequent visitor to the house in the early years of the war was the philosopher Ludwig Wittgenstein. My father had been to his lectures before the war, and there was an annotated copy of the *Blue Book* among his papers when my father died. I also found notes on a series of conversations on philosophical and scientific matters between Wittgenstein, my father and Cyril Waddington. These were published in 2003 by James Klagge and Alfred Nordmann in the book *Ludwig Wittgenstein: Public and Private Occasions*."

Winchester

St Faith's School in Cambridge encouraged David to compete for a scholarship to Winchester College. He was not sure he wanted to go, so his parents made alternative arrangements in case it should prove too stressful for him. He won the top scholarship of the year with an outstanding mathematics result and very good English and Latin. David had not studied Greek, so he did not attempt that paper. He was the first ever student to come top in the scholarship exam having only done three out of the four papers.

> It was also decided that I should take the School Certificate at the end of the first year, despite the fact that I would still be thirteen, because in preparation for the introduction of O-levels in 1950, there would be a minimum age for School Certificate in 1949.
>
> As a result of this I took the exams in English Language, English Literature, Mathematics, Further Mathematics, History (Ancient Greece), Latin, Greek and Divinity (including St. Luke in Greek). I was still struggling with Greek, quite enjoying the struggle, and got Credit in Greek, but got Very Good in all the other subjects. I had no official science background, nor any modern foreign language qualification. I did not take any serious external examination in foreign languages until I entered the Cornell Graduate School on my 22nd birthday.

David got an excellent education in science at Winchester. "In all subjects there was a lot of emphasis on private study and assignments, and we spent

Figure 2. Left to right, Ronald Peierls, Marvin Litvak and David Thouless, Cornell graduate students, 1958.

relatively little time in class, perhaps less than eighteen hours a week, over about 36 weeks a year." Time was found for a broad rounded education in addition to science. For example, "One of the joys of my second year was that the form-master was Harold Walker, the head of the history department. His one-term course on American history left enough in my memory that I found no need to revise when I took the test for US citizenship. His scholarly but sceptical teaching of divinity was challenging and refreshing, particularly to someone like me who took religion rather too seriously."

During David's time at Winchester his termly reports did comment on his mathematical ability but expended far more space on his untidy work and handwriting. This may have led to his excellent habit of developing his equations on scrap paper and when he was satisfied with them copying them into hard backed numbered page note books. These have now been deposited in the archives of the Royal Society.

Cambridge University

David was fortunate that he did not get the scholarship he wanted to Trinity College, but did get one to Trinity Hall next door. Trinity Hall was a much smaller college, better suited to his introverted personality. He made a number of really good friends while there, some of whom he sees to this day. He became an honorary fellow of Trinity Hall in 2014 and enjoys participating in some of their activities.

Describing his undergraduate experience, he said:

> I knew the Senior Tutor Charles Crawley well, as his son John was and is a good friend of mine. My own Tutor was the distinguished historian and theologian Owen Chadwick, and the other Tutor was Shaun Wylie, who supervised me in mathematics, and was probably the single most important influence on my academic development as an undergraduate. None of us were supposed to know the significance of Bletchley during the war and of Cheltenham later, but somehow or other, from various different sources, I had picked up a fair idea of what Shaun and his colleagues had been up to in those places.

The other piece of good fortune was that Trinity Hall did allow David to defer military service until after he graduated, which Trinity College would not have done. This led into his later studies with Hans Bethe (Nobel Laureate, 1967). In June 1955, the Cavendish Professor Mott (later Sir Nevill) called David into his

Figure 3. Hans Bethe and David Thouless on a picnic, May 1958.

office and asked him what he was doing next. David said that he was going to do his military service, as he did not wish to defer until after graduate school because he did not want to do military research, which would have been the likely outcome once he had a graduate degree. Mott told him that he could continue to get a deferral as long as he continued his scientific work and that the requirement for compulsory military service was likely to be discontinued. This led to an interesting situation in which no one had time to take David on as a doctoral student but the Cavendish had money for a stipend for him. Professor Mott suggested that he work with Hans Bethe, who was on sabbatical at the Cavendish Laboratory, Cambridge University. After a year, Hans offered David the opportunity to go to Ithaca with him and study for a Cornell PhD, which David accepted.

Cornell University

David obtained a Fulbright Foundation scholarship, which paid for his Atlantic trip on the Queen Elizabeth ocean liner and a train trip on the Lehigh Valley railroad to Ithaca. He travelled with Ronnie Peierls, son of Professor Peierls (later Sir Rudolph) of Birmingham University, who was also going to Cornell to study with Hans Bethe. There were various students sitting at their dining table on the liner who were going to a variety of different universities. One of these kindly sent congratulations to David after the award of the Nobel Prize; even though they had not seen or contacted each other in the 60 years since the journey.

In David's first week at Cornell he passed his modern language exams in French and German. He also passed the departmental qualifying examination with "flying colours." He was particularly pleased as there were no required courses for physics graduate students in Cambridge and "Hans Bethe had been complaining about the poor knowledge of general physics shown by PhD students he had met in England."

Cornell was unusual in having no graduate school course requirements, but the Physics Department required all its students to do two semesters of an experimental physics course. "I do not think any of my experiments came out right, but apparently the explanations I gave of what had gone wrong and what I needed to do about it were sufficiently convincing that I got the highest grade in the course, and was excused taking a second semester of experimental physics."

While at Cornell, David met Margaret Scrase, a biology student in the College of Agriculture. They married and have now been together for 60 years.

Mathematician Mark Kac was on David's doctoral committee and David said that "Getting to know Kac and to learn from him was one of the unexpected benefits of going to Cornell. I treasured his explanation of the difference between a physicist and a mathematician: that a physicist was interested in the simple properties of complicated systems, but a mathematician was interested in the complicated properties of simple systems."

In the 1960s (according to John Rehr) a story went around the Cornell physics department that David asked Hans Bethe for a topic for his PhD and then showed up two years later with a completed thesis. The fact is that Bethe was a scientific advisor to President Eisenhower. He travelled back and forth by train from Ithaca to Washington D.C. every week, so it was hard for them to have regular meetings. David's remark about this topic was, "if I had a good talk with him once a month, he left me with enough to think about for the next three months." However there was some truth to the rumour. David did produce a finished thesis and ask for a year's postdoctoral fellowship so that Margaret could finish her degree at Cornell. If David had shown Hans the thesis earlier the answer would have been yes, but as it was all of Hans' money was committed so David had to look elsewhere. Instead he obtained a postdoctoral fellowship for a year

Figure 4. David Thouless' parents: Robert and Priscilla Thouless, early 1960s.

at the Lawrence Radiation Laboratory in Berkeley, which allowed Margaret to complete her undergraduate degree.

POSTDOCTORAL FELLOWSHIPS

Cornell University had been such a marvellous experience for David and Margaret that whatever came next was bound to be a disappointment. David did not appear to have a preceptor in the Radiation Lab, but he did publish two papers and taught a course on atomic physics on the Berkeley campus which went quite well. Living in Berkeley was a pleasure and David and Margaret's explorations of San Francisco, the surrounding hills and beaches and the Sierra Nevada would not have happened if either of them had been taking work more seriously.

David moved to the Department of Mathematical Physics at Birmingham University for two more years of postdoctoral research. He worked under Rudolf Peierls from 1959 to 1961. David was working very hard because there were a great many interesting physicists in Professor Peierls' department in Birmingham University. David recollects, "I was probably more interactive with my colleagues during these two postdoctoral years in Birmingham than I was at any other period in my professional career." David and Margaret's two sons, Michael and Christopher, were born during this time.

David spent the summers of 1960 and 1961 at the Niels Bohr Institute and Nordita in Copenhagen. These were a pleasure for all concerned and helped

Figure 5. David Thouless' children: Michael, Christopher and Helen, December 1972.

Figure 6. Novosibirsk Conference March 1965. Thouless top right. Alexei Abrikosov fourth from left.

with the parlous financial state that resulted from trying to support a family on a British postdoctoral salary.

CAMBRIDGE: CHURCHILL COLLEGE FELLOW, UNIVERSITY LECTURER

David went to Churchill College as a Director of Studies and a Fellow in 1961, the first year the college took undergraduates. He also became a lecturer in the Department of Mathematics and Theoretical Physics. He learned a lot, particularly about teaching undergraduates, but he said there was less to show researchwise for the four years in Cambridge than in his previous positions. This had something to do with the intensive Cambridge 8 week term. He would get exhausted and spend much of the vacation recovering from respiratory diseases rather than doing research. His health only improved after the family moved into a centrally heated house in Birmingham in 1966.

In March 1965, David went to an interesting conference in Novosibirsk. Russia had temporarily opened up and there were no Intourist guides in Novosibirsk. This allowed the Russian physicists to talk freely with the Western physicists.

In David's own words:

> In the early spring of 1965 the most memorable scientific meeting I have ever attended took place. This was a conference on many-body

problems, which was held in Akademgorodok, about 20 km south of Novosibirsk. The town was the centre of work on nuclear physics, and had been closed to outsiders until that year. Teachers at the local English language school had been invited to translate for us, but Bogoliubov told them they were not wanted, because it was better for the Russians to practice their bad English rather than to rely on teachers with good English and no understanding of physics. The teachers sat in on the sessions and in the intervals tried to talk to the few of us who spoke English from the right side of the Atlantic.

We were able to go for walks in small groups, unobserved by security people. We met with people like Abrikosov, Gorkov and Dzyaloshinsky, whose book was making my own book out of date. This meeting was the first occasion on which I met Vitaly Ginzburg, who later spent time in Cambridge, and came, with his wife, to visit us in Birmingham. I also got to know A. B. Migdal and V. M. Galitskii, who were the authors of the Paper on Green's function that had been so in influential on my work at Cornell in 1958. An outing led, I think, by Migdal, was my first experience of cross-country skiing, in bright sunshine, but with crisp spring snow. The only unfortunate thing about this trip was that I had a bad cough, perhaps the remains of a pneumonia attack I had during the winter. I flew back as far as Sverdlovsk with a couple of young Russian physicists, but when we came back to the plane after a short walk I was accosted by a furious Intourist official, who was supposed to have been escorting me back to Moscow.

I spent two days in Moscow, visiting the Landau Institute and Moscow State University, hosted by Pitaevskii and by Abrikosov. One morning I wandered round the Kremlin by myself, and I was stopped by a guard, probably offended by my scruffy duffel coat. When I said I was "angliskii" he smiled broadly, waved his arms, and told me to look around. Unfortunately my first visit to Russia was probably also my last.

A couple of months later I happened to see Ginzburg and, if my memory is correct, Khalatnikov wandering around the Cambridge market place, during a break from a relativity conference. We invited them both home for dinner and got my colleague Roger Tayler to meet them. The third guest was a lucky choice, as he had translated one of Ginzburg's books.

UNIVERSITY OF BIRMINGHAM PROFESSOR

In 1965 David was appointed professor of physics at the University of Birmingham.

He has left six pages of detailed notes about the development of his research during the first three years at the University of Birmingham and one year of sabbatical leave when he visited Chalk River, Cornell, Stony Brook and several places in Australia.

Before leaving Birmingham on sabbatical in 1968 Margaret loaned their only car, a Bedford camper van, to neighbours. When Margaret and David got back a year later the friends had moved to Bristol, taking the van with them. The husband was in South America and his wife had a new baby just when Margaret and David needed their car back, so David had to go and fetch it. He stopped and had lunch with John Ziman, then a professor at the University of Bristol, which changed the future of his physics research. Two of Ziman's students said they had disproved Philip Warren Anderson and Nevill Francis Mott's 1958 theory of electron localisation disorder, so David said he would look at their papers. In the end he convinced himself that Anderson and Mott were right; the Nobel Committee for Physics came to the same conclusion in 1977 when they awarded them the prize. The exercise of reading, analysing and rewriting Anderson and Mott's work gave David opportunities to think about a topic that he had not thought about before and opened up connections within the physics world. David later thanked Margaret for changing the direction of his research life by lending their car.

Around 1970 Michael Kosterlitz, a research fellow whose funding was not tied to any particular project, began to work with David on the interaction energy of a pair of vortices in a two-dimensional neutral superfluid. David commented on their relationship, "We worked well together, since I had the broad ideas and tried to understand the big picture, whereas Mike would find the holes in my arguments and ways to solve the problems I had ignored." This collaboration resulted in Kosterlitz-Thouless transition theory, described in their 1972 paper, which is one of two cited for the 2016 Nobel Prize in Physics.

The other events of significance in 1972 for Michael and David were the births of their daughters Karin Kosterlitz and Helen Thouless.

REASONS FOR LEAVING BIRMINGHAM UNIVERSITY

There has been a lot of discussion of the "brain drain" of the 1970s, which is often attributed to a lack of money for academics. However, David did not leave the UK for money, but because of difficulties with the university administration. When David arrived back in Birmingham from sabbatical leave in 1969 he had a meeting with the new Vice Chancellor, who asked David what he would be doing next. David gave the true but impolitic answer that he did not have any definite plans. This led to an ongoing saga which resulted in the Vice Chancellor eventually telling David that if he had a chance to leave the university he should do it.

Although David did not have any definite plans on returning from sabbatical, his curiosity and openness to new topics led to an extraordinarily fruitful period from 1970 through 1978. He published 16 of his most important papers in five distinct topics, including the work for which he was eventually awarded the Nobel Prize. As noted by Ana Mari Cauce, President of the University of Washington, David was known for his curiosity-driven research which, decades after the initial research, has led to many practical uses.

There were no theoretical physics chairs vacant in the UK at that time so David left the UK, much as he did not want to. David went briefly to Yale but clearly he did not talk adequately to whoever was in charge of making the appointment because Yale wanted David to build a research group, whereas David had always preferred to work with colleagues rather than being a group leader.

University of Washington Seattle USA

David's life and work up to the year 1972 is known from his own detailed autobiographical notes. His story from that year forward is told without the benefit of such a first-hand account.

The University of Washington did not need David to build a big research group. There were enough other independent theoretical physics professors there to whom David's students and postdoctoral fellows could talk if he were away. He mostly taught graduate students and upper class undergraduate courses. He had many graduate students from around the world, but never an American-born student.

Shortly after arriving in Washington in 1980, David wrote a grant proposal in which he described the work he intended to do, but also suggested he might investigate some entirely different topic if a more interesting one came along. David's reputation for producing interesting work meant that he was awarded this grant despite the vagueness of the grant proposal; President Ana Mari Cauce has observed that this would be unlikely to be funded today, when grants driven mainly by curiosity do not get much support.

In 1982, David published a paper called *Quantized Hall conductance in a two-dimensional periodic potential* with research fellows Kohmoto, Nightingale and den Nijs (TKN^2), which is the second paper cited for the Nobel Prize. The word topology is not mentioned in the title of the 1982 paper and does not appear in his titles until 1985. However, when David Thouless wrote the book *Topological numbers in nonrelativistic physics* in 1998 he said "Topological numbers crept up on the physics community before the community was aware of them. I did not think in these terms until I started working on the topological aspects on long range order in the 1970s, although I had been working on aspects of superfluidity that are not topological for several years before that."

Marcel den Nijs has remained in Seattle and has been a great supporter of David but they have not published any more papers together.

In 1990 David was awarded the Wolf Prize in physics with Pierre-Gilles de Jennes (Nobel Laureate, 1991). Over the years he has received a number of other awards and honours. For example, he was elected a Fellow of the Royal Society (FRS) in 1979, a Fellow of the American Academy of Arts and Sciences (1981), a Fellow of the American Physical Society (1987) and a member of the US National Academy of Sciences (1995).

David enjoyed working and living in Seattle. He has never had many hobbies but he loved to hike in the mountains, camp, cross country ski and occasionally sail. His house had a 180-degree view of Lake Washington and mountains,

Figure 7. David Thouless, 1987. Distinguished Visiting Scientist, Brookhaven National Laboratories. Photograph courtesy of Brookhaven National Laboratories.

including Mt. Rainier. Even though some of the surrounding trees have grown, a marvellous view remains. The garden faces southeast and has excellent soil for gardening. David's biggest hobby over the years was reading. He read widely, but history interested him most. He was very happy in retirement reading in his chair and then resting his eyes on the view.

Assessing his own work, David wrote:

> My scientific accomplishments as a graduate student and in postdoctoral positions had been very solid, with a successful book published, and several papers which were either novel or at least close to the latest work in the field. There was less to show for my four years in Cambridge. The work on the exchange mechanism for nuclear magnetism in solid 3He, inspired by Phil Anderson, was a substantial contribution, and taught me a lot, but it was more than

Figure 8. David and Margaret Thouless at the Nobel Foundation, December 12, 2016.

twelve years after publication that the theoretical work became relevant to experimental measurements. The situation did not change immediately after I went to Birmingham, but my work, for various reasons, blossomed after I had been there for two years, and continued flourishing for the next twenty years or so.

The following are the papers David judged to be his most important work.

A. Papers on nuclear matter

D.J. Thouless. Application of perturbation methods to the theory of nuclear matter. *Phys. Rev.* **112** (1958) 906–22.

M.A. Thorpe and D.J. Thouless. Oscillations of the nuclear density. *Nucl.Phys.* **A156** (1970) 225–41.

B. Papers on collective motion in nuclei

D.J. Thouless. Stability conditions and nuclear rotations in the Hartree-Fock theory. *Nucl. Phys.* **21** (1960) 225–32.

D.J. Thouless. Vibrational states of nuclei in the random phase approximation. *Nucl. Phys.* **22** (1961) 78–85.

D.J. Thouless and J.G. Valatin. Time-dependent Hartree-Fock equations and rotational states of nuclei. *Nucl. Phys.* **31** (1962) 211–30.

R.E. Peierls and D.J. Thouless. Variational approach to collective motion. *Nucl. Phys.* **38** (1962) 154–76.

C. Papers on statistical mechanics

D.J. Thouless. Critical region for the Ising model with a long range interaction. *Phys. Rev.* **181** (1969) 954–68.

D.J. Thouless. Long range order in one-dimensional Ising systems. *Phys. Rev.* **187** (1969) 732–3.

J.M. Kosterlitz and D.J. Thouless. Long range order and metastability in two dimensional solids and superfluids. *J. Phys.* **C 5** (1972) L124–6.

J.M. Kosterlitz and D.J. Thouless. Ordering, metastability and phase transitions in two-dimensional systems. *J. Phys.* **C 6** (1973) 1181–1203.

D. Papers on superconductivity and superfluidity

D.J. Thouless. Strong coupling limit in the theory of superconductivity. *Phys. Rev.* **117** (1960) 1256–60.

D.J. Thouless. Perturbation theory in statistical mechanics and the theory of superconductivity. *Annals of Phys.* **10** (1960) 553–88.

D.J. Thouless. Critical fluctuations of a type II superconductor in a magnetic field. *Phys. Rev. Lett.* **34** (1975) 946–9.

G. Ruggeri and D.J. Thouless. Perturbation series for the critical behavior of type II superconductors near HC2. *J. Phys.* **F 6** (1976) 2063–79.

D.J. Thouless, P. Ao and Q. Niu. Vortex dynamics in superfluids and the Berry phase. *Physica* **A 200** (1993) 42–9.

D.J. Thouless, P. Ao and Q. Niu. Transverse force on a quantized vortex in a superfluid. *Phys. Rev. Lett.* **76** (1996) 3758–61.

M.R. Geller, C. Wexler and D.J. Thouless. Transverse Force on a Quantized Vortex in a Superconductor. *Phys. Rev.* **B 57** (1998) R8119–22.

D.J. Thouless, M.R. Geller, W.F. Vinen, J.-Y. Fortin and S.W. Rhee. Vortex dynamics in the two-fluid model. *Phys. Rev.* **B 63** (2001) 224504.

E. Papers on magnetism

D.J. Thouless. Exchange in solid 3He and the Heisenberg Hamiltonian. *Proc. Phys. Soc.* **86** (1965) 893–904.

D.J. Thouless, P.W. Anderson and R.G. Palmer. Solution of 'Solvable model of a spin glass'. *Phil. Mag.* **35** (1977) 593–601.

J. R. L. de Almeida and D. J. Thouless. Stability of the Sherrington-Kirkpatrick solution of a spin glass model. *J. Phys.* **A 11** (1978) 983–90.

D.J. Thouless, J.R.L. de Almeida and J.M. Kosterlitz. Stability and susceptibility in Parisi's solution of a spin glass model. *J. Phys.* **C 13** (1980) 3271–80.

D.J. Thouless. Spin glass on a Bethe lattice. *Phys. Rev. Lett.* **56** (1986) 1082–5.

F. Papers on electrons in disordered systems

D.J. Thouless. Anderson's theory of localized states. *J. Phys.* **C 4** (1970) 1559–66.

J.T. Edwards and D.J. Thouless. Regularity of the density of states in Anderson's localized electron model. *J. Phys.* **C 4** (1971) 453–7.

B.J. Last and D.J. Thouless. Percolation theory and electrical conductivity. *Phys. Rev. Lett.* **27** (1971) 1719–21.

D.J. Thouless. A relation between the density of states and range of localization for one dimensional random systems. *J. Phys.* **C 5** (1972) 77–81.

J.T. Edwards and D.J. Thouless. Numerical studies of localization in disordered systems. *J. Phys.* **C 5** (1972) 807–20.

D.J. Thouless. Localization distance and mean free path in one-dimensional disordered systems. *J. Phys.* **C 6** (1973) L49–51.

R. Abou-Chacra, P.W. Anderson and D.J. Thouless. A self-consistent theory of localization. *J. Phys.* **C 7** (1974) 1734–52.

R. Abou-Chacra and D.J. Thouless. Self-consistent theory of localization: II. Localization near the band edges. *J. Phys.* **C 7** (1974) 65–75.

D.J. Thouless. Electrons in disordered systems and the theory of localization. *Phys. Reports* **13 C** (1974) 93–142.

D.C. Licciardello and D.J. Thouless. Constancy of minimum metallic conductivity in two dimensions. *Phys. Rev. Lett.* **35** (1975) 1475–8.

D.C. Licciardello and D.J. Thouless. Conductivity and mobility edges for two-dimensional disordered systems. *J. Phys.* **C 8** (1975) 1803–12.
D.J. Thouless. Maximum metallic resistance in thin wires. *Phys. Rev. Lett.* **39** (1977) 1167–9.
D.J. Thouless. Percolation and localization. In *Ill condensed matter*, ed. R.Balian, R.Maynard and G.Toulouse (North-Holland 1979), pp 1–62.
D.J. Thouless. The effect of inelastic electron scattering on the conductivity of very thin wires. *Solid State Commun.* **34** (1980) 683–5.

G. Quantum Hall effect and related topics

D.J. Thouless, M. Kohmoto, M.P. Nightingale and M. den Nijs. Quantized Hall conductance in a two-dimensional periodic potential. *Phys. Rev. Lett.* **49** (1982) 405–8.
D.J. Thouless. Quantization of particle transport. *Phys Rev.* **B27** (1983) 6083–7.
D.J. Thouless. Band widths for a quasiperiodic tight-binding model. *Phys. Rev.* **B28** (1983) 4272–6.
D.J. Thouless. Wannier functions for magnetic sub-bands. *J. Phys.* **C 17** (1984) L325–7.
Q. Niu, D.J. Thouless and Y.S. Wu. Quantized Hall conductance as a topological invariant. *Phys. Rev.* **B 31** (1985) 3372–7.
Q. Niu and D.J. Thouless. Quantum Hall effect with realistic boundary conditions. *Phys. Rev.* **B 35** (1987) 2188–97.
D.J. Thouless. Scaling for the discrete Mathieu equation. *Commun. Math. Phys.* **127** (1990) 187–93.
D.J. Thouless and Y. Gefen. Fractional quantum Hall effect and multiple Aharonov-Bohm periods. *Phys. Rev. Lett.* **66** (1991) 806–9.
D.J. Thouless. Edge voltages and distributed currents in the quantum Hall effect. *Phys. Rev. Lett.* **71** (1993) 1879–82.

H. Mesoscopic systems

Y. Gefen and D.J. Thouless. Zener transitions and energy dissipation in driven systems. *Phys. Rev. Lett.* **59** (1987) 1752–5.

Books

D.J.Thouless. *The quantum mechanics of many-particle systems*. Academic Press, New York and London, 1961; second edition, 1972.
D.J. Thouless. *Topological quantum numbers in nonrelativistic physics*. World Scientific Publishing Company, Singapore, 1998.

Note: Much of this biography is based on David's words written in his own detailed autobiographical notes.

Thank you to Christopher Thouless, Michael Thouless and Peet Sasaki for making substantive comments and editing.

Editor's note: David Thouless did not deliver his Nobel Lecture.

F. Duncan M. Haldane. © Nobel Prize Outreach AB. Photo: A. Mahmoud

F. Duncan M. Haldane

Biography

I WAS BORN IN LONDON IN 1951, in a medical family who greatly valued science and education in general, but never tried to push their children to go into medicine, although my younger brother did choose that path. My father was a psychiatrist working in the newly-created National Health Service, and came from Scotland. He had wanted to become a psychoanalyst, but the war had prevented his planned training under Freud's pupil Melanie Klein, and he was trying to find some way to apply techniques or insights inspired by psychoanalytic theory to the much more limited possibilities for psychotherapy in an NHS practice. My mother was a Carinthian Slovene from a bilingual region in southern Austria, who had met my father when he was an army doctor in the British Occupation Forces there. She was a medical student working in a hospital when she met him, but never managed to complete her studies after coming to Britain, because all the exams she had passed in wartime Vienna would not have been recognized, and she would have had to restart all the medical training from scratch, in what was, to her, a foreign language. Instead, she had a family. My parents' backgrounds gave me a multicultural heritage, with relatives in both Scotland and in Austria, where we often visited for summer holidays, so I became reasonably fluent in German, but sadly my command of Slovenian remained very basic indeed. My mother was proud of her heritage, as was my father of his, and he would wear a kilt on formal occasions, so although I grew up in London, without a trace of a Scottish accent, I self-identified as half-Scot, half-Slovenian.

I was sent to private schools, first a mixed elementary school a short walk from our house in Bedford Park, in west London, where I appear to have excelled in subjects like arithmetic and spelling, but always lost out on my handwriting skills, which remained messy and irregular, despite my being made to copy out pages of text again and again (or so it seems in my memory). When I was ten, I was sent to the "preparatory school" (Colet Court) for St. Pauls School, and then to St. Pauls itself, which is a well-known "public" (i.e., private) boy's school noted for a rigorous educational curriculum. The school was very cosmopolitan, and was mainly a day-school with pupils coming from many parts of London, with a small boarding component. I was one of the one hundred and fifty-three scholars at the school (the number has biblical significance as the number of fish miraculously caught by the apostles), and because of this, I wore a little silver badge in the shape of a fish.

I always remember being interested in mathematics and science. In English schools, at least at that time, one had to specialize early. Looking at the list of General Certificate of Education "O levels" that I took, they were English, Latin, French, mathematics, "physics-with-chemistry," with the only unusual one being "physical geography and elementary geology." At some point I had to choose between continuing with history or geography, and the rocks and minerals seemed interesting and I was fascinated by the crystal collection the school had (perhaps an early attraction to "condensed matter"?).

For "A" levels, I just have mathematics, physics, chemistry, so somehow, I never studied biology (I think one only studied it if one was planning to go to medical school?). Of course, as these last years of school were during the late sixties, there were lots of distractions for teenagers in London during that period, but I managed to keep my academics on track. In my final year at school, I had a very enthusiastic and inspiring physics teacher who got me interested in the subject, while previously I had found chemistry definitely more interesting.

Somehow, I managed to combine interest in science with interests in rock music and sixties counterculture. I had a gap of nine months after leaving school, and before starting University, and decided to travel. I worked for a while at a book publishers' organization extracting data on names and fields of study of faculty members from German university catalogs, and with my savings, and a large backpack, then set off on the then-well-traveled overland trail to India and Nepal via Iran and Afghanistan (and back!)—a journey impossible today! (I would later get to see India (and Nepal) from a rather different perspective during visits as a professional Physicist.)

I was admitted to "read" Natural Sciences at Christ's College, Cambridge where I "matriculated" in October 1970. Three subjects plus mathematics were required, so I finally had the chance to learn some cell biology as well as physics and chemistry, but I found I was not so gifted in the laboratory, and after an experience when I accidentally swallowed some nasty chemical I was supposed to measure out a small dose of using a "mouth pipette" (I do not believe such things still exist with today's work-safety rules) I decided I should opt for prudence and focus on theory!

In my third and last Cambridge undergraduate year, 1973, I took a class called something like "advanced quantum mechanics" taught by Phil Anderson, where, if I remember correctly, he talked about the problem of localization by disorder, the Kondo effect, and other inspiring things. These were deeply conceptual quantum problems different from the diet of scattering problems which seemed like mathematical exercises in partial wave expansions and spherical harmonics that the more conventional classes had been feeding us. I was hooked and decided that if I was accepted to stay on at Cambridge as a graduate student in the Cavendish Laboratory (which happened), I would like to work with Anderson. I also considered working with Michael Green on an intriguing problem of "massless spinning relativistic strings": since string theory as a model for the hadrons was abandoned shortly thereafter, and took ten years "in the wilderness" till it was repopularized as a possible theory of quantum gravity, my choice to work with Phil seems a fortunate one, at least for one made in 1973! It is probably the case that any successful research career can be traced to "accidentally" making a series of non-obvious choices at the right time, and various chance events. I think it was the concreteness of condensed matter, in that it was much easier to experimentally realize systems that exhibit all sorts of remarkable effects, that kept me on the condensed matter theory trail. In some sense, particle theorists have only one physical vacuum, with its beautiful but highly constrained Lorentz

Figure 1. The Author in his office, with a working desk, after returning to Princeton as a faculty member (1980).

point-symmetry group, to play with, while condensed matter physics can "build" a huge variety of model vacua with different symmetry groups and "elementary particles" (elementary excitations), and play with them experimentally.

In the TCM (Theory of Condensed Matter) group at the Cavendish, Phil gave me the problem of "valence fluctuations" in the Anderson model of a magnetic impurities to look at and a reprint of his Les Houches lectures about the Kondo impurity spin model, including the "Anderson-Yuval-Hamann" renormalization group treatment of the mapping that turns the path-integral of spin-flips of the impurity into a coulomb gas of charges of alternating sign that interact with a logarithmic potential. This had a "renormalization group" (RG) treatment that provided the precursor for the method developed by Kosterlitz and Thouless for the Nobel-prize-winning solution of their famous problem. I also had to study Phil's less-complicated "poor-man's method" that rederived the same RG scaling equations for the Kondo model. Phil spent part of the year in Cambridge and the rest at Bell Laboratories, so I had to work through these mysterious texts by myself. The majority of the TCM group were interested in accurate computation of material properties, especially surface properties of metals with different kind of atoms or molecules absorbed on them for catalysis, so in my advisor's absence, I tried to learn from them and did not understand his "toy model" approach, which was that accurate details really do not matter if one is trying to understand the essence of some phenomenon, provided that the ingredients retained in the model are indeed the ones that matter.

I remember puzzling over the Kondo, Anderson and Wolff models which were all representations of something like a transition-metal d-orbital deep inside the core region of a transition metal atom, in which there are strong electron-electron interactions, mixing with a weakly-interacting metallic conduction band derived from outer s-like orbitals. I even got hold of a self-consistent Hartree-Fock program written in FORTRAN-66 line by line on a huge stack of IBM punched cards that had to be fed into a card-reader hopper to submit the job to a mainframe computer, and tried to puzzle out how the real orbitals of the notional metal atoms would behave as charge leaked off or onto the impurity atom from the metal background. Needless to say, all this was quite pointless, even though it was some kind of learning experience. When Phil returned again, I still had not figured out what the toy models really meant physically. For example the Wolf and Anderson models seemed to be mathematically equivalent, depending on whether the extra "d" orbital was interpreted as being part of the conduction band or orthogonal to it.

But instead of helping me struggle with these niggling details, when Phil returned, he gave a marvelous course of lectures that became his book "Basic

Notions of Condensed Matter Physics" where he sketched his ideas of "adiabatic continuity" within phases until critical points were reached, and that all points within the same phase shared the same essential "fixed-point" independent of all the fine details. Through hearing him flesh out his ways of thinking, and going to see him about some details I was missing, and instead having him share with me his interesting thoughts about some apparently quite different but essentially related issue, I began to see his point of view that tries to identify what is needed to see the "big picture," when trying to understand the physics of strongly-correlated systems. Somehow, that was what having a "mentor" was all about.

In the middle of my second year as a graduate student Phil announced that he would be exchanging his half-a-year at Cambridge, half-a-year at Bell Laboratories position for a similar one that replaced Cambridge University with Princeton University. I and Phil's other student, Ali Alpar, who was working on pion superfluidity in neutron stars, never learned the reason for the move. This was in any case a very interesting change for us both, as Phil arranged to take us with him to the very different world of Princeton, New Jersey, starting with a few summer months at Murray Hill, New Jersey, the location of Bell Telephone Laboratories, then in its heyday. This was a tremendous privilege for a graduate student.

In September 1975, I moved down from Murray Hill to Princeton, and Ali Alpar and I shared an office on the fourth floor of Jadwin Hall, which was a larger

Figure 2. With Eduardo Fradkin, at a tea house in Tbilisi, Georgia, 1988, during the visit to the USSR sponsored by the US National Academy of Science. Fradkin's paper with Dagotto and Boyanovsky played a key role in the Author's discovery of a model exhibiting a "quantum anomalous Hall effect."

office divided in two by partition, on the other side of which was Natan Andrei, working on particle theory with David Gross, and not yet on the Bethe Ansatz that he would go on to use to unexpectedly find the exact solution to problems I worked on such as the Kondo model, which I would tell him about. Other contemporary students on our floor included Ed Witten and Steve Girvin, who was working with John Hopfield on the "X-ray edge singularity" problem (which like the Kondo problem involved singular behavior at the Fermi level, especially the "orthogonality catastrophe" discovered by Anderson that affects dynamical degrees of freedom that excite particle-hole pairs in a metal that they couple to).

There were many blackboards on the fourth floor. One slightly disconcerting feature of the environment was that John Nash, the future Economics Nobel Laureate, who was in the middle of his illness, would gain access to the building at nights or weekends and systematically cover all the blackboards with mysterious equations connecting politics, pop culture, and numbers. Frank Wilczek had just become an Assistant Professor, and he gave a many-body class about the 3D interacting Bose fluid that I took. Barry Simon and Elliot Lieb were working on the stability of matter, which I also took a class on. Through Princeton and Phil Anderson, I was privileged to meet so many of the leading theorists who were at Princeton, or visiting; for example I was invited to dinner at the Andersons when Phil's old friend David Thouless visited, meeting him for the first time.

It was an intellectually exciting time to be at Princeton, and in that atmosphere, I finally understood what I was trying to achieve with my extension to Anderson's treatment of the Kondo problem that allowed valence (charge) fluctuations as well as spin fluctuations of an Anderson model impurity. The renormalization-group treatment showed a novel effect that there was a logarithmic temperature dependence of the energy level of the impurity orbital as a consequence of the interaction.

The renormalization group can be viewed as a way to resum a divergent series derived by perturbation theory, in this case in the mixing between the impurity orbital and the metal in which it is embedded. This means that the results can be validated by a detailed examination of the structure of the perturbation series. The test required that the sum of pieces of each of about forty distinct fourth-order terms in the series should exactly cancel. With the aid of the huge table of integrals by Gradsteyn and Ryzhik, I set out to do the test, but it did not quite work, the cancellation was just not happening. I checked and rechecked each of the forty terms time and time again, to no avail. Finally, after about two months of intense struggle, and being convinced that my results were correct, I realized that one of the complicated integrals I was taking from G&R could not possibly be right, because a simple approximation produced a lower bound that the printed result violated. When I worked out the integral for myself, there was a missing factor of two in the formula given in the tables, and the correction finally made everything work as expected. (When the next edition of G&R was published, there was indeed an erratum that corrected the printed formula!) The experience gave me confidence in standing by results I believed to be true, as well as a lifelong antipathy to doing high-order perturbation theory!

During my last year of graduate studies, the French physicist Philippe Nozières came to give a seminar, and Phil introduced me to him. Later, just when I was wondering where to apply for a postdoctoral position, I got a letter offering me a five-year position in France, at the Institut Laue-Langevin in Grenoble, a city with a large number of research laboratories. The ILL is a neutron-scattering facility, a joint consortium between France, Germany, and Britain, but had a theory group as well as experimental groups who used the neutron source. The idea of experiencing a new country, France, was very appealing, and especially as the dollar was at a low point of the exchange rate, the job looked very attractive, so I accepted. I finished writing up my thesis, and before leaving for France, I had the great opportunity to attend, with Phil's recommendation, a workshop on strongly-correlated electron systems at the Aspen Center for Physics, in Aspen, Colorado, and then spend a month with Sebastian Doniach at Stanford University.

My brother came to visit, and shared the driving in my old VW beetle from Princeton to Aspen, and it was quite amazing to experience transcontinental driving! That year was a special year at Aspen, as a high-powered delegation from the Landau Institute in the USSR, led by Lev Gorkov, was also attending the workshop. My future colleague Sasha Migdal was among the Soviet party, and it was very interesting to witness the internationalism of science. (There was also a lot of speculation about who was acting as the KGB minder who it was assumed had to be there to keep an eye on the rest of the delegation!) This was followed by a further drive through the spectacular western scenery to Stanford, where I met my future long-term collaborator Ed Rezayi, then a graduate student with Doniach, and then another transcontinental drive back across the US to New Jersey, from where I left for France.

I had perhaps foolishly shipped my American-model Volkswagen to France, but picked it up at the port of Le Havre, and was driving down the autoroute to Grenoble when I heard on the car radio that Phil Anderson was to share that year's Nobel Prize for Physics with his advisor John van Vleck and Nevil Mott, who had brought him to Cambridge in the sixties.

I soon found that while my years of French language studies at school had prepared me to decipher street signs and read menus, understanding what people were saying was another matter. On the other hand, the multinational work environment at the ILL was mainly English-speaking, which did not help to improve my French. This was remedied after I met Odile Belmont, a native of the Grenoble region who would later become my wife.

In learning about the Anderson-Yuval-Hamann treatment of the Kondo model I learned about the X-ray-edge singularity problem, which Nozières and de Domincis (ND) had solved in terms of singular integral equations, and the much simpler later variant treatment by Schotte and Schotte using "bosonization," a remarkable and mysterious representation of electron creation operators apparently just using harmonic oscillator variables, related to those used by Tomonaga in his 1950 treatment of sound waves in a one-dimensional Fermi gas. The two treatments agreed at weak coupling, but differed at strong coupling, where the ND treatment seemed more complete, but in fact the model assumptions used in the two treatments were slightly different, so the models were different at strong couplings. I became aware that Daniel Mattis had claimed to solve the Wolff model exactly using bosonization techniques, but I knew that, at least formally, the Wolff and Anderson magnetic impurity models were equivalent, and from my thesis work, felt something was not right with the proposed bosonization solution. One of my new colleagues at ILL, Hans Fogedby, was also working on the Wolff model with Mattis' technique, and I determined to try an understand the bosonization technique, and find out why it was giving results I disagreed with, including a phase transition to a magnetic state as the short-range (contact) interaction strength (usually denoted "*U*," by analogy to the Hubbard model, a widely-used "toy model" for studying magnetism) was increased.

The Anderson and Wolff models feature a single "impurity orbital" in which there is a Hubbard "*U*" coupling. Because the Pauli principle prevents two electrons with the same spin from being present in the single impurity orbital, there is no direct interaction between electrons with the same spin, and these were explicitly discarded in the bosonization treatment. However, while the ND treatment of the X-ray edge problem, preserved the contact-type nature of the interaction, the bosonization treatment was secretly treating a long-range interaction which could couple electrons of the same spin, so it was not valid to simply discard same-spin interactions. This subtlety was hidden in the now-explicitly-specified "ultra-violet cutoff" structure, invalidating the bosonization treatment of the Wolff model, but I wanted to "clean up" aspects of the bosonization technique, which had recently also been used to great effect by Luther and Emery for one-dimensional metals, and by Luther and Peschel for the spin-1/2 easy-plane spin chain.

At this time, the correctness of the Kosterlitz-Thouless treatment of the topological phase transition had not yet been universally acknowledged, and there was a counterproposal by Luther and Scalapino based on bosonization of a 1D quantum spin chain. I attended a workshop at NORDITA in Copenhagen, where Luther had moved to, where this was a heated subject of discussion. At that meeting I also first met my future colleague Kazumi Maki, and there was also a Soviet contingent, including Igor Dzyaloshinksky of the famous Landau-Institute AGD (Abrikosov-Gorkov-Dzyaloshinsky) triumvirate, who had produced the foremost text on diagrammatic perturbation methods in condensed matter theory. Igor was an old friend of Philippe Nozières, and I got to know him well when he subsequently came for an extended visit to Grenoble. He had produced an interpretation (with Anatoly Larkin) of bosonization in terms of standard diagrammatic perturbation theory, which was a useful alternative viewpoint.

In my investigation of bosonization, I found that its exact formulation needed two action-angle variables to replace the absent zero-wavelength sound-wave mode, and the lack of this in the earlier formulations such as Luther's had been "patched up" with a cutoff that was not really consistent. The new variables added topological winding-number excitations with their own distinctive energies to the well-known Tomonaga sound waves, and allowed me to formulate what I called "Luttinger liquid theory," first as a replacement for Landau Fermi-liquid theory in one dimensional electron systems. However, because "$2k_F$" for a spinless Fermi fluid would also be the Bragg vector if the fluid crystallized, it also applies to 1D Bose fluids and gapless uniaxially-anisotropic spin chains. As I described in my Nobel lecture, this led to a deeper understanding of spin chains, including my very expected discovery in early 1981 that the spin-1 isotropic antiferromagnetic chain had a gapped spin-liquid state that is now recognized as an early example of topological quantum matter.

I was quite surprised when analysis starting with the Luttinger-liquid approach, supplemented with the mapping of the Kosterlitz-Thouless transition to (1+1) dimensional quantum mechanics, led inescapably to my surprising conclusion. I was even more surprised at the resistance this received from the quantum magnetism community when I submitted the paper for publication: it was rejected by multiple journals, and was labeled a "conjecture" even though it was, in my mind, a clear prediction. I recall that one referee pontificated that my claims "were in manifest contradiction to fundamental principles such as renormalization and continuity"! Of course, my predictions were later vindicated both by numerical studies and experiments.

While in France, I received an unexpected invitation to visit the University of Southern California in Los Angeles for a job interview. It turned out that Kazumi

Figure 3. A Euoropean-style dinner at the 1980 Taniguchi Symposium in Japan. The Author is in the corner; David Edwards of Imperial College, the expert on magnetism who had been the external examiner of the Author's Ph.D. thesis, (and who has insisted on revisions to expand the explanations), is in the center.

Maki had written to Philippe Nozières asking for suggestions for candidates. I visited, and was seduced by the beach and palm trees. I had not yet actively started to look for a faculty position, but at that time, the news I was hearing from British friends was anecdotally rather pessimistic about the UK physics job market, and government research funding. So by default, I inadvertently joined the "brain drain" to the US. A lasting legacy from my time in France was my French life-partner Odile, who agreed to try out the California lifestyle with me.

In my last year at the ILL, I was fortunate, perhaps as a result of my "Luttinger liquid" work, to be invited to one of a small group of "promising young scholars" invited to a Taniguchi Symposium in Japan where the Japanese philanthropist Toyosaburo Taniguchi envisaged they would come together to interact in ideal and luxurious surroundings, in this case a lodge next to Mt. Fuji. Not all the "scholars" were that young, and I had the chance to meet and discuss with John Hubbard, who had introduced a key "toy model," the Hubbard model for strongly-interacting electrons, who was also there, and seemed to be enjoying the meeting. (Tragically, it was his last meeting, as he died just after returning home.)

By the time I got an extensively rewritten paper on the spin-chain finally published (in Physics Letters A) I had been in California for over a year. During that time I received two papers (from the same journal) to referee (first by Takhtajan, then by Babujian) both describing an exactly solvable gapless $S = 1$ spin chain with a Bethe Ansatz solution very similar to the gapless $S = 1/2$ chain. The exact solutions were claimed to represent the generic behavior of arbitrary-spin Heisenbeg antiferromagnets, and they apparently completely contradicted my theory! I must admit I had about ten minutes of self-doubt when I received the first of these papers, but soon saw that the solvable model was a modified model with a large non-Heisenberg unphysical "biquadratic exchange" term, and did not represent the standard Heisenberg model I had treated. Furthermore, though they were gapless, I could not fit them into my "Luttinger-liquid" picture. Around this time, (1+1)-d conformal field theory was starting to be developed. Eventually it emerged that "Luttinger liquids" were related to Abelian conformal field theories, that can have continuously tunable critical exponents. The new $S > 1/2$ Takhtajan-Babujian solvable models were critical, but correspond to non-Abelian conformal theories that require fine-tuning the couplings to exactly cancel "relevant" perturbations, so do not represent generic spin chains. The $S = 1$ case represents a critical point between the generic "Haldane-gap" non-degenerate symmetry-protected topological (SPT) state, and a non-topological gapped broken-symmetry two-fold-degenerate dimerized state.

An apparently unconnected series of surprises were independently discovered in those years. First, Klaus von Klitzing discovered the integer quantum Hall effect (QHE). As soon as it had been concluded that, in two dimensions, localization by a disordered potential would always lead to integer quantization

Figure 4. The Taniguchi Symposium participants in front of Mt. Fuji in November 1980. The Luttinger liquid approach to the spin chain was presented at this meeting.

Figure 5. At a formal Japanese dinner at a Buddhist temple: magnetism experts Toru Moriya, John Hertz and Richard Prange are in the foreground.

of the Hall conductivity, Dan Tsui, Horst Störmer and Art Gossard discovered the *fractional* quantum Hall effect. This was far more of a shock for theorists, as the understanding of the integer QHE showed a fractional effect could only occur as a consequence of interactions. Furthermore, at the time it was generally believed that second quantization and diagrammatic perturbation theory was the principal tool for understanding interaction effects. In fact these techniques are only useful if some adiabatic connection can be found between a non-interacting system and the interacting one, which was not the case for this problem. The Soviet physicists at the Landau Institute outside Moscow were the world's leading practitioners of diagrammatic techniques in condensed-matter physics, and interestingly, the fractional QHE was the first problem to which they were unable to make many contributions.

Of course, the key breakthrough was Laughlin's discovery of his eponymous state, apparently through carrying out a numerical diagonalization of a three-particle system projected into the lowest Landau level. Perhaps his training in band-structure calculation allowed him to take this direct route to investigate the problem. The key experimental clue was that the QHE states occurred at Landau-level filling $\nu = 1/3$ but not at $\nu = 1/2$. I had been thinking about some kind of "supersolid" picture, when in early 1983 I received Laughlin's paper to referee. Within ten minutes I knew he had found the right (Nobel prize-winning) explanation, an incompressible quantum fluid with fractionally-charged excitations, that was later realized to be topologically ordered. In addition, it was fundamentally disconnected from free-particle Slater-determinant states, so there seemed to be no hope of understanding it based on diagrammatic perturbation theory. The most convincing detail was that it provided a natural explanation based on Fermi statistics for why it occurred at $\nu = 1/3$ but not $\nu = 1/2$. The wavefunction also provides a clear picture of what was later called "flux attachment." The Laughlin state had a huge effect on the way I thought about condensed matter physics.

Later that year, there was a meeting at Bell Laboratories to celebrate Phil Anderson's sixtieth birthday, and I stopped over in New Jersey on my way to France, for a summer collaboration with Rémi Jullien, Robert Botet, and Max Kolb, who done the first numerical studies to test my claims about the integer-S antiferromagnetic spin chains, and had also attracted skepticism when they reported results supporting my predictions. I had a very interesting discussion about the Laughlin state with Phil, who noted that if three units of flux were injected at a point to create three concentric quasiholes, the resulting state was the same as that resulting from locally removing an electron from the Laughlin state. Thus adding three units of flux plus one electron would just change the N-particle state to the $(N + 1)$-particle state, in analogy to a Bose condensate where particles were composite objects. Independent of Laughlin's work, a numerical exact-diagonalization study had also independently been carried out with (quasi)periodic boundary conditions by Yoshioka, Halperin and Lee (YHL) in an anisotropic basis which seemed to suggest a liquid state with a three-fold degenerate ground state, but was not as revealing as Laughlin's picture.

I had been wondering how to do a numerical calculation that incorporated isotropy, without the problem of boundaries, which YHL avoids. That night, I was staying as a guest in Chandra Varma's house, and woke from a dream in the middle of the night with the image of a spherical surface around a magnetic monopole, which solved the problem, and turned up to be an incredibly powerful tool for numerical investigation of the fractional QHE. I suppose my brain had been churning over my discussion with Phil Anderson earlier that day, to produce this Kekulé-like experience. Having woken up, I worked out all the details there and then.

I flew on to France, but instead of working on spin-chains with Rémi Jullien and his group, I found that their spin-chain programs were built with arbitrary-range exchange, which allowed me to use them "as-is" (for bosonic Laughlin states) to test ideas suggested by the spherical geometry, such as the powerful pseudo-potential idea, and the idea that the Laughlin state was an exact eigenstate of a "toy model" that retained only short-range components of the interaction potential, analogous to the Hubbard model, except without a background lattice.

In France, I learned the basic techniques of the Lanczos sparse-matrix diagonalization method. When I got back to Los Angeles, I was coincidentally contacted by Ed Rezayi who had just moved there, and we began a fruitful collaboration on numerical studies of the fractional QHE. Because of the inapplicability of diagrammatic methods for this problem, these have been the only quantitative source of information about energies and stability in the problem, to date.

The next year I received an interesting job offer to join Bell Laboratories as a member of technical staff. I took a leave of absence from USC, and we moved to New Jersey, one month after our son was born. It was just the time of telephone deregulation, and of the split between AT&T Bell Laboratories, and BellCore, the part of the research division going to the new local telephone companies, and who were still in the same building as us for several months more. There was fantastic research going on at the Bell Labs, but in the end I decided that I missed the academic environment of a university and accepted a position at the University of California, San Diego, starting at the beginning of 1987, where I stayed until mid-1990. The effect of breaking up the Bell telephone monopoly inevitably led Bell Labs to decline to a shadow of its former self over the succeeding thirty years.

In January 1987, we moved to La Jolla, California, with its beautiful weather and beaches. At that time I was working on both quantum magnetism and the fractional quantum Hall effect. While at Bell Labs, as soon as I heard a rumor that Ian Affleck, with Tom Kennedy, Eliot Lieb and Hal Tasaki (AKLT) had come up with a variant spin-1 magnetic chain model for which the ground state could be exactly found (the AKLT state), I correctly immediately knew, with no further details, that it had to work by the same "pseudopotential" idea that made the Laughlin state an exact eigenstate of a truncated short range interaction. At UCSD, Assa Auerbach and Daniel Arovas, who were postdocs at the University of Chicago had asked to come to visit La Jolla during the Chicago winter, and do some work on quantum magnetism. They found a beach motel to stay for a month, and we were able to get very nice insights into the excitation spectrum of the AKLT model using methods borrowed from the fractional quantum Hall effect, in the process starting a lifelong friendship. Interestingly, this work provided the first clue that there could be some relation between the spin-chain and the quantum Hall effect: this is now clearer, as both are now recognized as forms of topological quantum matter.

In this period at UCSD, I came across various interesting results, such as an exactly-solvable spin chain model with long-range exchange (independently discovered simultaneously by Sriram Shastry, and now called the "Haldane-Shastry" model), in which the "spinons" of the spin-1/2 chain are especially simple.

Figure 6. With long-time collaborator Edward Rezayi and Nicholas Read (co-discoverer of non-Abelian statistics in a fractional quantum Hall state) at a restaurant in Aspen, Colorado, while participating in a workshop at the Aspen Center for Physics. Non-Abelian statistics are now seen as the key ingredient for topologically-protected quantum computing.

I also came up with the second discovery that the Nobel committee mentioned: I called it the "zero-field quantum Hall effect," but it is now usually called the "quantum anomalous Hall effect" or the "Chern Insulator," and is the first member of the topological insulator family, but one with broken time-reversal symmetry, unlike the later time-reversal-invariant topological insulators. The idea was started when I read a 1986 paper in *Physical Review Letters* (PRL) by Eduardo Fradkin, Elbio Dagotto, and Daniel Boyanovsky (FDB), called "Physical Realization of the Parity Anomaly in Condensed Matter Physics." I am not sure if I understood it properly, but it seemed to propose a quantum Hall effect in the absence of a magnetic field *and* with unbroken time-reversal symmetry, on a domain wall in a semiconductor with strong spin-orbit coupling. This interesting paper also stimulated Frank Wilczek to think about axion electrodynamics in a condensed-matter context. But thinking about it, I realized that there was no problem with a QHE in the absence of magnetic field, *provided time-reversal symmetry was broken*, which explicitly was not the case in the FDB paper. I tried to make this point by submitting a "comment" to PRL on the FDB paper, but as is often typical in this kind of "Comment/Response" dialog, it really became two monologs, where neither side understands what the other is saying.

Figure 7. The Author in front of a whiteboard describing the "toy-model" for a quantum anomalous Hall effect, or "Chern Insulator" thirty minutes after the early-morning phone call from the Nobel Prize committee.

In the course of sharpening my arguments, I looked for as simple and transparent a model as possible with which to make my point, and since Gordon Semenoff has used a "graphite monolayer" (*i.e.*, graphene) as the condensed-matter backdrop for Dirac points, I used that too. The 2D Dirac points are stable if both spatial inversion and time-reversal symmetries are unbroken: Semenoff broke inversion symmetry to get an entirely-unremarkable insulator that had a field-theoretic description as two copies of a massive Dirac equation. With a bit of magic involving complex second-neighbor bonds, I broke time-reversal symmetry and ended up with a topologically-non-trivial state exhibiting a "quantum anomalous Hall effect" where "anomalous" in this context means that the Hall effect is not driven by a uniform magnetic flux density, but arises from magnetization. At this point I realized that this effect was extremely interesting in its own right, especially if could be realized experimentally in a real material. I dropped out of the Comment/Response cycle, which in any case was getting nowhere, and published the result in its own right.

The model of graphene with a "mass gap" due to breaking of time-reversal symmetry, conceptually provided by an additional ferromagnetic degree of freedom with a magnetic moment normal to the graphene sheet, was a simple and transparent enough "toy model" to be used for a number of model calculations. As well as the gapped quantum anomalous Hall regime, it had a metallic regime, with the Fermi level inside a band, which could model a 2D version of a metallic (unquantized) anomalous Hall effect, and David Vanderbilt and coworkers later put it to good use to find and test a general Berry-like formula for the magnetization of a material in terms of its bulk bandstructure. I later also used it to guide me to new expressions for the anomalous Hall effect in 2D and 3D metals as a pure Fermi-surface formula, which is relevant to the currently-highly-studied "Weyl semi-metals."

Later in 1988, I had two very interesting foreign trips, one to the People's Republic of China, where T. D. Lee organized a meeting at Beijing University with a cast of colleagues such as Bob Laughlin, Steve Kivelson, Ganapathy Baskaran, Dung-Hai Lee, and others. This was when it was just becoming possible to travel to the PRC, and the Beijing streets were still rivers of bicycles, unlike today. In the second trip, I was invited by David Pines, to join a party sponsored by the National Academy of Sciences to visit the USSR, in particular the Landau Institute at Chernogolovka, which had long been off-limits to westerners (we were in fact the second group of western visitors to visit). There I met such future condensed colleagues such as Paul Wiegmann, who also independently solved the Kondo problem (and who a year later was able under Perestroika to get a passport to come with his family to a visiting position at UCSD), and my future Princeton colleague Sasha Migdal, as well as meeting senior Soviet physicists such as Gorkov, Abrikosov, and Khalatnikov. Everyone in the visiting party at the Landau Institute also wanted to meet my renowned future Princeton colleague Sasha Polyakov, but then, as now, he was a fanatical jogger, and was out running somewhere in the woods and could not be found! After a day at the Landau Institute, we went on to a meeting in Tbilisi, which was greatly enjoyed by our hosts, as the alcohol ban that Gorbachev had decreed in Moscow did not extend to Georgia. While meetings between physicists from the US with those from Russia and China are commonplace today, at the time these were quite exceptional experiences.

In 1990, Princeton University successfully enticed me away from California, and with a new baby daughter, we moved back to the East Coast. Princeton, long known for elementary particle physics, was building up its condensed matter group. In 1992, I spent a half-year sabbatical at the École Normale in Paris, and after giving a seminar on the mysterious symmetries of the Haldane-Shastry model, which a year earlier had led me to formulate a novel "fractional exclusion statistics" a suggestion from Vincent Pasquier and Denis Bernard led the

Figure 8. At the Author's 60th birthday "fest," with previous Nobel Laureates Robert B. Laughlin and Philip W. Anderson, the Author's mentor. Anderson is showing the proof copy of his latest book "More and Different."

identification of an unusual form of the "Yangian quantum group." In 1993, while attending a workshop, I unexpectedly learned from Steve Kivelson that I was that year's Oliver Buckley Prize winner for the old quantum spin chain work: it turned out that that had been announced a few weeks earlier, but I had mistaken the large white envelope with the APS letter for some routine circular, and left it unopened, and no-one else had told me. This must have been the days before email was widespread! A few years earlier, David Thouless had told me he was nominating me for Fellowship of the Royal Society of London, and (perhaps because of the Buckley Prize) I was finally elected in 1996 and had the honor of signing the parchment Charter Book, with entries going back to Newton.

For a long time, nothing had happened with my 1988 graphene-like toy-model for the zero-field quantum Hall effect. In 1999, work by Ganesh Sundaram and Qian Niu (a former graduate student of David Thouless) revived the long-ignored work of Karplus and Luttinger on the anomalous Hall effect in ferromagnetic metals, showing that it had a modern interpretation in terms of Berry curvature. This re-energized the study of Berry curvature effects in band structures.

My 1988 model satisfied the "TKNN" topological result of David Thouless, with coworkers Mahito Kohmoto, Marcel den Nijs, and Peter Nightingale, that was cited by the Nobel committee as David Thouless's seminal contribution to topological matter. When the gap was opened by breaking time-reversal invariance the conduction and valence bands had Chern invariants ±1 respectively.

In the early 2000s after attending a seminar by John Joanopoulos on the new subject of "photonic crystals" where the flow of light is modified by passing it through engineered spatially-periodic "metamaterials," I realized that, at least as far as "one-way" edge states were concerned, some of the physics of the quantum anomalous Hall effect could be transplanted into the field of photonic crystals, which could also have Chern invariants. Still it took some time to come up with an explicit photonic bandstructure that would this. Eventually, in early 2004, while I was on sabbatical at UC Santa Barbara, my student Srinivas Raghu, who came with me, found a candidate structure inspired by the same hexagonal graphene structure that exhibited the electronic effect in my 1988 model. A calculation confirmed that it indeed would show the effect, and the new field of "topological photonics" was born.

At that time, there was also a lot of discussion about a "spin Hall effect" in systems with spin-orbit coupling and unbroken time-reversal symmetry. As a toy model, it was natural to combine conjugate copies of the 1988 model for what could now be called the "quantum anomalous Hall effect" to form a time-reversal invariant structure that would exhibit a "quantum spin-Hall effect." While at UCSB, I played with this model, but because it had edge modes that traveled in opposite directions, I assumed that it could not represent a true stable topological phase because spin-non-conservation by generic Rashba spin-orbit coupling would surely mix and destroy the edge modes because the total Chern invariant satisfied $1 - 1 = 0$. This is a good lesson for not assuming things without actually doing a calculation! Charles Kane and Eugene Mele has the same idea, but actually tested it with a numerical calculation, and realized that the quantum spin-Hall state was indeed topologically stable because of a previously-unrecognized topological invariant. Furthermore, a few years later, in 2007, it was simultaneously realized by a number of groups that this new invariant could be extended to three dimensional materials, now called "topological insulators." This was shortly followed by the discovery by Liang Fu and Charles Kane of an extremely simple formula for determining whether such insulators with additional inversion symmetry were "topological" or not, leading to many experimental discoveries of topological materials, and an explosion of interest in the field.

In this period of the discovery of time-reversal-invariant topological insulators, my own work focused on rather different problems of the role of geometry rather than topology in the fractional quantum Hall effect, but in 2008, my student Hui Li and I discovered remarkable topological features in what we called the "entanglement spectrum" of quantum states, showing how the detailed structure of the entanglement revealed by its Schmidt decomposition contained far more information than just the single number characterizing entanglement entropy. This has turned into a widely-used diagnostic for studying the topology of entanglement.

In 2012, I was very gratified when the role that the 1988 "zero-field Hall effect" model had played in the topological insulator was recognized when I shared the prestigious International Centre for Theoretical Physics Dirac Medal with Charles Kane as well as Shoucheng Zhang, whose work with Laurens Molenkamp had led to a physical realization of the 2D quantum spin-Hall effect.

Finally in 2013, Shoucheng Zhang's collaboration with the experimental group at Tsinghua University in Beijing, where magnetic material was deposited on the surface of a layer of 3D topological insulator, finally led to the experimental realization of the quantum anomalous Hall effect envisaged in my 1988 paper. Because of the robustness of the unidirectional edge states, these materials are potentially even more useful than the time-reversal invariant topological insulators.

Finally, this chapter of my story ends in October 2016, when I was awakened by the 5:00 a.m. phone call from Stockholm, followed by the magnificent ceremony and banquet there on the 10th of December. While my mentor Phil Anderson was not able to travel to be in Stockholm in person, he passed on tips and observations he had made when he received his own Nobel prize in 1977. John van Vleck, Phil Anderson's thesis advisor, who shared the 1977 Nobel Prize for Physics with him, had as thesis advisor Edwin Kemble, who, while he himself did not win the Nobel Prize, had an advisor Percy Bridgeman who was the sole physics laureate in 1946. So I discovered I have an illustrious "academic gene line," stemming from fortunate choices I made back in 1973!

Topological Quantum Matter

Nobel Lecture, December 8, 2016 by
F. Duncan M. Haldane
Department of Physics, Princeton University.

ABSTRACT

Nobel Lecture, presented December 8, 2016, Aula Magna, Stockholm University. I will describe the history and background of three discoveries cited in this Nobel Prize: The "TKNN" topological formula for the integer quantum Hall effect found by David Thouless and collaborators, the Chern Insulator or quantum anomalous Hall effect, and its role in the later discovery of time-reversal-invariant topological insulators, and the unexpected topological spin-liquid state of the spin-1 quantum antiferromagnetic chain, which provided an initial example of topological quantum matter. I will summarize how these early beginnings have led to the exciting, and currently extremely active, field of "topological matter."

What we now know as "Topological quantum states" of condensed matter were first encountered around 1980, with the experimental discovery of the integer (Klitzing *et al.*, 1980), and later fractional (Tsui *et al.*, 1982) quantum Hall effects in the two-dimensional electron systems in semiconductor devices, and the theoretical discovery of the entangled gapped quantum spin-liquid state of integer-spin "quantum spin chains"(Haldane, 1981a, 1983a,b), which was later experimentally confirmed (Buyers *et al.*, 1986) in crystals of the organic chain molecule NENP. The common feature of these discoveries was their unexpectedness and the surprise that they engendered: they did not fit into the then-established paradigms of "condensed matter physics" (previously known as "solid state physics”). It was not at the time apparent that there could be any connection between these two surprises, but now, especially following the classification work of Xiao-Gang Wen (Chen *et al.*, 2013), we understand that their common feature is that they involve “topologically non-trivial” entangled states of matter that are fundamentally different from the previously-known “topologically trivial” states, and this lies at the heart of their unexpected properties.

Topology is the branch of mathematics originally used to classify the shapes of three-dimensional objects such as soccer balls, rugby (or American football) balls and coffee mugs (without a handle), which are “topologically trivial” surfaces without holes, and bagels, doughnuts, pretzels, and coffee cups with a handle, which are “non-trivial surfaces” with one or more holes. An ant crawling on such a “non-trivial” surface could walk around a closed path (one that ends at the same point that it started) that cannot be smoothly shrunk to a tiny circle around a point on the surface. These original ideas of topology were greatly generalized and made abstract by mathematicians, but the central idea, that things are only “topologically equivalent” if they can smoothly be transformed into each other, remains as its key idea. The essential feature is that different topologies are classified by whole numbers, like the number of holes in a surface, which cannot change gradually.

Entanglement is a central property of quantum mechanics whereby, if the state of a system is described in terms of the quantum state of its parts (typically if it is spatially separated into two halves), a measurement of a property localized in one of the two halves affects the state of the other half of the system. The “topology” of the “topological states of matter” celebrated in this Nobel Prize is more abstract than that of the shapes of everyday objects such as soccer balls and coffee cups, but distinguishes different types of “quantum entanglement” that cannot smoothly be transformed into one another, perhaps while some protective symmetries are being respected. In this case, a quantum state has “topologically trivial” entanglement if it can be smoothly transformed to a state where each part of the system is in an independent state where a measurement on that part has no effect on other parts of the system (this is called a “product state”). In the case of quantum spin systems (descriptions of non-metallic magnets), it turned out that almost all previously theoretically-described states were “topologically trivial,” so there was no precedent for the surprising properties of a non-trivial “topological state.”

It took some time for the general understanding that there was was a large class of new “topological states of matter” to emerge. An early milestone was the discovery (Thouless *et al.*, 1982) by David Thouless, and collaborators Mahito Kohmoto, Marcel den Nijs and Peter Nightingale (TKNN) of a remarkable formula that was soon recognized by the mathematical physicist Barry Simon (Simon, 1983) as just being the "first Chern class invariant" from the abstract mathematical topology of so-called " $U(1)$ fiber bundles," with an essential connection to a contemporaneous development, the "adiabatic quantum phase" discovered in 1983 by Michael Berry (Berry, 1984). As I am also presenting part of David Thouless' Nobel-Prize-winning work, I will describe this first in my lecture, and begin with the quantum Hall effect, for which two Nobel Prizes (1985 and 1998) have already been awarded.

In the presence of a uniform magnetic field with flux density B, charge-e electrons bound to a two-dimensional surface through which the magnetic field passes move in circular "Landau orbits." According to quantum mechanics, this periodic motion gives rise to a discrete set of positive energy levels of the electrons called "Landau levels." In the simplest model for these Landau levels, the period $T = 2\pi/\omega_c$ of the circular motion is independent of the radius of the circular motion, and the allowed energies of the Landau levels are those of a harmonic oscillator, $(n + \frac{1}{2})\hbar\omega_c$, where $\omega_c = eB/m_e$ is the so-called "cyclotron frequency." Assuming that the surface has translational symmetry, so all points on the surface are equivalent, the energy of each state in a Landau level is independent of the position of the center of the orbit, and the Landau level is highly (macroscopically) degenerate. The number of independent one-particle states in the Landau level is proportional to the area A of the system, in fact there are BA/Φ_0 states in each Landau level, where $\Phi_0 = h/e$ is the (London) quantum of magnetic flux.

The Pauli principle says that not more that one electron can "occupy" any independent one-particle state, and the Landau levels are somewhat analogous to the levels ($1s$, $2p$, $3d$ . . .) of the simple quantum mechanical model of the atom, familiar from high-school chemistry. However, instead of these levels accommodating finite and fixed numbers (2, 6,10. . .) of states available to be "filled," the number of states in a Landau level is huge (perhaps of order 10^{12} in a real sample) and varies with the magnetic field. Since the number of mobile electrons of the 2D surface is essentially fixed, it could in principle be possible to get things "just right" by "fine-tuning" the magnetic field so that in the ground state of the system, one or more Landau levels are completely filled, and the rest are completely empty, so that an energy gap separates the energy of the "highest occupied state" (the "HOMO" in quantum chemistry) and the "lowest unoccupied state" (or "LUMO"), making the system analogous to an intrinsic (undoped) semiconductor. Under these artificial "toy model" conditions, a simple calculation of the Hall conductivity σ^{xy} of the the system would indeed reproduce the quantum Hall effect with the universal value $\sigma^{xy} = ne^2/h$ (that depends only on material-independent fundamental constants and a whole number n, which is the number of occupied Landau levels), that would correspond to the results measured by von Klitzing.

The flaw in this näive explanation of the integer QHE is that it requires exquisite fine-tuning of the strength of the magnetic field. In contrast, it was the *insensitivity* to the fine-tuning of the magnetic field strength that alerted von Klitzing to the effect. He "switched on" the field to apply it to a device through which a fixed current was flowing stabilized by a constant current source, and observed that when things stabilized, a digital voltmeter always showed the same Hall voltage across the sample to many significant figures. (The story is told that he first thought the voltmeter was broken!) Of course, each time the magnetic field was "turned on" was different, so the final field would never have been the same on each run of the experiment, and certainly would never have "accidentally" taken the precise "magic value" of the näive explanation. It is fortunate that von Klitzing switched on the magnetic field with a fixed current through the sample, rather than switched on the current at fixed field, as the coincidence of the unchanged digital voltmeter readings would then never have happened!

The real samples, though comparatively clean, do not have the translational invariance that makes each state in a given Landau level have exactly the same energy. A local electric potential at the center of a given circular Landau orbit varies randomly from point to point, sometimes raising and sometimes lowering the energy, "broadening" the Landau level. The initial attempts to explain the effect focused on this effect of disorder, and found that, while two-dimensional electron systems with disorder generally have "localized" states, this is modified in a magnetic field. In this case, the centers of the Landau orbits slowly precess (in opposite senses) around either local minima or local maxima of the potential, corresponding to localized states, but there is an energy at the center of the broadened Landau level at which the centers of the orbits move along open snakelike paths, and the states at that energy are "extended" as opposed to "localized." In this picture, there is no gap between the "HOMO" and the "LUMO" which have equal energies (now called the "Fermi energy"), and, as the magnetic field strength is changed, the Fermi energy moves to keep the number of occupied states constant, but the integer n measured by von Klitzing only changes when the Fermi energy goes through the special energy at which extended states exist. This provided an explanation in terms of the somewhat arcane theory of localization that at first sight is not obviously topological, but what is now obvious as a very characteristically topological property emerged when Bert Halperin pointed out the importance of edge states (Halperin, 1982).

These edge states are easily seen as semiclassically as counterpropagating "skipping orbits" that precess around the boundary of the system in the opposite sense to that of the Landau orbits, when a particle in a Landau orbit intersects the boundary, and bounces off it (see Figure 1). Even without disorder in the interior of the disk, so that the energy gap between Landau levels remains, there is a continuous distribution of energy levels at the edge of the system that pins the Fermi level and accommodates the "spectral flow" of states between Landau levels as the field magnetic strength is changed, and removes the need for "fine tuning" of the magnetic field to have an energy gap in the interior of the sample. While the number of states in a Landau level changes with magnetic field strength, the number of states cannot change, so the states must flow between Landau levels: the gapless edge states provide the necessary "plumbing" connections between the Landau levels so states can be redistributed between them as the magnetic field changes.

The unavoidable edge states that transport particles in one direction only around the edge allow the robustness of the QHE to be understood in terms of the boundary of the system, but it is also valuable to understand it in terms of the bulk properties of the interior of the system. This is where the TKNN formula found by David Thouless and collaborators (Thouless *et al.*, 1982) enters the story. Thouless was inspired by the famous "Hoftstadter butterfly'" spectrum (Hofstadter, 1976) that results when there is a periodic potential on the 2D surface as well as magnetic flux (Figure 2). In this case, the energy band structure can be solved when the magnetic flux through the unit cell of the periodic

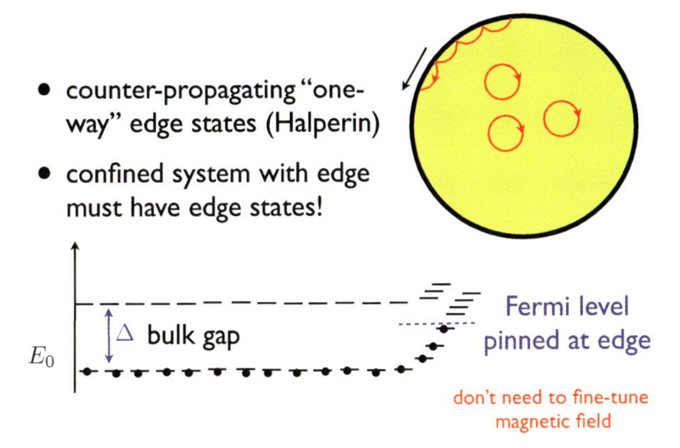

Figure 1. Simple energy-level picture for the integer quantum Hall effect, with an energy gap in the bulk stabilized by pinning of the Fermi level by gapless edge states. (The energy levels are show as as function of radius in a disk-shape sample.)

potential is a rational number p/q, where p and q are relative prime numbers with no common factors. The solution depends very delicately on the precise values of p and q, as it must be solved in an enlarged "magnetic unit cell" through which the magnetic flux must be an integer in units of Φ_0. The effect of the magnetic field is that each zero-field energy band that occurs in the absence of a magnetic field splits up into q energy bands, so that in going from a flux of 1/3 per unit cell to 100/301, what is one band at flux 1/3 splits up into 100 much narrower bands even though the flux change is very small!

A very clear argument formulated by Robert Laughlin (Laughlin, 1981) had already shown that in the absence of electron-electron interactions, if the Fermi level is inside a gap of the bulk electronic spectrum, the Hall conductivity σ^{xy} in the low-temperature limit $T \to 0$ had to be an integer multiple of $e/\Phi_0 = e^2/2\pi\hbar$. In the bottom left-hand corner of the "Hofstadter butterfly," where the magnetic flux through the unit cell is very small, the spectrum resembles that of simple Landau levels, with extremely narrow flat bands corresponding to a slightly widened Landau level, separated by large gaps. In this limit, the integer is just given by the number of filled Landau levels. But as the flux increases, the

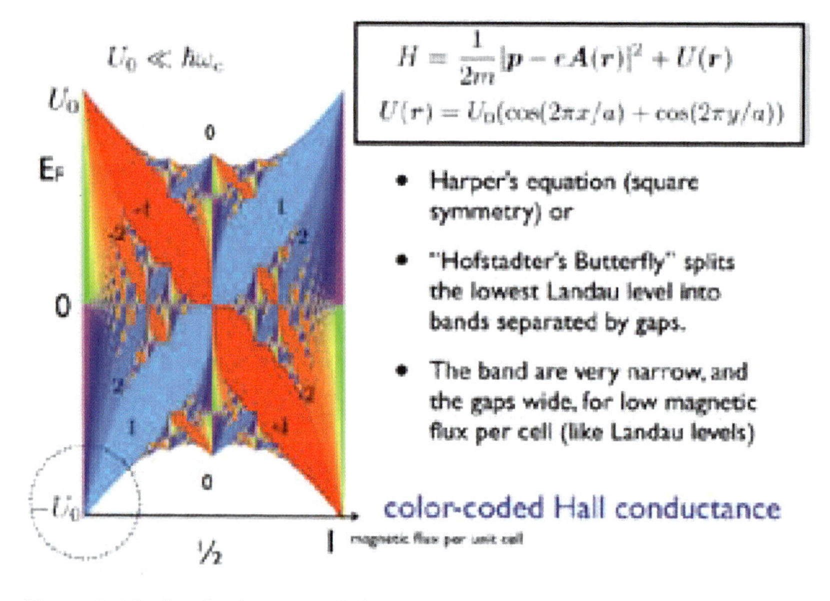

Figure 2. The "Hofstadter Butterfly" spectrum of electrons on a periodic lattice plus a uniform magnetic field, showing energy levels as a function of magnetic flux through a unit cell. The structure in the lower left corner becomes that of a system of simple Landau levels. Colors in gaps between subbands represent the different integer quantizations of the Hall effect if the Fermi level is in that gap. (Colored spectrum provided by D. Osadchy and J. Avron.)

Landau levels split up into an intricate pattern of sub-bands which are separated by many more gaps, which open and close as the magnetic field changes. When the Fermi level is in one of these new gaps, the question posed by TKNN was, what is the integer that defines the low-temperature Hall conductivity?

Even though TKNN were working in the enlarged "magnetic unit cell," the Bloch theorem remained valid, and showing the electronic wavefunctions had the form

$$\Psi_{kn}(\boldsymbol{r}) = u_n(\boldsymbol{k},\boldsymbol{r})e^{k\cdot r} \tag{1}$$

where $u_n(\boldsymbol{k},\boldsymbol{r})$ is a periodic function of $\boldsymbol{r}$ defined in the magnetic unit cell (MUC). Here $\boldsymbol{k}$ is a "Bloch vector" defined in the (magnetic) "Brillouin zone" (BZ) which in 2D is topologically equivalent to a torus, or doughnut shape. Using the fundamental Kubo formula for electrical conductivity, they found the formula

$$\sigma^{xy} = \frac{e^2}{2\pi\hbar}\sum_n\left(\frac{1}{2\pi}\int_{\mathrm{BZ}} d^2\boldsymbol{k}F_n^{xy}(\boldsymbol{k})\right)$$

$$F_n^{xy}(\boldsymbol{k}) = \frac{1}{2i}\int_{\mathrm{MUC}} d^2\boldsymbol{r}\left(\frac{\partial u_n^*}{\partial k_x}\frac{\partial u_n}{\partial k_y} - \frac{\partial u_n^*}{\partial k_y}\frac{\partial u_n}{\partial k_x}\right)$$

Here n labeled the occupied electronic bands below the Fermi level. The remarkable property was that the integral of each periodic function $F_n^{xy}(\boldsymbol{k})$ over the magnetic Brillouin zone was 2π times an integer, in agreement with Laughlin's result. TKNN realized that this had to be so, because $F_n^{xy}(\boldsymbol{k})$ could be written in the form

$$F_n^{xy}(\boldsymbol{k}) = \frac{\partial}{\partial k_x}\boldsymbol{A}_n^y(\boldsymbol{k}) - \frac{\partial}{\partial k_y}\boldsymbol{A}_n^x(\boldsymbol{k})$$

$$\boldsymbol{A}_n^i = \frac{1}{2i}\int_{\mathrm{MUC}} d^2\boldsymbol{r}\left(u_n^*\frac{\partial u_n}{\partial k_i} - u_n\frac{\partial u_n^*}{\partial k_i}\right)$$

leading to the key expression, as an integral around the Brillouin zone boundary (BZB) :

$$\sigma^{xy} = \frac{e^2}{2\pi\hbar}\sum_n\frac{1}{2\pi}\oint_{\mathrm{BZB}} dk_i\boldsymbol{A}_n^i(\boldsymbol{k}) \tag{2}$$

I learned from Marcel den Nijs and Peter Nightingale that their recollection is that the inclusion (in a single paragraph) of this remarkable explicit general formula in the paper (rather than formulas very specific to the Hofstadter model, which were the main aim of the paper) emerged as an "afterthought" while writing the paper! Another quote from den Nijs is that it was "the genius of David Thouless to choose the periodic potential generalization [to broaden the Landau level] rather than the random one, that was the essential step." This shows the power of choosing the right (and tractable) toy model for which a full and explicit calculation can be done. While there has been continuing interest to date in achieving a physical realization of the Hofstadter model, it had no relation whatsoever to the physical samples in which the integer quantum Hall effect was seen, for which the essentially intractable random potential was the physically-appropriate model, and the apparently-natural problem to study.

Shortly after the TKNN paper was published, Michael Berry discovered his famous geometric phase (Berry, 1984) of adiabatic quantum mechanics. In Berry's classic example, a spin with quantum number S is aligned along an axis represented by a unit vector $\hat{\Omega}$, with a direction that is slowly changed with time, defining a closed path on the unit sphere that finally returns to its original direction. Berry's result was that, in addition to the expected change of phase of the state with a rate proportional to its energy, there is an additional "geometric" change of phase that depends only on the geometry of the path, in this case given by the solid angle ω enclosed by the path (the area "enclosed" by the closed path on the surface of the unit sphere) times S. Looking at this more carefully, one sees that the notion of the area enclosed by the path is ambiguous, and the solid angle ω that it subtends is ambiguous up to multiples of 4π, but the physically-meaningful Berry phase factor exp $iS\omega$ is itself unambiguous because $2S$ is an integer. The influence of Berry's discovery of the geometric phase on modern developments in quantum theory cannot be overemphasized, and many consider that it deserves to get a fuller exposition in a future lecture in this series.

Both Berry's work and the TKNN formula were then brought to the attention of the mathematical physicist Berry Simon, who recognized (Simon, 1983) the connection between these formulas whereby the Berry phase could either be viewed as the integral of a "Berry connection" (analogous to the vector potential of electromagnetism) around a path, or by Stokes' theorem, as the integral of a "Berry flux" or "Berry curvature" through a surface bounded by the path. Furthermore, if the surface is a closed surface with no boundaries, its total Berry curvature or flux must be an integer multiple of 2π, and this integer is a topological invariant, the "first Chern class," technically of a "$U(1)$ fiber bundle" (the mathematical characterization of a quantum mechanical wavefunction) on a closed 2D manifold. This theorem is the close analog of the original Gauss-Bonnet theorem for integrals of the intrinsic (Gaussian) curvature over a 2D

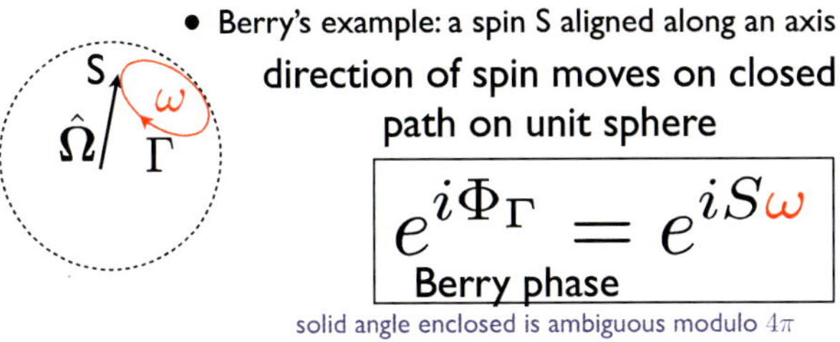

Figure 3. Berry phase for the adiabatic evolution of the state of a quantum spin aligned along a moving axis. The Berry phase Φ_Γ is the spin quantum number S times the solid angle subtended by the closed path Γ of the alignment axis $\hat{\Omega}$. After the axis returns to its inital orientation, the final quantum state is the initial state times the factor exp $i\Phi_\Gamma$ that depends geometrically on the path taken. Since the "solid angle subtended by the path" is ambiguous modulo 4π, $2S$ is topologically required to be an integer (which it is).

surface. If the surface is closed, like a sphere or a doughnut, the Gauss-Bonnet topological invariant counts the number of holes (the "genus" of the surface): it is this precise mathematical analogy that has given rise to the ubiquitous use of the movie showing a bagel or doughnut's topological equivalence to a coffee cup.

The identification of the TKNN formula as a topological invariant marked the beginning of the recognition that topology would play an important role in classifying quantum states themselves, in addition to the early discovery of the importance of topological excitations in the classical physics of the Berezinsky-Kosterlitz-Thouless transition (see J. Michael Kosterlitz's Nobel Lecture in this book) This invariant (the "Chern number" or "first Chern class," given by $\frac{1}{2}\pi$ times the integral of a Berry curvature over a 2D manifold) would remain the only known invariant in quantum condensed matter systems until the 2004 discovery by Kane and Mele (Kane and Mele, 2005) of a new "Z_2" invariant in time-reversal-invariant topological insulators, that led to the current explosion of new experimental and theoretical discoveries about topological states of matter.

The TKNN result was obtained for the bandstructure of electrons in uniform magnetic field with Landau levels that were split into Bloch bands by a periodic potential. In 1988, while analyzing a proposed realization of the "parity anomaly" by Fradkin, Dagotto and Boyanowsky (Fradkin *et al.*, 1986) I realized that the necessary condition for a quantum Hall effect was not a magnetic field, but just broken time-reversal invariance. This perhaps should have been seen as implicit in the TKNN result, but had not apparently been previously noted. I came up with a very simple model (Haldane, 1988) (see Figure 4) based on "a two-dimensional single sheet of graphite" (purely a "toy model" at that time, as the possibility that one day graphene sheets would be made then seemed like "science fiction") which I called a model for the "quantum Hall effect without Landau levels," based on standard Bloch states unlike the esoteric field-dependent ones of the Hofstadter model. This is now called the "quantum anomalous Hall effect" or "Chern insulator."

This state may also be called the first topological insulator, albeit one with broken time-reversal symmetry. It turns out that in 2D graphene, the "Dirac points" at the corners of the Brillouin zone where the conduction and valence band touch are stable only if both time-reversal and spatial inversion symmetry are unbroken, in which case the Berry curvature vanishes identically, and Berry phase factors for closed paths in the Brillouin zone are topological, with values $\exp i\varphi = \pm 1$, depending on whether their winding numbers around the Dirac points are even or odd. A gapped non-topological insulator state, investigated previously by Semenoff (Semenoff, 1984), results if *spatial inversion symmetry* is broken. In contrast the toy model I devised opens up a gap at the Dirac points to give a quantum Hall state by breaking *time-reversal symmetry*, through giving a chiral phase to second-neighbor hopping between states on the same sublattice. Once the gap opens and breaks the connection between conduction and valence

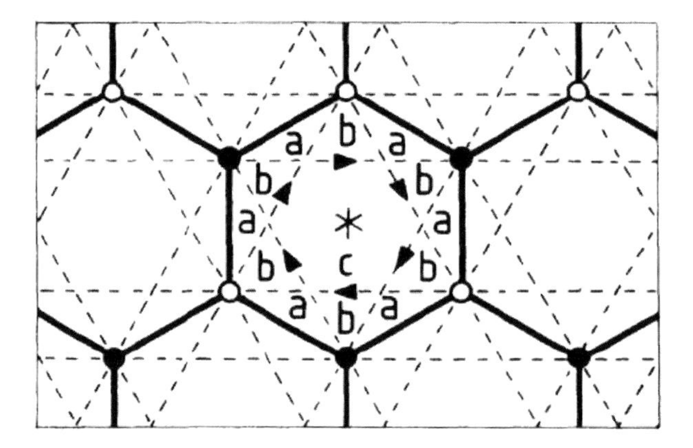

Figure 4. The simple graphene-like tight-binding "toy model"(Haldane, 1988) for the "broken-time-reversal topological insulator" or "Chern Insulator" that exhibits a zero-field "quantum anomalous Hall effect." Electrons "hop" along nearest-neighbor bonds (solid lines) with a real matrix element, and along second-neighbor bonds (dashed lines) with a complex matrix element, which has a positive phase for hopping in the direction of the arrow. Two conjugate copies (one for up-spin, one for down-spin electrons) were later combined by Kane and Mele to model a time-reversal-invariant topological insulator. The complex phases for hopping between second-neighbors introduces broken-time-reversal symmetry, which could come from a ferromagnetically-ordered magnetic dipole at the center (*) of each hexagonal cell, pointing normal to the 2D plane. The dipoles give rise to different magnetic flux through regions *a*, *b*, and *c* of the unit cell, but no net magnetic flux, leaving the standard Bloch structure intact.

bands in the interior of the system, they individually have opposite-sign Chern numbers ±1, and undirectional edge states are present. This is conveniently seen on the "zig-zag" edge, where in the absence of second-neighbor hopping, an zero-energy edge state spans one third of the surface Brillouin zone, connecting the projected Dirac points in a way reminiscent of the recently discovered "Fermi arc" surface states that connect the projected Dirac points of 3D Weyl semimetals found recently by Ashvin Vishwanath and coworkers (Wan *et al.*, 2011). When a gap opens, whether by breaking inversion symmetry, time-reversal symmetry, or both, the edge states must connect to either the valence or conduction band at each of the now gapped or "massive" Dirac point, leading to four possible outcomes (see Figure 5).

This simple toy model has proved very fruitful: rather surprisingly, while the original model was for charged fermions, it was translated from the language of electrons to that of neutral bosons and a photonic crystal (Haldane and Raghu, 2008), showing how topological "one-way" edge states could occur there too, initiating the growing field of topological photonics, and has been implemented experimentally with microwave-scale photonics.

In 2004, the possibility of a time-reversal-invariant analog of the Hall effect (the "spin-Hall effect") was under discussion, and a time-reversal invariant (TRI) model was considered by Charles Kane and Eugene Mele (Kane and Mele, 2005), who combined two conjugate copies of my model, one for spin-up electrons for which the valence band had Chern number ±1 and one for spin-down electrons where the valence band had the opposite value ∓1; on the edges, spin-up and spin-down edge modes propagated in opposite directions. Since the total Chern

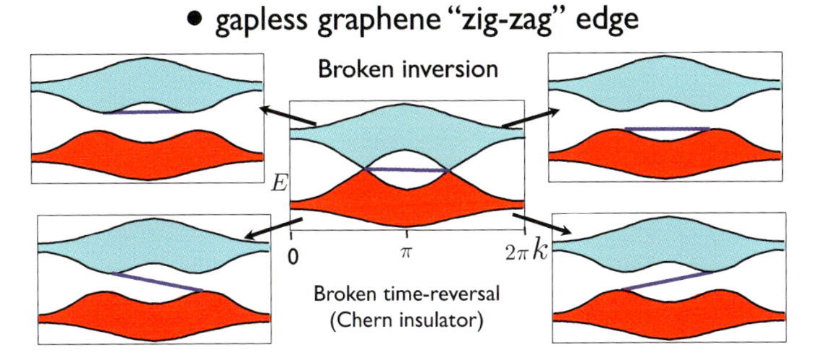

Figure 5. "Zig-Zag" edge of graphene after perturbation by terms that break inversion or time-reversal symmetry. The unperturbed edge has an edge state joining the projections of the two Dirac points where the filled valence band (red) touches the empty conduction bands (green). When a gap is induced by the perturbation, there are four ways the edge-state can be connected, two of which are topological, and connect conduction and valence bands.

number of the valence band vanished, there was no quantum Hall effect. Naively, it might have been expected that the gapless edge modes were not protected from backscattering and mixing, thus becoming gapped, if spin-non-conserving Rashba-type spin-orbit coupling was added to the system. However Kane and Mele discovered by a numerical calculation that, so long as time-reversal invariance was unbroken, the edge modes were in fact protected by a previously-unexpected "Z_2" topological invariant related to Kramers degeneracy. This new invariant had a 3D generalization discovered independently and simultaneousy in 2007 by Joel Moore and Leon Balents (Moore and Balents, 2007), Rahul Roy (Roy, 2009), and Liang Fu, Kane and Mele (Fu *et al.*, 2007), which led to the experimental discovery of the 3D time-reversal-invariant topological insulators (TI). This finally led to the reported experimental realization (Chang *et al.*, 2013) by Qikun Xue's group at Tsinghua University, Beijing, of the quantum anomalous Hall effect in thin films of TRI TI's which had been doped with magnetic material.

I now turn to the other (1981) discovery recognized by this Nobel prize: the novel quantum spin liquid states of the one-dimensional integer-spin antiferromagnets, which (for odd integral spin) have recently been classified by Xiao-Gang Wen and collaborators (Chen *et al.*, 2013) as "symmetry protected topological states" (SPT states), where the protective symmetries are time-reversal invariance and spatial inversion symmetry. The conventional magnetic ground states generally studied prior to 1981 were typically unentangled states, usually with long-range magnetic order, that could be modeled as a direct product of independent states on each sites, such as a ferromagnet (. . .↑↑↑↑↑↑. . .) or a Néel antiferromagnet (. . .↑↓↑↓↑↓↑. . .). The spin configurations shown have spins aligned parallel or antiparallel to the z-axis, but in the case of Heisenberg (isotropic) magnets these states spontaneously break rotational symmetry, and the alignment axis can point in any direction. In the case of the Heisenberg ferromagnet, the alignment direction is the direction of a macroscopic conserved angular momentum vector, and the conservation law for angular momentum of a rotationally-invariant system protects the ferromagnetic "order parameter" (the alignment direction) against deviation by zero-point fluctuations.

However, in the antiferromagnetic case, there is no conservation law to give protection against zero-point fluctuations, The celebrated Mermin-Wagner theorem that posed the key paradox in the case-of the finite temperature Kosterlitz-Thouless transition provides a similar result for quantum systems in one spatial dimension: without protection by a conservation law, the ground state of a quantum system with a continuous symmetry cannot exhibit long-range order of an order-parameter that breaks that symmetry. In higher dimensions, Heisenberg systems can exhibit antiferromagnetic broken symmetry ground states with gapless collective Goldstone-mode excitations known as (antiferromagnetic) spin waves that are small harmonic fluctuations of the Néel order-parameter around its uniform ground state configuration. But, if the assumption of long-range Néel antiferromagnetic order is made in the case of the one-dimensional spin-S antiferromagnet, it is easily found that the effect of the harmonic zero-point fluctuations would be to destroy the assumed long-range order.

At this point the power of exact (but not fully understood) mathematical results to sow confusion enters the story! In 1931, before he went on to discover how nuclear fusion powered the sun (and later to become David Thouless's thesis advisor at Cornell), Hans Bethe also worked on the one-dimensional Heisenberg chain as a "toy model" for magnetism, and discovered a remarkable "Ansatz"(Bethe, 1931) that provided exact solutions for eigenstates of the 1D model with $S = \frac{1}{2}$ and nearest-neighbor exchange, allowing the eigenvalue spectrum to be explicitly obtained. Unfortunately, it took more than sixty years for the underlying special mathematical structure of the model to be understood, and in the 1970s, only energy levels and thermodynamic properties could be extracted from the exact solutions, but not the correlation functions. However, the spectrum of low-energy eigenvalues superficially resembled the predictions of spin-wave theory with the only apparent change being that the speed of long-wavelength spin waves differed from the predictions of spin-wave theory by a factor of $\frac{1}{2}\pi$.

While the details of Bethe's Ansatz were somewhat arcane and mysterious, this was generally taken as confirmation that the spin-wave description was more-or-less correct despite the known destruction of true long-range order by quantum effects. In fact, we now know that the elementary excitations of the model that Bethe solves have *no relation whatsoever to spin waves*: they are spin-$\frac{1}{2}$ topological excitations (Faddeev and Takhtajan, 1981) that are created in pairs, and now known as "spinons," but even in the 1970s it ought to have been noticed that, when expressed in terms of the velocity of long wavelength excitations, the specific heat predicted by spin-wave theory was exactly twice the exact result extracted from the Bethe Ansatz, implying no relation of any kind between the spin-wave theory and low-energy excitations of Bethe's solvable model.

To get around the long-standing intractability of the problem of extracting correlation functions from Bethe's solution, new techniques for treating the problem emerged in the early 1970's from the work of Alan Luther and Ingo Peschel. Again old work (even older than Bethe's!) was important: they used the Jordan-Wigner (Jordan and Wigner, 1928) transformation that maps the one-dimensional magnet with nearest-neighbor exchange into a model of spinless

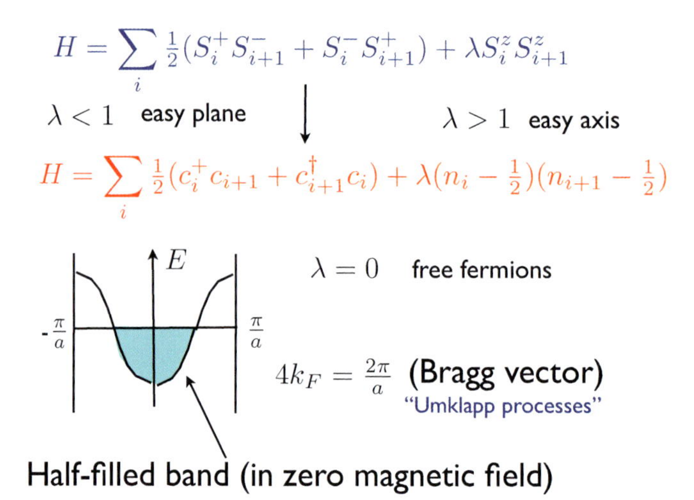

Figure 6. The Jordan-Wigner transformation maps the $S = \frac{1}{2}$ Heisenberg chain with zero magnetization into a half-filled interacting band of spinless fermions, where $4k_F$ is a Bragg vector.

fermions that move in one dimension by hopping between nearest neighbor sites on the lattice, with interactions between particles on neighbor sites. When the Heisenberg exchange coupling $J\vec{S}_n \cdot \vec{S}_{n+1}$ is decomposed into $J^x S_n^x S_{n+1}^x + J^y S_n^y S_{n+1}^y + J^z S_n^y S_{n+1}^y$, with $J^x = J^y = J^{xy}$, the $S = \frac{1}{2}$ "quantum XY" model with $J^z = 0$ is mapped into a non-interacting free-fermion model that can be completely and explicitly solved to extract all physical properties.

In the mid-1960s, Joaquin Luttinger (Luttinger, 1963) had noticed that a "toy model" of interacting spinless fermions with a linear Dirac-like dispersion and an interaction restricted to low momentum-transfer forward scattering should be solvable using the "Tomonaga bosons" found by Sin-itiro Tomonaga (Tomonaga, 1950). There were problems with Luttinger's solution, which was subsequently elucidated by Daniel Mattis and Elliott Lieb (Mattis and Lieb, 1965), and from this came the remarkable "bosonization" technique (representation of one-dimensional fermions in terms of Tomonaga's harmonic oscillator modes) explicitly formulated by Schotte and Schotte (Schotte and Schotte, 1969) in their 1969 simplified treatment of the "X-ray edge singularity" problem.

In 1975, Luther and Peschel (Luther and Peschel, 1975) adapted the new "bosonization" techniques to treat the easy-plane antiferromagnet with non-zero $J^z = \lambda|J^{xy}|$, with $|\lambda| < 1$, which they mapped into a "(1+1)-dimensional" effective quantum field theory could be treated by the "bosonization" mapping to a harmonic oscillator problem. This treatment was precisely equivalent (after a "Wick rotation" from (1+1)-dimensional Lorentz-invariant space-time to 2-dimensional Euclidean space) to the low-temperature "topologically-ordered" phase of the classical 2D XY model which Kosterlitz and Thouless were also studying at that time, with Néel correlations that decayed algebraically with non-universal power laws, where for large $|n - n'|$,

$$\langle S_n^x S_{n'}^x \rangle = \langle S_n^y S_{n'}^y \rangle \propto (-1)^{n-n'} |n-n'|^{-\eta}, \ \langle S_n^z S_{n'}^z \rangle = \propto (-1)^{n-n'} |n-n'|^{-\eta^{-1}} \quad (3)$$

where η varied with the coupling-constant ratio λ. Furthermore, introducing full "XYZ" anisotropy ($J^x \neq J^y$) maps the model to a massive field theory (the "sine-Gordon" model) with a excitation gap that depends algebraically on $J^x - J^y$ with an exponent fixed by η.

By that time Bethe's exact solution of the $S = \frac{1}{2}$ isotropic Heisenberg "XXX" model ($J^x = J^y = J^z$) had been extended to the full XYZ model by Rodney Baxter, following the identification of the Yang-Baxter algebra as the key ingredient that allowed Bethe's Ansatz to solve the model. Luther and Peschel were able to use this to indirectly obtain the value of the correlation exponent η as a function of λ for the easy plane "XXZ" model ($|\lambda| \leq 1$). They found that for positive (antiferromagnetic) J^z, η increases from $\frac{1}{2}$ at the fully-solvable "free-fermion" XY point with $J^z = 0$, reaching the consistent value $\eta = \eta^{-1} = 1$ at the antiferromagnetic Heisenberg "XXX" point $\lambda = 1$, while for negative (ferromagnetic) J^z, it decreases to zero when $\lambda = -1$, where the ground state develops long-range order with a conserved order parameter. Notably, the Luther-Peschel field-theory treatment failed to explain the gap that opens for $\lambda > 1$, when the model changes from an an easy-plane to an easy-axis antiferromagnet.

In 1979 I was working on the precise formulation of the bosonization method and found (Haldane, 1981b) that the zero-momentum modes of the fermion density needed to represented by action-angle variables as opposed to Tomonaga's harmonic oscillator modes which represented the modes carrying finite momentum. These action-angle degrees of freedom are topological in nature, and resolved the "mystery" of how one-dimensional fermions could apparently be represented in terms of "bosons" (harmonic oscillator modes): the representation in fact is constructed using harmonic oscillators *plus* topological winding-number degrees of freedom. This meant that the detailed structure of the excitation spectrum of a spinless fermion model with periodic boundary conditions contained two types of topological excitations (separate winding numbers of left- and right-moving fermion fields) as well as Tomonaga's sound-wave modes.

Knowledge of the energies of the two topological excitations fixed not only the speed of sound, which could be independently checked, but also the correlation exponent η, and applying this to the Bethe Ansatz solution of the XXZ model in zero field (or the equivalent Jordan-Wigner fermion model with a half-filled band) for which Luther and Peschel had indirectly found the exact value of the exponent η as a function of the couplings, I was able to confirm that the new expressions in terms of winding-number energies were also consistent, correct, and quite general.

This opened up by the possibility of extracting exact correlation exponents from Bethe Ansatz solutions of some models exhibiting one-dimension criticality by using the energies of their various topological excitations to fit them to what I called an effective "Luttinger liquid"(Haldane, 1981b) (or perhaps more properly a "Tomonaga-Luttinger liquid") modeled by a Luttinger model. These developments occurred before the later appearance of more powerful (1+1)-dimensional conformal field theory methods, and "Luttinger liquids" turn out to be systems decomposable into Abelian representations of the Virasoro algebra, with the constraints of Lorentz invariance removed.

When I applied this new picture to the full parameter space of the Bethe Ansatz solutions of the XXZ spin chain (which required numerical solution of the Bethe Ansatz integral equations away from half-filling of the fermion bands) it became immediately obvious from inspection of the results that the missing ingredient in Luther and Peschel's work was the omission of the "Umklapp" process by which, at half-filling of the band (where $4k_F$ is a Bragg vector), so scattering processes where the momentum changes by $4k_F$ allow two low-energy "left-moving" electrons (each with momentum near $-k_F$) to scatter into two low-energy right-moving electron states, each with momentum near k_F.

At first sight this should be represented by a term $\Psi^\dagger_R(x)\Psi^\dagger_R(x)\Psi_L(x)\Psi_L(x)$, but this is ruled out by the Pauli principle, which is presumably why Umklapp was not considered in the original work by Luther and Peschel, but the next-order term $(\Psi^\dagger_R(x)\partial_x\Psi^\dagger_R(x))(\Psi_L(x)\partial_x\Psi_L(x))$ is allowed, and when "bosonized" becomes $\cos 2\theta \equiv \cos 2(\varphi R(x) - \varphi L(\mathbf{x}))$. In the quantum analog of of the Berezinsky-Kosterlitz-Thouless (BKT) transition, this is a *double-vortex unbinding transition*, which is allowed, but the standard *single-vortex unbinding transition* is forbidden by momentum conservation. The translation of the usual single-vortex BKT process from classical 2D to quantum (1+1)D would be represented by a term $\cos\theta$ which becomes "relevant" (causing a gap to open) when $\eta > \frac{1}{4}$. The generalization of this is that a $\cos m\theta$ term becomes relevant when $\eta > \frac{1}{4}m^2$, which is perfectly consistent with the double-vortex term $\cos 2\theta$ becoming relevant (in the absence of the single-vortex term) exactly at the isotropic XXX point when $\eta = 1$.

This missing ingredient completed the field-theoretic picture of the $S = \frac{1}{2}$ begun by Luther and Peschel. It also removed the apparent "special" nature of the $S = \frac{1}{2}$ model which seemed to come from its mapping to a fermion model. The bosonization now provided a representation in terms of two "chiral" (left-moving and right-moving) topological winding number fields $\varphi_L(x)$ and $\varphi_R(x)$, without any obvious relation to the value of the spin S of the underlying spin chain.

A planar "XY" spin can be visualized as a "compass needle" that points in a 2D direction $(\cos(\varphi(x)), \sin(\varphi(x))$, and if it obeys a periodic boundary condition on a circle of circumference L, then $\varphi(x + L) = \varphi(x) + 2\pi W$ where the "winding number" W is a topological invariant that cannot change if the field $\varphi(x)$ varies smoothly with x. In the classical 2D XY model, $\varphi(x,y)$ is a smooth function except at singular points (x_0,y_0) which are the centers of vortices. In the quantum (1+1)D model these become space-time points (x_0, t_0) representing tunneling events (which have been called "instantons") at which the winding number changes, through a singular process that occurs briefly at a 1D space point x_0 and during an instant of time near t_0.

It turned out that for a spin-S easy-plane spin chain with zero magnetization along the z-axis, the usual "single-vortex" BKT "instanton" process is generically present, but is forbidden by an exact *quantum interference* process if $2S$ is odd. This highlights a difference between the classical statistical mechanics of the 2D BKT transition and the (1+1)D quantum version. In the classical 2D model, the

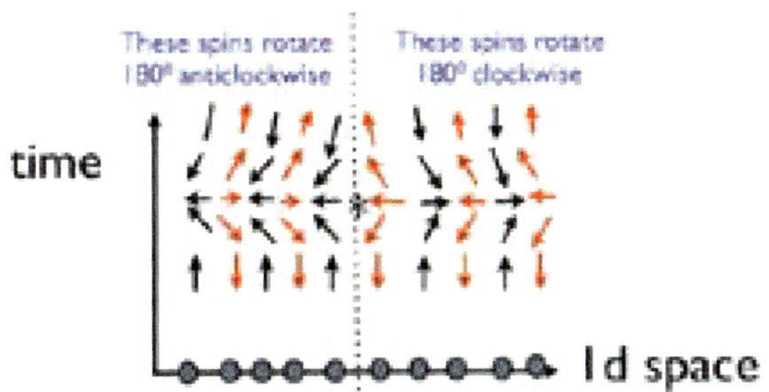

Figure 7. In (1+1)D space-time, the analog of the 2D vortex is an "instanton" tunneling process where the topological winding number of the easy-plane spin-chain changes. This process is centered on a "bond" between consecutive sites on which the local Néel order breaks down for a short time interval.

strength of the vortex term in the Boltzmann factor is a real positive fugacity factor, but in the quantum (1+1)D model, it is a complex amplitude for tunneling between topologically-different configurations with different winding number, and is real-positive or negative in time-reversal-invariant models. This means that quantum interference between competing instanton processes can occur.

In this case the tunneling process is centered at the midpoint of a "bond" between two neighboring spins. Assuming the spin chain is invariant under spatial translation by one site, the *magnitude* of the amplitude for the tunneling process must be the same independent of which bond it is centered on. But when two such processes on consecutive bonds are compared, the main difference is that one spin that rotated 180° clockwise now rotates 180° anticlockwise, so the two processes differ by a net rotation of one spin by 360°, with the histories of all other spins essentially identical. The fundamental difference between a spin where $2S$ is even and one where $2S$ is odd is that in the latter case, there is quantum state has a sign change as a result of the rotation. This means that, providing the exchange energy is the same on all bonds, there is destructive interference between instanton tunneling events on neighboring bonds is $2S$ is odd, but constructive interference if $2S$ is even.

This provides the "topological" explanation of why the instanton process that becomes relevant as the anisotropy of the spin-$\frac{1}{2}$ XXZ chain changes from easy-plane to easy axis corresponds to a *double vortex* of the BKT transition. It only drives the instability of the topologically-ordered easy-plane phase because the dominant *single vortex* process is canceled by destructive interference when $2S$ is odd. However, for integer S it is present, and the BKT transition will occur once the correlation exponent rise to the limiting value $\eta = \frac{1}{4}$ when tunneling between states with different winding number becomes relevant, topological order breaks down, and a gap in the excitation spectrum opens up. At this critical point the Néel correlations of $(S^x_n S^x_{n'})$ and $(S^y_n S^y_{n'})$ fall off much slower than than those of $(S^z_n S^z_{n'})$ implying that the transition happens before the isotropic Heisenberg point is reached. It is also a transition to a *unique* (singlet) ground state, while the double-vortex process conserves winding-number modulo 2, and leads to a two-fold degenerate (doublet) ground state when it becomes relevant.

From these results, it became clear that the progression from easy-plane to easy-axis models was quite different in the two cases of integer-S and half-odd-integer-S antiferromagnets. As λ increases, the chain with $2S$ odd has a direct "double-BKT" transition at $\lambda = 1$ from the topologically-ordered gapless easy-plane antiferromagnet with $\lambda < 1$ to the gapped easy-axis antiferromagnet with a doublet broken-symmetry Ising-Néel ground state. In contrast, the chain with even $2S$ has an a standard BKT transition at $\lambda = \lambda^c < 1$to a singlet gapped spin-liquid state with no broken symmetry, followed by a second Ising-type transition at $\lambda = \lambda^c > 1$ to the easy-axis Ising-Néel state.

These arguments exposed a fundamental topological difference between antiferromagetic Heisenberg (isotropic) quantum spin-S chains with $2S$ even and those with $2S$ odd, which contradicted the then-prevailing belief that the value of S entered as a continuous parameter as an expansion in powers of S^{-1} analogous to a semiclassical expansion in powers of $\hbar$. In this view, the asymptotic long-distance behavior of $\langle \vec{S}_n \cdot \vec{S}_{n'} \rangle$ would behave as $(-1)^{n-n'} |n - n'|^{-\eta}$, where $\eta(S^{-1})$ was a smooth function of S^{-1} that vanishes as $S^{-1} \to 0$.

My apparently-heretical 1981 claim, that there was a fundamental difference between one-dimensional quantum antiferromagnets with integer and half-odd-integer S, was presumably not well-enough explained, and the original paper (Haldane, 1981a) was rejected by a number of journals, and referred to by sceptics as a "conjecture," a description that seems to have stuck! By the time the paper was finally published (Haldane, 1983a), it had been significantly rewritten to emphasise the isotropic Heisenberg case, and the original preprint was eventually apparently lost, as this was years before preprints were stored on the internet. Happily, I recently recovered a copy that had been preserved by Jenő Sólyom, and placed it in the arXiv repository (Haldane, 1981a) for historical interest. Subsequently numerical exact diagonalization studies by Botet and Jullien (Botet and Jullien, 1983) found evidence for it, and finally, neutron scattering studies by Bill Buyers (Buyers *et al.*, 1986) on the quasi-one dimensional organic Nickel compound NENP provided experimental confirmation that the ground state of the spin-1 antiferromagnet was a singlet with an excitation gap.

The underlying reason that my 1981 result was so unexpected was that the spin-liquid state of the integer spin-1 chain was an early example of "topological quantum matter." The discovery predated Berry's 1983 discovery of the Berry phase, which in spin systems confirmed that the spin quantum number S had a topological role which relied on the value of $2S$ being an integer. Initially, from the standard Hamiltonian formulation used by condensed-matter physicists, it seemed mysterious that there seemed to be two distinct ways to apply quantum mechanics to a continuum field theory description of quantum antiferromagnetic spin chains, the "$O(3)$ non-linear sigma model," one for half-odd-integer spins, and the other for integer spins. In 1983, a very useful lead came from a discussion I had with Edward Witten, who mentioned that in the Lagrangian formulation favored by particle physicists, the sigma model could have an additional "topological term," which disappeared in the Hamiltonian formulation, and had no effect in the classical limit. This term is parameterized by an angle θ; and it was easy to use a formulation in terms of the Berry phases of the paths traced out by individual spins to show that this angle was $2\pi S$, taking the value 0 modulo 2π for integers spins, and π modulo 2π for half-odd-integer spins (these are the only two values compatible with time-reversal symmetry). This angle parameter is related to the "axion angle" introduced in high-energy physics in connection with the "strong-CP-violation" problem, and more recently in the electrodynamic description of "strong topological insulators" by Xiao-Liang Qi, Taylor Hughes and Shoucheng Zhang (Qi *et al.*, 2008), where the analogous "topological angle" takes the value $\theta = 0$ for non-topological TRI insulators, and $\theta = \pi$ for the strong 3D TRI topological insulators. The discovery of the "theta-term" in the Lagrangian form of the field theory of the one-dimensional anti-ferromagnets seems to mark the time after which the Lagrangian formulation started to become ubiquitous in theoretical quantum condensed-matter physics, and it is now a standard tool that complements Hamiltonian descriptions.

A simple model state that captures the essence of the gapped integer-S 1D antiferro-magnet was subsequently discovered by Ian Affleck, Tom Kennedy, Hal Tasaki, and Elliot Lieb (Affleck *et al.*, 1987), which is also the exact ground state of a modified "toy model" (the "AKLT model), which is particularly revealing, as it shows up the novel nature of quantum entanglement in the topological state. In this picture, a spin-1 object is viewed as a symmetric state of two spin-$\frac{1}{2}$ "half-spins," each of which can form an entangled singlet "valence bond" state

$$|\Psi\rangle = \frac{1}{\sqrt{2}}\left(|\uparrow\downarrow\rangle - |\downarrow\uparrow\rangle\right) \tag{4}$$

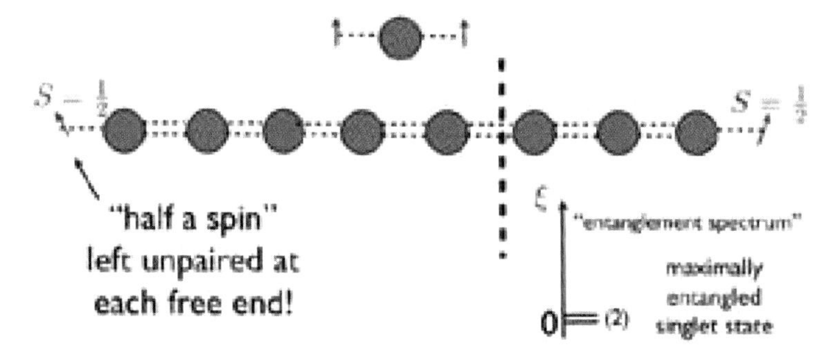

Figure 8. The $S = 1$ AKLT state treats each spin as a symmetric combination of two $S = \frac{1}{2}$ "half-spins," one of whch forms a singlet valence bond with a "half-spin" of the neighbor to the right, the other with the neighbor to the left. An unused spin-$\frac{1}{2}$ is left at each open end of the chain, and the "entanglement spectrum" consists of a single doublet.

by pairing with one of the half-spins of each neighbor. If the magnetic chain has free ends (*i.e.*, is "open"), this leaves an unpaired spin-$\frac{1}{2}$ at each end of the chain. This model also reveals the essentially "entangled" nature of the state: if the chain is cut in two, unpaired spin-$\frac{1}{2}$ degrees of freedom appear on either side of the cut, and the model state has a very simple characteristic "entanglement spectrum" (Li and Haldane, 2008) of a single spin-$\frac{1}{2}$ doublet. The feature that that all states in the entanglement spectrum are doublets, and that free ends of a long open spin-1 chain carry local spin-$\frac{1}{2}$ degrees of freedom is true for all states in the same topological class as the AKLT model, including the standard spin-1 Heisenberg antiferromagnet that I originally studied. (See Figure 8.)

The edges of the (integer) spin-S chain have local spin-$\frac{1}{2}S$ degrees of freedom, but since the elementary gapped bulk excitations are spin-1 magnons which can bind to the edge, the edge spins are topologically protected only when S is an *odd* integer. The final classification (Chen *et al.*, 2013) is that only the odd-integer-S state is a "symmetry protected topological state" (SPT state) protected by either time-reversal symmetry or spatial inversion, with a generic two-fold degeneracy of states in the entanglement spectrum.

Over the years, studies of topological state of the $S = 1$ Heisenberg antiferromagnet have been remarkable fruitful. The detailed study of its topological stability was the starting point that led a unified classification of SPT states in both one dimension and higher dimensions by Xiao-Gang Wen and collaborators (Chen *et al.*, 2013). In addition, its entanglement spectrum lies at the heart of the "density-matrix renormalization group" (White, 1992) and "matrix-product state" techniques that were in part developed for testing and verifying the so-called "Haldane conjecture." The features of unexpected topologically-protected edge states recur again and again in connection with "topological state of matter," for example in the "Majorana modes" that appear at the edge of topological superconducting wires, where the simple "toy model" introduced by Kitaev (Kitaev, 2001) plays a similar role to the AKLT model, and are now considered to be a possible platform for future topological quantum information processing. It is surprising how rich the developments stemming from the surprise discovery of topological phases of matter around 1980 have been.

Looking back at how this new field of topological quantum matter has developed since the initial discoveries in about 1980, I am struck by how important the use of stripped down "toy models" has been in discovering new physics. It also used to be thought that one-dimensional models were just "homework exercises" to be carried out before tackling the "real" three dimensional systems. In fact, partly because the effects of quantum fluctuations are more dramatic in low dimensions, we have found many interesting phenomena, in in doing so, a whole new way to look at condensed matter, and the exotic "topological states" that quantum mechanics make possible.

It has been my privilege to have been able to participate in opening up this field, to which many others have added amazing discoveries, and which has led to dreams of new quantum information technologies. I thank the Royal Swedish Academy of Sciences for honoring my co-Laureates and myself, and indeed our exciting subfield of physics.

REFERENCES

1. Affleck, I., T. Kennedy, E. H. Lieb, and H. Tasaki (1987), *Phys. Rev. Lett.* **59**, 799.
2. Berry, M. V. (1984), Proc. R. Soc. Lond. A **392** (1802), 45.
3. Bethe, H. (1931), *Z.Phys.* **71** (3), 205.
4. Botet, R., and R. Jullien (1983), *Phys. Rev.* B **27**, 613.
5. Buyers, W. J. L., R. M. Morra, R. L. Armstrong, M. J. Hogan, P. Gerlach, and K. Hirakawa. (1986), *Phys. Rev. Lett.* **56**, 371.
6. Chang, C.-Z., J. Zhang, X. Feng, J. Shen, Z. Zhang, M. Guo, K. Li, Y. Ou, P. Wei, L.-L. Wang, Z.-Q. Ji, Y. Feng, S. Ji, X. Chen, J. Jia, X. Dai, Z. Fang, S.-C. Zhang, K. He, Y. Wang, L. Lu, X.-C. Ma, and Q.-K. Xue (2013), *Science* **340** (6129), 167.
7. Chen, X., Z.-C. Gu, Z.-X. Liu, and X.-G. Wen (2013), *Phys. Rev.* B **87**, 155114.
8. Faddeev, L., and L. Takhtajan (1981), *Physics Letters* A **85** (6), 375.
9. Fradkin, E., E. Dagotto, and D. Boyanovsky (1986), *Phys. Rev. Lett.* **57**, 2967.
10. Fu, L., C. L. Kane, and E. J. Mele (2007), *Phys. Rev. Lett.* **98**, 106803.
11. Haldane, F. D. M. (1981a), ILL preprint SP81/95, arXiv:1612.00076.
12. Haldane, F. D. M. (1981b), *Journal of Physics C: Solid State Physics* **14** (19), 2585.
13. Haldane, F. D. M. (1983a), *Phys.Lett.* **93A**, 464.
14. Haldane, F. D. M. (1983b), *Phys. Rev. Lett.* **50**, 1153.
15. Haldane, F. D. M. (1988), *Phys. Rev. Lett.* **61**, 2015.
16. Haldane, F. D. M., and S. Raghu (2008), *Phys. Rev. Lett.* **100**, 013904.
17. Halperin, B. I. (1982), *Phys. Rev.* B **25**, 2185.
18. Hofstadter, D. R. (1976), *Phys. Rev.* B **14**, 2239.
19. Jordan, P., and E. Wigner (1928), *Z. Phys.* **47** (9), 631.
20. Kane, C. L., and E. J. Mele (2005), *Phys. Rev. Lett.* **95**, 146802.
21. Kitaev, A. Y. (2001), *Physics-Uspekhi* **44** (10S), 131.
22. Klitzing, K. v., G. Dorda, and M. Pepper (1980), Phys. Rev. Lett. **45**, 494.
23. Laughlin, R. B. (1981), *Phys. Rev.* B **23**, 5632.
24. Li, H., and F. D. M. Haldane (2008), *Phys. Rev. Lett.* **101**, 010504.
25. Luther, A., and I. Peschel (1975), *Phys. Rev.* B **12**, 3908.
26. Luttinger, J. M. (1963), *Journal of Mathematical Physics* **4** (9), 1154.
27. Mattis, D. C. M., and E. H. Lieb (1965), *Journal of Mathematical Physics* **6** (2), 304.
28. Moore, J. E., and L. Balents (2007), *Phys. Rev.* B **75**, 121306.
29. Qi, X.-L., T. L. Hughes, and S.-C. Zhang (2008), *Phys. Rev.* B **78**, 195424.
30. Roy, R. (2009), *Phys. Rev.* B **79**, 195322.

31. Schotte, K. D., and U. Schotte (1969), *Phys. Rev.* **182**, 479.
32. Semenoff, G. W. (1984), *Phys. Rev. Lett.* **53**, 2449.
33. Simon, B. (1983), *Phys. Rev. Lett.* **51**, 2167.
34. Thouless, D. J., M. Kohmoto, M. P. Nightingale, and M. den Nijs (1982), *Phys. Rev. Lett.* **49**, 405.
35. Tomonaga, S. (1950), *Progress of Theoretical Physics* **5** (4), 544.
36. Tsui, D. C., H. L. Stormer, and A. C. Gossard (1982), *Phys. Rev. Lett.* **48**, 1559.
37. Wan, X., A. M. Turner, A. Vishwanath, and S. Y. Savrasov (2011), *Phys. Rev.* B **83**, 205101.
38. White, S. R. (1992), *Phys. Rev. Lett.* **69**, 2863.

J. Michael Kosterlitz. © Nobel Prize Outreach AB. Photo: A. Mahmoud

J. Michael Kosterlitz

Biography

CHILDHOOD

I was born on June 22, 1943 in wartime Aberdeen, Scotland and lived there for the first sixteen years of my life. My parents, Hans Walter and Johanna Maria Kosterlitz (Gresshöner) had fled Hitler's Germany in 1934 because my father, a non-practicing Jew, came from a Jewish family and was forbidden to marry a non-Jewish woman like my mother or to be paid as a medical doctor in Berlin. Under the circumstances, my father decided that it would be in his best interests to leave Germany and accept the offer of a lectureship at Aberdeen University. My mother, who came from a conventional Christian German family, decided that she would follow my father to Britain so that they could be married, which they were in Glasgow.

I had a happy childhood in Cults which used to be a small village just outside Aberdeen and separated by farms and fields from the city but now is just another suburb of a much larger city. I was raised as a British child unaware of my German origins, although I was aware that my parents were different from those of my friends because they had a secret language used to communicate when they wanted to exclude me. My parents would have nothing to do with Germany and spoke only English at home except occasionally under special circumstances. As a result, I grew up speaking only English and had to wait several years before learning some basic German at school. In fact, for several years I did not know I was actually of German Jewish origin nor did I know what being Jewish meant. The only thing I knew was that a boy in my class got some extra vacation because of his Jewish religion. When this became known among the class, everyone wanted to change their religion for the extra holidays. My father had no interest in religion and left all instruction in these matters to my mother, who was a devout Christian. I was a nominal church going Christian until I left home for Cambridge University on a scholarship when, to my great relief, I could drop all religion and become my natural atheist self.

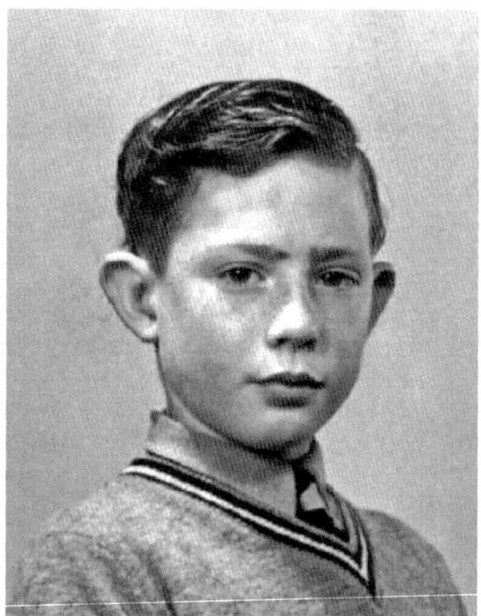

Figure 1. Left: my father and a very young Mike outside our granite house in Aberdeen, 1946. Right: Mike as a schoolboy, 1949.

EDUCATION

My early schooling was in Aberdeen at a semi private school, Robert Gordon's College, which I attended from kindergarten up to age sixteen. There I had a broad education including the sciences, mathematics, history, geography, Latin and French with a strong Aberdonian accent. Some of the teaching left much to be desired particularly in physics where even I, at the age of fourteen, could tell that the teacher did not understand the subject. My parents decided that I showed some talent for academics and that I was worth grooming for Cambridge or Oxford University. In 1959, I went to Edinburgh Academy where the English A and S level subjects were taught. There I was able to specialize in the sciences and mathematics. With the improved teaching, I found I could do best in physics and mathematics. Eventually, I concluded that the reason for this was that my ability to make logical deductions compensated for my unreliable memory.

At school, I was fairly average at the humanities but excelled at mathematics and the sciences. Chemistry was the science I enjoyed most because we were allowed a lot of freedom in the laboratory at school, where I would carry out various forbidden syntheses of explosives and other noxious substances. I remember a few occasions where the lab had to be evacuated when one of these experiments went wrong and a noxious gas escaped. Despite the enjoyment these chemistry "experiments" gave me, I was not very good at the subject because of

Figure 2. Left: Mike and parents in the Austrian Alps, 1958; Right: Mike with mother and paternal grandmother, Scotland, 1961.

the memory required and I had to make too many guesses, especially when we studied organic chemistry where the chemical formulae were too complicated to memorize. However, despite the rather boring experimental part, physics was where I excelled at school. It satisfied my six fact memory limit because I was able to deduce correct results more often than not. Also, about this time in my life, I discovered that I am red green blind and that this disability does not fit well with chemistry. Some of the experiments required me to distinguish between several test tubes each containing a different reddish fluid. They all looked the same to me, but my classmates assured me that they were all quite different. At this point, I decided that chemistry was not for me despite all the fun I had mixing the various chemicals I could access.

While I was at Edinburgh Academy, along with several fellow students, I sat the Cambridge University scholarship examination and, to my great pleasure, I was awarded a major scholarship in the Natural Sciences to Gonville and Caius College. A condition for attending Edinburgh Academy was that one joined the army cadet corps. Once every week in Edinburgh, I wore my kilt feeling foolish. Despite being born and raised in Aberdeen, Scotland, I did not have a drop of Scottish blood in me and had never worn the kilt before. I came to understand Edinburgh's reputation as a windy city by parading through the streets in winter in my kilt in the usual cold horizontal rain which is an experience never to be forgotten.

Figure 3. Mike in Aberdeen, 1959.

Undergraduate Years

As an undergraduate at Cambridge from 1961 to 1965, I did the natural sciences tripos which covered most of the science subjects of that time. I chose to do physics, mathematics, chemistry and biochemistry because I enjoyed the chemistry at school. My color blindness again made this very frustrating and my poor memory made organic chemistry a bit of a nightmare as there were many situations where guessing did not work. I remember one situation in a biochemical experiment, where I ended up with a test tube containing some nondescript fluid. I was staring at it wondering what I should see in it or if I should do something else to the contents when, suddenly, it started to change color. Instinctively, I averted my gaze just before the test tube exploded. To this day, I have no idea what happened but that episode confirmed that chemistry was not for me and that the less dangerous and less memory intensive subject of physics would be my best bet. For fun, I joined the Cambridge Climbing Club which ran a bus or minibus to Derbyshire or North Wales every weekend. I discovered I was good at rock climbing and enjoyed the thrill of being high on a cliff with almost nothing to stop a fatal fall. From that moment, climbing at weekends became like a drug and I became obsessed by the sport.

During this period, I met the girl who is now my wife. One day, I woke from sleep with a terrible sore throat which refused to get better. In desperation, I went to a doctor who peered at my throat, took a swab and ordered me to rest. A few days later, I was informed that I had contracted mononucleosis and that I should take things easy until further notice. Horrified, I asked if that meant that I was not allowed to go rock climbing. The doctor stared at me as if I were mad and wearily explained that mononucleosis meant that my spleen was enlarged and, if I took even a small fall on to the rope, it would probably rupture resulting in death. This sounded like a good reason to stay in Cambridge for the next few weekends rather than courting a messy unpleasant death, so I did. Of course, my weekends became very boring without the thrill of rock climbing so, instead of climbing, I tried dancing at a nearby nightclub where I met a girl from Stockholm who was in Cambridge to learn English. She and I got on well and we spent a lot of time together. Slowly, our relationship developed into something more serious and we became an inseparable couple.

About this time my grandmother died at age ninety-two and left me a small bequest which I promptly spent on a car and the necessary insurance. With the help of my little car a typical weekend would be scheduled as: Friday evening—drive as fast as possible to Llanberis pass in North Wales or to the Lake District or, in the cold winter, to Glencoe or Ben Nevis in Scotland for the ice climbing. Climb on Saturday and Sunday, if not raining, and drive back to Cambridge late

Figure 4. Left: Harvey Court, Gonville & Caius College, Cambridge, 1965.

Figure 5.. Mike bouldering.

Sunday night and early Monday morning. Then I would sleep until I woke too late to go to class, thus not having any lecture notes. The weekly tutorials kept my nose to the grindstone and rescued my academic career.

My life settled into a routine which was close to my father's dictum of work hard and play hard. By now Berit had become part of my weekly routine. Although she did not climb herself, she enjoyed being with me in the mountains. On Friday evening, Berit and I would jump into the car and drive as fast as possible to North Wales and Berit blames this period in our lives for causing whiplash injuries to her neck. We would return to Cambridge or Oxford late on Sunday after a hair raising race through the dark and I would try to stay awake in class until the following Friday. I would often leave for an afternoon's climbing on some closer rocks once a week or perhaps do a bit of climbing on some nearby buildings. I had so much fun with these extra curricular activities that I seriously neglected my studies, especially in my final year. Also, in the Part 1 exams at the end of my second year at Cambridge, I did rather well in line with the expectations of me as major scholarship holder and I now expected to obtain a first-class degree in my third year with ease.

Figure 6. Mike graduating, June 1965.

However, I did not perform as expected in the all-important final year and ended with an upper second class degree. I am a bit surprised that I did so well because I hardly studied, nor did I attend many classes mostly because I had done so well in the first two years without studying much. I thought to myself "Michael, you are some sort of genius", but the final year taught me differently. Shortly before the final exam, I realized that I did not know what had been in the syllabus and panicked. I borrowed lecture notes from friends as I did not have any and read and read for most of each day as I watched time passing inexorably towards the final examinations. I struggled with questions I half understood, knowing that my enjoyable undergraduate days of climbing and pubs were now exacting their price. In the summer, I went on a climbing expedition to the Peruvian Andes where I could forget my Cambridge failures.

While I was in Peru, my father arranged for me to have an extra year, 1965–66, at Cambridge to do Part III mathematics to try to improve my disappointing performance. He was an eminent academic who well understood the importance of a Ph.D. degree. This year was a mixed success because I did not appreciate the almost rigorous approach of the applied mathematicians and physicists teaching the courses. Much of what I learned I found useful later. Unfortunately, I had not learned my lesson and my obsession with rock climbing again prevented me from spending the necessary time and effort my studies really needed. Again, I had to be satisfied with an upper second class performance. At the end of the year, I was lucky to have the necessary qualification for graduate school at Cambridge, but not in high energy theory which I was determined to do. I was offered a position with Neville Mott in experimental solid state physics but I turned this down in favor of an offer from Oxford in high energy theory.

GRADUATE AND POSTDOCTORAL YEARS

I spent the next three years, 1966–69, in Oxford sharing a rented house with several medical students and my future wife, who spent the time complaining that she had to work too hard keeping several messy males tidy, clean and fed while I worked on my Oxford D.Phil. We quickly fell into a routine where Berit left for work at 8:00 am and I at 10:30 am to arrive for the 11:00 am coffee at Rudolph Peierls' department of theoretical physics. My D.Phil. supervisor, John Taylor, left me alone to do my own research and find my own problems which upset me at the time but, in retrospect, this was excellent training for my later career. Whatever his reasons, I am eternally grateful to John for putting up with my foibles at Oxford. I managed to write three papers on Regge poles and the Veneziano model, a precursor to modern string theory, with other graduate students.

These were the subjects I would continue to work on in Torino and later at Birmingham until I changed fields. In 1969, I managed to write my thesis, imaginatively titled "Problems in strong interaction physics", which, I suspect, has never been read. Of course, the weekends and vacations were still reserved for my climbing obsession, which still occupied all my leisure time. At this time, I spent all summer vacations climbing in the French or Italian Alps and even Yosemite Valley in the USA and managed to get quite a reputation as a mountaineer.

Our next adventure was when I managed to obtain a Royal Society grant for a postdoctoral fellowship which I could use anywhere in Europe. I decided on the Istituto di Fisica Teorica, Torino, Italy because Sergio Fubini, one of the pioneers of modern string theory, was there but, more importantly, it was close to the Alps where the best mountains such as Mont Blanc are situated. Neither Berit nor I spoke any Italian but we were young enough that this was just a minor challenge to be overcome. We had many other interesting challenges to overcome of which renting an apartment and furnishing it, all in Italian, was one. Another was to deal with the local car drivers. I learned to love alpine skiing, made contact with outstanding local climbers and created a climb in the Val d'Orco which bears my name, "Fessura Kosterlitz" which was not repeated for a decade. My achievements in physics are somewhat less well known but I did do some very long calculations on a precursor to modern string theory. This resulted in one paper "The General N-Point Vertex in a Dual Model" with a fellow postdoc, Dennis Wray. I did not realize at the time that I was in the forefront doing research in what was later to become string theory.

While I was in Italy, there occurred a pivotal event which led to my Nobel Prize. I applied to CERN for a postdoctoral position for 1971–1972 but failed to submit the necessary paperwork in a timely fashion and was turned down. Panic set in as the prospect of unemployment hung over me. Berit walked to the main train station to buy a British newspaper which contained advertisements for academic jobs. There was one for a three-year postdoctoral position at Birmingham University in England and I dutifully applied for this although I did not really want to go there for high energy physics. I got offered the job and duly accepted.

Another, even more pivotal, event occurred in Italy. Berit and I had by now been an inseparable couple for seven years. The question of marriage arose on and off during those years, but we decided that I was too immature and the three-year difference between our ages was too great. By now we felt differently and I suggested that our relationship seemed very stable and we might as well make it legal. We married in Torino in September 1970 after a long battle with the local bureaucracy. Also, this avoided the problem of obtaining work permits for Berit. Since then, we have had more than a few adventures and three children

Figure 7. Berit and Mike's wedding, September 1970, Torino, Italy.

who are now settled in Boston and Providence in New England. We have also been blessed with five grandchildren.

I spent the next three years, 1970–1973, as a Research Fellow at the Department of Mathematical Physics at Birmingham University. I continued my calculations on the dual resonance model of Veneziano and was about to write up my calculations when a preprint by a group at Berkeley doing exactly what I had done appeared on my desk. Needless to say, I was rather annoyed but shrugged my shoulders and started a new long and laborious calculation. I completed this and started to write it up when another Berkeley preprint arrived on my desk. When this happened yet a third time, I did get rather upset and went from office to office asking the occupants if they had a problem I could look at or if I could help in some way. Eventually, I found myself in David Thouless' office listening as he described concepts and ideas I knew nothing about. He talked about superfluidity in ^{4}He films, crystals in two dimensions, vortices, dislocations, topology and many other related ideas.

Although this was all new to me, as was statistical mechanics which I had ignored as being unnecessary for high energy physics, the ideas made sense to

Figure 8. The Kosterlitz family, Lairhillock, 1991. Clockwise: my mother, my father, Elisabeth, Karin, Berit, Mike, Jonathan.

me. After I left David's office with my head spinning with all these new ideas and concepts, I returned to my own smaller office and began to work on these new wonderful ideas which David had introduced to me. The central idea was that the only way a flow in ^{4}He can dissipate is by the creation of vortices and their subsequent motion. A superfluid can be characterized by the absence of free vortices and a normal, dissipating fluid by the presence of a finite concentration in thermal equilibrium. In two dimensions, the problem becomes equivalent to the equilibrium statistical mechanics of a set of point charges interacting by a Coulomb potential. David and I introduced the concept of a vortex as a topological excitation or defect. The same ideas can be used to discuss the melting of a two-dimensional crystal with point dislocations playing the same role as vortices in a superfluid. David and I wrote two papers on this [1, 2] where we discussed the basic theory of defect mediated transitions. About this time, David casually directed my attention to some papers by Phil Anderson and coworkers on the Kondo problem and its mapping to a one dimensional $1/r^2$ Ising model [3] which introduced me to renormalization group methods, although this terminology only came into use later. I did nothing for six months but reading and re-reading this seminal paper and reproducing the calculations to try to understand it. During this time, Berit tried in vain to tidy my office but was firmly told to leave all papers alone. A year later, based on this research, I published a paper which discusses a renormalization group treatment of the two-dimensional planar rotor model of superfluid ^{4}He [4] which is the basis for the exact prediction for the superfluid density [5].

My next position arranged by David Thouless was as a Postdoctoral Fellow at LASSP at Cornell in 1973–74. There I met Michael Fisher and his young, very smart graduate student, David Nelson. Even at this early stage in his career, David demonstrated that he was going to become something special. I was excited by the prospect of learning about phase transitions and critical phenomena from the Cornell experts, Michael Fisher and Ken Wilson. Field theoretic methods and the epsilon expansion of Wilson and Fisher [6] had permitted enormous progress in understanding a huge variety of phase transitions and I badly wanted to be one of the pioneers in this. Working with Nelson and Fisher opened my eyes to what physics is all about, how important experimental data are and how to choose the problems to work on. In the 1970s, critical phenomena was a field which was at last opening out by Wilson and Fisher's epsilon expansion methods. With Nelson and Fisher, I worked on bicritical points using renormalization group methods. To our great pleasure, we were able to understand in great detail the shape of the phase diagram in the vicinity of a bicritical point and why the various phase boundaries had the shape of experiments on anisotropic antiferromagnets. This was at the height of the development of critical phenomena in $4 - \varepsilon$ dimensions and I was excited to be in the middle of it with the leading authorities in the field. I learned the importance of testing one's theory against the ultimate authority in physics, experiment.

During all my postdoctoral years I kept to my mantra: first climbing, then physics and last family. In fact, when I was in my twenties, I was one of the best climbers in Britain and even considered giving up physics in favor of a professional climbing career. My teaching duties prevented me from going on any of the Himalayan expeditions I could have joined. However, on thinking about the possible consequences of this choice, sanity and my wife finally prevailed. I realized that, although I was technically good enough, a career in academia and physics would allow me enough vacation time to indulge in my climbing obsession. Some of my climbing acquaintances had chosen to become professional mountaineers and a few succeeded but most did not. I decided that I would probably not succeed in this.

TENURED YEARS: BIRMINGHAM AND BROWN

I returned to Birmingham University in 1974 as a tenured lecturer, then was promoted to Senior Lecturer in 1978 and finally to Reader in 1980. I continued working on phase transitions and critical phenomena while teaching two courses

Figure 9. Dinner with our family and my wife's relatives at the Swedish summer house.

at the same time. David Nelson and I managed to produce our important prediction for the superfluid density of a thin film of ^{4}He [5], but my significant output slowed down although I produced several papers on critical phenomena. David Thouless was still at Birmingham, during which time we continued our collaboration on spin glasses until he moved to the USA. I spent a semester in 1978 as a visiting professor at Princeton, Bell Laboratories and Harvard respectively, bringing my family. My stay at Harvard was especially productive as David Nelson and I wrote our paper "Universal Jump in the Superfluid Density of Two-Dimensional Superfluids."

By 1978 we had two children in Birmingham schools and my wife and I were happy and thought we were settled there forever. I was doing what I loved, climbing, immersed in physics, and spending the remaining time with my growing family. However, this contented period of my life was not to last, because I contracted the nasty autoimmune disease of multiple sclerosis. I awoke one day in September 1978 and was unable to stand up because my balance did not function. I was admitted to hospital where I spent one week while the doctors tried to figure out what was wrong. Eventually, a solemn neurologist said that there were two possibilities, a brain tumor or multiple sclerosis, of which the latter was the better alternative. It turned out I did indeed suffer from MS and life as I knew it was forever changed. Needless to say, I did not react well to this news as I assumed it meant the mountaineering half of my life was over and I would have

Figure 10. Karin, Jonathan, Liz and Mike at Dunottar Castle in Aberdeenshire, Scotland, 1991.

to live the rest of my life without it. My wife was not as upset as I was because, by this time, a number of my climbing friends had died in climbing accidents and she was relieved that this would not happen to her husband. However, this thought was little consolation to me who could not envisage life without the mountains and I went into a deep depression which lasted for several years. The professor of neurology offered these kind words of encouragement, "There is no cure, some people live longer than others. If you can look back after 25 years you will know how bad a case you are." Needless to say, this information also affected my physics productivity for a few years.

In 1979 I was offered a position as a tenured full professor at Brown University and Birmingham counter offered a promotion to a research professor as an incentive to stay. This would be at a Center of Excellence centered at Birmingham, which I was inclined to accept. I was about to refuse the offer from Brown when Birmingham abruptly withdrew their offer. Combined with my illness, for which I subconsciously blamed Britain, this was the last straw and I immediately tendered my resignation and left for Brown, where we have been since 1982. My wife and I finally became citizens of the USA in 2004 because, in that year, Sweden permitted dual nationality and my wife did not wish to give up her Swedish nationality. As a British citizen, I had no difficulty because Britain has always permitted dual nationality. After 9/11, I felt that my wife and I and, especially, our children needed the protection of citizenship, so we paid a lot of money to an immigration lawyer and became US citizens in 2004.

At Brown, my interests changed somewhat and, with the help of a grant from NSF, I started to work on various effects in two dimensional arrays of Josephson junctions such as disorder and in a magnetic field. These can be represented by a frustrated planar rotor model, which is quite different from the original *2D* planar rotor model [4]. In this, I was greatly helped by a very good graduate student from Brazil, Enzo Granato. This system is an excellent system for the study of many variants of the original system of Kosterlitz and Thouless and is still under quite active theoretical and experimental investigation. We looked at some of the more elementary aspects of the system and slightly increased our understanding of it. These experimentally accessible variants of the model took us out of the realm of analytic work and my student and I turned to numerical simulations, which was the only way we could make any progress. This has turned into a more than twenty-year collaboration with Enzo at INPE in Brazil.

In 1985, I went on a sabbatical to France, bringing my family as I realized this might be the last opportunity for a family adventure. The children went to French schools and I spent six months at Saclay and Orsay with my Brown graduate students continuing the work on planar rotor models.

Figure 11. Putting on a new roof on our summer house in Sweden. Left to right: Elisabeth, Jonathan, Karin.

Figure 12. Visiting the Thouless family in Seattle. From left to right: David, Mike, Margaret.

On return to Brown I became interested in numerical work with a couple of graduate students from Korea. Our projects were to study the kinetics of growth of a surface by random deposition. We studied the scaling of the interface width with time and evaluated the exponent to a high degree of accuracy. However, we could not compete with the massive simulations from a group in Germany. The other project was to investigate if it was possible to identify a weak first order transition by purely numerical methods [7] which method is still being used in 2016. Jooyoung has turned his talents to the protein folding problem and his group is now recognized as a leader in this field as they consistently score very highly in the CASP competitions. A Japanese graduate student, Nobuhiko Akino, has been very successful in his numerical work on randomness in superconductors and in *XY* spin glasses which have been longstanding intractable problems. We concluded that an *XY* spin glass exists in three dimensions and above, which result can also be obtained via massive simulations.

For reasons which are still unclear to me, I lost my NSF funding over this and have never been able to get it back. However, the problem never stopped to intrigue me and over the last ten years I have doggedly pursued the solution although it has proven to be somewhat elusive. I had a brilliant graduate student given to me by Brown who was invaluable help to me doing difficult numerical work, and together we managed to get a paper accepted in 2010.

I also have had a longstanding collaboration for the last twenty-five years with my colleague Tapio Ala-Nissila in Finland working on phase field models of growth. This is a surprisingly successful method for the numerical study of growth in fluids and in solids which we recently applied to the hydrodynamics of crystals [8]. This collaboration has also included my colleague, Martin Grant, at McGill in Montreal, Canada and Ken Elder at Oakland University in Michigan, USA as well as my Brown colleague See-Chen Jing. I also started a collaboration at the Korea Institute of Advanced Study in Seoul, Korea where I am now a Distinguished Professor visiting for two months every summer. Even at my advanced age of 73, physics still fascinates me because there are so many problems waiting for a solution that, despite my increasing incompetence, I would like to see understood before I retire. Perhaps in this respect, I am like my father who refused to give up working until he was over 90! On reflection having produced nearly sixty papers in my time at Brown is not bad, but nothing will ever compare to the exhilaration of our 1977 paper [5] when theory agreed quantitatively with experiment [9]. Each summer, Berit and I travel a lot, spending time in Brazil, Finland and Korea but always keep four or more weeks sacrosanct for our Swedish summer house where we can relax completely by watching the grass

Figure 13. Eating chicken in a restaurant in chicken street, Dongdaemun market, Seoul, Korea.

grow. The only disadvantage is that it always does grow and then needs cutting, which gives me about the only exercise I have during the year.

Last but by no means least, I am happy that I have managed to work since that dreadful day in September 1978 when I was diagnosed with MS. The twenty-five years have gone and, as predicted by the neurologist then, I now know the outcome. I was not a bad case. I had attacks every 18 months from age 35 to 55, some quite bad, some small relapses. When I was 55 my neurologist put me into a trial for a new MS drug. This was very successful and opened up a whole new field of pharmacological drugs for the easing of MS. Since then, I have been lucky in that I have never had another attack. I only battle the deadly fatigue that comes with the disease. I want to take this space to tell any budding scientist that, however bleak the future may seem due to illness or other problems, one cannot say you will not be successful.

More people than I can list here have contributed in vital ways to my success. Those that are probably the most important are David Thouless whose friendship, patience and collaboration are central to my career, Berit, my wife, for her patience and forbearance with my peculiarities and absences when I was either climbing mountains or working too hard and my children, Karin, Jonathan and Elisabeth for putting up with and loving their strange father who was absent too often and too long. I also acknowledge the support and friendship of my colleagues at Birmingham and Brown.

REFERENCES

1. J.M. Kosterlitz and D.J. Thouless, *J Phys C: Solid State Phys* **5** L124–6 (1972).
2. J.M. Kosterlitz and D.J. Thouless, *J Phys C: Solid State Phys* **6** 1181–203 (1973).
3. P.W. Anderson, G. Yuval and D.R. Hammann, *Phys Rev B* **1** 4464 (1970).
4. J.M. Kosterlitz, *J Phys C: Solid State Phys* **7** 1046–60 (1974).
5. D.R. Nelson and J.M. Kosterlitz, *Phys Rev Lett* **39** 1201 (1977).
6. K.G. Wilson and M.E. Fisher, *Phys Rev Lett* **28** 240–3 (1972).
7. Jooyoung Lee and J.M. Kosterlitz, *Phys Rev Lett* **1990** 137 (1990).
8. V. Heinonen et al., *Phys Rev Lett* **116** 024303 (2016).
9. D.J. Bishop and J.D. Reppy, *Phys Rev Lett* **40** 1727–30 (1978).

Topological Defects and Phase Transitions

Nobel Lecture, December 8, 2016 by John Michael Kosterlitz
Brown University, Providence, Rhode Island, USA.

1. INTRODUCTION

It is a great honor to speak to you today about "theoretical discoveries of topological phase transitions and topological phases of matter." Since the main character, David Thouless, is not able to speak here, the two minor characters, Duncan Haldane and I, have been asked to speak for David. This is a very daunting task which I agonized over for a considerable period of time as I feel inadequate for this. Eventually, time ran out and I was forced to produce something relevant so I decided to start by talking my earliest experience of David and how we ended as collaborators on our prize winning work. Then I will summarize my understanding of his seminal contributions to his applications of topology to classical ($\hbar = 0$) Berezinskii-Kosterlitz-Thouless or BKT phase transition. David has worked on many more applications of topology to quantum mechanical systems such as the Quantum Hall effect and Duncan Haldane will talk about David's contributions to these.

My first experience of David Thouless took place in 1961 when I was a freshman at Cambridge University. I was in a large introductory class on mathematics for physics waiting for the instructor to appear to enlighten us when a young man who was clearly too young for this advanced science course walked in. Obviously, he had wandered into the wrong lecture hall. To our astonishment, he stopped in front of the class and proceeded to talk about various complicated pieces of mathematics which most of the class had either not met before or had not understood. It rapidly became clear that the class was in the presence of a mind which operated on a different level to those of the audience. My later experiences of David merely reinforced this early impression. My next meeting with him was in 1971 in the Department of Mathematical Physics at Birmingham University in England where I went by accident as a postdoctoral fellow in high energy physics. After being frustrated for a year, I looked for a new tractable problem and David introduced me to the new worlds of topology and phase transitions in two dimensional systems.

As far as I am concerned, the study of topological excitations started in 1970 when I was a postdoc in high energy physics at the Istituto di Fisica Teorica at Torino University, Italy. As a very disorganized person, I failed to submit my application for a position at CERN, Geneva in a timely fashion and, instead, found myself without a position for the following year. After replying to some advertisements in the British newspapers, I was offered a postdoctoral position in the department of Mathematical Physics at Birmingham University in England. I did not want to go to Birmingham which, at that time, was a large industrial city in the flat middle of England where a lot of cars and trucks were built. It was certainly not my ideal place to live, but my girlfriend and I decided that it was better than the alternative of unemployment. During my first year there, I continued some elaborate field theory calculations but I had an unhappy experience. I was about to write up my calculations for publication when a preprint from a group in Berkeley doing exactly the same thing appeared on my desk. After two or three repeats of this, I became very disillusioned. In desperation, I went round the department looking for a tractable problem in any branch of physics. I appeared in David Thouless' office listening to him describing several new and mysterious concepts such as topology, vortices in superfluids and dislocations in two dimensional crystals. To make matters even worse, my knowledge of statistical mechanics was almost non-existent as I had omitted that course as irrelevant to high energy physics which I considered to be the only field of physics of any interest. However, to my surprise, David's ideas made sense to me as being new and very different and they seemed worth considering. We began to work on the problem of phase transitions mediated by topological defects, which to my untutored mind seemed just another application of field theoretic ideas and was therefore worthy of consideration. Little did I know just how different and important these ideas and their applications would be in the following decades and where they would take us.

At this point, I would like to talk about David's vital contribution to our understanding of two dimensional phase transitions. In fact, one of our motivations for looking at two dimensions was that we thought that life was easier in two than in three dimensions. David had already done some work on the importance of topological defect driven phase transitions in the context of the one dimensional Ising chain with interactions between spins decaying as $1/r^2$. This model can be discussed in terms of topological defects, or domain walls, interacting as $\ln r/a$ [1] and David had shown that the magnetization dropped discontinuously to zero at T_c although it was not a first order transition. This was later made quantitative by Anderson and Yuval [2] who used an early version of the renormalization group. This work was very influential in our thinking about defect driven phase transitions because it led us to seek other systems in which there are point topological defects with a logarithmic interaction. Examples of this are point vortices in ^{4}He films, in superconducting films and point dislocations in $2D$ crystals. This, in turn, led us to the Coulomb gas description of such systems. However, those of you who are paying attention to the details will have noticed a serious flaw in this analogy since our basic $1D$ example is different from our other systems which are Coulomb gases in $2D$. The reason why the $1D$ system with logarithmic interactions works is because of the constraint that the charges or domain walls alternate in sign along the line. If this constraint is relaxed, the phase transition disappears. Of course, this is not the first time that a correct conclusion is arrived at for the wrong reason.

The first thing we had to understand was the role of long range order in crystals and superfluids, as the standard picture of a crystal in two dimensions is a system of molecules in which knowledge of the position of a single particle means that one knows the positions of all the others from the equation $\mathbf{r}(n,m) = n\mathbf{e}_1 + m\mathbf{e}_2$ where $\mathbf{e}_1,\mathbf{e}_2$ are the fundamental lattice vectors and n, $m = \pm 1, \pm 2, \ldots \pm \infty$. The problem here is the Peierls argument [3, 4] which says that long range order is not possible in two dimensional solids because low energy phonons give a mean square deviation of atoms from their equilibrium positions in an $L \times L$ system increasing logarithmically with the size of the system, L. A useful picture of a two dimensional crystal is to consider a flat elastic sheet on which is drawn a lattice of dots representing the atoms of a crystal. Now, stretch some regions and compress other regions of the sheet without tearing it, representing smooth elastic distortions of the crystal. Clearly, the dots (particles) will move far from their initial positions—in fact a distance proportional to $\sqrt{\ln L}$—although the lattice structure is preserved. The absence of long range order in this form has been shown rigorously by Mermin [5]. Similar arguments show that there is no spontaneous magnetization in a $2D$ Heisenberg magnet [6] and that the expectation value of the superfluid order parameter vanishes in a $2D$ Bose liquid is zero [7].

According to the conventional wisdom of the early 1970s, this implies that there can be no phase transition to an ordered state at any finite temperature because an ordered state does not exist! However, this minor contradiction did not deter David and myself because David understood the subtleties of the situation and could see a way out of the apparent contradiction while I was too ignorant to realize that there was any such contradiction. In hindsight, I understood that, *very* occasionally, being ignorant of the fact that a problem is insoluble, allows one to proceed and solve it anyway. As luck had it, this was one of those few occasions for me. Of course, it also helped that there existed some experimental and numerical evidence for transitions to more ordered low temperature phases in $2D$ crystals [8, 9], very thin films of ^{4}He [10, 11] and $2D$ models of magnets [12, 13, 14]. The most compelling piece of experimental evidence for us is shown in Figure 1 where the deviation of $-\Delta f$, the decrease in the resonant frequency f of the crystal with a film of ^{4}He adsorbed on the surface, from the

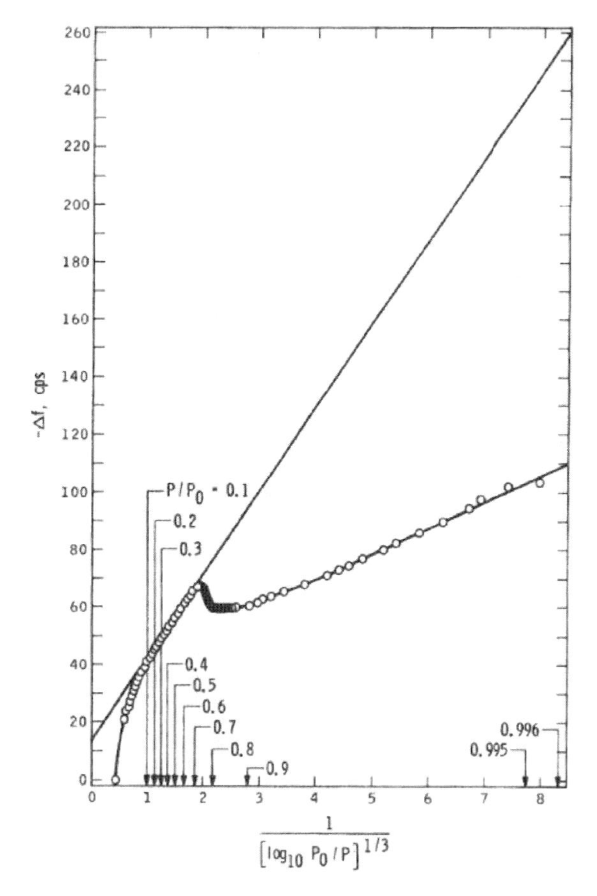

Figure 1. The horizontal axis is a measure of the total areal mass density of the adsorbed film and the vertical axis is $-\Delta f$, a measure of the adsorbed mass which decouples from the oscillating substrate. Reprinted from M. Chester, L. C. Yang and J. B. Stephens, *Phys. Rev. Lett.* **31** 211 (1972) with permission. Copyright American Physical Society.

straight line is a measure of the areal superfluid mass density. Clearly, the $2D$ film undergoes an abrupt transition as the adsorbed mass density increases with a probable finite discontinuity in $\rho_s(T)$. This behavior seemed very strange as conventional wisdom said that ρ_s would increase continuously from zero as the ordered phase is entered. This needed an explanation which, clearly, had to be rather different from anything known previously.

2. BREAKTHROUGH

The solution to this puzzle is that there can be a more subtle type of order called topological order in some two dimensional systems. The simplest example is the Ising ferromagnet which consists of a set of spins $S_\alpha = \pm 1$ on a D-dimensional cubic lattice. The rules of statistical mechanics are (i) any configuration of the system occurs with probability $e^{(-E/k_B T)}$ where E is the energy of that configuration and (ii) compute the partition function $Z(T) \equiv \sum_{\text{configs}} \exp(-E/k_B T)$, which gives all necessary thermodynamic information. The most probable excitations are the low energy ones which are responsible for the absence of true long range order but, otherwise, have no effect. To discuss the destruction of superfluidity and the melting of a $2D$ crystal, we have to include the very improbable high energy topological defects responsible for the destruction of a superfluid and of a crystal. These are the vortices in a superfluid and dislocations in a crystal [15, 16, 17]. I should point out that similar ideas had been proposed a bit earlier by Berezinskii [18, 19] but, when we did our work, we were not aware of this. For some reason, our work has received much more attention than that of Berezinskii.

Of course, you may well ask about the connection between topology which is the study of spheres with N holes while our physical systems all lie on a flat simply connected $2D$ surface with no holes. The topology we are considering is determined by the underlying physics and its corresponding energetics and a phase transition can be thought of as a transition between topological sectors defined by the topological invariants. We can discuss the importance of topology by comparing the $2D$ planar rotor magnet with two component spins and the Heisenberg model with three component spins. For the planar rotor model

$$\mathbf{s}_i = (s_{ix}, s_{iy}) = s(\cos\phi_i, \sin\phi_i)$$
$$\Psi_i = s_{ix} + i s_{iy} = s e^{i\phi_i}$$

where s denotes the length of the spins, usually taken as unity. Consider a large $L_x \times L_y$ system with periodic boundary conditions (similar considerations hold for other boundary conditions). In the planar rotor model, the direction of magnetization in a region is defined by the angle ϕ which varies slowly in space. Although the angle ϕ fluctuates by a large amount in a large system, the number of multiples of 2π it changes by on a path which goes completely round the system is a *topological* invariant, so that

$$\frac{1}{2\pi}\int_0^{L_x}\frac{\partial\phi}{\partial x}dx=n_x,$$

$$\frac{1}{2\pi}\int_0^{L_y}\frac{\partial\phi}{\partial y}dy=n_y,$$

are numbers defining a particular metastable state. Transitions can take place from one metastable state to another only if a vortex–antivortex pair is formed, separate and recombine after one has gone right round the system. This process causes n_x or n_y to change by one, but there is an energy barrier proportional to the logarithm of the system size to prevent such a transition.

The same system composed of three component spins

$$\mathbf{s}_i=\left(s_{ix},s_{iy},s_{iz}\right)=s\left(\sin\theta_i\cos\phi_i,\sin\theta_i\sin\phi_i,\cos\theta_i\right)$$

is called the Heisenberg model. A quantity such as $\frac{1}{2\pi}\int_0^{L_x}\frac{\partial\phi}{\partial x}$ is *not* a topological invariant. A twist of the azimuthal angle ϕ by 2π across the system can be continuously unwound by changing the polar angle θ, which we take to be the same everywhere from $\pi/2$ to zero. In fact, the Heisenberg model in two dimensions has a single topological invariant $N = 0, \pm1, \pm2, \ldots$ where

$$N=\frac{1}{4\pi}\int dx\,dy\sin\theta\left(\frac{\partial\theta}{\partial x}\frac{\partial\phi}{\partial y}-\frac{\partial\theta}{\partial y}\frac{\partial\phi}{\partial x}\right).$$

If we regard the direction of magnetization in space as giving a mapping of the space on to the surface of a unit sphere, the invariant N measures the number of times space encloses the unit sphere. This invariant is of no significance in statistical mechanics because the energy barrier separating configurations with different values of N is of order unity. Thus, there is no barrier between different topological sectors (different values of N) which implies that there is no ordered state for the $2D$ $n = 3$ Heisenberg magnet. In the $2D$ planar rotor model, there is an infinite energy barrier between different topological sectors parametrized by n_x and n_y and, in consequence, there is a phase transition when the system can fluctuate between different topological sectors.

We can show this by showing how a configuration with $N = 1$ can be continuously deformed into one with $N = 0$. A simple example of an $N = 1$ configuration is one where θ is a continuous function of $r = \sqrt{x^2 + y^2}$ and $\theta = \pi$ for $r > a$ and $\theta(r = 0) = 0$. The angle $\phi(x,y) = \tan^{-1}(y/x)$. The energy of a slowly varying configuration is

$$E = \frac{Js^2}{2}\int dx\,dy\left[(\nabla\theta)^2 + \sin^2\theta(\nabla\phi)^2\right] = \pi Js^2\int_0^a\left[\left(\frac{d\theta}{dr}\right)^2 + \frac{\sin^2\theta}{r^2}\right]rdr$$

for this configuration. Even if θ varies linearly between $r = 0$ and $r = a$, E is finite and independent of a. Of course, for small values of a this expression for the energy is invalid, but the number of spins in the disk of radius a is small so that any energy barrier is also small and the topological invariant N can be changed by small thermal fluctuations. The conclusion is that the $2D$ planar rotor and related models can have a finite temperature topologically ordered state while the three component Heisenberg model does not. This is consistent with numerical studies [14], a later renormalization group study by Polyakov [20] and experiments on superfluids [10, 11]. Note that the calculation by Polyakov is performed in a single topological sector $N = 0$ so that the absence of a phase transition in the $2D$ Heisenberg model is verified by both arguments separately.

3. VORTICES IN THE PLANAR ROTOR MODEL IN TWO DIMENSIONS

The importance of topological defects in phase transitions in these two dimensional systems was discussed in our 1972 paper [15] where our defect free energy argument was presented. The planar rotor and the superfluid film free energy can be written as

$$\frac{H}{k_B T} = \frac{K_0(T)}{2}\int \frac{d^2\mathbf{r}}{a_0^2}(\nabla\theta(\mathbf{r}))^2$$

where a_0 is the lattice spacing or some microscopic cut-off length scale and

$$K_0(T) = \begin{cases} \dfrac{J}{k_B T} & \text{planar rotor} \\[2ex] \dfrac{\hbar^2\rho_s^0(T)}{m^2 k_B T} & \text{superfluid film} \end{cases}.$$

Here, J is the exchange interaction between nearest neighbor unit length spins so that $H[\{\mathbf{s}\}] = \frac{J}{2}\sum_{<\mathbf{r},\mathbf{r}'>}[s(\mathbf{r})-s(\mathbf{r}')]^2 = J\sum_{<\mathbf{r},\mathbf{r}'>}[1-\cos(\theta(\mathbf{r})-\theta(\mathbf{r}'))]$. For a ^{4}He film,

$$H=\frac{1}{2}\int\frac{d^2\mathbf{r}}{a_0^2}\rho_s^0(T,\mathbf{r})\mathbf{v}_s^2$$

where $v_s=\frac{\hbar}{m}\nabla\theta$ is the superfluid velocity, $\theta(\mathbf{r})$ is the phase of the superfluid order parameter $\psi(\mathbf{r}) = |\psi(\mathbf{r})|e^{i\theta(\mathbf{r})}$ and $\rho_s^0(T,\mathbf{r})$ is the position dependent bare superfluid density. $\rho_s^0(T,\mathbf{r}) = 0$ at the vortex cores and constant elsewhere. It turns out that its exact spatial dependence is irrelevant as the only important consequence is that there is a finite energy E_c associated with each vortex core. The physical reason is that a vortex core costs a finite free energy because the vortex core is a region where the superfluid order parameter vanishes. Now we can see how the topology arises—each vortex corresponds to a hole in the surface and the superfluid lives on the $2D$ surface with a set of holes where $\oint_C d\theta = 2\pi n$ and a vortex can be called a topological defect.

Since vortices interact pairwise by a logarithmic energy

$$\frac{H}{k_BT}=-\pi K_0(T)\sum_{\mathbf{R},\mathbf{R}'}n(\mathbf{R})n(\mathbf{R}')\ln\left(\frac{|\mathbf{R}-\mathbf{R}'|}{a}\right)-\ln y_0\sum_{\mathbf{R}}n^2(\mathbf{R})$$

the Hamiltonian is exactly that of a neutral plasma of Coulomb charges. Also, one can restrict consideration to the lowest charges $n = 0, \pm 1$ since the larger values are suppressed by powers of the fugacity $y_0 = e^{-E_c/k_BT} << 1$. Our first attempt at solving this was to consider a single isolated vortex of unit circulation in a $L \times L$ system. The free energy of such a vortex is $\Delta F = \Delta E - T\Delta S = k_BT(\pi K_0(T) - 2)\ln(L/a)$ since $\Delta E/(k_BT) = \pi K_0(T)\ln L/a$ and the entropy $\Delta S = k_B \ln L^2/a^2$. Now, at low temperature T, $2k_BT < \pi J$, $\Delta F \to +\infty$ and the probability of having a vortex $P \propto e^{-\Delta F/k_BT} \to 0$ while, for $2k_BT > \pi J$, $P \to 1$ and there will be a finite concentration of free vortices. David and I realized that we could treat the Coulomb plasma of n charges $q = +1$ and $nq = -1$ charges by introducing a scale dependent dielectric function $\in(r)$ such that the force between a pair of test charges separated by a distance r is $2\pi K_0/r\in(r) = 2\pi K(r)/r$. The energy of this pair is $E(r) = \int_a^r dr'\frac{K(r')}{r'} = U(r)\ln(r/a)$. Our self consistent equation for $K(r = e^l)$ becomes

$$K^{-1}(l)=K^{-1}(0)+4\pi^3y_0^2\int_0^l dl' e^{4l'-2\pi U(l')}$$

Kosterlitz and Thouless [15] derived this self consistent integral equation for the effective interaction energy [16].

The central problem is to solve this equation since it is clear that a transition between a phase of bound dipoles and a phase of free charges will happen when $\pi K(\infty) = 2$. However, to find the behavior of the system near T_c requires solving the self consistent equation for $K(l)$. Unfortunately, KT made an unnecessary approximation by replacing $U(r)$ by $K(r)$ and solving self consistently for $K(r)$. The approximation was justified on the grounds that $U(r) - K(r) << 1$ but this led to incorrect results. A proper treatment has been given by Young [21] who showed that this is equivalent to the renormalization group equations of Kosterlitz [17].

$$\begin{aligned} \frac{dK^{-1}}{dl} &= 4\pi^3 y^2 + \mathcal{O}(y^4) \\ \frac{dy}{dl} &= (2 - \pi K)y + \mathcal{O}(y^3) \end{aligned} \tag{1}$$

Remarkably, these *approximate* RG equations to lowest order in the vortex fugacity y yield an *exact*, inescapable prediction for an experimentally measurable quantity. The flows are shown in Figure 2. If the experimental number is different from the theoretical prediction then, either the experiment is wrong or

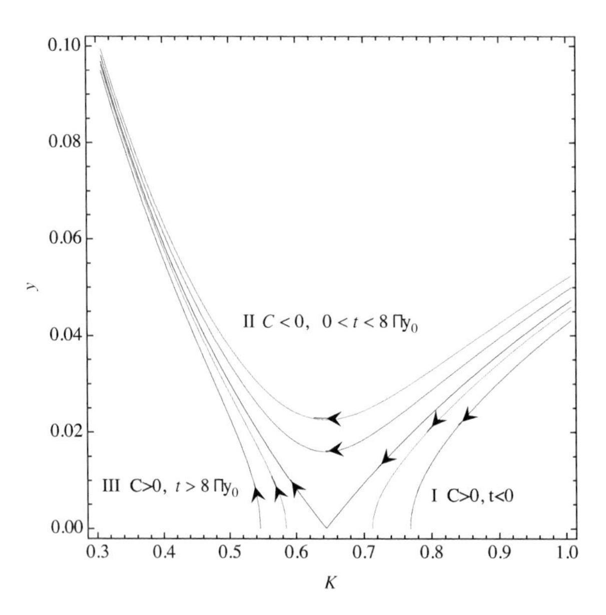

Figure 2. Renormalization group flows from equation (1) for the $2D$ planar rotor model. Note that for $T \leq T_c$, $y(\infty) = 0$ and $K^{-1}(\infty) = \pi/2$.

the whole theory is wrong. To our great relief and pleasure the key experiment by Bishop and Reppy was done in 1978 [23].

The theoretical prediction [24]

$$\frac{\rho_s^R(T_c^-)}{T_c} = \frac{2m^2k_B}{\pi\hbar^2} = 3.491\times10^{-8}\ \text{gm cm}^{-2}\ K^{-1}$$

has been checked experimentally [23, 25] and the data from several different experiments [26, 27, 28, 29, 30] is presented in Figure 3. It is of interest to note that the experimental data were obtained and plotted before the authors were aware of our theoretical prediction. This can be viewed as experimental confirmation of the BKT theory. There has also been extensive experimental investigation into melting in *2D* by the Maret group [32, 32, 33].

4. MELTING OF TWO DIMENSIONAL CRYSTALS

There is quantitative agreement with the theory of melting by topological defects due to Young, Halperin and Nelson [35, 36]. In the theory of melting of *2D*

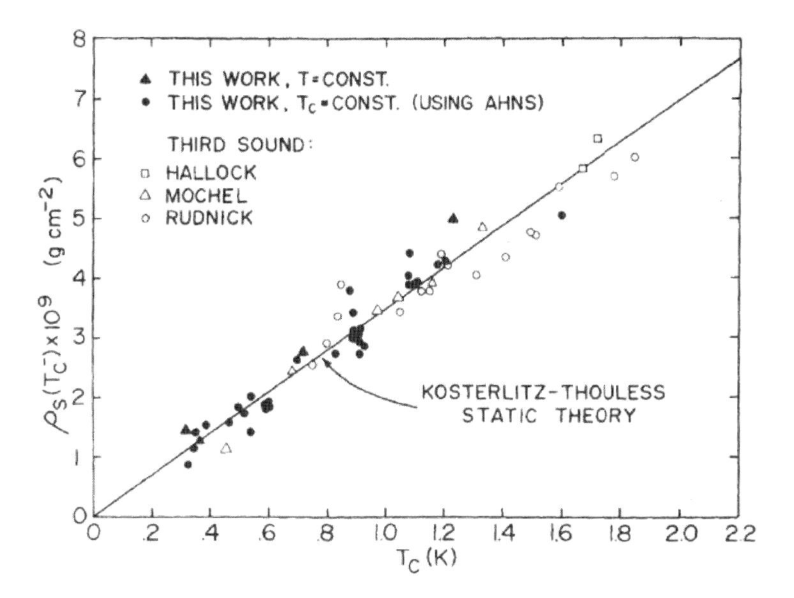

Figure 3. Results of third sound and torsional oscillator experiments for the superfluid density discontinuity $\rho_s(T_c^-)$ as a function of temperature. The solid line is the theoretical prediction for the static theory. Reprinted from D J Bishop and J D Reppy, *Phys Rev Lett* **40** 1727 (1978) with permission. Copyright American Physical Society.

crystals, one starts with the expression for linear elasticity of a triangular lattice, which is the usual lattice structure in $2D$.

$$F=\frac{1}{2}\int d^2\mathbf{r}\left(2\mu_0 u_{ij}^2+lu_{kk}^2\right),$$

$$u_{ij}=\frac{1}{2}\left(\frac{\partial u_i}{\partial r_j}+\frac{\partial u_j}{\partial r_i}\right),$$

where u_{ij} is the linear elastic strain tensor and u_i is the displacement field. The strain field can be decomposed into a smooth part ϕ_{ij} and a singular part $u^s(\mathbf{r})$ due to dislocations [37]. These are characterized by the integral of the displacement $\mathbf{u}(\mathbf{r})$ round a contour enclosing a topological defect or dislocation

$$\oint_C d\mathbf{u}=a_0\mathbf{b}(\mathbf{r})=a_0\left(n(\mathbf{r})\hat{\mathbf{e}}_1+m(\mathbf{r})\hat{\mathbf{e}}_2\right).$$

Here, $\mathbf{b}(\mathbf{r})$ is the dimensionless Burgers vector, a_0 is the crystal lattice spacing and n,m are integers. Within continuum elasticity theory, one can show that [36]

$$u_{ij}^s(\mathbf{r})=\frac{1}{2}\left(\frac{1}{2\mu}\in_{ik}\in_{jl}\frac{\partial^2}{\partial r_k\partial r_l}-\frac{\lambda}{4\mu(\lambda+\mu)}\delta_{ij}\nabla^2\right)a_0\sum_{\mathbf{r}'}b_m G_m(\mathbf{r},\mathbf{r}'),$$

$$G_m(\mathbf{r},\mathbf{r}')=-\frac{K_0}{4\pi}\sum_{n=1}^{2}\in_{nm}\left(r_n-r_n'\right)\left[\ln\left(\frac{|\mathbf{r}-\mathbf{r}'|}{a}\right)+C\right].$$

From this, one obtains the energy of an set of dislocations of Burgers vectors $\mathbf{b}(\mathbf{r})$ as

$$\frac{H_D}{k_BT}=-\frac{K_0(T)}{8\pi}\int d\mathbf{r}\,d\mathbf{r}'\left(\mathbf{b}(\mathbf{r})\cdot\mathbf{b}(\mathbf{r}')\ln\frac{|\mathbf{r}-\mathbf{r}'|}{a}-\frac{\mathbf{b}(\mathbf{r})\cdot(\mathbf{r}-\mathbf{r}')\mathbf{b}(\mathbf{r}')\cdot(\mathbf{r}-\mathbf{r}')}{(\mathbf{r}-\mathbf{r}')^2}\right).$$

In our paper, we ignored the second term in this equation on the grounds that it is less relevant than the logarithmic term, which was an unfortunate error. This was corrected by Halperin and Nelson who predicted the now famous hexatic fluid phase with 6-fold orientational symmetry. We assumed that dislocation unbinding led directly to an isotropic fluid which is now known to be wrong. Melting in $2D$ is a two stage process. At temperature T_m, the crystal melts by dislocation unbinding to an anisotropic hexatic fluid and, at $T_i > T_m$, this undergoes a transition where the algebraic orientational order is destroyed by disclination unbinding, resulting in the expected high temperature isotropic fluid [35, 36].

The predictions from this theory are similar to those for superfluid ^{4}He films and the corresponding universal jump is for the renormalized (measured) Young's modulus

$$\tilde{K}_R\left(T_m^-\right) = \lim_{T\to T_m^-} \frac{4\tilde{\mu}_R(T)\left(\tilde{\mu}_R(T)+\tilde{\lambda}_R(T)\right)}{2\tilde{\mu}_R(T)+\tilde{\lambda}_R(T)} = 16\pi$$

where $\tilde{\mu}_R(T)$ is the renormalized value of μ/k_BT One of the interesting but unmeasurable predictions of the dislocation theory is the X-ray structure function

$$S(\mathbf{q}) = \left\langle |\rho(\mathbf{q})|^2 \right\rangle = \sum_{\mathbf{r}} e^{i\mathbf{q}\cdot\mathbf{r}} \left\langle e^{i\mathbf{q}\cdot(\mathbf{u}(\mathbf{r})-\mathbf{u}(0))} \right\rangle \sim |\mathbf{q}-\mathbf{G}|^{-2+\eta_G(T)}$$

$$\eta_G(T) = \frac{k_BT|\mathbf{G}|^2}{4\pi} \frac{3\mu_R(T)+\lambda_R(T)}{\mu_R(T)\left(2\mu_R(T)+\lambda_R(T)\right)}$$

There are no δ-function Bragg peaks in the structure function but algebraic peaks behaving as

$$S(\mathbf{q}) \sim |\mathbf{q}-\mathbf{G}|^{-2+\eta_G(T)}$$

We see that this diverges at $\mathbf{q} = \mathbf{G}$ for small $|\mathbf{G}|$ so that the expected X-ray structure function looks like that sketched in Figure 4. This is one of the

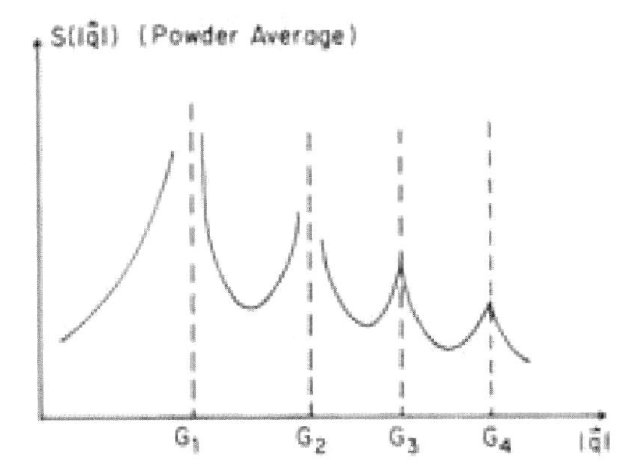

Figure 4. A schematic sketch of the structure function $S(\mathbf{q})$ of a $2D$ crystal. For $T \leq T_m$, peaks for small $\mathbf{G}$ diverge as $|\mathbf{q} - \mathbf{G}|^{-2+\eta G}$ but for larger $\mathbf{G}$ are finite cusps. For $T > T_m$, all peaks are finite with maximum height $\sim\xi_+(T)^{2-\eta G}$. Reprinted from [34] with permission. Copyright 2002 Cambridge University Press.

characteristic predictions of the dislocation theory of melting but, unfortunately, it is not measurable by experiment because the accessible system size and quality are not yet sufficient.

One of the main measurable predictions of the dislocation theory of melting is the renormalized (measured) Young's modulus for which there is remarkable agreement between experiment and theory as shown in Figure 5. Although the theoretical predictions were made in the 1970s [35, 36], experimental measurements [31, 32] were not done for several decades because of the difficulties of realizing a suitable experimental system. In general, these 2*D* systems are extremely sensitive to perturbations due to the supporting substrate and the theory assumes no substrate effects.

In our original papers, we did consider melting of a crystal by dislocations but we did not discuss the fluid phase described by a periodic lattice with a fine concentration of free dislocations. A periodic solid has *two* types of order—translational order and orientational order—describing the orientation of the crystal axes. These order parameters are the density $\rho(\mathbf{r})$ and the orientational order parameter $\psi_6(\mathbf{r}) = e^{6i\theta(\mathbf{r})}$

$$\rho(\mathbf{r}) = \rho_0(\mathbf{r}) + \sum_{\mathbf{G}} |\rho_G(\mathbf{r})| e^{i\mathbf{G}\cdot\mathbf{u}(\mathbf{r})}$$

$$\psi_6(\mathbf{r}) = e^{6i\theta(\mathbf{r})}$$

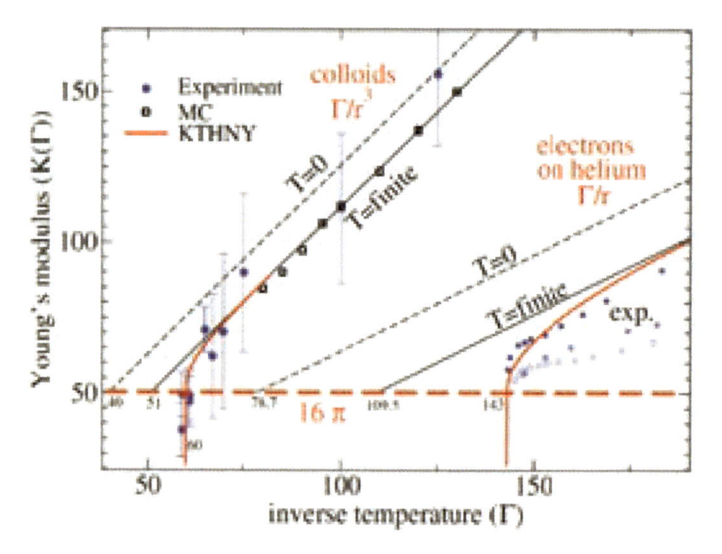

Figure 5. Young's modulus $K_R(T)$ as a function of the effective inverse temperature Γ. The solid line is the dislocation theory prediction of Halperin and Nelson and the symbols are the experimental points. Reprinted from Fig. (2) of *J. Phys.: Condens. Matter* **17** (2005) S3579-S3586 doi: 10.1088/0953-8984/17/45/051 with permission. Copyright 2005 IOP.

The topological defects are (i) dislocations which are responsible for the melting of the solid to an orientationally ordered hexatic fluid, and (ii) disclinations (vortices) responsible for the transition to a high temperature isotropic fluid [35, 36].

The theory has been worked out by Young [35] and Halperin and Nelson [36] with very detailed predictions which have been confirmed by experiment [31, 32] and summarized in Figure 5. One of the most sensitive tests of the theory to date are the numerical simulations by Kapfer and Krauth [38] who performed large scale simulations on up to 10^6 particles interacting by $V(r) = \in (\sigma/r)^n$ repulsive potentials. They found that melting does proceed via the KTHNY scenario with an intermediate hexatic fluid for long range ($n < 6$) potentials, which includes the colloid experiments with $n = 3$ [31, 32] and the electrons on the surface of ^{4}He [39] $n = 1$ while for $n > 6$, the hexatic-isotropic transition becomes first order, which agrees with the hard disk ($n = \infty$) simulations. Note that these simulations are on larger systems than the experimental ones.
BKT theory has also been applied to superconductivity in thin films. In our original paper, we stated that true superconductivity in a $2D$ superconducting film could not exist because of the finite penetration depth $\lambda(T)$ which limits the range of the logarithmic interaction between vortices. For separations $\lambda(T)$, the vortex-vortex interaction behaves as $1/r$ so that the vortices are always free at any $T > 0$ thus destroying superconductivity. Although our argument is correct, in many thin film superconductors, the penetration depth can be $\mathcal{O}(1\text{cm})$ which is a typical system size. For the small applied currents used, this is so large that its effects are smaller than that of the finite currents or the finite frequencies so that the behavior of the system is indistinguishable from that of the $\lambda = \infty$ limit [40, 41]. The theory has also been applied to $2D$ layers of cold atoms [42, 43] with reasonable agreement which may be improved in the future.

REFERENCES

1. D J Thouless *Phys Rev* **187** 732 (1969)
2. P W Anderson, G Yuval and D R Hamann *Phys Rev* B **1** 4464 (1970)
3. R E Peierls, *Helv. Phys. Acta.* **7** Suppl. II 81–3 (1934)
4. R E Peierls, it Ann. Inst. Henri Poincaré **5** 177–222 (1935)
5. N D Mermin *Phys Rev* **176** 250–4 (1968)
6. N D Mermin and H Wagner, *Phys Rev Lett* **17** 1133–6 (1966)
7. P C Hohenberg, *Phys Rev* **158** 383–6 (1967)
8. B J Alder and T E Wainwright, *J Chem Phys* **33** (1960)
9. B J Alder and T E Wainwright, *Phys Rev* **127** 359–61 (1962)
10. M Chester, L C Yang and J B Stephens *Phys Rev Lett* **29** 211–4 (1972)

11. M Chester and L C Yang *Phys Rev Lett* **31** 1377–80 (1973)
12. H E Stanley and T A Kaplan *Phys Rev Lett* **17** 913–5 (1966)
13. H E Stanley *Phys Rev Lett* **20** 589–92 (1968)
14. M A Moore *Phys Rev Lett* **23** 861–3 (1969)
15. J M Kosterlitz and D J Thouless *J Phys C: Solid State Phys* **5** L124–6 (1972)
16. J M Kosterlitz and D J Thouless *J Phys C: Solid State Phys* **6** 1181–203 (1973)
17. J M Kosterlitz *J Phys C: Solid State Phys* **7** 1046–60 (1974)
18. V L Berezinskii *Zh Eksp Teor Fiz* **59** 907 (1970); *JETP* **32** 493 (1971)
19. V L Berezinskii *Zh Eksp Teor Fiz* **61** 1144 (1971); *JETP* **34** 601 (1972)
20. A M Polyakov *Phys Lett* B **59** 79–81 (1975)
21. A P Young *J Phys C: Solid State Phys* **11** L453–5 (1978)
22. J M Kosterlitz *J Phys C: Solid State Phys* **7** 1046–60
23. D J Bishop and J D Reppy *Phys Rev Lett* **40** 1727–30 (1978)
24. D R Nelson and J M Kosterlitz *Phys Rev Lett* **39** 1201 (1977)
25. D J Bishop and J D Reppy *Phys Rev* B **22** 5171 (1980)
26. I Rudnick *Phys Rev Lett* **40** 1454 (1978)
27. J Maps and R B Hallock *Phys Rev Lett* **47** 1533 (1981)
28. J Maps and R B Hallock *Phys Rev* B **27** 5491 (1983)
29. J E Rutledge, W L McMillan and J M Mochel Phys Rev B **18** 2155 (1978)
30. A L Buck and J M Mochel *Physica* B **107** 403 (1980)]
31. J Zanghellini, P Keim and H H von Grünberg *J Phys: Condens Matter* **17** S3579 (2005)
32. H H von Grünberg, P Keim and G Maret *Soft Matter vol 3: Colloidal Order from Entropic and Surface Forces* eds G Gomper and M Schick (Weinheim: Wiley) pp 40–83 (2007)
33. U Gasser, C Eisenmann, G Maret and P Keim *ChemPhysChem* **11** 963 (2010)
34. D R Nelson *Defects and Geometry in Condensed Matter Physics* Cambridge University Press (2002)
35. A P Young *Phys Rev* B **19** 1855 (1979)
36. B I Halperin and D R Nelson *Phys Rev Lett* **41** 121 (1978); *Phys Rev* B **19** 2457 (1979)
37. F R N Nabarro *Theory of Dislocations* Oxford University Press (1967)
38. S C Kapfer and W Krauth *Phys Rev Lett* **114** 035702 (2015)
39. C C Grimes and G Adams *Phys Rev Lett* **42** 795 (1979)
40. D J Resnick, J C Garland, J T Boyd, S Shoemaker and R S Newrock *Phys Rev Lett* **47** 1542 (1981)
41. A F Hebard and A T Fiory *Phys Rev Lett* **50** 1603 (1983)
42. Z Hadzibabic, P Krüger, M Chenau, B Battelier and J Dalibard *Nature* **441** 1118 (2006)
43. M Holzmann, G Baym, J-P Blaizot and F Laloë *Proc Natl Acad Sci* **104** 1476 (2007)

Physics 2017

one half to

Rainer Weiss

and the other half jointly to

Barry C. Barish and Kip S. Thorne

"for decisive contributions to the LIGO detector and the observation of gravitational waves"

The Nobel Prize in Physics, 2017

Presentation speech by Professor Olga Botner, Member of the Royal Swedish Academy of Sciences; Member of the Nobel Committee for Physics, 10 December 2017.

Your Majesties, Your Royal Highnesses, Esteemed Laureates, Ladies and Gentlemen,

Gravitation is the weakest force we know – yet it is this force that keeps our feet on the ground, determines the orbits of the planets around our Sun and governs the violent encounters of black holes in the distant Universe. The 2017 Nobel Prize in Physics honors a discovery that reflects this apparent discord between strength and weakness in a subtle way: the observation of an incredibly weak distortion of space due to a passing gravitational wave created in a momentous collision between two black holes in a faraway galaxy.

About 1.3 billion years ago, at a time when the first multicellular life emerged on Earth, two black holes with masses close to 30 times that of our Sun entered the final stages of a deadly pas-de-deux. Whirling finally at half the speed of light, they collided and merged – an event which sent gravitational waves speeding through space and time, carrying information on what had just happened. Ever since then, the waves have traveled through the Universe, sweeping through the Earth on September 14, 2015, when they were discovered by the newly commissioned twin detectors of the Laser Interferometer Gravitational-Wave Observatory, designed to notice the tiniest vibrations. This was the first time ever that a passing gravitational wave had been directly observed on Earth. This amazing observation represents the dawn of a new era, opening up new vistas for astronomy and bringing about new possibilities to study gravitation close to black holes where it is at its strongest.

Gravitational waves were predicted about 100 years ago in the framework of Einstein's general theory of relativity, which connects gravitation with the geometry of spacetime. They arise as ripples in spacetime whenever objects accelerate, traveling outward from the source as ever weaker waves. When gravitational waves reach the Earth, they stretch and squeeze space by the tiniest amount, far too small to be detected by human senses.

Nevertheless, the team of scientists led by this year's Nobel Laureates succeeded in overcoming the challenge, constructing detectors called laser-interferometers, capable of measuring the microscopic distortions – billions of times smaller than the thickness of a strand of spider silk – in the distance between two mirrors. These interferometers use laser light, split along two arms, each four kilometers long. Mirrors at each end reflect the light back to a point where the two light beams are superimposed. The resulting light wave is channeled to a detector where a pattern of shadows is seen and registered. A passing gravitational wave stretches one of the arms and compresses the other, causing the pattern to shift minutely, which is what the scientists detect.

To exclude local disturbances, like passing trucks or tiny earthquakes causing the instrument to vibrate, the observatory comprises two identical detectors on opposite sides of the American continent, separated by about 3000 km. On September 14, at 9:50 UTC both detectors registered almost identical signals from the passing wave, with less than 7 milliseconds delay. This not only confirmed the signal but allowed localization of its source within an arc on the Southern sky.

The first direct observation of a gravitational wave is in more than one sense an earth-shaking discovery, a breakthrough which has been followed by additional significant observations. It opens unanticipated opportunities to explore hitherto invisible parts of the Universe where according to the renowned astronomer Carl Sagan "something incredible is waiting to be known".

Professor Weiss, Professor Barish, Professor Thorne:

You have been awarded the 2017 Nobel Prize for Physics for your decisive contributions to the detectors of the Laser Interferometer Gravitational-Wave Observatory and for the observation of gravitational waves. On behalf of the Royal Swedish Academy of Sciences it is my honor and great pleasure to convey to you our warmest congratulations. I now ask you to step forward to receive your Nobel Prizes from the hands of His Majesty the King.

Rainer Weiss. © Nobel Prize Outreach AB. Photo: A. Mahmoud

Rainer Weiss

Biography

MY FATHER CAME FROM a well-off German Jewish family in Berlin with connections to the Rathenau family that had begun the Allgemeine Electrische Gesellschaft (AEG). As a young man he had become an ardent and idealistic communist. After finishing medical school he worked in a communist workers' hospital as a neurologist in Berlin. My mother came from a Protestant family of government workers and lawyers on the Rhine near Koblenz. She had come to Berlin with ambitions to become a professional actress. I was born on September 29, 1932 in Berlin to this unlikely and unwed pair.

At the time Berlin was balkanized into sectors that were Communist, Nazi and Weimar. My father had gotten into difficulties with Nazis infiltrators at his hospital and had been taken "prisoner" by a Nazi gang. My mother's family, with still some connection to the civil authorities, managed to get him released and sent him to Prague, Czechoslovakia. After I was born she joined him in Prague. It was not a convivial relationship even though they had gotten married and had another child, Sybille, by 1937. A critical moment came in September of 1938 when Chamberlain gave Sudetenland to Hitler and effectively opened Czechoslovakia to the Nazis. We heard the decision on a radio while on vacation in Slovakia and joined a large group of people heading toward Prague to attempt to get a visa to emigrate to almost anywhere else in the world that would accept Jews. There were not many places to go. We were extremely lucky in gaining the support of the Stix family of St Louis who gave bond for about ten thousand Jews who were professionals to gain favored entry to the United States.

The family came to the United States in January 1939, landing in New York. It took my father about 4 years to pass the New York State examinations to practice as a medical doctor. During those years he worked as a medical aide in a New York City tuberculosis hospital and my mother took counter jobs at department and drug stores. The family stabilized by the late 1940s, with my father becoming a psychoanalyst in the Horney group.

My sister became an actress and is now a successful playwright at New York University.

Initially I went to New York City public school but through a refugee relief organization associated with a neighborhood church I received a scholarship to the Columbia Grammar School, a private school in mid-Manhattan at one time associated with preparing students for Columbia University. I started there in 5th grade and remained to graduate as a high school senior in 1950.

Music, science and history were my favorite courses. Mathematics had to be motivated by real problems. In part due to my father introducing me to magnets and batteries, I became interested in electricity and, especially, radio and electronics. By 13, at the end of the Second World War, I could go to Cortland Street in New York and for pennies buy vacuum tubes, transformers, capacitors, resistors and for a few dollars buy complete assemblies such as radar receivers with oscilloscopes, servo controllers for gun mounts, crystal oscillators. All of this war surplus was arriving by the truck load and effectively being dumped into the street. With these components and magazines such as *Popular Mechanics and the American Radio Relay League Handbook for guidance*, one could build ham transmitters, audio amplifiers, even an FM radio. I made pocket change by fixing radios and other broken electronics. During that time, I built an audio system for the school gymnasium and the transmitter for a school ham station, W2ZIQ. A significant opportunity for even grander electronics occurred when a Brooklyn movie theatre had a fire behind the screen and a large number of coaxial tweeter/woofer speakers were destroyed. A friend and I were allowed to unscrew six of the speakers and take them home via the subway. The heat of the fire was not enough to depolarize the magnets nor destroy the tweeters; all that was needed to restore the speakers was to buy new cones and voice coils from the manufacturer.

By 1948 I had assembled a high-fidelity audio system consisting of an FM tuner, a Williamson power amplifier and one of these movie house speakers. At that time the New York Philharmonic was being broadcast live from Carnegie Hall on an FM station. I invited some immigrant friends of my family with an interest in classical music to come listen. They were truly impressed at the sound, it was like being in Carnegie Hall for the concert. They asked, if they paid for the parts, would I make them such a system. That is how a small business began – they had friends and their friends had friends and so on. By the time I was a senior in high school I had more orders than I could handle but had run into a problem that challenged my "street" electronics knowledge.

At the time the very best phonograph records for classical music were being made by the British company Decca FFRR (Full Frequency Range Recordings); they were 78 RPM and made of the smoothest shellac available. Even so the roughness of the surface made it difficult to enjoy piano music due to the high frequency noise generated by the needle running in a rough groove. The noise was tolerable in a fast loud movement but it was all one heard in a quiet slow movement and spoiled the listening, especially in a wideband high fidelity system. To deal with this problem, the thought I had was to make a frequency variable filter that depended on the amplitude of the sound – to reduce the high frequency response at low sound volume and open the bandwidth when the music was louder. I never got the idea to work satisfactorily. It just seemed difficult to vary the bandwidth smoothly in such a way to avoid swishing noises and adding more distracting sounds. I didn't know enough about filter theory and mathematics and decided it would be useful to learn electronics more formally by going to college.

MIT accepted me as a Freshman in 1950. At the time MIT had a rigid curriculum for Freshmen consisting of what was considered a basis for all scientific and engineering disciplines. The choice of a major was deferred to the sophomore year. I chose Electrical Engineering as it seemed closest to the problems I was trying to solve. The Electrical Engineering curriculum was unfortunately also rigid. All had to first take a course in power engineering and mechanical structures followed by the elements of circuit theory. Electronics and more interesting courses in noise and signal processing were reserved for juniors and seniors. By the second term sophomore year I had shifted into Physics because it had a more flexible curriculum.

Furthermore, vinyl phonograph records with much smoother surfaces had begun to replace shellac records and the problem I was worrying about had been eliminated by this new technology.

The summer between sophomore and junior year I worked for a small-time entrepreneur who had a cost plus contract with the Air Force to design and construct an automatic blood cell counter. The idea was to enable quick triage in the event of a nuclear war with Russia, to save those people who still had reasonable red cell counts and viable white cells to survive. This was part of a national effort of civilian defense to allow misguided military planners to contemplate the use of nuclear weapons, at a time when some Americans who could afford it were being encouraged by our government to build fall-out shelters.

I designed pulse counter circuits with pulse width discrimination. Others were designing a microscope with a rotating stage and a fast photodetector at the optical output which would drive the electronics. I don't think the project was ever completed. When the summer ended I made a bicycling trip to Nantucket which changed my life.

On the trip I met a girl who literally swept me off my feet. She played the piano, folk danced and had a very sensible attitude to what was important in life. We spent several days together in Boston before she had to go to

Northwestern University in Chicago to continue her education. The relation was initially maintained by frequent letters and came to a high point during Christmas when we met each other's families. After she returned to Chicago and I to Boston the letters were less frequent and I went to Chicago in the middle of the school term to try understand why. To a more world wise person it would have been obvious. She had found a more interesting guy, and I went into what is best described by Schubert in the song cycle "Die Winterreise" as the disappointed rejected lover who could think only of "her" – saw and heard her in the trees, in the waterfalls, the sunsets ... The result was I failed all my courses at MIT and had to leave as a student.

In the spring of 1953 I became an electronics technician in the Atomic Beam Laboratory of Physics Professor Jerrold Zacharias in the Research Laboratory of Electronics (RLE) at MIT. I had a union card and punched a time clock. My colleagues were machinists and lab technicians, Frank O'Brien, John McClean, Mark Kelly and a collection of graduate students some of whom were still veterans from WWII. I learned how to machine, do sheet metal work, soft and hard solder, Heliarc weld and design equipment around those things available in metals stockrooms and hardware stores – the art of improvisation in experimental science. The science being done in that laboratory was exquisite. The experiments were looking at the properties of isolated single atoms and molecules unperturbed by neighboring systems. Each atom was the same as the next and it was possible to ask fundamental questions about their structure and the interactions that held them together.

I started by helping the graduate students design and build the electronics they needed for their thesis projects and eventually began working directly with Jerrold on the Cesium atomic beam clock. The laboratory developed a prototype clock with potential of a precision of 10^{-12} in one second of integration time. The clock was commercialized by the National Company and then became the standard of time for the Bureau of Standards (now the National Institute of Standards and Technology, NIST) and the United States Navy.

Jerrold had bigger ideas. He wanted next to make an atomic clock with about 100 times better precision so he could make a direct measurement of the Einstein gravitational red shift on the Earth. His idea was to increase the observing time of the atom in the region where the instrument translated the internal oscillations of the Cesium atom into a radio frequency signal. In the initial clock the atoms were flying through this region in milliseconds since they moved with the velocity of sound horizontally. His new idea was to make the atoms travel vertically so that the slower ones in the Maxwell distribution would be turned around by the gravitational field of the Earth – they would follow the same parabolic trajectory as a ball thrown vertically. The observation time could become

Figure 1. Jerrold Zacharias and me on the porch of his house on Monument Beach, Cape Cod, Massachusetts in the mid-1970s. The photograph was taken by Rebecca.

a decent fraction of a second. The concept was called the Zacharias atomic fountain. When the new clock was operating, Jerrold and I would go to Switzerland where we would put one clock in the laboratory on top of the Jungfrau and another one in the valley below and compare their rates by sending signals between them.

Unfortunately, the fountain clock did not work. The first attempt was made in a vertical vacuum system about 3 meters high. Although we were injecting about 10^{18} atoms/sec into the fountain we saw less than 10 background atoms/sec hitting a detector on the opposite side of the fountain. The same results when extending the height of the apparatus first to 6 meters and finally to 9 meters – there were just no slow atoms in the beam. It seemed the Maxwell distribution was not satisfied in a beam. In 1956 Jerrold began a project to revitalize secondary science and mathematics education in the United States and I had the free run of his laboratory. I did want to understand why the fountain had failed and set up a fast shutter near the source of atoms and a detector at the 6-meter-high point in the upward going beam. I found that the Maxwell distribution was already deficient at 1/3 the average velocity atoms and that there were simply no atoms at 1/20 of the average velocity we were hoping to use to make the fountain. The problem was the copious fast atoms in the beam were hitting the slow ones and throwing them out of the beam.

It is worth noting that now, with the ability to laser cool a gas of atoms, it is possible to make a Zacharias fountain with heights of less than a meter and clocks that can measure the Einstein gravitational red-shift over a height difference of a few cm.

With Jerrold's help I finished my undergraduate degree and became a graduate student working in the same laboratory. I kept trying to make better clocks. The next idea was to increase the standard frequency using molecular rotation states of light molecules at 50 to 100GHz rather than the hyperfine structure of Cesium at 10GHz. There I had to invent a way to detect all kinds of atoms and molecules rather than just alkali atoms which ionized easily on a hot wire with a work function higher than the ionization potential of the atom. I designed and built an electron impact ionizer with high current densities and a scheme to use the electron space charge to collect and focus the positive ions. The device was able to convert a neutral atomic or molecular beam to a collimated positive ion beam with 20% efficiency. Next I built an electric resonance molecular beam apparatus to use the rotation states of $C^{12}O^{18}$ as the basis of the new clock with the fancy ionizer as the molecule detector.

While I was waiting for the O^{18} enriched sample of carbon monoxide to be produced in an Israeli reactor, I worked with Lee Grodzins on a Mossbauer experiment. The Mossbauer effect had just been discovered, affording a way of measuring fractional energy shifts of 10^{-13} in a simple apparatus. The idea we had was to test a somewhat zany hypothesis of Finlay-Freundlich (this in the epoch of the also wild but seductive hypothesis of the steady state universe) who had noticed that spectral lines in bright stars were more red shifted than in dimmer ones. He attributed this to a photon/photon scattering (not predicted by quantum field theory) where a photon from a spectral line in the star was reduced in frequency by colliding with the background thermal photons generated by the star. He furthermore made an estimate of the average photon field in the universe and provocatively attributed the Hubble cosmological red shift to this new type of scattering which got called the "tired light" hypothesis. We built a Mossbauer apparatus where we passed the gamma rays through a hot oven to look for a frequency shift. By comparing the Mossbauer line shift with the oven hot and cold we established no frequency shift at a level that would have mattered for the Finlay-Freundlich hypothesis (several years later the experiment was done again with light and a microwave cavity again showing no frequency shift – this was the beginning of precision interferometry, more on this later).

Just as we were finishing the Mossbauer experiment, I was told that I had exhausted the funding for a PhD candidate and that I had to finally do a PhD thesis and graduate. Furthermore, my wife had become pregnant and a real income had become more important. (My wife, Rebecca, was a plant physiologist at Harvard and later became a children's librarian.) Jerrold managed to get me an instructor's job in the Physics department at Tufts University in Medford, Massachusetts. I taught in the day and worked on the thesis at MIT at night. The ambitious CO clock was dropped and instead I did a boring but useful measurement of the electric dipole moment of HF and its hyperfine structure in a set of low angular momentum rotational states.

All this was done by May 1962 when Sarah, our daughter, was born. (Sarah has become an ethnomusicologist. A son Benjamin was born in 1967, he is now an art historian.) Tufts had made me an Assistant Professor and it seemed I could have stayed as a faculty member, but I wanted to work with Professor Robert Dicke at Princeton who had become interested in gravitation. These were the years of a renaissance in General Relativity, in good measure due to the vast improvements in the technology since 1915. It was now possible to contemplate measuring the tiny deviations of Einstein's theory from Newton's. Gravitation, because of this, was making a transition from mathematics back into physics.

When I arrived at Princeton, Dicke and his group had just finished a modern version of the Eötvös experiment showing the equivalence of the inertial and gravitational mass, one of the cornerstones of general relativity (the weak principal of equivalence), to a part in 10^{-11}. Dicke was working on a new theory of gravitation which combined a scalar field to the tensor field of general relativity. The motivation was to better incorporate Mach's principle into gravitation. He asked another post doctoral scientist, Barry Block, and me to consider an experiment to measure the excitation of the Earth in the spherically symmetric ${}_0S_0$ mode by scalar gravitational waves coming from astrophysical sources. The mode has a period around 20 minutes with a Q about 3 000 as had been seen after some strong deep focus earthquakes. We made a quartz gravimeter and placed it in the same temperature regulated pit used earlier by the Eötvös experiment. We did not set interesting limits on the spectrum of scalar gravitational radiation. Early in our observing run, it became clear that geophysical excitations were going to severely limit the sensitivity of our measurements.

Even though the experiment was unsuccessful, the two years at Princeton were profoundly important in my scientific development. During my stay a range of experiments and experimental techniques were being tried. These included: a successful measurement of the Einstein gravitational red-shift in the sun (the first really believable measurement), an experiment that showed the equivalence between passive gravitational mass (ability of mass to respond to gravitational fields) and the active gravitational mass (ability of mass to make gravitational fields), an experiment to try to answer how round is the sun, and a precise absolute measurement of g, earth's gravitational acceleration, using a freely falling corner cube in an interferometer. Lots of new ideas were being talked about at the group meetings such as the notion of putting optical corner cubes on the moon to allow precision measurements of moon-earth dynamics as well as the early thoughts about the heat that might accompany the origin of the universe. (The actual work of looking at the microwave spectrum of the sky started shortly after I left.) The critical and lasting knowledge was how one designs an experiment to get to its fundamental limits. Dicke was a master at this. I tried to learn a little formal general relativity from courses taught by Dicke (very many diversions) and Wigner (too abstract) but have to say they were interesting but not successful.

In 1965 Jerrold invited me to come back to MIT as a faculty member in the Physics department, with research support through the Joint Services Research Program in the Research Laboratory of Electronics. The program was not fussy about the actual research topics it supported but was dedicated to training more scientists and engineers as a resource for the national defense. At the time MIT was a better place than Princeton to do experimental work, as there was still the legacy of machine shops and store rooms filled with equipment from the wartime radar lab. It was easier to start a new experimental program at MIT. I began a laboratory dedicated to Cosmology and Gravitation. One of the first research goals was to try to establish if G, the Newtonian constant, was varying in time by a fractional amount 10^{-10}/year. Both Dirac and Dicke had suggested that due to the expansion of the universe G was getting smaller with time. The way we were going to measure this was with an absolute gravimeter based on a plate held against Earth gravity by electric forces whose strength was determined by the Stark effect on molecular states in a beam. Thereby g would be turned into a frequency that could be compared to an atomic clock. It was also necessary to take sample measurements of the shape of the Earth to estimate its change in radius with time. The idea was to develop kilometer long laser strain gauges with absolute knowledge of the wavelength by comparison to an optical molecular resonance reference. The proposed program was long range and probably too ambitious for a starting (untenured) faculty member, although I felt it fit well into the capabilities of the MIT infrastructure.

We started with the laser frequency stabilization when Schaoul Ezekiel, an aeronautical instrumentation graduate student, became the first student to join the new group. We made an Argon ion laser in the RLE facilities and frequency stabilized it against narrow molecular iodine resonances to a relative frequency accuracy of 10^{-12}. At about the same time, two experiments were done to look at the quantum noise of the laser. One experiment measured the fundamental phase noise in a laser due to the spontaneous emission as had been predicted by Townes. The other (mentioned previously as a redo of the tired light hypothesis) was a table-top Michelson interferometer operating at significant power (~100mW) with the phase measurement limited by the shot noise (quantum noise) at 10's of KHz. The fringe was maintained by a servo system and the signal was translated from the 1/f region of the amplitude noise of the laser to higher frequency by modulation techniques – all direct applications of the Dicke methods for precision measurement.

The absolute gravimeter never got constructed as I began to realize that the lunar laser ranging observations and the solar system radar astronomy, which gave the critical radial dimension to the dynamics measurements, would lead to more reliable measurements of changes in g at the necessary precision.

In 1966 I was asked to teach a graduate general relativity course. I describe the difficulties and also some of the things learned in my Nobel lecture, especially, the beginnings of thinking about gravitational wave detection by interferometric methods. The other new topic for me was general relativity applied to cosmology. It was love at first sight (even though we all learn bits of this as we go along) finally understanding Bondi's book on Cosmology and working with the Friedmann-Robertson-Walker equations was magical. One of the students in the course was Dirk Muehlner, who had some experience with far-infrared physics and, furthermore, knew some astronomy. At the end of the course we began talking about the new measurements that had been made by Penzias and Wilson and their interpretation by Dicke and his group as the red shifted relic heat of the cosmic explosion. We both thought it would be critical to show that the radiation actually exhibited a Planck spectrum, but it was only talk until Bernard Burke, the head of my division in the Physics department, and I had a heart to heart conversation about my future in the department. Burke felt that these laser and gravity experiments would not lead to interesting results soon enough for decisions that had to be made in the department. He suggested why not really do cosmology and measure the spectrum of cosmic background radiation. His radio astronomy colleagues could help, since they had some experience with flying balloons in the stratosphere above most of the water in the atmosphere that would disturb such measurements.

Dirk joined the lab and we began to explore the possibility of making a measurement of the spectrum of the cosmic microwave background. At the time there were only measurements in the Rayleigh-Jeans low frequency part of a 3K thermal spectrum. The thermal peak is near 180GHZ, while the highest frequency measurements at the time was at 32GHz. There was an optical measurement made in the 1930s of the rotational excitation of CN (Cyanogen) molecules in stellar atmospheres, which could be interpreted as due to the molecules being in equilibrium with thermal radiation at 3K. The lowest rotational state energies were close to the thermal peak and could have been excited by the radiation but also by local charged particles. I describe our effort in a book edited by Jim Peebles, Lyman Page and Bruce Partridge, *Finding the Big Bang*[1].

The laser science and technology were taken over by Ezekiel as the research area in a new group in RLE. Starting in 1967 we began a program to measure the spectrum of cosmic background radiation from high altitude balloons. The research was supported by NASA.

Btween 1967 and 1982 we flew around 20 flights, first to measure the spectrum and then the isotropy of the cosmic background radiation. The program was the mainstay for graduate student theses as it had both significant technical development (new mm detectors and filters, cryogenic instrumentation) but also astrophysical science results that could be published. Initially the spectrum measurements were done in three frequency bands, one at low frequencies in the Rayleigh Jeans region of the black body spectrum, which overlapped with the ground-based measurements, a second embracing the black body peak and a third in the Wien part of the spectrum above the peak. Even at an altitude of 40km in the atmosphere the ozone and water emission lines had to be accounted for and modeling was necessary to recover the cosmic background spectrum. At the end more channels were added to specifically measure the atmospheric emission lines. Eventually we were able to establish the black body nature of the spectrum (there was a peak in the spectrum) but not with much precision. We began to realize that a precision measurement would require a satellite mission.

Next, we turned to measurements of the angular distribution of the radiation. The first goal for these measurements was to see a dipole distribution of cosmic background radiation in the sky determined by our motion relative to the average rest frame of the universe or, another way to think of it, as relative to the last scatterers of the radiation at a red shift of about 1000. The largest term would come from the rotation of our galaxy, which gives a v/c of about 10^{-3} producing a variation of the radiation temperature over the sky, being hottest in the direction of the velocity and coldest in the opposing direction.

High sensitivity to small changes in the temperature (in the intensity of the radiation) was required in these measurements, but one could relax the requirements for absolute calibration so important in the spectrum measurements. Our first flights made an unfortunate discovery and indicated a significant problem for the future of these measurements. In the channel embracing the black body peak and the high frequency channel we saw anisotropies easily 10 times larger than the expected dipole and also discrete sources tied to the sky and not to the atmosphere. Eventually we realized we were measuring dust emission from these discrete sources as well as broadly distributed dust throughout our galaxy with the strongest emission from the galactic plane. We had to become astronomers to get at the cosmology, In fact it took many flights viewing much of the sky to actually measure the dipole in the low frequency channel (least effected by the dust) using corrections from the higher frequency channels. The galaxy had replaced the atmosphere as the worst source of contamination. Other groups measuring the isotropy of the cosmic background from balloons and aircraft with channels at lower frequencies, less sensitive to the dust but more sensitive to free-free and synchrotron emission by electrons, were able to map the dipole better.

By the early 1970s it was clear that a satellite mission would be more definitive in making measurements of the spectrum and the isotropy of the cosmic background radiation. By placing the instruments outside of the atmosphere in Earth orbit, one could get long integration times to improve the signal to noise but also have time to test for systematics. Furthermore, without the absorption in the atmosphere, it would be possible to add enough wavelength coverage to separate the cosmic background radiation from the emission of nearer astronomical foregrounds. John Mather recognized this and acted on it. As a graduate student at Berkeley he conceived of COBE (Cosmic Background Explorer) and after graduating he pulled a team together including me to actually do it. He and John Boslough wrote a book about COBE, *The Very First Light*[2] which includes much of the story of the project.

I became the COBE science working group chairman, in part since I was the oldest but also because of the experience I had gained with being on many NASA advisory committees dealing with science policy and management, though COBE took close to 20 years from John's conception to results. The results were significant: the cosmic background spectrum was found to be thermal to a 10^{-4} between 90 to 600 GHz , an intrinsic anisotropy of the universe at a level of 10^{-5} K was discovered at angular scales 7 degrees and larger which indicated that there were quantum fluctuations in the beginning of the universe that created a structure maintained through the universal expansion by the gravitational interaction of dark matter, and it found that the galaxies were filled with dust at close to 5K. For these discoveries, two COBE scientists, Mather and George Smoot, won the Nobel Prize in Physics in 2006.

Research in experimental gravitation did not end in the laboratory once the cosmic background measurements began. The same graduate course in general relativity led to a gedanken experiment which became a real experiment in 1972 to try to detect gravitational waves from astronomical sources. The experiences of the earlier research on laser frequency stabilization and the characterization of the fundamental noise in laser interferometry found application in the design and construction of a prototype interferometric gravitational wave detector. By the mid-1980s, Stephan Meyer took over the cosmic background radiation research and I became more involved with the gravitational wave detection project, as is described in my Nobel Physics Prize Lecture.

REFERENCES

1. Peebles, P.J.E., Page, L.A., Partridge, R.B., *Finding the Big Bang*, Cambridge University Press (2009).
2. Mather, J.C., Boslough, J. *The Very First Light*, Basic Books (1996, 2008).

Ligo and the Discovery of Gravitational Waves, I

Nobel Lecture, December 8, 2017 by Rainer Weiss
Massachusetts Institute of Technology (MIT), Cambridge, MA, USA.

ALL THREE OF US – Barry Barish, Kip Thorne and I – want to recognize the critical role played by the scientists, engineers, students, technicians and administrators of the LIGO laboratory and LIGO Scientific Collaboration who are responsible for opening a new field of scientific research: Gravitational Wave Astronomy and Astrophysics. We are also deeply indebted to the United States National Science Foundation, which was willing to take a risk in supporting a new field that required significant technical development and with an uncertain knowledge of sources but certain that, should it succeed, it would have a profound influence on our understanding of physics and the universe.

The three of us will give talks with the same title, "LIGO and the Discovery of Gravitational Waves", but focus on different aspects. I will discuss some of the early history of gravitational waves and develop the concepts to understand the detectors as well as the challenges faced in measuring strains as small as 10^{-21}. Barry will show how the LIGO project organized to make steady improvements and ultimately carry out a successful scientific program. He will describe the detections as well as ideas to improve the detectors driven by the new science we hope to learn. Kip will look at the broader aspects of the new field of gravitational wave astronomy. He will tell of the critical role numerical relativity and understanding the quantum mechanics of precision measurements has played. He will also give a vision of the science that could come from an investigation of the gravitational wave sky from periods of fractions of milliseconds to tens of billions of years.

In 1915, a little over 100 years ago, Einstein published the General Theory of Relativity (GR)1,2, a new theory of gravitation which replaced the Newtonian force by the idea that mass distorts the geometry of space and the flow of time. Matter then moves in this new space-time along the shortest four-dimensional paths. (*Figure 1* and its caption give an impression of the idea.)

The new theory solved a puzzle remaining from Newton's theory for the motion of the planet Mercury, the planet in the strongest gravitational field of the sun, which after correction for the motions of other objects in the solar system, still did not seem to obey Newton's theory. Specifically,

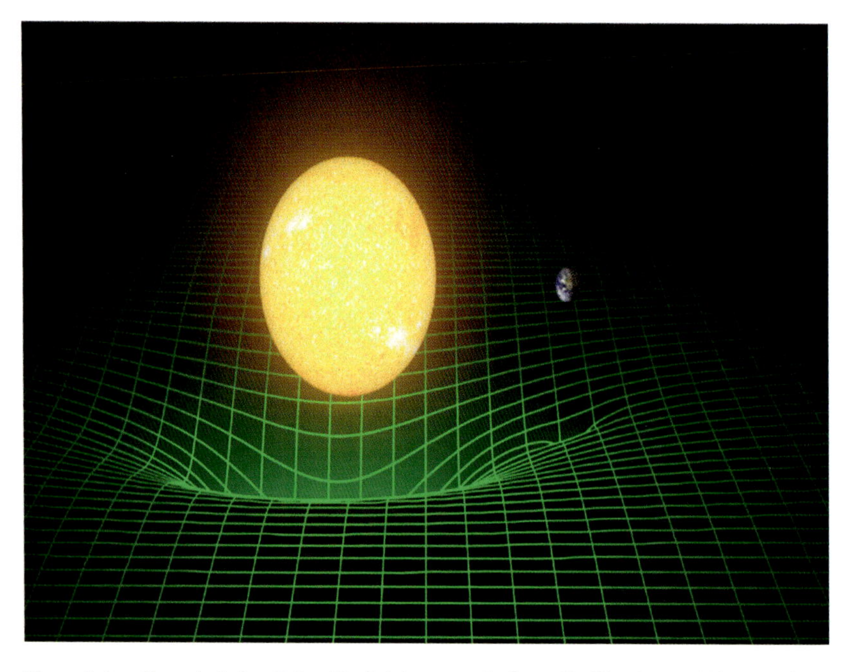

Figure 1. An attempt at visualizing Einstein's concept of gravity (the General Theory of Relativity) in which distortions in the geometry of space and time have replaced the gravitational force of Newton. Think of a jungle gym of rectangular bars in the three spatial dimensions. The intersection points of the bars are all evenly spaced and straight lines. Imagine also that a clock is placed at all the intersection points and that all the clocks read the same time at the same moment in time. The situation when there are no masses in the vicinity. Now, take a two-dimensional cut through the structure and place the sun and earth into the picture. The structure near the sun and to a smaller extent that near the earth is distorted. Also, though not shown, the clocks at the intersection points near the sun and to a lesser extent those near the Earth run a little more slowly than those in the less distorted parts. The orbit of the Earth around the sun is now dictated by the distortion of space and the time dilation – the path an object with no forces would follow in the new geometry.

Mercury was not moving along a path dictated by a pure $1/r^2$ force law. To save Newton, another planet between the Sun and Mercury was hypothesized but never found. To Einstein's enormous pleasure, GR gave the measured orbits. It was the first confirmation that he was on the right track. GR also provided some new phenomena open to measurement, small effects difficult to measure but profound in their importance. These were the prediction that clocks ran more slowly in strong gravitational fields than weak ones[3,4] and that starlight passing the limb of the sun would be deflected toward the sun, the so-called bending of light[5]. It is a tribute to the difficulty of the measurements that both of these effects were only really measured reliably about 50 years later.

GR had more in it still. In principle, as we have now have found directly, one could carry out calculations of massive systems moving at relativistic speeds. Furthermore, gravitational information – gravitational waves – did not travel at infinite speed as implied by Newton's gravity, but travelled at the speed of light as is necessary to be compatible with special relativity.

In 1916 Einstein[6] wrote a paper showing some of the ideas discussed above by perturbation calculations using linearized versions of his full field equations. It is in this paper he first describes gravitational waves. (*Figure* 2).

They travel at the speed of light and are transverse waves much as electromagnetic waves, but rather than exerting forces on charges, they distort space perpendicular to the direction along which they propagate. Alternately stretching space in the east-west direction while simultaneously compressing space in the north-south direction. The distortion has the special property that the change in separation of two points is proportional to their separation, the strain $h = \Delta l/l$ in the east-west direction is the same over the entire wavefront, while strain is equal but opposite in the north-south direction. (For completeness it is necessary to at least say there is another polarization for the waves rotated by 45 degrees to the east-west, north-south directions traveling along the same direction which acts independently [is orthogonal]). The gravitational waves carry energy as well as linear and angular momentum. In the 1916 paper, Einstein describes the wave kinematics well but runs into trouble showing how accelerated masses radiate gravitational waves. In a second paper dedicated entirely to gravitational waves in 1918[7], he derives the quadrupole formula (to within a factor of 2) relating the gravitational wave energy radiated by a mass distribution to its non-spherically symmetric accelerations. Despite the mistake in the 1916 paper, he makes a bold (and for 1916 correct) claim at the end of this paper that gravitational waves will never play a significant role in physics. I have asked the editors of the

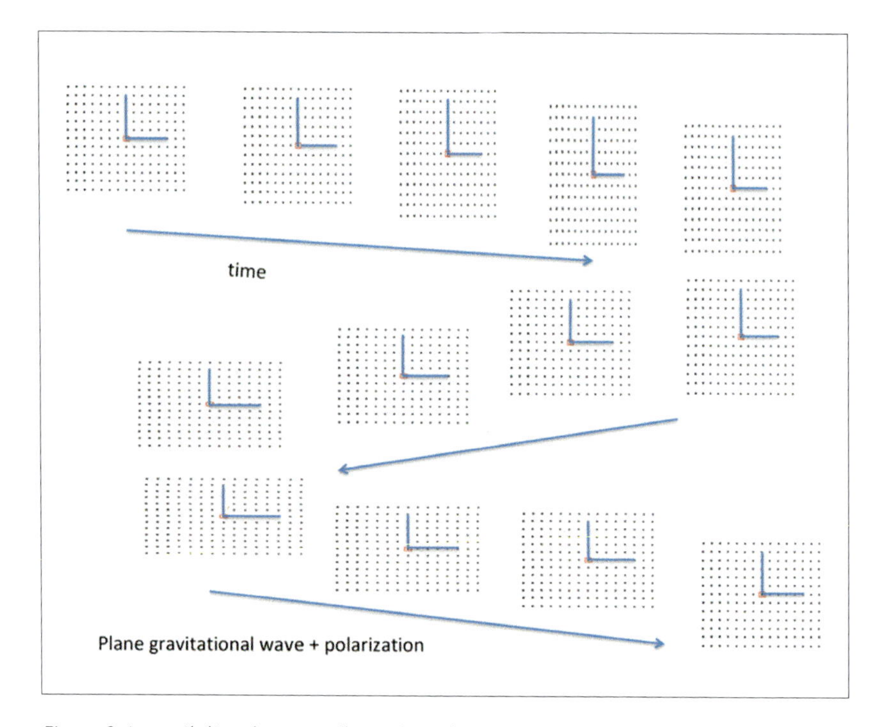

Figure 2. A gravitational wave with + polarization. Gravitational waves are emitted by accelerated masses. They propagate at the speed of light. Once they have become plane waves some number of wavelengths from their source they become transverse waves which cause a strain in space perpendicular to the direction in which they propagate. The strain stretches space in one dimension while simultaneously contracting space in a perpendicular dimension. The strain $h = \Delta l/l$ is the change in distance between two places divided by their separation. The figure tries to show this by distributing a set of probe masses throughout space. The gravitational wave propagates perpendicular to the plane of the figure and the masses. Time evolves from the top left to the bottom right by going through one complete cycle of an oscillating strain. L in the figure symbolizes the LIGO detector. The thing to notice is the change in length of the arms of the L which becomes measurable in the detector. The longer the arms the larger is the measurable displacement signal. The extension of one arm with the contraction of the other arm enables the use of a Michelson interferometer which is specifically sensitive to the difference in light travel time along the two arms.

Einstein Papers Project to find evidence in his notebooks and back of envelopes for the estimates that led him to this, but nothing has been found.

I will be presumptive and guess at what he might have considered, given the technology and knowledge of astronomy open to him in 1916. There are two equations that will help us with this: an estimate for the gravitational wave strain h that comes from the quadrupole formula and the relation between the power carried by the gravitational waves and the time dependence of the strain. A way of estimating the gravitational wave strain from the motion of the masses at the source of the waves is

$$h = \left(\frac{Gm}{Rc^2}\right)\left(\frac{v^2}{c^2}\right)$$

Here G is the Newtonian gravitational constant, m is the mass of the radiating system, R the distance from the source, c is the velocity of light and v is a non-spherically symmetric velocity of the radiating mass distribution. For example, if the system is composed of two orbiting objects, it becomes the relative tangential velocity in their orbits. The term in the first pair of brackets has a special meaning in general relativity, it is dimensionless and is a measure of the absolute strength of gravity in the situation. We who are standing on the surface of the Earth experience the strength of this term as 10^{-10}, at the surface of the sun it is about 10^{-6}, or the surface of a neutron star it is 10^{-1}, at the event horizon of a black hole, it is about 1. It is clear we live in very weak gravity.

The other relation we will find useful is the power per area carried away by the gravitational wave knowing the gravitational strain in the wave

$$S_g = \frac{c^3}{16\pi G}\left\langle\left(\frac{dh}{dt}\right)^2\right\rangle \text{ where } \frac{c^3}{16\pi G} = 7.8\times 10^{36}\, erg\sec/\,cm^2$$

The power per area in the wave is proportional to the square of the rate of change of the strain times a gigantic factor which tells that a small amount of strain in space is accompanied by a huge amount of energy. In other words, it takes enormous amounts of energy to distort space. One way to say it is, the stiffness (Young's modulus) of space at a distortion frequency of 100 Hz is 10^{20} larger than steel.

An example one might have expected Einstein to use is the gravitational radiation emitted by the collision of two trains, a man-made event of significant energy transfer. Using parameters such as m = 10^5 kg, v = 100km/hr, collision times of 1/3 second and a distance of R = 300km (far enough away to be dominated by the radiation rather than Newtonian interaction), one gets a strain h =10^{-42} truly too small to have any physical effect. Another possibility might have been to estimate the change one might be able to detect in observing a binary star system through a telescope. The two stars orbiting each other would lead to the loss of energy by gravitational radiation and would be changing their period as they fall toward each other. Some typical numbers that might have been available in 1916, the two stars both have a solar mass and an orbital period of a day. The energy lost to gravitational waves each orbit relative to the kinetic energy stored in the orbit is about 10^{-15} so that it would take about 10^{13} years to see the orbit collapse, not a practical astronomical observation.

As with many of the other effects predicted by general relativity, it took the development of technology as well as the improved knowledge of the universe through improved technology to make reliable measurements and observations.

It took until the mid-1970s with discovery of pulsars and vastly improved time keeping to perform the measurements of a binary star system which showed evidence for energy loss due to the radiation of gravitational waves. The definitive measurements were made after the discovery of a binary neutron star system by Russell Hulse and Joseph Taylor (*Figure 3*).

Hulse and Taylor tracked the pulsations from a neutron star using the **radio telescope at Arecibo** (Puerto Rico) beginning in 1972. The frequency of the pulsar was nominally 17Hz but they noticed that pulsation frequency was frequency modulated with a period of 8 hours. After a con-

Figure 3. Russell Hulse and Joseph Taylor who were awarded the Nobel Prize in Physics in 1993 for the discovery of the binary pulsar system, which became a remarkable test laboratory for general relativity. They were able to measure relativistic periastron precession of 4 degrees/year and many other relativistic dynamical effects which allowed them to solve for the individual masses of the neutron stars. As shown in Figure 4 they solved for the dynamics of this system and minimized the residual between their data and the full relativistic model of the motion; to do this it was necessary to include the loss of energy by the system to gravitational radiation.

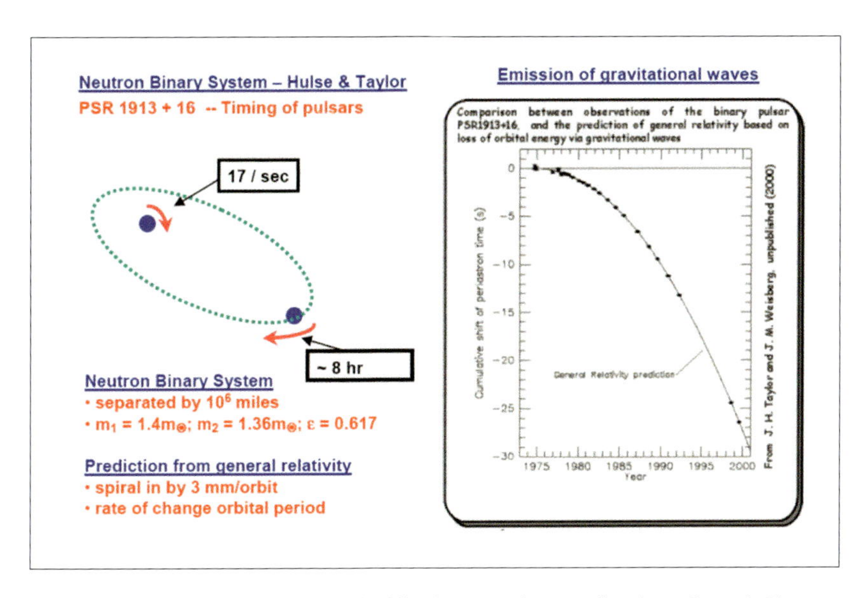

Figure 4. The change in orbital period of the binary pulsar as a function of epoch. The dots indicate the measured reduction of the orbital period as the neutron stars lose total energy by radiation of gravitational waves. The line is the expectation from the General Theory of Relativity. The system was the first to show the effects of gravitational radiation and also provided the first demonstrated source of gravitational waves for direct detection. As Barry will show, the coalescence of a pair of neutron stars has been detected by LIGO and VIRGO.

siderable effort to establish that there were no other stars in the vicinity, they came to a model of a pair of neutron stars orbiting each other. They established quickly that the dimensionless gravitational field strength of one neutron star at the position of the other was about 10^{-6} and that they had come upon a wonderful laboratory to test general relativity. The General Relativistic periastron advance was close to 4 degrees per year, enormous compared to the 43 seconds of arc per century in the Mercury Sun system, and through other relativistic terms it was possible to solve for the mass of the two objects separately. One of the more stunning measurements was the change of the orbital period as a function of epoch. *Figure 4* from Taylor and Weisberg8 shows the period was getting shorter systematically and was consistent with being due to energy being lost by the system to gravitational waves using the quadrupole formula. The explorations of gravitation with this really remarkable laboratory led to Hulse and Taylor receiving the 1993 Nobel Prize in Physics. The discovery not only confirmed the existence of gravitational waves but also identified a source for the eventual direct detection of gravitational waves.

The first experiments that attempted to directly detect gravitational waves from astrophysical sources were an outcome of the Chapel Hill Conference on Gravitation in 1957[9]. At this conference John Wheeler and

Joseph Weber presented the idea of writing the gravitational wave as a tidal force transverse to the propagation direction and treating the detection of weak gravitational waves as a Newtonian interaction of these forces exciting a mechanical resonator. The idea was to measure a gravitational wave arising from an astrophysical event such as a supernova explosion in our own or a neighboring galaxy. The event involved the conversion of a decent fraction of the rest mass of the exploding star into gravitational waves. The frequencies of the motion at the star are in the low kHz band. The waves, when incident on an aluminum bar with longitudinal resonances in the same band, would drive the bar into oscillation10. The bar would ring on after the wave had passed through, making it easier to measure.

Figure 5 shows Weber mounting strain gauges on one of these detectors. A critical idea from the beginning was to look for coincident excitations in several bars as a means of making a detection over the noise in the device which was dominated by Brownian motion (the thermally induced random motions of the bar's longitudinal modes). All through the mid- to late 1960s, Weber kept improving the experiment. In 1969 with a detector at the University of Maryland, another at a golf course about 8 miles from the university and a remote detector at the Argonne Laboratory in Chicago, he wrote a discovery paper in *Physical Review Letters*[11] showing a few coincident pulses per day at a peak strain level of $h \sim 10^{-15}$ in the KHz band. It soon became evident that if the pulses were gravitational waves from the region around the center of our galaxy, this implied an extremely efficient conversion of rest mass to gravitational waves, in fact the conversion of all the rest mass in the galaxy into gravitational waves in a few million years. Despite this unlikely possibility, about a dozen experimental groups throughout the world began the development of Weber-like detectors. By the early 1970s it was becoming clear that no one was able to confirm the Weber measurements.

In 1966 I was asked by the MIT Physics department to teach a course in General Relativity. It was at the time of the revived interest in GR in the physics community which had begun with the Chapel Hill meeting and with the increased research in experimental work in gravitation due to Dicke and Wheeler at Princeton and Schiff at Stanford and others. I had just returned to MIT from a post doctoral position with Dicke and had started a new group in Cosmology and Gravitation. For the prior 40 years GR had not been taught in physics at MIT but had occasionally been taught in the mathematics department as part of differential geometry. What little I knew of the physics of GR had come from conversations with colleagues at Princeton and simply from reading. The formal mathematics of differential geometry and tensor analysis was entirely new to me and at the time there were not yet really good text books on the subject. It

Figure 5. Joseph Weber was the first to attempt to detect gravitational waves from astrophysical sources directly. The concept for doing this was generated by John Wheeler and Weber at the 1957 Chapel Hill conference on gravitation. The idea was to measure the distortions induced by a gravitational wave in a cylinder. Instead of thinking of the gravitational wave as a strain in space, they thought of the wave as exerting tidal forces on the cylinder and calculated the motion of the bar as an oscillator in inertial space. The tidal force was just another Newtonian force on the bar. Weber imagined short pulses of gravitational waves with some Fourier components at the resonant frequency of the cylinder. The wave would pass through the cylinder and leave it ringing. In the picture at the right, Weber is attaching some strain gauges to the cylinder to measure the excitation. The cylinder is placed in the vacuum chamber (behind him) when searching for gravitational waves.

was a hard term, often learning from the students and just catching up to them. The students in the course were aware of the Weber experiments and asked for lectures about them. By that time in the course, I had become a convert to Einstein's geometric view and had a hard time looking at gravitational waves as a tidal force and, especially, the idea of a metal bar interacting with the wave.

I spent the weekend before the lecture trying to apply F.A.E. Pirani's[12] (*Figure 7*) approach of the geodesic deviation to a simple gedanken experiment to measure a gravitational wave. Eventually it turned into a straightforward idea of measuring the time it took light to travel back and forth between two free masses. With a light source mounted on one of the free masses, one set a clock also mounted on the mass to start as the light left to go to the distant mass. At the distant mass the light was reflected by a mirror back to the source and the clock was stopped by the arrival of the return light. One did this measurement when no gravitational wave passed between the two masses and when a wave did pass and compared the times. It turns out an easy problem to set up in GR and since all the measurements were made at the same mass there was no question of the conversion of coordinate time to proper time. After this we went on to cosmology and more interesting problems. Unknown to me two Russians, Gertsenshtein and Pustovoit[13], in 1963 had come up with a similar idea.

By early in the 1970s, when it became evident that there were no confirmations of the Weber experiments, I returned to thinking whether it was actually possible to convert the gedanken experiment in the GR course to a real experiment. (*Figure* 6) The real experiment would use a Michelson interferometer configuration to exploit the symmetry of the gravitational waves.

The interferometer would need some unusual properties: the mirror masses were suspended so they could move in response to the wave, the light was to be reflected back and forth many times along the interferometer arms, and to get to the shot noise limit (quantum limit) one needed a method to high frequency modulate the light above the excess noise of the available laser sources. (*Figure* 7) A study of the various fundamental physics noises in such an instrument and of the environmental noise sources showed by making the instrument on km scales there was a chance that one could intersect the strain sensitivity required for detection of some astrophysical sources. I put the calculations in a Quarterly Progress Report of the Research Laboratory for Electronics14 and asked the Laboratory management to help fund the construction of a 1.5-meter prototype using military research funds. (*Figure* 7).

At that time in the early 1970s the primary condition required by the military to support research was to train scientists and engineers. By 1974 one of the corrosive effects of the Vietnam War was the demand by many of the anti-war demonstrators to starve the war effort by insisting the military only support research essential for their mission. This was also urged by the supporters of the war in the administration who distrusted academic scientists in general. As judged by the Director of the Research Laboratory of Electronics, cosmology and gravitation research was not relevant to the military's mission and the lab support was terminated. I then began a multi-year effort to gain support from other federal agencies but ran against a skepticism among peer reviewers of whether interferometric methods using free masses could be used to detect gravitational waves as well as increased pessimism about gravitational wave research in general.

Scientists in Europe who had been involved in attempting to confirm the Weber results had become interested in the interferometric free mass detector. The group at the Max Planck Institute of Astrophysics in Garch-

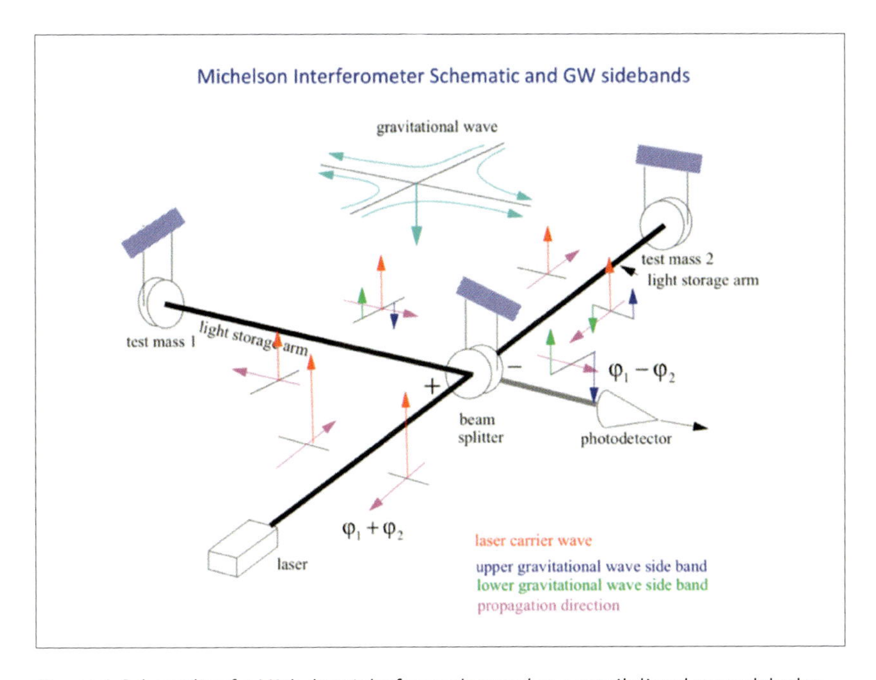

Figure 6. Schematic of a Michelson interferometer used as a gravitational wave detector. The gravitational wave at a low frequency f_g is incident from above the plane of the interferometer. Start by following the light from the laser to the symmetric port of the beam splitter (+ no phase inversion of the light on reflection). The laser generates carrier light (red amplitude). The violet vectors indicate the propagation direction of the light. The laser light that gets reflected by the beam splitter heads to the mirror on test mass1 reduced by the reflectivity of the beam splitter. On reflection from test mass1, which is moving due to the gravitational wave, the carrier generates two sidebands one at a frequency f_g above the carrier (blue) and another f_g below (green). The beam heads back to the beam splitter. The beam from the laser that gets transmitted by the beam splitter heads toward test mass 2; the carrier is reduced by the transmission of the beam splitter. That beam on reflection from test mass 2 also gets two sidebands from the motion of the test mass 2, but they have the opposite sign because the gravitational wave is compressing space on the test mass 1 side while expanding space on the test mass 2 side. The beam from test mass 2 gets reflected toward the detector from the antisymmetric port of the beam splitter (the phases of the carrier and the sidebands are inverted). The beam from test mass 1 is transmitted by the beam splitter toward the detector. The beams from the two sides of the interferometer are added. If the paths on the two sides take equal time from the first encounter with the beam splitter to the second one (or take a difference in time that is an integral number of periods of the light wave), the carrier cancels while the sidebands double at the detector. The sidebands carry the information about the gravitational wave both the wave amplitude and the phase. To make the sidebands detectable as a current in the photodetector requires a small amount of carrier to beat against the sidebands.

ing led by Heinz Billing had collaborated with an Italian group in Frascati to run coincident bar detectors. They had done a thorough analysis of the noise in their detectors and designed a data acquisition system with adequate bandwidth to fully characterize their signals and also had developed software algorithms that provided meaningful false alarm probabilities. They saw no coincident pulses. They were deciding whether to drop

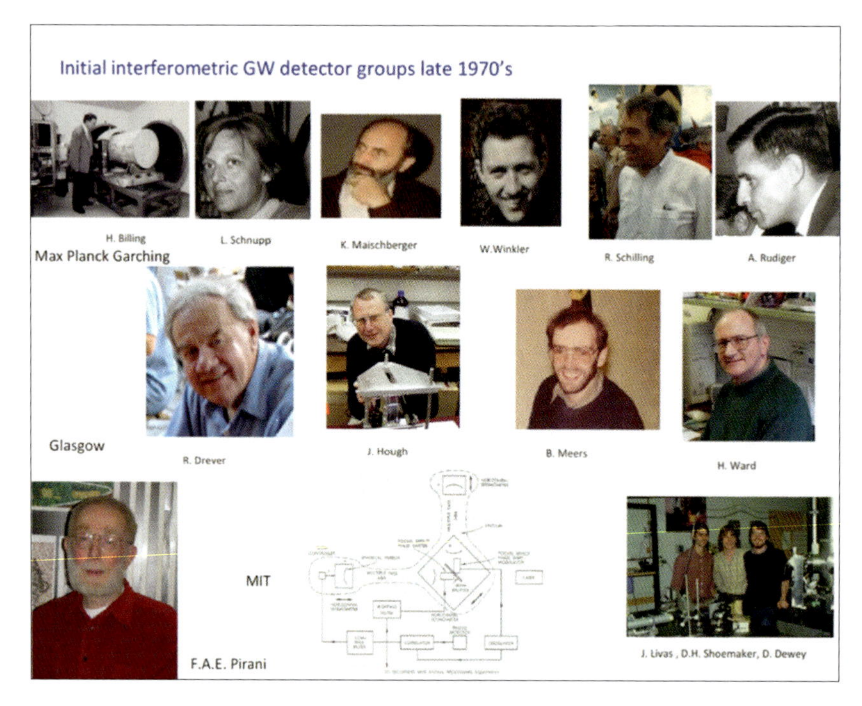

Figure 7. Bottom left, F.A.E. Pirani, who showed that it is possible to measure the relative motion of two free masses traveling through a gravitational wave (so-called geodesic deviation) in a coordinate-independent manner. A critical idea that was not fully accepted until the 1957 Chapel Hill Conference. *Bottom middle*, a schematic diagram of an interferometric gravitational wave detector with suspended masses, multiple passes along the interferometer arms and a technique to reduce the amplitude noise of the laser light. *Bottom right*, the 1.5-meter prototype at MIT and the three graduate students who constructed and operated it. *Top line*, the gravity research group at the Max Planck Institute of astrophysics in Garching, Germany, who built a 3-meter interferometer prototype and then a 30-meter one to show that the idea scaled properly. The group invented many of the solutions for the noise in these systems including the idea of suspending all the optics in the phase sensitive part of the interferometer and the need to frequency stabilize the laser to deal with scattering by stray light. Schnupp came up with the idea to make a slight imbalance in the interferometer arms to use external phase modulation. Schilling invented the idea of power recycling to increase the modulated light in the interferometer. *Middle line*, the gravity research group in Glasgow, Scotland, Drever developed Fabry-Perot cavities as multipass elements in the interferometer arms and also invented the idea of power recycling[16]. Brian Meers[17] was the first to suggest putting another partially reflecting mirror between the photo detector and the beam splitter to tailor the interferometer spectral response by reflecting the sidebands back into the main interferometer arms. Ward developed a method to align the interferometer.

out or develop cryogenic detectors or to try free mass interferometers. They decided to build a 3-meter interferometric prototype and quickly established the fundamental noise processes that would limit performance, some of which I had estimated in my study but also others which I had neglected. (*Figure 7*). They worked on this prototype systematically diagnosing the noise and designing methods and concepts to circumvent

PHYSICAL REVIEW D

PARTICLES AND FIELDS

THIRD SERIES, VOLUME 38, NUMBER 2 15 JULY 1988

Noise behavior of the Garching 30-meter prototype gravitational-wave detector

D. Shoemaker,* R. Schilling, L. Schnupp, W. Winkler, K. Maischberger, and A. Rüdiger

Max-Planck-Institut für Quantenoptik, D-8046 Garching bei München, Federal Republic of Germany

(Received 22 February 1988)

The prototype gravitational-wave detector at Garching is described. In a laser-illuminated Michelson interferometer having arms 30 m in length, a folded optical path of 3 km is realized. The origin, action, and magnitude of possible noise sources are given. The agreement between the expected and measured noise is good. For a band of astrophysical interest, extending from 1 to 6 kHz, the quantum shot noise corresponding to a light power of $P = 0.23$ W is dominant. In terms of the dimensionless strain h the best sensitivity in a 1-kHz bandwidth is $h = 3 \times 10^{-18}$, comparable to the most sensitive Weber-bar-type antennas.

Figure 8. The abstract of the paper on the 30-meter interferometer operated by the Max Planck Group in Garching showing that an interferometer had attained better sensitivity than the best cryogenic Weber bar detectors. The 30-meter results were important in helping to make the case for long baseline instruments such as LIGO. The top two photos are of the 40-meter instrument constructed at Caltech by Whitcomb and Drever. The 40-meter at Caltech and the 5-meter at MIT (*Figure 13*) were the final test beds for the first LIGO detector.

the noise. Eventually they were limited by the fundamental quantum and thermal noise. At that point in their research they designed and constructed a 30-meter prototype[15] to test the scaling laws toward km scale instruments. (*Figure 8*).

Another group in Glasgow Scotland, led by Ronald Drever (*Figure 7*) also became interested in interferometric free mass detectors after having worked with bar detectors. The experience here was a little different. They had redesigned the bar detector to be broad band and were now more disturbed by noise generated in the motion transducer. They became interested in the possibility of using an interferometric rather than a piezoelectric sensor as readout. They also eventually went to a free mass design but decided on optical cavities (Fabry-Perot interferometers) as the means of bouncing the light back and forth in the interferometers arms instead of the Herriot delay lines (discrete spots on the mirrors) I had proposed. They came to this design by noting that delay lines were more prone to phase noise from optical scattering than a Fabry-Perot.

An important turning point occurred in 1975 when Kip Thorne (*Figure 9*) and I met in Washington D.C. to work on a committee to study the possible role of the space program in research on gravitation and cosmology.

At the time Kip had established one of the premier groups in theoretical gravitation at Caltech and was thinking of encouraging Caltech to start a new complementary program in experimental gravitation. The issue we discussed was the nature of such a program. Kip had been thinking of gravitational wave research and had come to realize that a program ultimately going to be able to measure gravitational waves from "allowed" astrophysical sources would require sensitivities $h \leq 10^{-21}$ in the 10 to 1000Hz band, a million times more sensitive than the Weber bars. Kip had been much influenced by Vladimir Braginsky (Moscow State University), whose notion was to develop bars operating near the quantum limit. I suggested he think about long baseline interferometers and look into bringing people from the Max Planck or Glasgow groups to Caltech to start such a program. In 1978 Caltech did make the decision to form a new group with Ronald Drever as its leader and with a significant internal Caltech investment. In 1980, Stan Whitcomb was brought to Caltech to help lead the new Caltech Group.

The combined facts of the success of the Max Planck group in demonstrating the scaling relations for the interferometers, the progress in gaining sensitivity in the Max Planck as well as the MIT prototypes and the significant investment in the field being made by Caltech led me to make a proposal to the NSF to study what was involved in actually making an interferometric detector with adequate sensitivity to detect astrophysical sources. Our study involved several scientists working (*Figure 10*) with

The measurement challenge

Kip Thorne

$$h = \frac{\Delta L}{L} \leq 10^{-21}$$

$$L = 4\text{km} \quad \Delta L \leq 4\text{x}10^{-18}\,\text{meters}$$

$$\Delta L \sim 10^{-12}\,\text{wavelength of light}$$

$$\Delta L \sim 10^{-12}\,\text{vibrations at earth's surface}$$

Figure 9. Kip Thorne around 1980 with his measurement challenge to the experimenters based on his and others' estimates of possible astronomical gravitational wave source dynamics and populations.

industrial engineering consultants to look at designs and costs of the vacuum system, the large-scale optics and lasers, the siting for the construction of two multikilometer interferometric detectors separated by continental distance. The aim of the study was to establish feasibility and cost scaling relations for the large infrastructure and to look at the experimental challenges to gain the two large factors in the detector design itself. With multi kilometer long arms measuring a strain of 10^{-21} requires a displacement sensitivity of 10^{-18} meters, about 1/1000 of the size of a proton. Using light with a wavelength of 10^{-6} meters, one needs to develop optical techniques that can measure 10^{-12} of a wavelength. Furthermore, typical ground vibrations even in quiet places are broadly around 10^{-6} meters, so that to be able to sense a gravitational wave above just ground noise one would need to isolate the mirrors that determine the end points of the interferometer to another factor of 10^{-12}. Reducing the noise in the system by these two factors of 10^{-12} was known at the outset to be the primary challenge. The intent of the study was to serve as the basis for a collaborative proposal by all groups interested in the construction of a Laser Interferometer Gravitational-wave Observatory (LIGO).

The results of the study were presented to an NSF committee considering large new projects in physics in 1983. The presentations were given jointly by the Caltech and MIT research groups who had begun to form a collaboration. The committee was remarkably encouraging in their evaluation:

The committee is impressed with the long-range scientific potential of gravitational wave detection. It will not only test our basic understanding of gravitation, but provide an entirely new window on the universe.

We have considered the major interferometric laser detection system now being developed by the Caltech and MIT groups. We note that not only is this an outstanding scientific opportunity, but the Foundation is the only source of support for ground based gravitational physics.

As with any attempt at a qualitative advance, there are risks: here the uncertainties involve both the magnitude of the signals to be detected and the large extrapolation of known experimental technique inherent in the proposed scale. We find, however, the fundamental scientific merits of such an investigation so important as to be worth a substantial investment.

(National Science Foundation, Advisory Committee for Physics, December 12–13, 1983).

Following this encouragement from the NSF committee we began making the transition from independent investigator small-scale science to the project organization required for a large-scale coordinated scientific effort. At first, we did not succeed; it took until 1994 with Barry Barish becoming Director of LIGO to truly make this transition. In the interim there was progress in the technology in the various laboratories but only

A STUDY OF A LONG BASELINE

GRAVITATIONAL WAVE ANTENNA SYSTEM

Prepared for the National Science Foundation
under NSF Grant PHY-8109581
to the Massachusetts Institute of Technology

Prepared By:

Paul Linsay MIT
Peter Saulson MIT
Rainer Weiss MIT

With Contributions By:

Stan Whitcomb CalTech

Industrial Consultants:

Arthur D. Little Corporation Cambridge, Massachusetts
Stone & Webster Engineering Corporation Boston, Massachusetts

OCTOBER 1983

Figure 10. In 1979 the NSF supported a study[18] with industry to look at the feasibility of building a multi-kilometer baseline interferometric gravitational wave detector. The study summarized the current state of prototype research and estimates for gravitational wave sources. It looked at the ability of industry combined with academic research to develop adequate optics, lasers, vibration isolation and control systems to meet the scientific requirements. The designs and costs of the vacuum system and the buildings as well cost scaling relations were developed. Possible sites for up to 10km baseline systems both above and below ground were investigated. One of the intents of the study was to be the factual basis for a proposal to the NSF by a consortium of scientists and institutions to build a pair of detectors in the United States.

halting progress in major project decisions such as choosing sites or more than a conceptual design for the interferometer configuration. In 1986 we began an effort to seek significant funding for the interferometric gravitational wave detection program, Richard Garwin suggested that the NSF carry out a summer study of the field by an independent group to both reevaluate its importance as well as to assess readiness and cost. A "Panel on Interferometric Observatories for Gravitational Waves" was established consisting of established scientists with experience in large projects, experts in the various technologies required and scientists with knowledge of the relevant aspects of gravitation and astrophysics. In January 1987 they issued their report, which encouraged the NSF to build two full scale interferometric detectors at widely separated sites and insisted that the project find a single Director before moving forward.

Robbie Vogt became the first Director of LIGO in 1987. He introduced a structure into the laboratory research program assigning responsibilities

Figure 11. The 1989 proposal[19] to build and operate LIGO. Rochus (Robbie) Vogt, who had become Director of the LIGO project in 1987, organized a joint Caltech/MIT collaboration to write the proposal. The proposal specified specific parameters for the project: two sites run in coincidence – one in Hanford, Washington and the other in Livingston, Louisiana – a 4 km baseline, an initial detector design and strain sensitivity spectrum, an environmental monitoring system, the vacuum system for the instrumentation and the vacuum system for the beamtubes, the buildings and permanent infrastructure. The proposal also provided a plan for a developmental approach for the detector with an initial detector using then available technology, having a possibility for detection; and a later advanced detector with not yet fully available technology, in the development stages at the various LIGO research laboratories, having a good chance for detection. The facilities were designed to accommodate both the initial and the advanced detector. I worked with engineers Boude Moore and later Larry Jones on the design, construction and qualification of the 4 km beam tubes.

for both detector development and infrastructure planning and estimation. His major effort was to guide us, both Caltech and MIT participants, scientists and engineers, to write a construction proposal for LIGO with sufficient detail to estimate schedules and costs. The proposal also served to pull together in one place the knowledge of gravitational wave sources, the noise budgets for the detectors and the plans for data analysis to establish confidence of detection. It became a source book for the field[19] (*Figure 11*). The proposal also laid out the plans for the field with a staged strategy. The infrastructure would be designed and built to be adequate for both an initial detector using technology we now had (almost) in hand and would not compromise a future advanced detector we could envision

Figure 12. By 1994 the NSF as well as both the Caltech and MIT administrations felt the management of the LIGO project needed strengthening. In 1994 Professor Barry Barish of Caltech became Director of LIGO. He substantially increased the management and technical staff of the project. He brought in Dr. Gary Sanders as the Project Manager and engaged Dr. Albert Lazzarini and Dennis Coyne as system scientist and engineer. He asked Dr. Stan Whitcomb to manage the design and construction of the first interferometer. Barish also realized that a larger group than only the Caltech and MIT groups would be needed to analyze the data and to publish results. He was the architect for the LIGO Scientific Collaboration. Barry asked me to be the first spokesperson for the Collaboration.

operating at the quantum limit of a 1 ton test mass. In the process of writing the proposal, Vogt paired an engineer with a scientist to assure that the planned infrastructure would be able to serve the experiment requirements. Vogt also oversaw the site selection.

After a stormy period in the project, Barry Barish became the second Director of LIGO in 1994. He instituted significant changes in the project organization by broadening the leadership from a skunk works to a larger managerial group (*Figure 12*). He also conceived and organized the LIGO Scientific Collaboration, which included both LIGO project members as well as groups outside of Caltech and MIT who were interested in the new science to come from LIGO. The Collaboration has major roles in the data analysis and publications of the science. Some groups in the Collaboration are involved in basic detector research as well.

By the mid 1980s the technical development to accomplish the two factors of 10^{-12} was advancing in several research laboratories: a 40-meter prototype at Caltech, a 5-meter prototype at MIT (*Figure 13*), a 10-meter prototype in Glasgow, at a new laboratory in Hannover where the Max

Planck group had moved from Garching and a new program joint between Italy and France in Cascina, Italy (VIRGO). Instead of continuing with the historical development I will turn to the major ideas to gain the two factors. The optical arrangement of the advanced detector, for which Barry will show some of the results, is schematized in Figure 14. The initial LIGO detector was the same, except there was no signal recycling mirror. The caption explains some of the basic concepts. The noise budget for the initial long baseline detector in the LIGO facilities is shown in *Figure 15*, giving both the estimated contributions from fundamental physical principles and contributions from technical noise that can be reduced by further engineering. The noise budget is an excellent way to keep track of our progress in accomplishing the two factors of 10^{-12}, and, as Barry will show with real data from the instrument, the slow but steady progress made as we learned more about the instrument in the commissioning.

5 Meter Prototype at MIT

Test of LIGO alignment system
Test of power recycled Michelson-Fabry Perot interferometer @ 0.5 and 1.06 microns
Test of suspended power recycled Michelson
Perturbed Optics Modeling

Figure 13. The MIT group in the 1990s and some of the research that was done. *Front row* left to right: Michael Zucker, Nergis Mavalvala, Peter Csatorday, Peter Fritschel, Joseph Kovalik, R. Weiss, *next row back*: Yaron Hefetz, David Shoemaker, Brett Bochner, *in the back*: Brian Lantz.

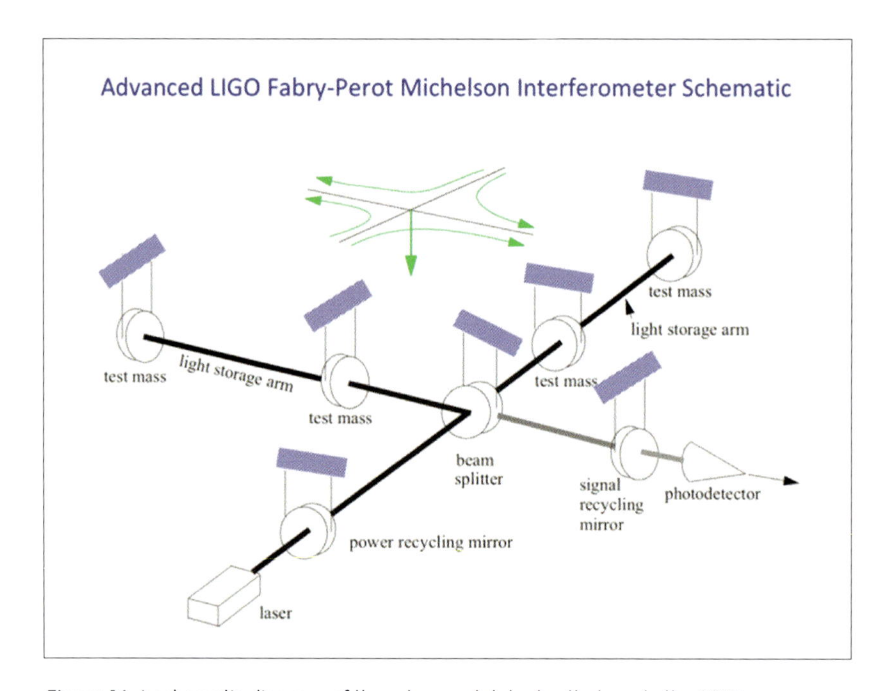

Figure 14. A schematic diagram of the advanced detector that made the 2015 measurement of the gravitational waves from a binary black hole collision. The interferometer has all the ingredients needed to make the factor of 10–12 sensitivity improvement in measuring the small displacements of the test masses. To the main elements of the interferometer presented in *Figure 6* are added input test masses with partially transmitting coatings between the beam splitter and the end test masses. The combination of the input and end test mass comprise an optical resonator (Fabry-Perot cavities) which can be thought of as bouncing the light between them many times (typically several hundred times) to increase the gravitational wave induced sidebands by the number of bounces before returning the light to the beam splitter. One attempts to make the number of bounces the same in both arms. As in *Figure 6* the sidebands from the two arms add going toward the photodetector and the carrier cancels. No sidebands return to the laser, but all the carrier does. Between the beam splitter and the laser a new partially reflecting mirror is inserted, the power recycling mirror. The position of this mirror and its transmission is arranged to make another interferometer that cancels the carrier from the laser reflected by the power recycling mirror with the carrier transmitted back by the recycling mirror from the beam splitter. This eliminates the carrier being reflected by the interferometer to the laser and builds up the carrier power between the beam splitter and the input test masses by several hundred. This is the equivalent of using a more powerful laser. The optical configuration described was used in the initial LIGO interferometer. Finally, the advanced detector includes another partially reflecting mirror between the beam splitter and the photodetector, the signal recycling mirror. This mirror reflects the sidebands back into the interferometer and modifies the spectral response of the entire interferometer to the sidebands, thereby tuning the spectral response of the detector to the gravitational waves being sought.

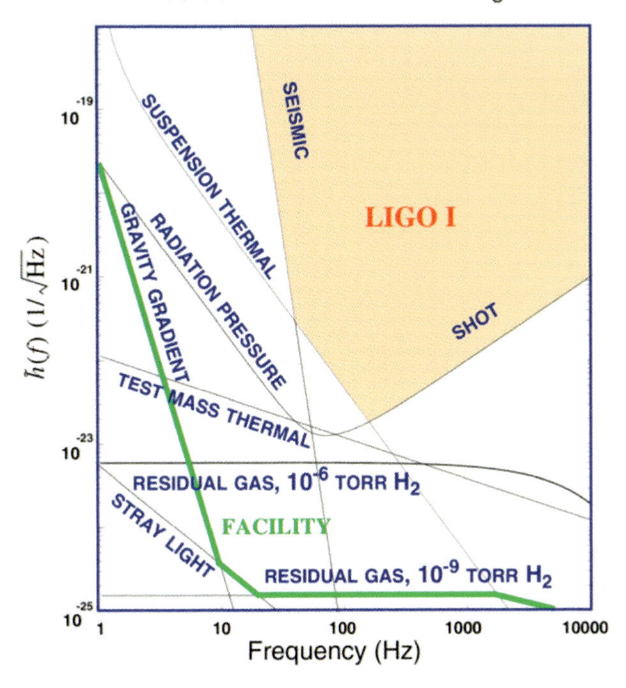

Figure 15. The noise budget of the initial LIGO interferometer as described in the 1989 proposal. Note the vertical axis is the spectral amplitude strain/sqrt(Hz). To convert to the rms strain, which has been used in other parts of this paper, multiply the plotted value by the square root of the detection bandwidth. At approximately 100 Hz with a 100 Hz detection bandwidth, the rms detector sensitivity is about 10^{-22} strain. The plot shows the principal noise terms that limit the performance of the detector. At frequencies above 200 Hz, the quantum noise (shot noise) limits the performance. The line labeled radiation pressure, which does not limit this detector, is the other part of the quantum noise due to the fluctuations in radiation pressure on the mirror. The two terms together constitute the quantum limit and are the equivalent of the noise in a Heisenberg microscope but here with a macroscopic object such as a 10kg mirror rather than an electron. In this naïve rendering one can reduce the shot noise by increasing the optical power but at the expense of increasing the radiation pressure noise. The noise near 100Hz, the sweet spot for this detector, is limited by the thermal noise in the pendulum suspension, in this case phonon excitations at the end points of the pendulum. At frequencies below 70Hz the detector is limited by incomplete removal of the ground vibrations (seismic noise) at the mirror. Seismic noise and pendulum thermal noise were greatly reduced for the advanced detector (*Figure 16*). The broad noise spectra called residual gas are the phase fluctuations induced by forward scattering of residual gas molecules traveling through the optical beams in the long arms, this term is the primary reason for ultra-high vacuum in the beam path. The sharply rising curve at low frequencies labeled gravity gradients is due to the fluctuating Newtonian gravitational forces on the end test masses caused by time dependent density fluctuations in both the atmosphere and the ground. Even though there is some hope in being able to measure these fluctuations or reduce them by burying the instrument, these constitute a severe limit to measure low frequency, less than a few Hz, gravitational waves on the ground.

A significant step in achieving the improvement against seismic and pendulum thermal noise was made in the development and construction of the advanced detector, *Figure 16*. The two major improvements were a new suspension system with four pendula in series each stage providing a filter against seismic displacement noise varying as 1/frequency². The final stage supporting the test mass using very low mechanical loss fused silica fibers (rather than steel fibers) to reduce the suspension thermal noise. The suspension system was provided by the Glasgow group. The second improvement was a multi stage active vibration isolation system to further reduce the seismic motion, especially at low frequencies where

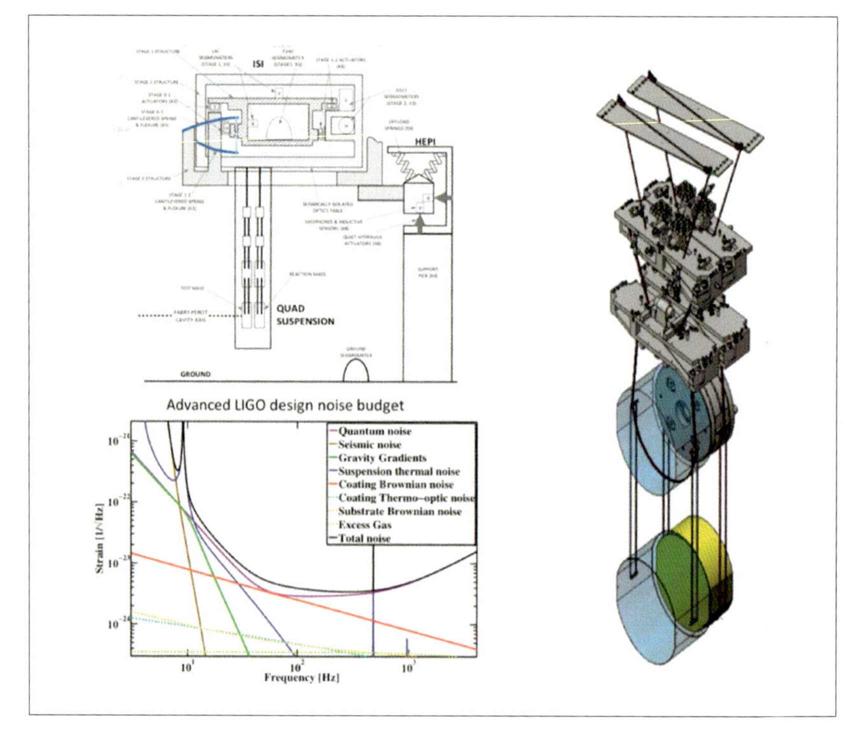

Figure 16. Some properties of the advanced detector. *Right*, a quadruple pendulum system developed at Glasgow University which provides four stages of high frequency ground noise isolation as well as low pendulum thermal noise by using fused silica fiber supports (pioneered by Braginsky's Moscow group) in the final pendulum stage. *Top left*, a dual active seismic isolation system, initially developed at Stanford University, which becomes a platform for the quadruple suspension. The active system measures the motion of a platform with a three-dimensional seismometer and then feeds this back to a set of controllers to null the seismometer. Two such active systems are placed in series. *Bottom left*, the noise budget for the advanced detector. The dominant noise has become the quantum noise at both high and low frequencies (discussed in Kip's lecture). The noise limiting the performance at the sweet spot near 100 Hz has become the thermal noise generated in the mirror coatings. At this writing in February 2018, the noise in the advanced detector, as Barry will show, is about a factor 2.5 worse at the sweet spot than this projected noise curve.

PHYSICAL REVIEW VOLUME 166, NUMBER 5 25 FEBRUARY 1968

Gravitational Radiation in the Limit of High Frequency.
II. Nonlinear Terms and the Effective Stress Tensor*

RICHARD A. ISAACSON†
Department of Physics and Astronomy, University of Maryland, College Park, Maryland
(Received 14 July 1967)

The high-frequency expansion of a vacuum gravitational field in powers of its small wavelength is continued. We go beyond the previously discussed linearization of the field equations to consider the lowest-order nonlinearities. These are shown to provide a natural, gauge-invariant, averaged stress tensor for the effective energy localized in the high-frequency gravitational waves. Under the assumption of the WKB form for the field, this stress tensor is found to have the same algebraic structure as that for an electromagnetic null field. A Poynting vector is used to investigate the flow of energy and momentum by gravitational waves, and it is seen that high-frequency waves propagate along null hypersurfaces and are not backscattered by the lowest-order nonlinearities. Expressions for the total energy and momentum carried by the field to flat null infinity are given in terms of coordinate-independent hypersurface integrals valid within regions of high field strength. The formalism is applied to the case of spherical gravitational waves where a news function is obtained and where the source is found to lose exactly the energy and momentum contained in the radiation field. Second-order terms in the metric are found to be finite and free of divergences of the $(\ln r)/r$ variety.

Figure 17. A key individual at the National Science Foundation (NSF) was Dr. Richard Isaacson, who became discipline chief for Gravity in the Division of Physics in the mid-1970s. He had received his Ph.D. from the University of Maryland in Charles Misner's group and had written a paper that rigorously showed gravitational waves carried energy and momentum away from their sources – an issue that was still controversial in the 1960s. Isaacson sensed the scientific promise in a vital program of research in experimental and theoretical gravitation and became a powerful advocate for the field at the NSF. He convinced the Director of the Physics Division, Dr. Marcel Bardon, that gravitational wave detection, though risky since the technology needed development and the sources were not certain, was the type of science which could produce transformative results and was well suited to the NSF. In addition to his advocacy within the Foundation, Isaacson guided the scientists in the field with strategic advice and wisdom about what was possible. After he retired from the NSF, he made a profession of a significant hobby in a study of textiles and the cultures that produce them in Asia. He curated an exhibit of the tent bands of yurts for the Washington Textile Museum.

the pendula are not effective. The results of these developments will be described by Barry in the next talk.

I started this talk with an acknowledgment of the remarkable role the National Science Foundation played in fostering and supporting LIGO over the 45 years from an idea to the first detections and the opening of a new field of science. All the more remarkable as it was risky science and the NSF saw the project through significant external criticism and our internal disorganization. Much of the responsibility and foresight for this is due to Dr. Richard Isaacson (*Figure 17*), who was both an advocate for the field of gravitation within the Foundation and a strategist for the possible.

Once an internal NSF decision was made to construct LIGO, the strong

NSF Directors critical for LIGO

Eric Block Walter Massey Neal Lane

Figure 18. Three Directors of the National Science Foundation who played critical roles in the evolution of the LIGO project: Eric Block was the first Director to consider going forward with the large baseline LIGO project. Walter Massey defended the program during the time when astronomers considered LIGO premature and a possible waste of funds. Neal Lane was a strong advocate for a line item in the congressional NSF budget to support Major Research Equipment Funds for Construction (MREFC). LIGO was one of the first projects to take advantage of these funds, and since then the funds have been used to begin many other large NSF projects.

support of three NSF Directors (*Figure 18*) was essential to keep the project moving at critical junctures with Congress and amidst criticism in a skeptical subset of the scientific community.

A personal reflection. Science reporters who have written about the LIGO project have noted the 45 years from its inception to the first measurements. They comment on our perseverance and the novelty of a group of scientists working so long together without the satisfaction of a significant scientific result. This might be the way a person outside of the project views it. From the inside it is a very different story. What held the group together were the daily little triumphs or puzzles generated in the development of the ideas and technology (both hardware and software) as well as the collegiality of a dedicated group of people.

REFERENCES

1. Einstein, A. (1915) "Zur Allgemeinen Relativitätstheorie", *Sitzunsberichte der Königlich Preussichen Akademie der Wissenschaften,* 778 & 799.
2. Einstein, A. (1916) "Die Grundlage der Allgemeinen Relativitätstheorie", *Annalen der Physik,* **49**, 769.
3. Pound, R.V. and Rebka, G.A. (1960) "Apparent Weight of Photons", *Physical Review Letters* 4, 337.
4. Brault, J (1963) "Gravitational Redshift of Solar Lines", *Bulletin of the American Astronomical Society* **8**, 28.
5. Fomalont, E.B. and Sramek, R.A. (1975) "A Confirmation of Einstein's General Theory of Relativity by Measuring the Bending of Microwave Radiation in the Gravitational Field of the Sun", *Astrophysical Journal* **199**, 749.
6. Einstein, A. (1916) "Näherungsweise integration der feldgleichungen der gravitation", *Sitzungsberichte der Königlich Preussichen Akademie der Wissenschaften* XXXIII, 688.
7. Einstein, A. (1918) "Über gravitationswellen", *Sitzungsberichte der Königlich Preussichen Akademie der Wissenschaften,* **8**, 154.
8. Taylor, J. H. and Weisberg, J.M. (1982) "A new test of general relativity: Gravitational radiation and the binary pulsar PSR 1913+16", *Astrophysical Journal* **253**, 908.
9. "The Role of Gravitation in Physics", Report from the 1957 Chapel Hill Conference, Reprinted Max Planck Research Library for the History and Development of Knowledge Sources 5.
10. Weber, J. (1960) "Detection and generation of gravitational waves", *Physical Review* **117**, 306.
11. Weber, J. (1969) "Evidence for the Discovery of Gravitational Waves", *Physical Review Letters* **22**,1320.
12. Pirani, F.A.E. (1956) "On the physical significance of the Riemann tensor", *Acta Physica Polonica,* **15**, 389 and (1957) "Measurement of classical gravitation fields", Chapel Hill Conference (Reference 9, 61).
13. Gertsenshtein, M.E. and Pustovoit, V.I. (1963) "On the Detection of Low Frequency Gravitational Waves", *Soviet Physics – JETP* **16**, 433.
14. Weiss, R. (1972) "Electromagnetically coupled broadband gravitational antenna", *Research Laboratory for Electronics*, MIT, Quarterly Progress Report, No 105, 54.
15. Shoemaker, D., Schilling, R., Schnupp, L., Winkler, W., Maischberger, K., and Rüdiger, A. (1988). "Noise behavior of the Garching 30-meter prototype gravitational wave detector", *Physical Review* D **38**, 423.
16. Drever, R.W.P. (1983). "Interferometric detectors for gravitational radiation", *Gravitational Radiation*, NATO Advanced Physics Institute, Les Houches, ed N. Deruelle and T. Piran, (North Holland Publishing), 321.
17. Meers, B.J. (1988). "Recycling in laser-interferometric gravitational-wave detectors", *Physical Review* D **38**, 2317.
18. NSF Bluebook (1983). https://emvogil-3.mit.edu/~weiss/ligo_history_documents/NSF_bluebook_1983.pdf
19. Caltech/MIT NSF proposal (1989). https://dcc.ligo.org/LIGO-M890001/public/main/

Barry C. Barish. © Nobel Prize Outreach AB. Photo: A. Mahmoud

Barry C. Barish

Biography

REFLECTING ON MY BACKGROUND, it is not at all apparent why my life did not go in other directions, rather than my becoming a physicist, and one dedicated to an academic career and pursuing research on some of the most fundamental problems in nature. My families' backgrounds are not very well documented, but I do know that both my parents' families immigrated to the U.S. from small shtetls in Eastern Europe, in the general area of Ukraine and Belarus. I know little about my ancestors there. They apparently were not very distinguished, and I do not know what they did or under what circumstances they immigrated to the U.S. at about the turn of the 20th century.

My mother's family (maiden name Shames) settled in St Joseph, Missouri, where she was born. They moved while she was a child to Omaha, Nebraska, where her father, Max, opened a body and fender shop, where they fixed cars. My mother had one sister and they lived a middle-class life with their social life centered around the small, but tight knit, Jewish community in Omaha. My grandmother died at a young age and my grandfather, Max, remarried. My mother graduated from high school and had aspirations to go to college at the University of Nebraska, even earned a scholarship, but her father would not let her go. This was a disappointment she expressed her whole life.

My father's father, two brothers and grandfather immigrated to the U.S. in 1901, and became homesteaders in North Dakota, under a program where the U.S. government gave away land to be farmed and developed. They did this for 18 months, sold the land for a good profit and moved to Sioux City, Iowa, where they eventually owned and ran a Ford agency until they had a falling-out with Henry Ford over his anti-Semitic attitudes. In 1929, just before the U.S. depression, the Barish family moved to Omaha, Nebraska. My grandfather had died in Sioux City when my father was only 10 years old. Although his mother remarried, my father had to work from a young age to help support the family. As a result, my father also never went to college.

Figure 1. Photo as a young boy.

I was born in Omaha in 1936 and we spent the first 10 years of my life there. I had a brother, born in 1940. I was a quiet child, but a very good student from a young age. My father worked for his father-in-law in his automobile repair business. During World War II, my father worked at a factory near Omaha that built military bombers for the war. He was in charge of an assembly line. As a result of his war work, he did not serve in the military during the war.

After the war, my father did not want to work any longer for his father-in-law and, instead, decided to move the family to California, where he would work for his uncle. His uncle had moved his car business to Los Angeles. We moved to the Los Feliz area, a quiet neighborhood adjacent to Hollywood. I went to public schools through high school. As a young child, I loved to read and after consuming many mystery and science fiction books and I began reading good literature, including the classics. In my young teenage years, in addition to liking story-telling, I enjoyed writing. When was asked what I wanted to do when I grew up, it was to become a novelist. At this age, I did not have any special interest in science or mathematics but was always very good in math and won some contests testing skills.

I actually spent most of my non-school hours playing sports – football, basketball and tennis. Most of my friends were made through sports. By the time I entered high school, my sports interests focused on tennis and I played on my high school tennis team for the next three years, during which time, I won some honors.

Academically, I was always at or near the top of my class, but I had no real direction. Neither of my parents had gone to college, but they were convinced of the importance of education and were committed to my brother and me getting a college education. However, they were not able to give us much guidance in terms of what we might study. I also never happened to have a good mentor in high school and, as a result, I had little idea what to study. There was a general expectation in my family that my brother and I would become either doctors or lawyers. I was not very interested, but my brother did become a lawyer!

Lacking much good guidance, I decided I would study engineering, based on the fact that I had good math skills and liked what I knew about engineering. I entered the University of California, Berkeley, where as a Freshman I took two Engineering courses, Inorganic Chemistry, Physics, Mathematics and a French Literature course. This was my first physics course, because we did not have one in my high school, and I began my love affair with physics, immediately. At the same time, my engineering courses, drafting and surveying were too tedious for me. So, I switched to becoming a physics major, not really understanding what it meant to be a physicist.

I loved Berkeley from the beginning. I thrived in the academic part, made many new friends, including having my first serious girlfriend. I emerged from being a terribly shy young boy to being a socially (if not outgoing) active college student. My only disappointment was that I dropped tennis. I had aspirations to play tennis at Berkeley, but I was only

Figure 2. My days as a tennis star.

marginally good enough to be on the tennis team, and the large demands for practice time interfered with my many labs in physics, chemistry and engineering.

In undergraduate school, I became interested in particle physics, stimulated by all the new elementary particles being discovered on the particle accelerators at the "Rad Lab" (now Lawrence Berkeley National Laboratory, LBNL) above the campus. I did some research there as an undergrad and spent time at the 184-inch cyclotron. As I approached getting my BA, I was very attracted to doing my graduate work in Berkeley on the large accelerators. The physics department discouraged their own undergrads from going to graduate school in Berkeley, but in the end they accepted me to their graduate school.

In graduate school, after my course work and candidacy exams, I went into particle physics and did my thesis on the 184-in cyclotron studying single pion production in pion-proton collisions tracing the production of the $\Delta 33$ resonance in the final state. My thesis advisor was A. Carl Helmholz, who was chairman of the physics department. During this period I learned not just about particle physics detectors, but also the accelerators in Berkeley, the 184-in Cyclotron and the Bevatron. In fact, when I received my PhD, my wife had a job she liked and we decided to stay for an extra year, me on a postdoctoral appointment working at the Bevatron.

Figure 3. Marriage to Samoan in 1960.

My years in Berkeley were transformative. I entered as a shy young boy, good academically, but otherwise pretty lost. I found and developed the professional love of my life, physics, as an undergraduate in Berkeley. I did well enough in graduate school to land a really good postdoctoral job at California Institute of Technology (Caltech), which has been my professional home ever since. I also met the other love of my life, my wife Samoan, in Berkeley and we have made our lives together ever since, she as a psychoanalyst, me as a scientist, and with a daughter and son, and three grandchildren. I left Berkeley with happiness and confidence to take on the world!

I was hired at Caltech in the fall of 1963 as a postdoc for Alvin Tollestrup to work on a new experiment at the Brookhaven Alternating Gradient Synchrotron to study the annihilations of antiproton-proton → electron-positron pairs. This reaction probed the time-like form factors and we were searching for heavy resonances in that system, but found none. For the experiment, I designed and built an intense separated anti-proton beam as my primary contribution to the experiment. The experiment worked well and we made important time-like measurements of the proton structure for the first time. For me, it provided for the first-time, my identification with one of the major experiments in particle physics.

Then, in about 1965, I went to work on the new two-mile long linear electron accelerator at the Stanford Linear Accelerator Center (SLAC), working with Henry Kendall, Jerry Friedman of Massachusetts Institute of Technology (MIT) and Richard Taylor of SLAC. This long-term project involved designing and building a large 6-GeV spectrometer and preparing a three part program: 1) to measure the proton form factors at high momentum transfer; 2) to make a comparison of electron and positron scattering to probe two-photon exchange effects; 3) and finally, to measure electron proton inelastic scattering. These were the greatly anticipated initial experiments to be performed at the new SLAC accelerator. I participated in both the electron elastic scattering and electron-positron comparisons – highly anticipated and important studies of the proton form-factors to large momentum transfer. We found no big surprises, and I made a fateful decision not to participate in the inelastic scattering experiment, expecting that this long-term program would again provide nice measurements, but no big surprises. How wrong I was! The experiment surprisingly (at least to me) found very large scattering cross sections, which became pivotal evidence for the quark structure of the proton. My colleagues, Kendall, Friedman and Taylor, won the Nobel Prize in 1990 for this important discovery.

Figure 4. Working with Marty Breidenbach at SLAC.

I left the SLAC experiment to return to the Brookhaven AGS for a follow-up experiment from our earlier anti-proton proton annihilation experiment this time to measure final pion production states, which had been an important background in the original experiment. I organized a small group from Caltech and collaborated with Adrian Melissinos and his group from Rochester University. It was the first experiment where I was the leader (or spokesperson). The experiment was successful in that we made a series of nice measurements of high energy antiproton-proton annihilation cross sections.

On the personal side, during this period my wife and I had two children and our lives became complicated by our trying to balance my demanding professional life that required lots of travel and a home life that involved raising small children and my wife, Samoan, pursuing her own professional life as a mental health professional. I also was promoted from post-doc to Assistant Professor at Caltech.

This was a very exciting time in high energy physics. The new high energy accelerator at the National Accelerator Laboratory (now called Fermilab) was nearing completion and proposals were being made for the initial scientific program. I teamed with my Assistant Professor colleague at Caltech, Frank Sciulli, to develop a proposal to make a narrow band neutrino beam and a detector to exploit it. The high energy protons at Fermilab provided the opportunity to produce intense high energy neutrino beams. Our experiment was approved, along with a wide-band neutrino beam proposal by David Cline, Al Mann and Carlo Rubbia.

We succeeded in measuring the first high energy scattering distributions and cross sections for both neutrinos and antineutrinos. These results were complementary to the SLAC electron deep inelastic scattering experiments in revealing the quark structure of the proton. But, even more importantly, Electroweak Symmetry Breaking or the, so-called, Standard Model for particle physics had been proposed by Sheldon Glashow, Stephen Weinberg and Abdus Salam. A unique prediction of that model was that it predicted the existence of weak neutral currents, or in other words, neutrino scattering without charge exchange ($\nu + N \rightarrow \nu + X$). An event had been observed in the Gargamelle heavy liquid Bubble Chamber at the European Organization for Nuclear Research (CERN) that provided initial evidence for neutral currents, and our experiment at Fermilab provided the definitive proof. This was the lynchpin that established the Standard Model of particle physics, still the best description of high energy physics that we have today.

In the subsequent years, I was promoted and settled into having my career based at Caltech. Interestingly, I had considered going to Caltech as an undergraduate, also as a graduate student, then I finally went as a postdoc, and have made my entire career since then at Caltech. It has been a great place to me and for me! The smallness and dedication to being a place where great science is done has provided the encouragement and support for me to address the most fundamental problems experimentally possible at the different stages of my career. I have led a series of forefront particle physics experiments, including: the Neutrino Experiment at Fermilab (with Frank Sciulli), where we established the weak neutral current; definitive evidence for the Z^0, the carrier of the weak neutral current at the SLAC Linear Collider; measurements of the fundamental parameters of the τ-lepton (our heaviest lepton) at the Cornell Electronic Storage Ring (CESR) electron accelerator; the most sensitive search for Grand Unified magnetic monopoles at the Gran Sasso Laboratory in Italy in the MACRO experiment, as well as experimental confirmatory evidence for the existence of atmospheric neutrino oscillations discovered by Takaaki Kajita and collaborators in the Super-Kamiokande experiment in Japan.

In 1990, I joined with Bill Willis from Columbia to co-lead the design of one of the two large efforts (Gammas, Electrons, Muons = GEM) for the Super Collider in Texas. We did detector development R&D and simulation work to optimize a detector for the main particle physics goals in the multi-TeV range. The GEM collaboration was highly international and we developed a detector design that was very well reviewed shortly before the Superconducting Super Collider (SSC) was cancelled by Congress in the fall of 1993. This was a huge disappointment for those of us in particle physics in the U.S. A large contingent of my collaborators in GEM joined the CERN Large Hadron Collider (LHC) experiments, and both the U.S. Department of Energy (DoE) and National Science Foundation (NSF) approved funding for U.S. participation at CERN. In fact, some central design features and technologies from GEM have been implemented in the LHC detectors.

Following the cancellation of the SSC, I made a personal decision not to join the CERN experiments, but was not in a rush to decide what I wanted to do next. Being a professor at Caltech, I had no job issue and my main preoccupation in the months immediately following the SSC cancellation was finding jobs for some of the very talented scientists I had hired at the SSC and who now had no job. Some of them were absorbed in the broader U.S. high energy physics program, some left high energy physics for technical jobs or Wall Street or elsewhere, and some were unfortunately lost to science.

About two months after the SSC cancellation, I received a phone call from Charles Peck, the chair of Physics, Math and Astronomy at Caltech. Charlie was a long-time colleague and friend, who I had worked with on and off through the years. He asked if he could come out and visit me in Santa Monica, where we live (about an hour drive from Caltech) to take a walk with me on the beach. This was completely out of character for Charlie, and I had no idea what he wanted to talk about (maybe to commiserate over the demise of the SSC?), but I didn't ask, just set a time for him to come. Charlie came and we took a walk during which he asked me if I would be willing to take over the leadership of the Laser Interferometer Gravitational-Wave Observatory (LIGO). This was completely out of the blue and a surprise to me. Charlie explained the problems that existed in LIGO and that NSF wanted a change of leadership, before they would consider funding the project. He told me that the NSF Director, the MIT and Caltech Presidents, etc. had conferred and they wanted him to ask me to take the project over. I told Charlie that I was flattered, supported Caltech's and MIT's initiative toward gravitational wave detection, but that I hadn't yet considered what I wanted to do next. I said that to decide on LIGO, I would need a month to do my homework to determine whether I thought I could make LIGO a success. After a month, I was not able to convince myself that I could succeed, but on the other hand, I couldn't convince myself I couldn't. As a consequence, I accepted and became the LIGO Principal Investigator, and I plunged into this great challenge.

In the spring of 1994, we revised the LIGO proposal in ways that raised the costs substantially. The reasons were to strengthen the technical team, have larger staffs at the sites to operate LIGO, and to invest more in making the technical infrastructure as robust, flexible and forward-looking as was practical. We made some significant design changes, including

Figure 5. Gary Sanders (center), Kip Thorne and myself (left) in 2000.

investing in some modern developing technologies like solid state lasers, digital controls, computer interferometer simulations, etc. The revised proposal was submitted to the NSF and was reviewed by a distinguished panel by summer of 1994. The new proposal was strongly endorsed by the review committee.

The next step was for NSF to make a decision whether to fund our proposal. We had strong support in Physics from Rich Isaacson and Dave Berley, who were responsible for the NSF gravity program and for LIGO and from the physics leadership, Marcel Bardon and Bob Eisenstein. The Math and Physical Sciences Assistant Director at that time was Bill Harris. I met with this group several times over this transition period. Bill became convinced that LIGO was potentially a transformative project and that we were on a path to succeed.

Bill Harris convinced Neal Lane, Director of the NSF at that time and the National Science Board (NSB) chairperson that Kip Thorne and I should be invited to make a presentation directly to the NSB, to help them be able to make an informed decision whether to fund and proceed with LIGO. Kip Thorne and I were invited to a meeting of the NSB during the summer of 1994. I note that it was unprecedented for scientific proposals to be presented to the NSB by the proposers. In any case, the NSB agreed and Kip made an inspired presentation on the science potential of LIGO. In my opinion, this made funding LIGO irresistible, if the board could only be convinced the investment would be used in a way such that the project could succeed. I followed Kip's talk by laying out our plan for how we would build and evolve LIGO over the coming decade, as well as how to have a clear path in the longer term to Advanced LIGO, a yet more sensitive detector. The NSB accepted our arguments and approved funding to construct LIGO.

We then organized the LIGO project and began to move toward construction. The project was (and is today) a Caltech-MIT collaboration, funded through a "Cooperative Agreement" between NSF and Caltech, and with MIT as a subcontractor. Although administratively MIT is a subcontractor, in fact operationally LIGO and Advanced LIGO were built and are operated by the LIGO Laboratory, consisting of staffs and Caltech, MIT, Hanford Washington and Livingston Louisiana. Some other university groups participate in particular technical areas under subcontracts with the LIGO Laboratory and non-U.S. collaborators thorough memorandums of understanding (MOUs).

The construction of Initial LIGO took place between 1994 and 1999. Some key members of LIGO who joined the project early during the construction period included: Gary Sanders, whom I had worked with at the SSC and who had experience with large projects, became our project manager; I recruited Albert Lazzarini from industry, and he took on the responsibility of being our systems engineer and integration manager; Dennis Coyne, who became our chief engineer, and Mark Coles, who became head of the LIGO Livingston Observatory (LLO). They joined the very strong team who were working on LIGO before I took it over, including Stan Whitcomb, who has been our chief scientist; Fred Raab, who became the head of the LIGO Hanford Observatory (LHO), Mike Zucker who led LLO for a few years, David Shoemaker and Peter Fritschel, who worked with Rai Weiss on the many responsibilities assumed by the MIT group.

Two important events occurred in 1997, while LIGO was in the midst of construction.

First, I invited Benoit Mours from Laboratoire d'Annecy and a senior member of Virgo to visit Caltech on sabbatical, and during that visit he worked with LIGO scientists to agree on a data format structure that would be used by both LIGO and Virgo. This was the earliest step toward these two collaborations working together to analyze data. The MOU to collaborate between LIGO and Virgo that has led to joint authorship of our discovery paper was formalized ten years later.

The second important event that took place in 1997 was the creation of the LIGO Scientific Collaboration (LSC), which brought in a broader group than Caltech and MIT to carry out the science from LIGO. The LSC was made by design to be a separate organization from the LIGO Laboratory, an organization where LIGO Laboratory scientists and scientists from other U.S. or non-U.S. universities and laboratories would work on equal footing with other individuals and groups that joined to do LIGO science. This has evolved over 20 years into the present LSC of over 1,200 members from 18 countries. The data pipelines for the various searches, calibration of data, data quality studies, scientific computing and the responsibility for scientific papers, talks etc. are carried out. The LSC has been a great success in enabling scientists from around the world to join together to carry out the science of LIGO.

In about 2000, we began commissioning LIGO and it became more sensitive than any previous instrument within about one year. Over the next decade we went through six cycles of 'observing,' which we called runs S1 through S6, each taking months of data, and each searching for gravitational waves and not finding any. Each of these observational runs was at significantly better sensitivity than the previous, enabling us to set better and better limits on gravitational waves. Finally, the last data runs reached our design sensitivities. One very important feature of this period is that there we had found no serious backgrounds or rejected possible candidates for gravitational waves. This meant that we were limited by our instrument sensitivity, not backgrounds that had features that could mimic gravitational waves.

At the time I proposed our plan for LIGO to the National Science Board in 1994, I asked for support to keep the key technical experts who developed Initial LIGO to do the R&D to develop the technologies for Advanced LIGO. By 2003, we had developed the basic technological concepts for Advanced LIGO and that conceptual design was reviewed and approved by the National Science Foundation. We developed an engineering design and developed the project plan and responsibilities over the next few years. Advanced LIGO was funded and we began construction soon after completion of our S5 data runs, when we had pretty much achieved the best sensitivity possible before a major upgrade to Advanced LIGO. A few changes for Advanced LIGO were implemented immediately and a final data run we called Enhanced LIGO, S6, was taken before the major construction was undertaken for Advanced LIGO.

During much of the period of Advanced LIGO construction, I was involved part-time on LIGO. I had been recruited to lead the design of the International Linear Collider, the leading concept for the next generation particle physics accelerator. For three generations, particle physics had advanced to new higher energy regimes through the complementarity of two accelerators, a proton-proton accelerator with good rates and ability to survey, and an electron-positron accelerator with capability of doing more precise measurements. Vigorous R&D programs were carried out at the Japanese High-Energy Accelerator Research Organization Laboratory (KEK), SLAC in the U.S. and Deutsches Elektronen-Synchroton (DESY) in Germany to develop a design for an electron-positron machine to complement the Large Hadron Collider (LHC) at CERN. The fundamental problem is that traditional circular machines at such high energies for electrons and positrons radiate away too much energy for a practical size machine. This motivated the development of a new type of accelerator, a linear collider.

A linear collider is a very challenging concept, because it is a single pass machine, while a conventional circular collider has many cycles for the beam particles to collide. Therefore, to achieve comparable event rate (luminosity) the beams have to be focused into very small spots. Two possible solutions were pursued in major R&D programs in the 1990s, one using conventional room temperature cavities at SLAC and KEK, and

Figure 6. Inspecting superconducting cavities for the International Linear Collider.

one using superconducting cavities at DESY. I chaired a committee in 2004 that chose the superconducting radiofrequency (RF) cavity concept, based to a large extent on the promise of a developing technology, other applications, and some technical advantages. I was then asked by the International Committee on Future Accelerators (ICFA) to lead the design effort for the International Linear Collider. I accepted and led an international R&D and design effort resulting in a detailed technical design that was completed in 2012 and is under consideration by the Japanese government with a decision expected during 2018.

In 2002, I was appointed by President Bush to the National Science Board (NSB). The NSB has 24 members, 8 appointed by the sitting President every two years. I served from 2002 to 2008, then as a consultant for two more years. The NSB advises both Congress and the President on science issues and serves as the oversight board for the National Science Foundation. Just as examples, during my tenure we did a study and advised Congress on the issue of 'rare materials,' many of which have become essential for high technology electronics, etc., yet, are not being mined in the U.S. Other issues that we grappled with continuously included diversity issues in science education and science professions. As the oversight board for the National Science Foundation we advised on how to do the best science within a system that has become more and more accountable to the government and has a smaller fraction of successful grant applications. Both trends lead to more conservatism and lack of risk taking in the NSF grant program, while doing the best science inherently involves risk taking and willingness to fail or find nothing that is interesting. My tenure on the NSB was very broadening for me, working closely with very talented and knowledgeable colleagues with very different backgrounds and skills on interesting and important issues.

I returned full-time to LIGO as the construction was complete on Advanced LIGO and commissioning was underway. The commissioning went well and when an improvement of a factor of three or more over the best achieved sensitivity for Initial LIGO was achieved, we were ready to perform the first Advanced LIGO data run. It began in September of 2015 and we observed the merger of two Black Holes on September 14, 2015. This discovery was a direct result of the new technology we had developed to improve our isolation from the ground motion by adding seismic isolation to the passive isolation in LIGO. This technology was developed in the Advanced LIGO R&D program that immediately followed the construction of Initial LIGO.

My first reaction to the dramatic event of Sept 14, 2015 was concern that we might be either fooling ourselves or were being fooled. The Advanced LIGO interferometer was basically a new instrument and we had only officially begun to operate it a few days before the discovery. So, maybe there were ways that such event candidates could be generated in the new detector itself or by something external. The main point is that we could measure the improved sensitivity but needed to run the detector for some length of time to determine whether events as observed could be some form of background. To test that required running for about one month and looking at off coincidence bins to see if such events are generated accidently. After doing this test for one month, we determined the probability of the event being an accidental was infinitesimal. This relieved one source of concern. The other was "how could we be fooled." In this case, the worry we had was that maybe there was some way that someone had 'injected' a fake event into our data stream. We created a 'tiger team' of experts to investigate this, led by Matt Evans of MIT. They did a thorough job of tracing the signal back to each detector (meaning that it would have had to be generated and injected into the data at our two LIGO detector sites within 6.9 milliseconds of each other. They showed that this could not have happened in any way they could conceive.

At this point, everyone in LIGO, including me, decided the event was real and we should proceed toward publication, keeping the discovery totally internal until we published, so that we could present it as professionally as possible. We spent another month doing the physics analysis for fitting to general relativity, parameter estimation, etc. By early November, we began writing the discovery paper. It was completed in early December and Physical Review Letters said they could not do an expedited review with Christmas vacation coming so soon. So, we held the paper until January. In the interim, on December 26 we saw our second 'five-sigma' event. The black holes for this event were somewhat lighter and therefore the event went to higher frequencies and had many cycles in our frequency band. Any doubt that might have lingered in my colleague and my mind were set to rest with this observation.

For me, despite the fact that the first event made an incredibly strong case, I let out a final sigh of relief. This was not due to any lingering doubts about what we had done, but a history in physics of the difficulty in claiming a discovery based on one event. I recall the discovery of the Ω- based on a single event in a hydrogen bubble chamber beautifully established the quark model of particle physics. On the other hand, a claimed discovery of the magnetic monopole, again based on a single event, appeared convincing, but was not confirmed in much more sensitive experiments.

The announcement of our observation was made on February 11, 2016. The LIGO press conferences was in Washington DC and simultaneously I gave the first scientific seminar on our result at CERN. We were, of course, very pleased by the acceptance and enthusiasm for our discovery by the scientific community. That has continued through our observations of several more black hole mergers, and more recently, a binary neutron star merger. The neutron star merger event had electromagnetic counterparts. This combined science has been amazing and is the beginning of a new astronomy – multimessenger astronomy, which I believe will lead to exciting new understanding about our universe in the coming decades and beyond.

As I write this, we are working hard to make LIGO an even more sensitive instrument, as well as doing R&D and developing concepts for next generation gravitational wave detectors. This will keep me busy for the rest of my career, and I eagerly embrace my future!

Ligo and the Discovery of Gravitational Waves, II

Nobel Lecture, December 8, 2017 by Barry C. Barish
California Institute of Technology, Pasadena, CA, USA.

INTRODUCTION

The observation of gravitational waves in the Laser Interferometer Gravitational-Wave Observatory (LIGO) was announced on February 11, 2016 [1], one hundred years after Einstein proposed the existence of gravitational waves [2][3]. This observation came after more than fifty years of experimental efforts to develop sensitive enough detectors to observe the tiny distortions in spacetime from gravitational waves. The Nobel Prize for 2017 was awarded to Rainer ("Rai") Weiss, Kip Thorne and myself *"for decisive contributions to the LIGO detector and the observation of gravitational waves."* In fact, the success of LIGO follows from decades of R&D on the concept and techniques, which were covered in Rai Weiss' Nobel Lecture, followed by the design, construction and evolving the LIGO large-scale interferometers to be more and more sensitive to gravitational waves. This work has been carried through the LIGO Laboratory and the scientific exploitation through the LIGO Scientific Collaboration, having more than a thousand scientists, who author the gravitational wave observational papers. In addition, many others made important contributions to the science of black holes, numerical relativity, etc.

In these three lectures, Rai, Kip and I tell the story of LIGO and gravitational waves in three parts. Rai covers the physics of gravitational waves, the experimental challenges and some of the pioneering interferometer work. He highlights the experimental challenges and some important early innovations that were proposed, tested at small-scale and have been incorporated in the LIGO interferometers. In this lecture, I describe the LIGO project and, the improvements that led to detection of merging black holes in Advanced LIGO. I also describe some key features of the interferometers, some implications of the discoveries, and finally, how we envision the evolution of LIGO over the coming decade. Kip will talk about some early personal history, theoretical advances that were crucial to making and interpreting the LIGO, and finally, his vision of the future opportunities in this new field of science.

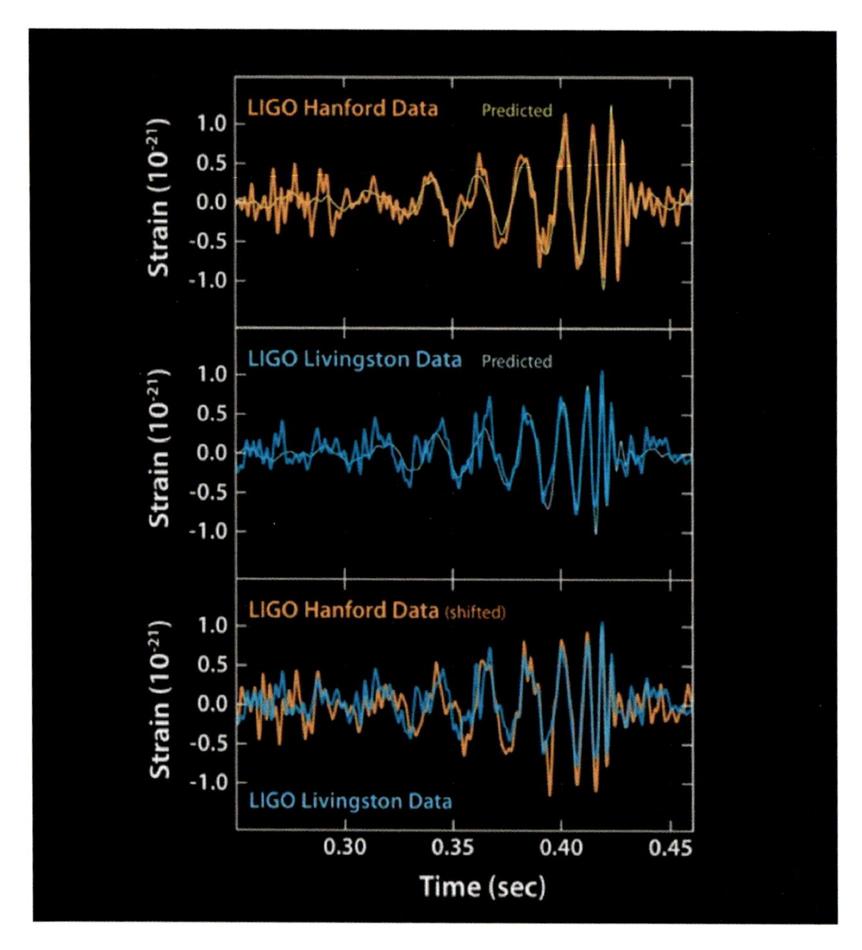

Figure 1. This is the gravitational wave discovery figure that was observed by members of the LIGO Scientific Collaboration within minutes after the event was recorded in Advanced LIGO. Each of the three figures shows the detected "strain" signals in units of 10^{-21} vs time. The top trace is the observed waveform detected in the Hanford, Washington interferometer, the middle trace is the observed waveform in Livingston, Louisiana The two signals are almost identical and, but are shifted by 6.9 msec and are superposed in the bottom trace. Image Credit: Caltech/MIT/LIGO Lab

Our three talks can be read as a series having some overlaps, or can each be read individually. In our lectures, we single out some individuals who played key roles in the discovery of gravitational waves, but by necessity, we have left out many other very important contributors.

THE LIGO LABORATORY

I became Principal Investigator of LIGO in early 1994 and, consequently, my part of our story begins at that point. A few months earlier, I had become 'available,' due to the unfortunate cancellation of the Superconducting Super Collider (SSC) in Texas by the U.S. Congress [4]. The SSC was the realization of a long process by the U.S. and world particle physics communities, conceptually and technically, to develop a facility that would open particle physics to a new energy regime, where there were strong arguments that science beyond the Standard Model of particle physics would become accessible. The Higgs phenomenon had been proposed to explain the origin of mass for the elementary particles, and the search for the associated Higgs particle was to be the first major goal of the SSC. It was eventually discovered at the Large Hadron Collider at the European Organization for Nuclear Research (CERN) Laboratory in Geneva, Switzerland that has some of the capability of the SSC.

The SSC was designed to have two complementary detectors with different features that would complement each other and confirm discoveries. I was co-spokesperson with Bill Willis of Columbia University for the Gammas, Electronics, Muons (GEM) detector, one of two large detector facilities. We had spent several years in the early 1990s developing concepts, technologies and design for GEM, and in 1993, the conceptual design for GEM had been successfully reviewed. We were just embarking on the final technical design and preparing for the beginning of construction when the SSC was cancelled by Congress. This was obviously extremely disruptive to all involved in the SSC. From GEM, a significant contingent of the physicists joined the CERN Large Hadron Collider experiments, and many of our technical developments were incorporated in the CMS and Atlas detectors at CERN.

I had decided not to join the CERN experiments, at least right away, because I preferred to take my time to decide what I wanted to do next. In fact, I was extremely busy on all the tasks involved in closing down the GEM facilities, and especially, in helping many colleagues I had hired to the SSC find jobs. However, I was approached by California Institute of Technology (Caltech) to become the LIGO Principal Investigator in early winter 1994. My previous roles on LIGO had been advisory, as I had been a strong supporter for developing the experimental gravitational wave effort at Caltech. I agreed to take over the LIGO leadership, and my first task was to strengthen the LIGO team and to evaluate and revise the LIGO proposal to the National Science Foundation (NSF). The NSF was on a tight schedule to make a decision whether to move forward with LIGO. They wanted an external review of the revised proposal by early summer 1994 and a discussion at the National Science Board later in the summer. They needed a decision whether to go forward with the LIGO construction project by fall.

To organize the LIGO effort, I created a new structure for the project built around the LIGO Laboratory that unified the Caltech and Massachusetts Institute of Technology (MIT) efforts. We divided responsibilities and created an overall concept for developing the two distant instruments in Hanford, Washington and Livingston, Louisiana. The decision to build LIGO at those two sites had already been made, following an evaluation by the NSF of about twenty proposals.

After becoming LIGO PI, my first priority was to make some key hires to strengthen the project, especially from the SSC where some extremely talented individuals had become available. Most importantly, I hired Gary Sanders as project manager, the same position he had with me for GEM at the SSC. Together, we rapidly filled out key positions on LIGO with other senior members, including Albert Lazzarini, as integration manager, John Worden for the large vacuum system, Jay Heefner (now deceased) and Rolf Bork to develop digital controls, Dennis Coyne as chief engineer, and others to fill out the initial team.

The next task was to integrate the new members of LIGO with the very talented existing LIGO staff members that included Stan Whitcomb, Robbie Vogt, Bill Althouse, Mike Zucker, Fred Raab at Caltech, and the MIT group under Rai's leadership that included David Shoemaker and Peter Fritschel. The combined Caltech-MIT effort became the LIGO Laboratory, which was to be the organization responsible for the construction and operation of LIGO. LIGO Laboratory is jointly operated by Caltech and MIT through a Cooperative Agreement between Caltech and NSF. LIGO Laboratory includes LIGO Hanford and Livingston Observatories, Caltech and the MIT LIGO facilities. There are currently 178 staff: scientific (including academic staff, postdocs, grad students, engineers, and technicians), and administrative support staff. I stepped down from being LIGO Director in 2006 and Jay Marx became the second LIGO Lab Director. Jay very capably led further improvements of initial LIGO to design sensitivity, as well as the early developments of Advanced LIGO. In 2011, Dave Reitze became LIGO Executive Director and has done a superb job of leading LIGO through the construction and commissioning of Advanced LIGO, and most importantly, the first gravitational wave detections.

In early 1994, we rapidly revised and re-costed the LIGO construction proposal to NSF to account for the larger team, increasing the planned staffing for the Hanford and Louisiana sites, and incorporating a more ambitious technical infrastructure. The plan was to make the initial implementation more robust, as much as possible to be able to accommodate building a second improved version (Advanced LIGO) within the same infrastructure. The total cost of the increases amounted to about \$100M, bringing the total Initial LIGO construction costs to almost \$300M. At the time, this corresponded to our asking for funding for the largest project the NSF had ever undertaken.

The NSF conducted the external review of our revised proposal in late spring 1994 and we received a very encouraging and strongly positive review. Before making a decision, however, Kip Thorne and I were invited to present LIGO to the National Science Board (NSB). This was very unusual at the NSF, as the NSB does not normally interact directly with proponents. At the NSB meeting, Kip presented the theoretical underpinnings of gravitational waves, as well as giving a description of the physics we would be able to produce with detections. I presented our plans for the project, which involved first building and exploiting Initial LIGO, which would be based as much as possible on technologies we had already demonstrated. This approach was hardly conservative, because LIGO was such a huge extrapolation from the R&D prototypes. It was a factor of a 100 in size and at least as large a factor in technical performance. A key feature of our plan was to initiate an ambitious R&D program to develop and test Advanced LIGO technologies, immediately following the completion of Initial LIGO construction. This would be carried out by keeping the key technical staff that had developed Initial LIGO. This was an unusual request, because Advanced LIGO was only a strategic concept and had not been proposed at this stage. Following our presentations to the NSB, we received formal approval at the full requested funding for Initial LIGO, with a commitment to support the crucial R&D program for Advanced LIGO.

The basic scheme for LIGO was to use a special high power stabilized single-line laser (neodymium-doped yttrium aluminum garnet = Nd:YAG) that entered the interferometer and was split into two beams transported

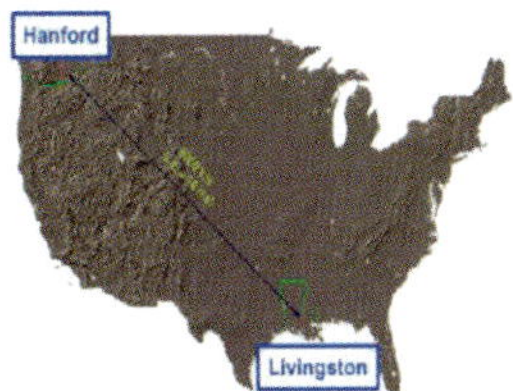

Figure 2. The LIGO Interferometers in Hanford, Washington and Livingston, Louisiana. Credit: Caltech/MIT/LIGO Lab

in perpendicular directions. The LIGO vacuum pipe is 1.2 m in diameter and is kept at high vacuum (10^{-9} torr). The 'test' masses are very high-quality mirrors that are suspended, in order to keep them isolated from the earth. They are made of fused Silica and hung in a four-stage pendulum for Advanced LIGO. In the simplest version of the interferometer, the equal length arms are adjusted such that the reflected light from mirrors at the far ends arrive back at the same time, and inverting one, the two beams cancel each other and no light is recorded in the photodetector. This is the normal state of the interferometer working at the 'dark port.' Many effects make the beams not completely cancel, and the actual optical configuration is more sophisticated.

When a gravitational wave crosses the interferometer, it stretches one arm and compresses the other, at the frequency of the gravitational wave. Consequently, the light from the two arms returns at slightly different times (or phase) and the two beams no longer completely cancel. This process reverses itself, stretching the other arm and squeezing the initial arm at the frequency of the gravitational wave. The resulting frequency and time-dependent amount of light is recorded by a photo-sensor and recorded as the waveform from the passage of a gravitational wave. The experimental challenge is to make the interferometer sensitive to the incredibly tiny distortions of spacetime that come from a gravitational wave, while at the same time, suppressing the various background noise sources.

The spacetime distortions from the passage of an astrophysical source are expected to be of the order of $h = \Delta L/L \sim 10^{-21}$, a difference in length of a small fraction of the size of a proton. In LIGO, we have made the length of the interferometer arms as long as is practical, in our case four kilometers, and this results in a difference in length we must be sensitive to that is still incredibly small, about 10^{-18} meters. For reference, that is about a 1,000 times smaller than the size of a proton! If that sounds very hard, it is!! Skipping the details, what enables us to achieve this precision is the sophisticated instrumentation that reduces seismic and thermal noise sources, effectively making the statistics very high by having many photons traversing the interferometer arms.

The initial version of LIGO was constructed during the period from 1994 to 1999, employing technologies that represented a balance between being capable of achieving sensitivity levels where the detections of gravitational waves might be 'possible,' and using techniques that we had fully demonstrated in our laboratories. LIGO was a huge extrapolation from the 30m prototype interferometer [5] in Garching, Germany and the 40m prototype [6] at Caltech interferometers that preceded it, and especially considering the very large NSF investment, we needed to be confident of technical success. In reality, from the best theoretical estimates at the time, we anticipated that we would likely need to achieve sensitivities well beyond those of Initial LIGO before achieving detections. So, developing the techniques and building Advanced LIGO was always an integral feature of our plans.

THE LIGO SCIENTIFIC COLLABORATION (LSC)

LIGO Laboratory, even after the strengthening in 1994, was relatively small for building such a large, ambitious and challenging construction project. The key hires that were added to the original Caltech and MIT LIGO teams were focused in areas needed for the construction of this sophisticated and challenging project. There were other crucial areas where we were weak. These were basically in areas that would be needed to extract the science, building computing facilities, data analysis infrastructure, search algorithms and pipelines, etc. In addition, there was expertise in hardware areas outside Caltech and MIT that could be strengthened by involving expertise from the larger worldwide community.

Gary Sanders and I both had worked within high energy physics collaborations. We appreciated the value of such collaborations, but believed LIGO needed a different model. By 1997, in the middle of LIGO construction I made a proposal to the NSF through a review committee chaired by Boyce McDaniel of Cornell, who knew how high energy collaborations worked. Our model for LIGO was somewhat different. We wanted to create a collaboration for LIGO that was focused on the science, an 'open' collaboration where individuals or groups could join if they could make significant contributions to LIGO science. They would not be expected to necessarily provide resources or hardware, as is done in high energy physics collaborations. In order to make joining the LSC as attractive as possible, we took a further step to insure that Caltech and MIT LIGO Laboratory scientists would not have a strong built in advantage to do LIGO science. The step was to make LIGO Laboratory individual scientists join the LSC to do their science and that the science component of LIGO would be carried out through that organization. This concept was endorsed by the McDaniel committee, the NSF approved, and we initiated a fledgling collaboration.

It took some effort, but I managed to convince the Caltech and MIT administrations and LIGO Laboratory scientific staff of this plan. I asked Rai Weiss to become the first spokesperson of the LSC and his credibility and approach helped to get the collaboration off the ground. Rai was succeeded by Peter Saulson of Syracuse, who also had a long history in LIGO, was much respected, and was the first spokesperson from outside Caltech and MIT. He was subsequently followed by David Reitze (now LIGO Laboratory Executive Director), Gabriela Gonzalez of Louisiana

State University (LSU) and now David Shoemaker (MIT). As the LSC matured, it has become more democratic in involving collaborators, publication policies, etc.

The LSC has all the responsibilities to produce the science with calibrated strain data. The data pipelines have been developed through the LSC and the various analysis groups on different science are organized and run by the LSC. LSC members also participate in some hardware detector areas and future detector R&D and planning. Also, LSC members participate in calibration of the data and have played the major role in detector characterization of the data. We have a program for young scientists to spend time at the LIGO sites, a "fellows" program.

The LSC has been extremely successful, the proof being the strong and effective role it played in making the discovery of gravitational waves, analyzing the data, interpreting the results, and in writing up and presenting the results. The strong and effective LSC role has continued to the present and the announcement of detection of a neutron star binary merger and all the follow-up measurements with various astronomical instruments.

The LIGO Scientific Collaboration and its 1200 members deserve shared credit for the discoveries reported here, resulting in the Nobel Prize, which must be given to no more than three individuals. The LSC has grown to more than 1200+ scientific collaborators, from 108 institutions and 18 countries! We conducted a study last year to figure out how to evolve the LSC for the gravitational wave observational era we are entering. We expect to institute some changes by the time of the next data run schedule to begin about the end of this calendar year, and we fully expect the LSC to play the central role in producing the future science with LIGO.

INITIAL LIGO PERFORMANCE AND SCIENCE

We had a two-step concept for LIGO, since beginning the project. Initial LIGO [7], as much as possible, was based on demonstrated methods and technologies, while the second stage, Advanced LIGO, was to achieve significantly improved sensitivities, through implementing methods and technologies that we would develop through an ambitious post construction R&D program. It was from that perspective that we proposed to NSF that while we would be commissioning, running and learning from Initial LIGO, we be funded to simultaneously carry out an ambitious R&D program to develop the techniques that would improve LIGO in a second step to a sensitivity where detections would become 'probable.' The NSF approved that plan and funded what was to become a successful R&D program that began in about 2000 and led to the Advanced LIGO concept that was proposed and approved by the NSF in 2003. The actual project funding was awarded several years later. I emphasize that our being able to carry out this R&D/design program, keeping key individuals who developed and built Initial LIGO, was crucial to our eventual success in detecting gravitational waves with Advanced LIGO. Another important point is that the Initial LIGO infrastructure was designed such that the interferometer subsystems could be evolved or replaced inside the same infrastructure (vacuum vessels).

After the completion of construction of Initial LIGO, we began commissioning and rapidly achieved better sensitivity than any previous gravitational wave detectors. As a result, we embarked on our first gravitational wave search data run. We did not detect gravitational waves, but we did set new astronomical limits on a variety of possible gravitational wave sources. Following the first data run, we made some technical improvements that reduced the background noise levels, both planned improvements and others from what we had learned in the first data run, and then we embarked on a second data run. Again, we did not detect gravitational waves, and again we achieved new limits and published them on various possible sources. All together, we repeated this basic cycle for over a decade, improving sensitivity and taking data for a total of six data runs at ever-increasing sensitivity (see Figure 3). For the final data runs, the interferometer sensitivities achieved very close to our original Initial LIGO design goals.

We searched for gravitational waves from several potential sources: mergers of binary black holes, a black hole and a neutron star, and binary neutron star systems. We also searched for signals from continuous sources, such as known and unknown pulsars, possible stochastic background signals; and unmodeled signals from some new source we had not specifically targeted. Unfortunately, even with the impressive improve-

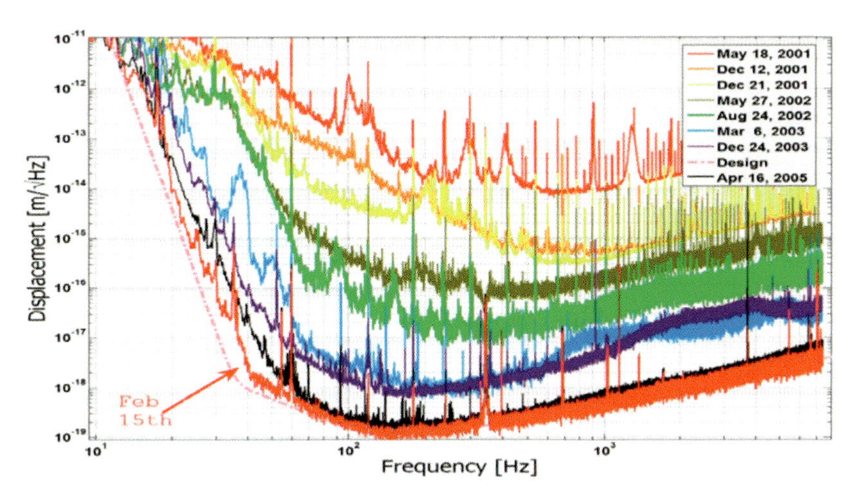

Figure 3. The evolving improved sensitivity of Initial LIGO.

ments in interferometer sensitivity, we did not detect gravitational waves. The resulting limits on various sources of gravitational waves were constraining for some of the published models of gravitational wave production from astrophysical phenomena.

The final Initial LIGO searches for black hole binary systems were performed in collaboration with the Virgo detector [8]. Although we had not detected signals, we were cautiously confident that the technical improvements envisioned for Advanced LIGO would be sufficient to finally achieve detection.

ADVANCED LIGO

By about 2004, the improved technologies developed for Advanced LIGO [9] were mature enough to propose Advanced LIGO to the National Science Foundation. After reviewing the proposal, NSF continued to support the technical developments in parallel with the continued running of Initial LIGO. The Advanced LIGO project received major funding through the NSF Major Research Equipment and Construction (MREFC), leading us to end the Initial LIGO scientific program and begin construction of Advanced LIGO. Additional significant contributions to Advanced LIGO included; a pre-stabilized laser system from the Max Planck Institute (Germany); Test Mass Suspension systems from the Science and Technology Facilities Council (UK); and thermal compensation wavefront sensors and interferometer controls components from the Australian Research Council:

The basic goal of Advanced LIGO is to improve the sensitivity from Initial LIGO by at least a factor of ten over the entire frequency range of

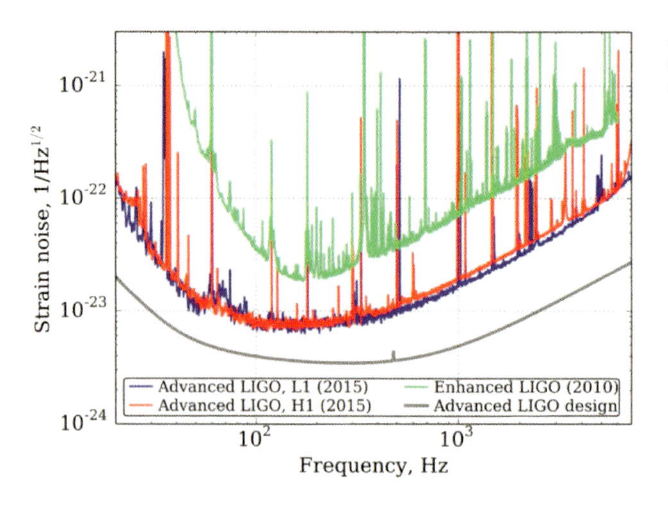

Figure 4. Advanced LIGO Sensitivity Goal.

the interferometer (see Figure 4). It is important to note that a factor of x10 improvement in sensitivity increases the distance we can search by that factor, since we measure an amplitude. It thus increases the volume of the universe (or rate for most sources) searched for by a factor of x1000. (The sensitivity to most sources is proportional to the volume we search) Therefore, there is a very high premium in LIGO on increasing the range we can search, and consequently, we spend a good fraction of our time improving the sensitivity, rather than taking very long data runs.

The Initial and Advanced LIGO gravitational wave detectors are Michelson interferometers with 4 km long arms. Both use Fabry-Perot cavities to increase the interaction time with a gravitational wave, and power recycling to increase the effective laser power. Signal recycling at the output (dark) port is a new feature of Advanced LIGO, and this changes the control and readout systems. Signal recycling enables tuning the sensitivity response to the physics goals, presently black hole and neutron star mergers. For Advanced LIGO, the design sensitivity is moved to lower frequencies (e.g. 40 Hz down to 10 Hz).

The improved seismic isolation system uses both passive and active isolation, and the improved test mass suspensions use a quadruple pendulum. Higher laser power, larger test masses and improved mirror coatings have been incorporated. The Advanced LIGO interferometers are installed in the same infrastructure, including the same vacuum system as used for Initial LIGO.

The Advanced LIGO laser is a multi-stage Nd:YAG laser. Our goal is to raise the power from ~18 W in Initial LIGO to 180 W for Advanced LIGO, improving the high frequency sensitivity, accordingly. The pre-stabilized laser system consists of the laser and a control system to stabilize the laser in frequency, beam direction, and intensity. For the results presented here, due to stability issues, heating and scattered-light effects, the laser power has only modestly been increased. We plan to bring the power up systematically in steps, studying these effects for the next data runs.

The key improvement in Advanced LIGO that enabled the detection of the black hole merger was implementing active seismic isolation and a quadruple suspension system (see figure 5).

The multiple suspension system moved all active components off the final test masses and gives better isolation. Initial LIGO used 25-cm, 11-kg, fused-silica test masses, while for Advanced LIGO the test masses are 34 cm in diameter to reduce thermal noise contributions and are 40 kg, which reduces the radiation pressure noise to a level comparable to the suspension thermal noise. The test mass is suspended by fused silica fibers, rather than the steel wires used in initial LIGO. The complete suspension system has four pendulum stages, increasing the seismic isolation and providing multiple points for actuation.

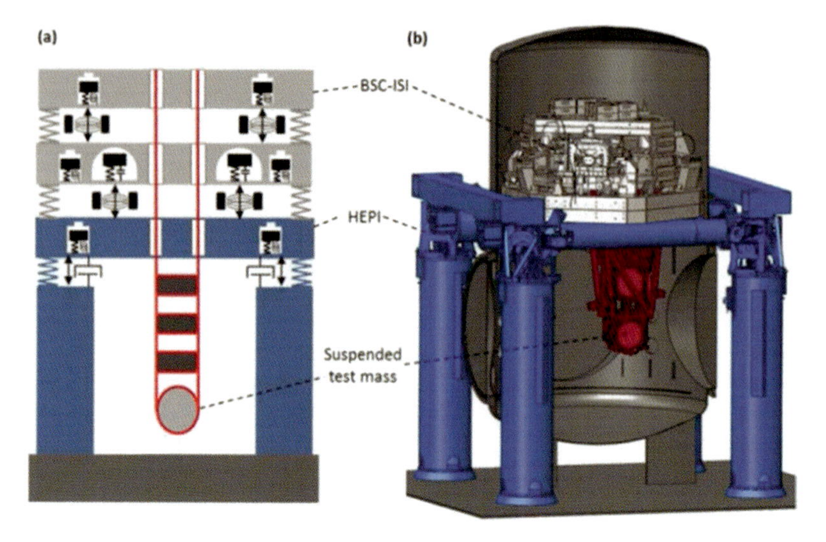

Figure 5. Advanced LIGO multi-stage suspension system for the test masses with active-passive seismic isolation.

The active seismic isolation senses motion and is combined with the passive seismic isolation using servo techniques to improve the low frequency sensitivity by a factor of x100. Since the rate for gravitational events scales with the volume, this improvement increases the rate for events by 10^6. This improvement enabled Advanced LIGO to make a detection of a black hole merger in days, while Initial LIGO failed to detect gravitational waves in years of data taking.

THE BLACK HOLE MERGER EVENT (GW150914)

The observation of the first Black Hole merger by Advanced LIGO [11], [12], [13], [14] was made on September 14, 2015. Figure 1 shows the data, and Figure 6 reveals the key features of the observed compact binary merger. These are a result of the analysis of the observed event shown in Figure 1. At the top of Figure 6, the three phases of the coalescence (in-spiral, merger, and ringdown) are indicated above the wave forms. As the objects in-spiral together, more and more gravitational waves are emitted and the frequency and amplitude of the signal increases (the characteristic chirp signal). This is following by the final merger, and then, the merged single object rings down. The bottom pane shows on the left scale that the objects are highly relativistic and are moving at more than 0.5 the speed of light by the time of the final coalescence. On the right side, the scale is units of Schwarzschild radii and indicate that the objects are very compact, only a few hundred kilometers apart when they enter our frequency band.

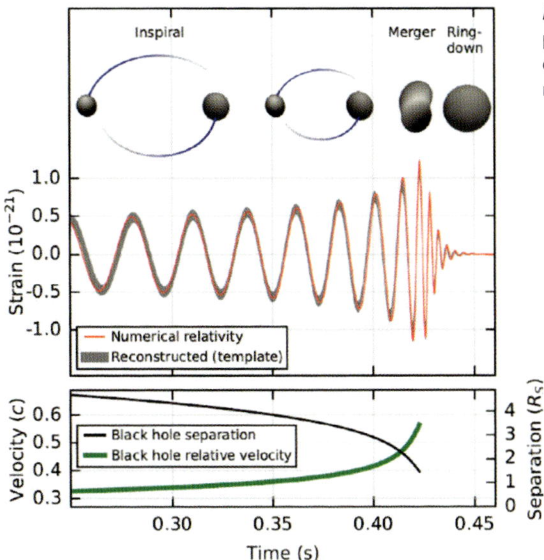

Figure 6. The physics interpretation of the observed event as a binary black hole merger.

By comparing and fitting our waveform to general relativity, we have concluded that we have observed the merger of two heavy compact objects (black holes), each ~30 times the mass of the sun, and going around each other separated by only a few hundred kilometers and moving at relativistic velocities.

On the top of the figure, we see the three different phases of the merger: the inspiral phase of the binary black holes system, then their merger of the two compact objects, and finally, there is a ringdown. The characteristic increasing frequency and amplitude with time (the so-called chirp signal) can be seen in the inspiral phase. The largest amplitude is during the final merger, and finally, there is a characteristic ringdown frequency. The two black holes inspiral and merge together due to the emission of gravitational radiation coming from the accelerations.

The bottom of Figure 6 is even more revealing. The right-hand axis is basically in units of about a hundred kilometers. That means the separation between the two merging objects at the beginning was only ~ 400 kilometers, and at the end ~100 kilometers. Therefore, these ~ 30 Solar Mass objects are confined in a volume only about twice the size of Stockholm, yet the final black hole has 60 times the mass of our sun, or 10 million times as massive as the earth. From the axis on the left, we see that when we first observed this event, the two objects were moving at about three-tenths the speed of light and that increases to over half the speed of light by the time of the final merges!

In order to be confident that what we observed was a real event and not a background fluctuation, we directly measure the background

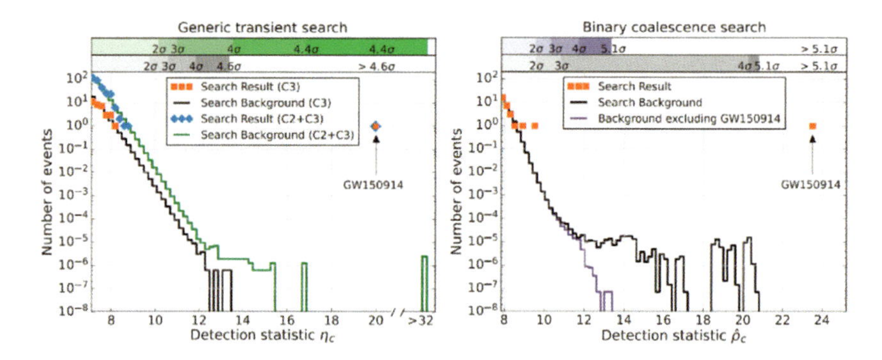

Figure 7. The statistical significance of the event as a generic transient event on the left, and directly searching for events from a merger on the right.

probability by comparing coincidence time slices for the two detectors, both in time (e.g. ±10 msec) and out of time! Since GW150914 occurred only days after we began data taking with Advanced LIGO, this required taking about one month of data before we could quantitatively establish the probability that the event was real. In other words, in addition to searching for coincidence in-time signals, we look for coincidences between all the out-of-time slices during our data taking. These background slices could not have come from any physical phenomena traveling at relativistic speeds, like gravitational waves. The total number of time slices we compared was equivalent to an in-time background levels equivalent to over 67,000 years of data taking. Taking into account the different event classes in our search, we reduce the limit on the false alarm rate to 1 in over 22,500 years. This corresponds to a probability that our observed event is accidental to $< 2 \times 10^{-6}$, establishing a significance level of 4.6σ. I emphasize that the measured significance level is set from the number of bins compared from 16 live days of data taking. This represents a lower limit on the actual significance of GW150914.

Figure 7 shows the statistical significance as described above for the GW150914 event, compared to the measured background levels, under two different assumptions. The horizontal axis is a measure of the significance of the events and the vertical access is the rate. The left-hand plot shows the observed GW150914 event at the level of one event level, and having statistical significance $\sigma > 4.6$, as described above. This plot assumes a generic signal shape for the event. The right-hand plot shows a significance of > 5σ, when a binary coalescence form is assumed. Note that the second most significant event in this data is about 2σ, which may well also be a binary black hole merger, but at this early stage in LIGO, we are only declaring 5σ events as gravitational wave binary mergers.

The shape of the waveforms that describe the merger, coalescence and ringdown reveal the detailed parameters of the merger. The orbits decay as the two black holes accelerate around each other and emit energy into gravitational waves determined by the 'chirp mass,' as defined below, at leading order in the strength of the binary's gravity.

$$\mathcal{M} = \frac{(m_1 m_2)^{3/5}}{M^{1/5}} \simeq \frac{c^3}{G}\left[\frac{5}{96}\pi^{-8/3} f^{-11/3} \dot{f}\right]^{3/5}$$

The next order terms enable the measurement of the mass ratios and spins; the red-shifted masses are directly measured; and the amplitude is inversely proportional to the luminosity distance. Orbital precession occurs when spins are misaligned with the orbital angular momentum. GW150914 shows no evidence for precession. The sky location is extracted from the time-delay between detectors and the differences in the amplitude and phase in the detectors.

Using numerical simulations to fit for the black hole merger parameters, we determine that the total energy radiated into gravitational waves is $3.0 \pm 0.5\ M_{\odot}\ c^2$. The system reached a peak luminosity of ~3.6×10^{56} ergs/s, and the spin of the final black hole < 0.7 of the maximal black hole spin.

The main parameters of the black hole merger are summarized below: *Table 1.*

Primary black hole mass	$36^{+5}_{-4}\ M_{\odot}$
Secondary black hole mass	$29^{+4}_{-4}\ M_{\odot}$
Final black hole mass	$62^{+4}_{-4}\ M_{\odot}$
Final black hole spin	$0.67^{+0.05}_{-0.07}$
Luminosity distance	410^{+160}_{-180} Mpc
Source redshift, z	$0.09^{+0.03}_{-0.04}$

With only two detectors we cannot locate the direction very well, but comparing the time, amplitude and phase in the Livingston and Hanford interferometers, we are able to locate the gravitational wave as coming up from the Southern Hemisphere within an area of about 600 square degrees. Our most recent observations, discussed later, include Virgo and with three detectors we determine the sky location to tens of square degrees.

MORE BLACK HOLE MERGERS

The first Advanced LIGO data run (O1) continued for four months, from September 2015 to January 2016. Our second data run (O2) ran from December 2016 to the end of August 2017. Much like the evolution described for Initial LIGO, we are making improvements between Advanced LIGO data runs and expect to achieve design sensitivity within a few years. We are actively searching for other signals, besides binary merges, including bust signals from phenomena like supernova explosions or gamma ray bursts, continuous wave signals from spinning neutron stars (pulsars), stochastic background signals, etc. So far, we have only detected binary merger signals, but hope to detect others as our sensitivity improves.

We search for transient or burst events in two different ways. First, we use an unmodeled search for a power excess, using a wavelet technique, which makes no assumptions about the expected binary merger waveform. The second search method employs a matched template technique using binary merger waveforms. We have populated the available mass 1 vs mass 2 plane with templates that describe the waveform, calculated using post-Newtonian calculations and/or numerical relativity, as required. There are several hundred thousand templates, which are each multiplied by the observed noise in each interval of time (matched template technique).

Since announcing our observation of GW150914, we have reported

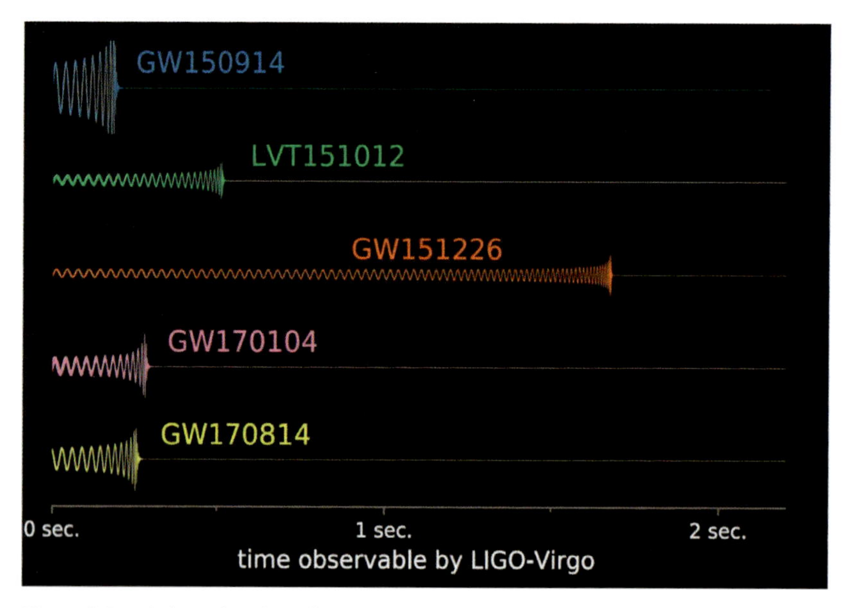

Figure 8. Event characteristics of reported black hole mergers in LIGO.

[15], [16] several more observations of black hole mergers. As shown in figure 8, we have so far reported four black hole merger events having high statistical significance (> 5σ), and one of 2σ, which we do not declare as an event. However, all the characteristics of that event are similar to the other black hole mergers, except that it is further away, making the signal weaker and the signal to noise less significant.

It is interesting that we have only one such event candidate, having marginal signal to noise. This is a result of the extremely steep cutoff of the noise background when requiring a coincidence between the two sites, as is evident by the extremely sharp fall-off of the background noise in Figure 7. We can conclude that there is very little correlation between the noisy events observed at the two sites. This was not necessarily anticipated, as we worried about correlated noise. However, this result is both more convincing, in terms of the events we have observed and bodes very well for the cleanliness of events in future observations by using the coincidence technique.

For the black hole merger events shown in Figure 8, notice that the amount of time and number of cycles observable in LIGO is very dependent on the mass of the black hole system. The heaviest black hole we observed was the first one and we only observed a few cycles, while for the lightest one (GW151226) we have many cycles. Also, I repeat that the candidate event, LVT151-12, has characteristics completely consistent with the other events, but is about twice as far away and therefore has lower statistical significance.

SCIENCE IMPLICATIONS OF THE OBSERVED BLACK HOLE MERGERS

Gravitational waves represent a completely new way to view the universe. We have every reason to expect that we will discover new phenomena and learn 'new' astrophysics from gravitational waves. This has already been realized from the very first observations of gravitational events.

Conclusions from the first observations of black hole merger observations from gravitational waves include the following:

- Stellar binary black holes exist
- They form into binary pairs
- They merge within the lifetime of the universe
- Masses ($M > 20\, M_\odot$) are considerably heavier than what was known or expected of stellar mass black holes.

The fact that the observed black hole mergers in LIGO are so massive has opened the question of how they were produced. LIGO is strongly biased to detecting heavy masses, because the stronger signal provides sensitivity to greater distance and the volume observed grows like the cube of the distance. Nevertheless, the creation of such large masses needs explanation. If these stellar mass black holes come from the gravitational collapse of heavy stars, it requires special conditions for the parent star to survive, such as low-metallicity regions of the universe. Another possibility is that the heavy black holes were produced in dense clusters, and a third is that they are primordial and may even be associated with the dark matter.

The next challenge will be to distinguish between these or other possibilities for the origin of such heavy stellar black holes. More events will give us distributions of masses and other parameters, while larger signal/noise events will enable determining the other feature of these mergers. For example, are the spins of the merging black holes aligned, anti-aligned or is there no correlation in the spins?

Our other major scientific goal is to test general relativity in the important regime of strong field gravity. The merger of black holes presents just such a new 'laboratory' for these studies. We can compare the observed waveform with the predictions of general relativity looking for deviations. So far, all tests are in good agreement with the theory.

The first test is just how well general relativity fits the details of our data, or if there are any other features in the data. At the present level, the deviations we see are all consistent with the instrumental noise in our interferometers.

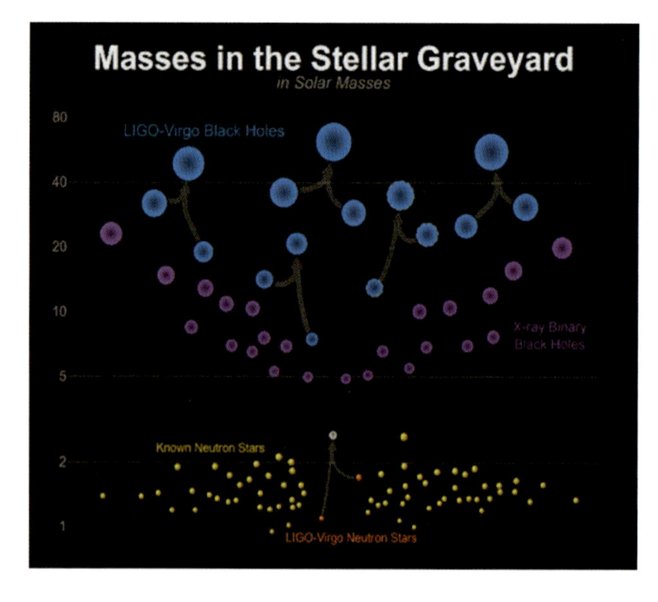

Figure 9. Masses of observed black hole mergers in LIGO.

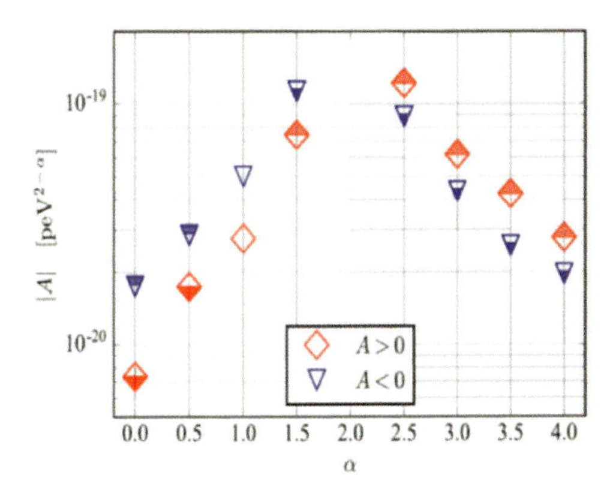

Figure 10. Tests of the inclusion of a dispersion term in general relativity. Upper limits are shown on the presence of a dispersion term.

We have tried to quantify our level of agreement by adding a dispersion term, which would indicate that some components might not travel at the speed of light and are dispersed. The large distance to the black hole merger events provides good sensitivity for such tests. From Figure 10 we see that very little dispersion is allowed. The limit $\alpha = 0$ corresponds to an upper limit on the mass of gravitons (a hypothetical particle mediating the force of gravity). Assuming that gravitons are dispersed in vacuum like massive particles, then the bound on graviton mass is $M_g < 7.7 \times 10^{-23}$ eV/c^2

More stringent tests of general relativity will be possible as we collect more events, and in particular we can test alternate formulations of general relativity.

VIRGO DETECTS GRAVITATIONAL WAVES

Finally, very recently we reported [17] one more black hole merger event (GW170814) and for the first time this was also observed in the Virgo detector near Pisa in Italy. Virgo is a collaboration of France, Italy, Netherlands, Poland and Hungary. The founders of Virgo were Adalberto Giazotto and Alain Brillet and the present leadership is Federico Ferrini as the European Gravitational Observatory (EGO) director and Jo van de Brand as Virgo spokesperson. Virgo is a 3 km interferometer, similar to LIGO, but with some technical differences.

Adding Virgo, in this detection, not only gives independent confirmation of the LIGO black hole merger detections, but improves markedly the ability to triangulate. This is a precursor to also adding KAGRA in Japan and LIGO-India detectors to the network. The combination will give good sky coverage and more than an order-of-magnitude improvement in located the direction of the source. This can be seen in Figure 12, where the sky location of this last observed event is dramatically better than the previous observed black hole merger events.

Figure 11. The Virgo Suspended Mass Interferometer in Cascina, Italy

The first three-way coincidence enabled a test of general relativity that could not be carried out with LIGO alone, because the two LIGO detectors are almost co-aligned. Including the non-aligned Virgo data, the polarization of the gravitational waves can be studied to determine whether the waves are transverse as predicted in general relativity. GW170814 enabled the first tests for non-transverse polarizations. The first such analysis was performed using polarizations that are forbidden by general relativity and these were disfavored in the analysis. More precise tests and analysis will be possible in the future using events detected in both LIGO and Virgo.

NEUTRON STAR BINARY MERGER

Two weeks after the Nobel announcement in October and almost two months before this lecture, we announced the first observation of a merger of a neutron star binary system [18]. This was also the first gravi-

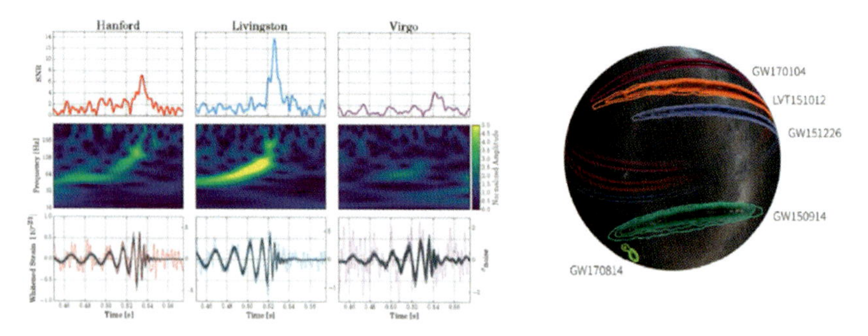

Figure 12. Black hole merger 170814 showing LIGO Hanford, LIGO Livingston and Virgo signals on the left, and the reconstructed direction of the gravitational wave on the right.

tational wave event to have electromagnetic counterparts observed in a large variety of astronomical instruments. This event initiates the eagerly sought after new field of multi-messenger astronomy.

On August 17, 2017, LIGO and Virgo registered a gravitational wave signal from the inspiral of two neutron stars. Many of us expected the first source we would observe in LIGO would be from the merger of binary neutron stars, because there is a known rate of detected binary neutron stars in our own galaxies by radio telescopes. Although accurately predicting the rate of mergers detectable by LIGO from this data is not very precise, we knew it to be in the range of detectability by LIGO. No such definite predictions were possible for binary black hole mergers or for black hole – neutron star mergers (which we have not yet detected).

Neutron stars are very dense nuclear matter and one of the main goals studying such describe super-dense matter by a relationship called the "equation of state." The neutron stars mass and equation of state determines size and tidal deformation from the gravitational pull of the companion neutron star. We cannot determine the equation of state from this event, but this is a future goal from these detections.

From our estimation of the parameters, the masses of the two compact objects turn out to be between 0.86 and 2.26 solar masses as shown in Figure 14. In fact, restricting the spins to the range for binary neutron stars, limits the mass determination to between 1.17 and 1.60 solar masses, consistent with being neutron stars. The distance to the source is well determined from the amplitude of the merger gravitational wave signal to be about 40 Mpc (130 million light years).

This event occurred only a few days after Virgo had joined the gravitational wave network and greatly improved localization for the black hole merger event discussed above. The localization of the neutron merger event, including Virgo, is an oblong about 2 degrees across, and 15 degrees long, covering about 28 square degrees.

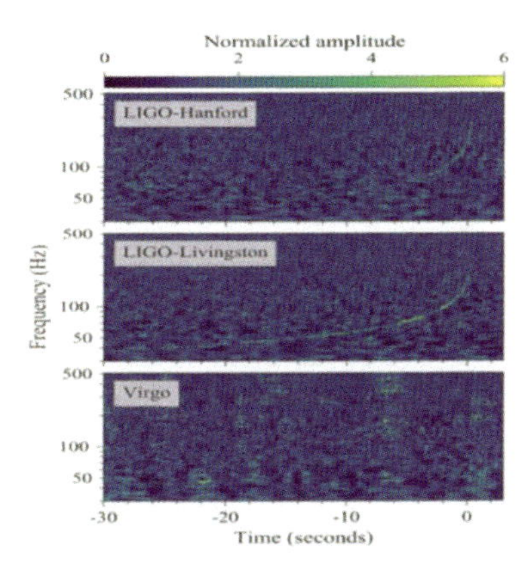

Figure 13. The first observation of binary neutron star merger. Time-frequency representations of gravitational-wave event GW170817, observed by the LIGO-Hanford (top), LIGO-Livingston (middle), and Virgo (bottom) detectors.

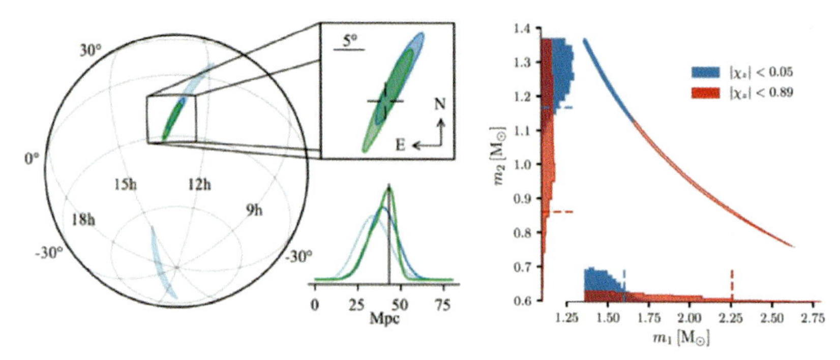

Figure 14. The main parameters of GW170817.

The Fermi satellite observed a gamma-ray burst GRB170817A in the same region of the sky and the triggered follow-ups identified the fading light from the event from near NGC 4993, first by the 1m Swope optical telescope [19]. The gravitational wave observation of the neutron star merger provides a new independent method of measuring the Hubble constant, the rate of expansion of the universe, using a method proposed by Bernard Schutz, called a 'standard siren' measurement [20]. The idea of is to use the distances of galaxies determined just from the gravitational-wave observations, by determining the luminosity (e.g. distance) directly from the observations. Standard sirens are compact binary systems, consisting of neutron stars or black holes, whose gravitational waveform as the compact objects inspiral towards merger carries information about the distance of the source – as well as the masses of the compact objects and other parameters of the system. In this case, we use

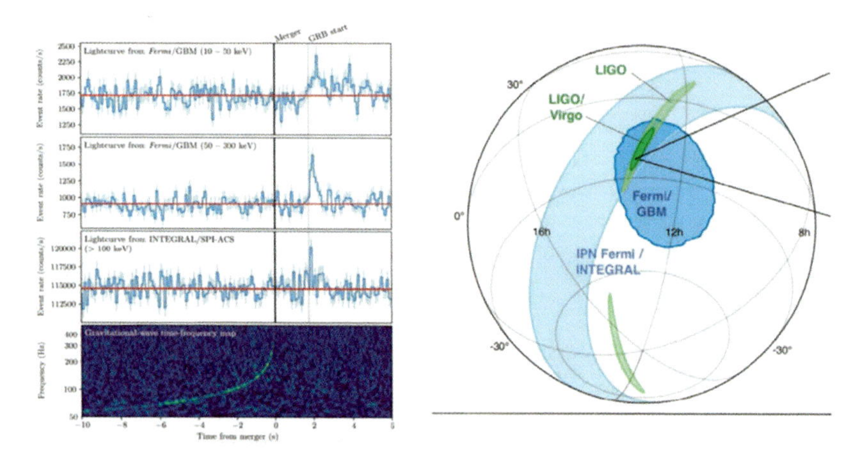

Figure 15. Fermi short gamma ray burst signal observed 1.7 sec after gravitational wave signal in the same region of the sky.

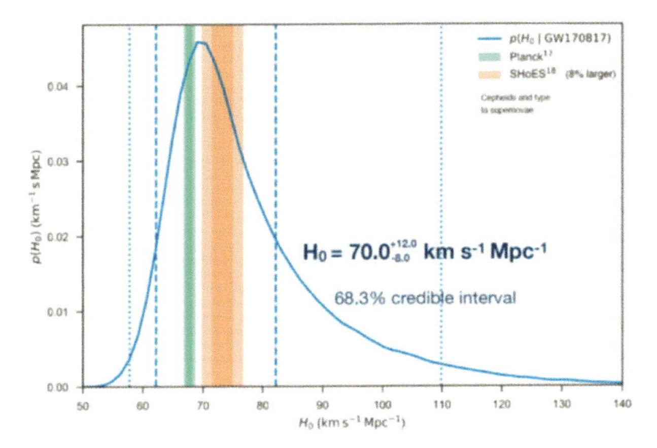

Figure 16. Determination of the Hubble Constant from gravitational wave data. The blue curve shows the relative probability of different values of H0 with the peak at 70 kms^{-1} Mpc^{-1}. The dashed and dotted blue vertical lines are the limits of 68.3% and 95.4% credible intervals for H0. Our gravitational-wave results are in agreement with both Planck and SHoES. Note, that we cannot distinguish from just one event between these two, which disagree with each other at the 95.4% probability level.

the precise information of which galaxy the event occurred in from the optical observations.

GW170817 represents the first joint detection of gravitational and electromagnetic waves from the same astrophysical source. All the data agrees with the hypothesis that the source is the merger of two neutron stars in the host galaxy NGC4993 in the constellation of Hydra. Analysis of the waveform of GW170817 yielded a distance estimate of about 44 Mpc, assuming that the sky position of GW170817 was exactly coincident with its optical counterpart. The gravitational-wave distance estimate is completely independent of the Cosmic Distance Ladder derived from electromagnetic observations, so that future measurements with more statistics from gravitational waves can help distinguish the uncertainties between the electromagnetic observations.

The observation in LIGO and companion observation of a short gamma ray burst opened a full astronomical campaign to observe this event in different wavelengths and devices, including the large neutrino detectors. I will not summarize those results here [21], [22], [23], as that is not the subject of this Nobel Lecture, but the rich observations gave strong support to the 'kilonova' picture of a binary neutron star merger, as well as that neutron star binary mergers are a significant source of the heavy elements in nature. Even more exciting is the fact that the long anticipated idea of doing multimessenger astronomy – using the complementary information from electromagnetic, neutrinos and gravitational waves to study the same phenomena – has become a reality.

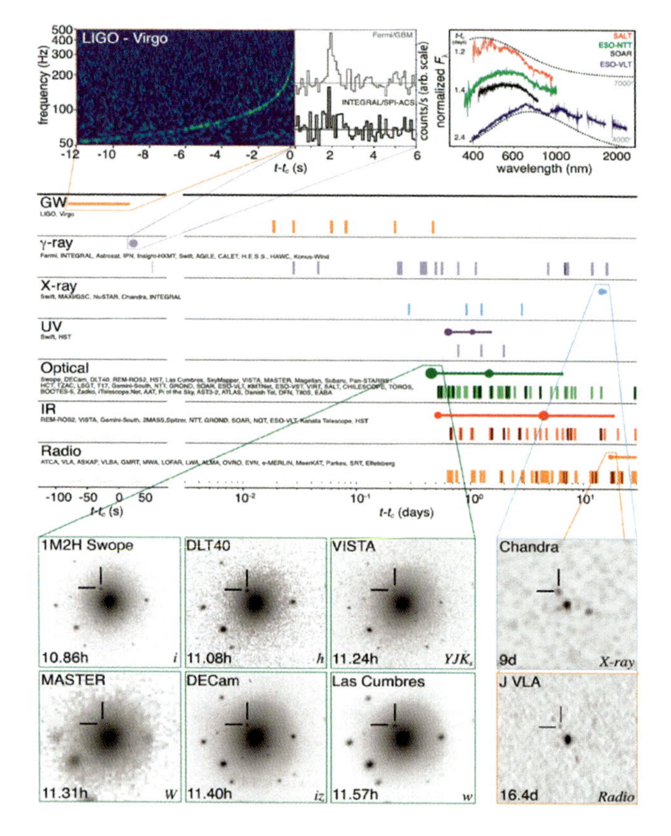

Figure 17. The electromagnetic campaign for counterparts to GW170817.

FUTURE DIRECTIONS FOR LIGO AND GRAVITATIONAL WAVES

As discussed above, we have completed two data taking runs (O1 and O2) with Advanced LIGO. In the first data run, we made the first direct observation of gravitational waves, resulting from a black hole merger ~1.3 billion years ago. The waveform data fit the predictions of Einstein's general relativity remarkably well and all detailed analyses are in good agreement with general relativity. The merging black holes were each about 30 solar masses, which is surprisingly heavy. During the same data run, we observed a second black hole merger about three months later, again having more than 5 sigma significance. During this data run, we also observed a weaker event, having about 2 sigma significance, which we do not claim as a gravitational wave event, but again appears to fit the black hole merger hypothesis very well, however occurred about twice as far away as the other two events and therefore has a weaker signal and is less significant.

Following O1, we spent about nine months calibrating the interferometers and improving them. The sensitivity improvements for O2 were within the range of our planning, but technical problems, especially in Hanford, limited the improvements. We could hardly expect O2 to be as spectacular as O1, especially since our sensitivity improvements were less than we had hoped for. However, O2, has turned out to be comparably exciting to O1. We recorded several more black hole mergers and are starting to get enough to get some statistical information on mass distributions, etc. Near the end of the data taking, in August 2017, the Virgo detector in Italy had good enough sensitivity to join LIGO as part of a network. A short time later, we observed another black hole merger, but this time with a three-fold coincidence. As expected, the third detector dramatically improved our ability to locate the source on the sky from many hundreds of square degrees to a few tens of square degrees. This demonstrated our future ability to compare information with other astronomical instruments looking for signals from the same phenomena. No companion electromagnetic signals are expected for black hole mergers and none have been seen.

Finally, as discussed above, we have made the first observation of gravitational waves from a neutron star merger. Using the combined LIGO and Virgo observations produced much better source location, enabling all the related electromagnetic observations discussed above. This was a spectacular initiation of multimessenger astronomy and portents exciting future combined observations.

THE FUTURE

We are now in the process of preparing for our third observational run, O3, which will begin in about one year. The long down periods for LIGO are a direct result of the large premium in improving the interferometer sensitivities. We are presently about a factor of two away from Advanced LIGO design sensitivity. Figure 18 shows the performance improvement planned for Advanced LIGO and what was achieved for the O1 observing run. We made some modest improvements for O2 and expect to make more for O3, the next data run. In steps, we expect to reach the design sensitivity over the next few years.

The worldwide gravitational wave network will add the KAGRA interferometer in Japan and the LIGO-India interferometer over the next decade, giving both improved location information and broad sky coverage.

The rates for the black hole and neutron star mergers by more than an order of magnitude, compared to O1 and O2. In addition to compact binary mergers, over the next decade, we will be able to study the features and science of compact binary mergers, and it is possible that the improved sensitivity will open up other gravitational wave channels to observations.

In addition, we are working on a set of improvements beyond Advanced LIGO [24] that are to be implemented over the next decade. We call this improvement program A+, and we believe it is a realistic goal to do the enabling R&D and implement improvements over the coming decade that will improve the sensitivity by another factor of 2 beyond advanced LIGO. We are preparing to submit a specific proposal to the NSF in the near future.

In addition, we are working on technological improvements to fully exploit the potential of the present LIGO facilities and sites (Voyager), as well as investigating concepts for a third generation detector (Cosmic Explorer). Figure 18 shows our present view of improvements that could ultimately be made fully exploiting the present LIGO sites. The R&D to develop Voyager could be incorporated such a third generation detector. It appears possible to improve the sensitivity of LIGO by about a factor of ten beyond what has presently been achieved.

The past two years have seen an impressive beginning to this new field, based on gravitational wave observations. We hope to make a set of incremental improvements over the next decade, first achieving the Advanced LIGO and, hopefully, followed by continued improvements to fully exploit the sites, achieving x10 the present sensitivity.

This will be far from the end of the story. Conceptual studies for the

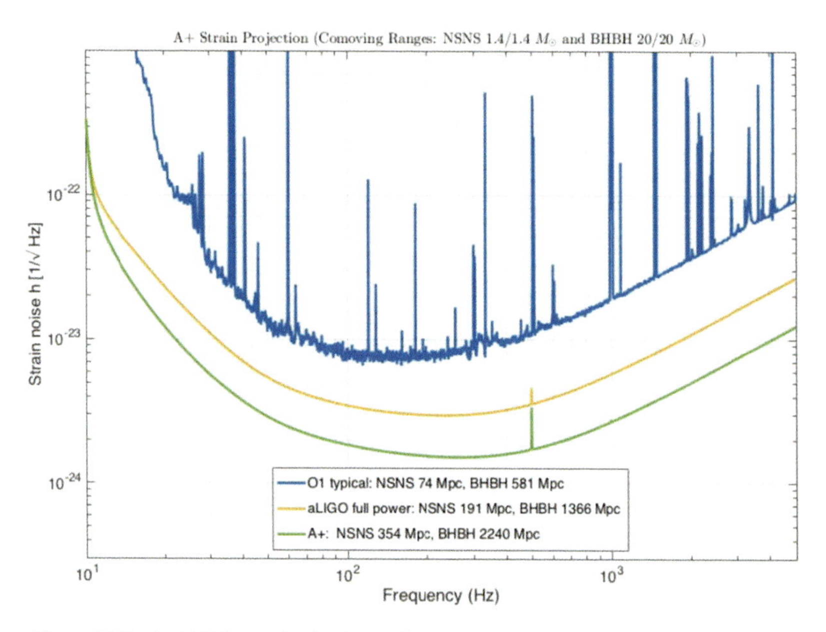

Figure 18. Typical LIGO sensitivity during the O1 data run, the Advanced LIGO design sensitivity, and the estimated sensitivity achievable with the A+ upgrades. A+ will involve a variety of improvements over the present LIGO, including implementing squeezed light, improved optical coatings and test masses, and possibly cryogenic cooling.

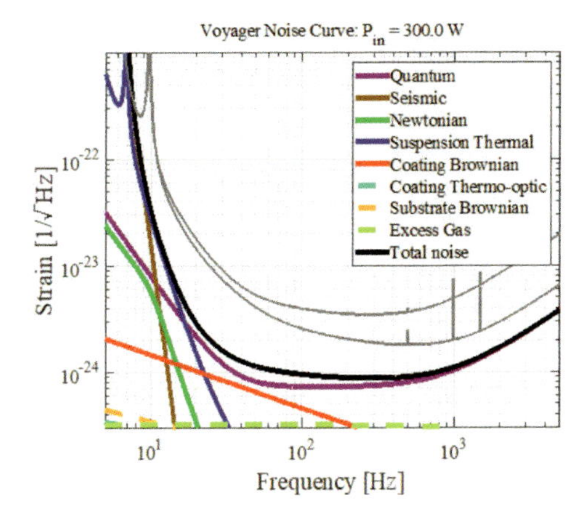

Figure 19. Fully exploiting the LIGO sites. Employing improved technologies, a factor of x10 improvement over O1 sensitivities appears possible.

Cosmic Explorer [25] in the U.S., and especially for the Einstein Telescope [26] in Europe illustrate that we should be able to make third generation ground based detectors that will be x10 times more sensitive than can be achieved at the present LIGO facilities. The Einstein Telescope: is designed to have sensitivity x10 better than LIGO. The key features in the concept are that it is deep underground, has 10 km arms and a triangular configuration, is cryogenic and has both a low frequency optimized and a high frequency optimized configuration. The Cosmic Explorer concept is less mature at this time, but is basically a 10x scale-up of LIGO and on the surface. Which is the better approach, how many detectors are needed in the next generation for pointing are unresolved. The community is now doing a study of the science potential for a new generation of detectors and this will help set the priorities in developing unified concept to pursue.

It is clear that it is possible to make future detectors that will open up impressive new science possibilities with gravitational waves, as are discussed in Kip Thorne's lecture. He also discusses the future science for LISA, the planned space-based gravitational wave experiment, as well as pulsar timing and other possible future gravitational wave probes.

We eagerly look forward to deeper understanding of our universe and likely, many exciting surprises, as we open and exploit this new field of gravitational wave science.

FINAL REMARKS AND REFLECTIONS

LIGO has been a remarkable adventure, beginning with Einstein's prediction in 1916, and some later wavering in his own belief in gravitational waves. There was a long road, bringing us into the 1950s before the theoretical community became fully convinced of gravitational waves. The experimental part of the story started in earnest in the 1950s when Joseph Weber began the first serious attempts to detect gravitational waves with resonant bars. Some of the early history was marred by false discoveries, but overall, much was learned from the resonant bar era that has been built into the way we approach using interferometers. The concept of doing coincidences from separated detectors to be confident of detections, the analysis of background noise sources is used extensively in LIGO, and finally analyzing off-coincident time slices to measure accidental background levels.

The early interferometer era is covered in Rai Weiss' lecture. It is characterized by a broad worldwide R&D effort, incorporating ideas and testing them in large scale prototypes. Most of the key concepts employed in Advanced LIGO were first suggested and tested several decades ago.

LIGO, itself, has been built and operated by a very strong collaboration between the Caltech and MIT groups, with some key contributions from collaborators in Scotland, Germany, Australia and several U.S. universities. The LIGO Scientific Collaboration has carried out the data analysis and written our series of discovery papers in a way that we can all be proud.

Finally, our home institutions, Caltech and MIT, have given us unusual support and encouragement through all the years we toiled, before achieving the ultimate success, detections of gravitational waves. But, most of all, we are indebted to the U.S. National Science Foundation, for being an amazing institution, where supporting and facilitating the best science is what they do. Thank you, all!

LIGO – SOME KEY MEMBERS

We are especially indebted to the LIGO Laboratory, who performed the enabling R&D, the interferometer designs, the construction of LIGO, the commissioning, the operations and upgrades. The LIGO Laboratory has worked closely with the 1200 member LIGO Scientific Collaboration, who have been responsible for the data analysis and science. By necessity, the key members of the LIGO Lab shown below, represent only a partial list.

LIGO Laboratory Directorate

Barry Barish
Former Director

Jay Marx
Former Director

Dave Reitze
Director

Albert Lazzarini
Deputy Director

LIGO Scientific Collaboration Spokespersons

Rai Weiss
Former
Spokesperson

Peter Saulson
Former
Spokesperson

Dave Reitze
Former
Spokesperson

Gabriela Gonzalez
Former
Spokesperson

David Shoemaker
Spokesperson

LIGO Site Heads

Fred Raab
Former
Hanford
Head

Mike Landry
Hanford
Head

Mark Coles
Former
Livingston
Head

Mike Zucker
Former
Livingston
Head

Joe Giaime
Livingston
Head

Advanced LIGO

Dennis Coyne
Chief Engineer

Peter Fritschel
Interferometer
Design

Daniel Sigg
Interferometer
Operations

Valery Frolav
Interferometer
Operations

LIGO Lab – Senior Scientists

Stan Whitcomb
Chief Scientist

Rana Adakari
Future Detectors

Matt Evans
Future Detectors

AlanWeinstein
Open Data

REFERENCES

1. B.P. Abbott et al., *Phys. Rev. Lett.* **116**, 061102 (2016).
2. A. Einstein, *Sitzungsber. K. Preuss. Akad. Wiss.* **1**, 688 (1916).
3. A. Einstein, *Sitzungsber. K. Preuss. Akad. Wiss.* **1**, 154 (1918).
4. M. Riordan, L. Hoddeson, and A. Kolb. (2015). *Tunnel Visions: The Rise and Fall of the Superconducting Super Collider.* U. of Chicago Press. ISBN 978-0-226-29479-7.
5. Shoemaker et al., *Phys. Rev.* **38**, 423 (1988).
6. Abramovici et al., *Physics Letters* A **218** 157-163 (1996).
7. Abbott et al., *Rep. Prog. Phys.* **72** (2009) 076901.
8. J. Aasi et al., *Phys. Rev.* D **87**, 022002 (2013).
9. Advanced LIGO, *Classical and Quantum Gravity*, Vol **32**, Number 5, 2015.
10. B.P. Abbott et al., *Phys. Rev. Lett.* **116**, 061102 (2016).
11. B.P. Abbott et al., *Phys. Rev. Lett.* **116**, 131103 (2016).
12. B.P. Abbott et al., *Phys. Rev* D **93**, 122003 (2016).
13. B.P. Abbott et al., *Phys. Rev Lett.* **116**, 221101 (2016).
14. B.P. Abbot et al., *Astrophysics Journal Letters* **833**, 1 (2016).
15. B.P. Abbot et al., *Phys. Rev. Lett.* **116**, 241103 (2016).
16. B.P. Abbot et al., *Phys. Rev. Lett.* **118**, 221101 (2017).
17. B.P. Abbot et al., *Phys. Rev. Lett.* **119**, 141101 (2017).
18. B.P. Abbot et al., *Phys. Rev. Lett.* 119, 161101 (2017).
19. LIGO, Virgo, 1M2H, Dark Energy Camera, DES, DLT40 las Cumbres Observatory, VINROUGE and MASTER Collaborations, *Nature* **551**, 85 (2017).
20. B.F. Schutz, *Nature* **323**, 310 (1986).
21. Multimessenger Collaboration *Astro Phys. Journal Letters* **848**:L12 (2017).
22. LIGO, Virgo, Fermi and Integral Collaborations *Astro Phys. Journal Letters,* **848**:L13 (2017).
23. LIGO and VIRGO Collaborations, *Astro Phys. Phys. Journal Letters,* **850**:L39 (2017).
24. Miller, John; Barsotti, Lisa; Vitale, Salvatore; Fritschel, Peter; Evans, Matthew; Sigg, Daniel (16 March 2015). *Physical Review* D. **91**.
25. McClelland, David; Evans, Matthew; Lantz, Brian; Martin, Ian; Quetschke, Volker; Schnabel, Roman (8 October 2015). Instrument Science White Paper (PDF) (Report). LIGO Scientific Collaboration. LIGO Document T1500290-v2.
26. M. Punturo, K. Somiya Int. *J. Mod. Phys.* D **22**, 1330010 (2013).

Kip S. Thorne. © Nobel Prize Outreach AB. Photo: A. Mahmoud

Kip S. Thorne

Biography

MY YOUTH

I was born in 1940 in Logan, Utah, USA, a college town of 16,000, nestled in a verdant valley in the Rocky Mountains.

My father, David Wynne Thorne, was a professor of soil chemistry at the Utah Agricultural College (since renamed Utah State University). Over his lifetime he had a major impact, through research and consulting, on arid-land agriculture, not only in the USA but also in the Middle East, Pakistan, and India. He was an intellectual inspiration to me.

My mother, Alison Comish Thorne, with a PhD in economics, aspired to be an academic, too. However, her career was thwarted by Utah's nepotism law that forbad the wife of a University employee from also working for the University; so she devoted most of her life to community organizing and community activism, and to raising and mentoring five children. Her lifetime impact on the community led the University to award her an honorary doctorate in 2000, when she was 86; and in 2004 when she died, a giant headline in the local newspaper, the Herald Journal, read "Old Radical Dies".

Our parents encouraged my siblings and me to pursue our own interests, treasure our individuality, think for ourselves, and not automatically accept the dictates of the culture in which we lived. This and much more about my youth are described in my Mother's autobiography, *Leave the Dishes in the Sink: Adventures of an Activist in Conservative Utah.*

As a small boy, watching plows create snow banks as high as 3 meters in front of our home, I aspired to become a snow plow driver. Then, when I was eight, my mother took me to a lecture about the solar system at the local Mormon church (Logan's fifth ward), a lecture by a professor from the University. I was enthralled, so my Mother suggested we make a model of the solar system on the sidewalk alongside our home. We drew the sun as a circle four and a half feet in diameter (about a meter), and then she showed me how, mathematically, to take the solar system's

Figure 1. Back row: Me (age 11), my father Wynne, and my sister Barrie; front row: my mother Alison and siblings Sandra, Lance (the baby) and Avril; in Logan, Utah, summer 1951.

actual dimensions and scale them down to this 4.5-foot sun. With our calculations completed we drew each planet as a circle at the appropriate distance from our sun. It was amazing to me: the Earth was a half-inch diameter circle a bit beyond the fourth home north of ours; and Pluto was a tiny circle about 3 miles away, in North Logan. I was hooked. I began to devour everything I could find about astronomy in local libraries and bookstores.

Five years later I discovered, in a bookstore in Salt Lake City, a paperback edition of *One, Two, Three, ..., Infinity* by the physicist George Gamow. It dazzled me. It revealed the role of astronomy as a subfield of physics, the role of mathematics as the language of physics, the beauty of Einstein's relativity, and the power of physical laws to explain the universe. I read it three times and decided I wanted to become a physicist, pursuing a quest to understand the universe. Fourteen years later, when I had started publishing my own research, George Gamow sent me a letter inquiring about ideas in one of my publications. Thrilled, I wrote back, telling him I was a physicist because of having read his book three times. In response, he sent me a copy of *One, Two, Three, ... Infinity* in Turkish, with an inscription "To Kip so that he would not be able to re-read it a 4th time". That book remains one of my most treasured possessions.

My mother encouraged each of her other children to pursue their own chosen dreams. My sister Barrie, two years younger than I, became a professor of sociology. My sister Sandra, eight years younger, became one of the first female forest rangers for the US Forest Service. My sister Avril, nine years younger, became a professor of psychology. And my brother, Lance, eleven years younger, became an artist in wood.

Our ancestors, on all genealogical lines, joined the Mormon church and migrated to Utah on foot, on horseback, or in covered wagons before the railroad arrived (1869). Throughout my youth, our parents, Alison and Wynne, taught an adult Sunday school class, focusing on comparisons between Mormon thought and culture, and other religions and the ideas of great philosophers.

For me as a youth, Logan and its Mormon culture and history provided an idyllic environment, and I still treasure my Mormon roots. However, in my teen age years, as I learned more and more about science and discovered its power for explaining Nature and the Universe in testable and tested ways, and for producing technology that can improve dramatically the lives of people, and as I contrasted this with the more magical and less verifiable character of religion, I gradually lost interest in religion and even in whether God exists. (Much later, when my mother was 75 years old, at her urging, she and all her children resigned our membership in the Mormon Church, because of the church's discrimination against women. My sister Barrie had already been excommunicated for her feminist activities.)

As a teenager in the 1950s, I had an active social life. I played saxophone and clarinet in a dance band, participated in exhibition dancing, edited the high school year book, and was on the high school debate squad, partnering with the future All-American and All-Pro football player Merlin Olsen. But my deepest passion continued to be physics. While others were building telescopes, I – having been captivated by Mr. Thomas's high school course on axiomatic Euclidean geometry in two and three dimensions – formulated it in four dimensions. I recall my excitement upon discovering, through a sequence of lemmas and a theorem, that in four dimensions the intersection of two planes is generically a point, not a plane.

In the summer of my eighth birthday I was at loose ends, so my mother sent me to typing school at Logan High School. "This may be useful to you someday," she said; and indeed it became very useful decades later, in the era of computers. In my fifteenth summer my parents enrolled me in a geology course and an analytic geometry course at the University, opening my eyes to phenomena I had not dreamed existed. Thereafter, throughout high school, I continued taking an occasional university course.

MY UNIVERSITY STUDENT YEARS

Despite my university experience as a teenager, when I arrived at Caltech as a freshman in September 1958, I found myself overwhelmed. I had had no calculus, I was a slow reader, and it quickly became evident that my thinking was slower than that of most other Caltech freshmen. I stumbled and struggled for a year and a half, but gradually developed my own ways of mastering the physics and mathematics that were coming at me like water from the proverbial fire hose. Most valuable of all was a series of notebooks that I developed for myself – one for each major class that I took. In each I wrote down the most important ideas and results I was learning, in my own words and equations, and formulated my own mathematical proof and/or physical explanation for each major result. I continued this through graduate school, then abandoned it for about 15 years, and then started up again in the late 1970s, when I was trying to master new topics and tools relevant to astrophysics and to gravitational-wave experiment. I still find myself consulting those old notebooks from time to time.

By the middle of my sophomore year at Caltech, I got my feet under myself and started enjoying my studies thoroughly, and started moving through difficult material at a reasonable pace.

In the summers before my first, second, and third years of college (1958, 1959, and 1960), I worked as an engineer's assistant in the Great Salt Lake Desert, designing solid propellant rocket engines for the Thiokol Chemical Corporation's Minute Man Intercontinental Ballistic Missile – engines that would later power the space shuttle. This gave me my first taste of "big science", it showed me how various components of an R&D program should come together on a predefined time schedule, and it showed me how Nature can confound a research program: hot, turbulent gas swirling near the entrance to the rocket nozzles kept eroding the slowest-burning solid propellant (the "inhibitor") so rapidly that the turbulence ate into the rocket casing, blowing the nozzles off the engine. The explosions, in test after test of our evolving design, were spectacular and frustrating.

In the summer before my fourth college year (1961), I got a job doing theoretical astrophysics research under the inspiring mentorship of the astronomer Jesse Greenstein. The result was my first published paper, on "The Theory of Synchrotron Radiation from Stars with Dipole Magnetic Fields".

Ever since reading *One, Two, Three, ... Infinity*, I had been fascinated by relativity. During my fourth year at Caltech I decided that was the direction I wanted to go for my PhD, so I spent many hours in the Caltech physics library trying to read relativity articles in research journals such as *Reviews of Modern Physics*. It soon became evident that by far the most interesting research on general relativity was being done by John Archibald Wheeler at Princeton University and his students, so I applied there for graduate school – despite Jesse Greenstein's warnings that the only significant application of relativity was the expansion of the universe. In Jesse's view, and that of many other eminent astronomers and physicists of the era, relativity was a dead end.

At Princeton, John Wheeler was an even more inspiring mentor than I expected, and his young associate Charles Misner added to the inspiration. From Wheeler and Misner I learned about black holes, neutron stars, singularities, and geometrodynamics (the ill-understood nonlinear dynamics of curved spacetime). In parallel, I sat in on the weekly research group meetings of Robert Dicke, whose focus was experiments to test general relativity; and there I met and admired postdoc Rainer ("Rai") Weiss.

In that era, when relativity theory was far ahead of experiment and was only weakly tested, I somehow understood that the interface of the theory with experiment could become a fruitful and exciting area of research, so I not only immersed myself in Dicke's experimental-gravity milieux; I also spent much of my first year at Princeton getting hands-on experience with experiment. In the bowels of the Princeton physics building there was a cyclotron (particle accelerator) on which, under the mentoring of assistant professor Edwin Kashy, I explored the internal structure of the nuclei of Rhodium atoms. This was rather far from relativity, but that experience (like my earlier experience with big science at Thiokol) would turn out to be extremely useful later, when I embarked on gravitational wave research.

In the summer of 1963, I spent eight weeks in a relativity summer school at the *École d'Été de Physique Theorique* in the French Alps. There Wheeler and Dicke gave inspiring lectures, and I met gravitational waves in depth for the first time, in lectures by Rainer Sachs (University of Texas) on the elegant, mathematical theory of the waves, and by Joseph Weber (University of Maryland) on his pioneering experimental effort to discover gravitational waves from the distant universe. I hiked with Weber in the surrounding Alps, we talked at length about his experimental program, I became a convert to the importance and possibilities of gravitational wave experiments, and I became rather fond of Weber himself.

I completed my PhD in June 1965 and spent one more postdoctoral year at Princeton, honing my theory research skills. In 1966 Willy Fowler (who would win the 1983 Nobel Prize for explaining the origin of the elements in stars) invited me back to Caltech as a postdoc, and I jumped at the opportunity. In May, while driving from Princeton to Caltech to start my new job, I stopped in Chicago for discussions with Subrahmanyan Chandrasekhar (who would share the 1983 Nobel Prize with Fowler). Over the following decade both Fowler and Chandrasekhar made major contributions to my chosen areas of research and influenced me substantially (Fowler on relativistic stars; Chandrasekhar on black holes and gravitational waves), and both became dear friends of mine.

EARLY YEARS AS A CALTECH PROFESSOR

When I arrived back at Caltech in 1966, there was a paucity of theoretical physics faculty working outside elementary particle theory. Particle theory was in the doldrums and I was bubbling over with research problems involving black holes, neutron stars, and gravitational waves, so a number of outstanding physics graduate students gravitated toward me, looking for interesting research problems. By late winter, although just a postdoc, I had built a research group of five graduate students and was having a wonderful time working with them. Then in the spring, to my great surprise, the University of Chicago – under Chandrasekhar's influence – offered me a tenured associate professorship. To my great joy, Caltech matched the offer, and almost overnight I was a tenured member of the Caltech faculty.

One of the great things about Caltech is the support that the administration and one's colleagues provide to young faculty members, to help them reach their potential. Maintaining a research group of five or six graduate students and several postdocs, as I was doing almost from the outset, is not cheap. Initially most of the expenses were covered by Fowler's research grants from the National Science Foundation (NSF) and from the Office of Naval Research. In 1968, when Fowler became a member of the National Science Board, which oversees NSF, he arranged for me to take over from him as the Principal Investigator on his large NSF grant. Under my leadership, that grant was renewed time and time again over the next forty years and remained my largest source of research funding until my formal retirement in 2009, whereupon my successor, Yanbei Chen, became the grant's Principal Investigator, and remains so today, after several renewals.

My group's initial research topics – black holes, neutron stars and gravitational waves – were all subtopics in a brand-new field called *relativistic astrophysics*. This new field grew out of the discoveries of quasars (1963; Maarten Schmidt at Caltech), pulsars (1967; Tony Hewish and Jocelyn Bell at the University of Cambridge), cosmic X-ray sources (1962; Ricardo Giacconi and colleagues at American Science and Engineering), and the cosmic microwave background radiation (CMB 1964; Arno Penzias and Robert Wilson at Bell Labs, and then Robert Dicke and his group at Princeton). Thanks to these observational discoveries, relativity was suddenly relevant to a whole lot more in the universe than just its expansion. The merger of these discoveries with the theoretical ideas of Wheeler (Princeton), Yakov Borisovich

Zel'dovich (Moscow), Dennis Sciama (Cambridge), Fowler, Chandrasekhar, and others, gave rise to relativistic astrophysics.

Very early in the development of this new field (summer 1965; before moving to Caltech), I attended the Fifth International Conference on General Relativity and Gravitation, in London. There I met and initiated close friendships with a few physicists who would profoundly influence my life and career. Most important, perhaps, were Stephen Hawking (a student of Sciama) and Igor Novikov (a young colleague of Zel'dovich).

Hawking had contracted Amyotrophic Lateral Sclerosis only two years earlier. In London, walking with a cane and talking with modestly mutilated enunciation, he lectured about his recent insights into the big bang. I was mesmerized by his science and also his personality. We talked in the conference corridors and found ourselves kindred spirits. Although, in the subsequent half century, Hawking's research on black holes and the big bang has greatly impacted my gravitational wave work, we have never collaborated on research, and when together we have spent more time discussing life and death and love, than physics; so I shall describe the details of our friendship elsewhere, not here.

In London, Igor Novikov lectured about new insights in relativistic astrophysics that he and Zel'dovich had been developing. I had studied the Russian language as a Caltech undergraduate, and in London I found that my Russian was about as good (or bad!) as Novikov's English, so we stumbled along in a semi-coherent mixture of the two languages, exchanging astrophysics ideas and initiating a friendship that would soon grow strong and deep.

In 1968, with my new Caltech research group beginning to make an impact, I was well prepared to take advantage of the next international conference on general relativity, this time in Tbilisi (Soviet Georgia). There I met Zel'dovich in person for the first time, and Zel'dovich introduced me to Vladimir Braginsky, who was building a research program in gravitational wave experiment at Moscow University in parallel with Joseph Weber's in America. This was the beginning of my career-long research collaboration with the groups of Braginsky (on gravitational waves and experimental tests of relativity) and of Zel'dovich and Novikov (on black holes and neutron stars, and later on wormholes and time travel). To facilitate our collaborations, Braginsky, Novikov and I began traveling back and forth between Moscow and Pasadena with typically one trip per year in one direction or the other – despite the raging cold war. For a few details, see my book *Black Holes and Time Warps: Einstein's Outrageous Legacy.*

During my first dozen years on the Caltech faculty, 1966–1978, gravitational waves were only a modest portion of my group's research portfolio. Our larger foci were black holes, and other astrophysical phenomena where gravity is so strong that it must be described by Einstein's relativity laws rather than Newton's laws – primarily neutron stars and dense, relativistic clusters of stars. My students and postdocs (sometimes with a little help from me) used general relativity to analyze the structures and astrophysical roles of these objects, and also how they would behave when disturbed – their pulsations and their emission of gravitational waves. This fed into the main thrust of our gravitational wave research: our evolving vision for the information that can be extracted from gravitational waves, when they are ultimately detected; and more broadly, our vision for the future of gravitational wave astronomy; see my Nobel Lecture.

In the next to last section of this biography, I describe the style in which we carried out this research. That style included extensive interactions with colleagues from other institutions, including Zel'dovich, Novikov, Braginsky, and also Leonid Grishchuk in Russia; Hawking and Brandon Carter in the UK; Wheeler, Chandrasekhar, Fowler, James Bardeen and James Hartle in the US; and many more.

CALTECH'S EARLY RESEARCH IN GRAVITATIONAL-WAVE EXPERIMENT

In his Part I of our joint Nobel Lecture, Rai Weiss describes the early history of experimental research on gravitational waves, including (very briefly) at Caltech. Here I shall add some details about the genesis and early years of the Caltech experimental effort.

My early ideas about gravitational-wave experiment were influenced profoundly by Vladimir Braginsky. After Weber's 1969 announcement that he might be seeing gravitational waves, Braginsky (1969–1972) was the first other experimenter to build and operate gravitational wave detectors using the "bar" technology that Weber had initiated, and was the first to fail to find the waves that Weber appeared to be detecting (1972), and among the first to move on toward second generation detectors (1974). In 1972, after Rai Weiss wrote his seminal paper proposing the gravitational wave detectors – "gravitational interferometers" – that would ultimately be used in LIGO (see my Nobel Lecture), I turned to Braginsky for insights and advice about future gravitational wave experiments.

It was my many discussions with Braginsky in 1972–1976, as well as those with Weiss, that convinced me gravitational wave detection was truly feasible and led me in 1976 to propose to Caltech that we create a research group working on gravitational wave experiment. My first choice to lead our Caltech group was Braginsky. After many months of struggling with the idea of moving from Moscow to Caltech, he told me *No*. Even if he managed to get himself and his family through the iron curtain to California, the consequences for his professional colleagues and friends left back in Moscow could be dire, he thought.

When I asked Braginsky whom we should go after to lead the Caltech effort, at the top of his list was the same person as Weiss suggested to me: Ronald Drever of the University of Glasgow. Why? Because of Drever's high creativity and his experimental insights. (For example, Drever had already proposed operating the arms of gravitational interferometers as Fabry-Perot cavities, which has turned out to be a major improvement on Weiss's original design.) So, I suggested Drever to the Caltech physics and astronomy faculty, and after many months of learning about him and other candidates, they chose him to initiate our new experimental effort. The Caltech administration made him an offer which after many many more months, in 1979, he ultimately accepted. The next year we recruited Stan Whitcomb from the University of Chicago to assist Drever in leading our experimental effort. (Today Whitcomb is the LIGO Laboratory's Chief Scientist.)

As a precursor to Drever's acceptance, the Caltech administration pledged roughly two million dollars of Caltech's own private funds for the construction of laboratories and equipment for the new experimental group, including, most importantly, funds toward a prototype gravitational interferometer with 40-meter arms.

This was the first substantial investment in gravitational interferometer research by any institution in the US: Neither MIT (Weiss's home institution) nor the National Science Foundation had yet been willing to commit significant funds for such research. With Caltech on board, Weiss, Drever, and I, working with NSF's Richard Isaacson, were able to trigger significant NSF funding from 1979 onward.

[I take great pride in Caltech's early and enthusiastic commitment to this field and unwavering support from the 1970s through today. Caltech's atmosphere of collegiality, intellectual ferment, and easy communication across fields of science, and our administration's enthusiastic efforts to help us find the funding needed for realizing our dreams, have anchored me to Caltech throughout my career, as they also anchored Richard Feynman and many others of my colleagues.]

For me, the late 1970s and early 1980s were a particularly exciting period:

Drever, commuting back and forth between Caltech and Glasgow, made several inventions that would significantly improve gravitational interferometers:

- *power recycling* (recycling unused light back into the interferometer – which was also invented independently by Roland Schilling in Garching, Germany).
- *resonant recycling* (tuning the response of the interferometer to waves of different frequencies by recycling some of the signal back into the interferometer before extracting it. A few years later, Brian Meers improved on Drever's version of this and it got renamed *signal recycling*).

- the *PDH technique for stabilizing the frequency of lasers* (adapted by Drever from an earlier microwave idea by Robert Pound, and then first demonstrated by John Hall and Drever in Hall's lab in Colorado). This is now widely used in other areas of science and technology.

While Drever was inventing and commuting, Whitcomb and the students and postdocs that he and Drever hired were focused on building and perfecting the 40-meter prototype interferometer on the Caltech campus, and with it exploring technical issues that had to be surmounted in any ultimately successful gravitational interferometer.

In parallel, Carlton Caves and my other theory students and I – with very helpful input from Drever and Whitcomb – embarked on *Quantum Nondemolition* research: an effort to devise ways to circumvent the Heisenberg uncertainty principle in gravitational interferometers and other gravitational wave detectors. This effort was triggered by insights from Braginsky, much of it was in collaboration with Braginsky and his group, and it continues to this day; see my Nobel Lecture for details.

LIGO

In 1984 – building on successes with the interferometer prototypes at MIT, Caltech, Glasgow and Garching, and building on a feasibility study for kilometer-sized interferometers that Weiss and his MIT group and Whitcomb had carried out – Drever, Weiss and I founded LIGO as a Caltech/MIT collaboration. MIT was unwilling to make any substantial institutional commitment to LIGO until a few years later, so Caltech became our collaboration's lead institution. Weiss and Barish sketch the subsequent history of LIGO in their parts of our joint Nobel Lecture.

From 1984 to 1987, I served as the "glue" that held our Caltech/MIT collaboration together, mediating between Weiss (who understood clearly that collaboration was essential for success) and Drever (who needed to be in complete control of all he did in order to remain creative and productive, and so had difficulty truly collaborating). It was with great relief that I relinquished my mediation role in 1987, when the three of us turned over the leadership of LIGO to our first director, Robbie Vogt, who quickly molded us into a truly functional, joint Caltech/MIT team.

In the meantime, Braginsky – despite having endorsed Weiss's gravitational-interferometer ideas in the 1970s – focused the energy of his research group unwaveringly on a variant of Weber's "bar" gravitational-wave technology. Braginsky was concerned that, to succeed, gravitational interferometers would have to become extremely complex (which they indeed are today, with 100,000 data channels that monitor their sub-

Figure 2. Me, Ron Drever and Robbie Vogt in 1978 [Credit: Caltech.]

systems and the environment); and he worried that this complexity might ultimately doom the interferometers to failure.

Throughout the late 1970s and the 1980s, Braginsky and I both commuted back and forth between Moscow and California, maintaining a tight collaboration (particularly on quantum nondemolition techniques and technology; see my Nobel Lecture). And throughout this period, Braginsky advised Drever, Weiss, and their colleagues about interferometer R&D and planning. In the late 1980s, when Braginsky saw the progress that was being made with the prototype interferometers and saw the Caltech/MIT plans for a proposal to the NSF to construct LIGO, he became convinced that the probability of success was reasonably high; so he went home to Moscow, shut down his bar-detector research, and initiated in its place a whole new research program in support of LIGO. This had a profound effect on me, bolstering my confidence at just the moment my Caltech/MIT colleagues and I were developing our proposal and plans for LIGO construction.

In the early 1990s, under Vogt's leadership, we secured approval from NSF for LIGO's construction and we took major steps toward construction. Then in 1994–2001, our second director, Barry Barish, transformed LIGO from a small Caltech/MIT project into a large international collaboration, and led us through the construction of LIGO's facilities, the installation of LIGO's first interferometers, and the writing of a proposal for the advanced interferometers that have now succeeded in discovering gravitational waves; see Barish's part II of our joint Nobel Lecture.

In 1992, with LIGO starting to move forward, I wound down other efforts (including the theory of time travel) that my theory research group was doing and I refocused our research almost completely onto theoretical support for LIGO. This included analyzing sources of noise in LIGO's interferometers and ways of controlling the noise (see my Nobel Lecture), a beefed-up effort on quantum nondemolition, and a renewed effort to understand sources of gravitational waves, and the shapes of their waves – their *waveforms.*

It was only then that I began to realize how difficult would be the analysis of LIGO's data – finding weak gravitational-wave signals amidst LIGO's noise, and extracting the information carried by the signals. Fortunately, my former student Bernard Schutz, at the University of Cardiff, UK, had recognized this as early as 1986 and had begun then to lay foundations for the data analysis (see my Nobel Lecture). To bring Caltech up to speed on the data analysis, we imported Bruce Allen from the University of Wisconsin; and he, together with a number of my students and postdocs, dove into the problem while I cheered them on. Soon thereafter, Barish, as LIGO's second director, created the LIGO Scientific Collaboration (LSC), which facilitated expanding the data analysis effort to scientists at many other institutions; see Barish's Part II of our joint Nobel Lecture.

To help educate the many hundreds of scientists who joined the LIGO effort in the late 1990s and the 2000s, I created in 2002 an online course in gravitational-wave physics that included videos of lectures about all aspects of the field, by the best experts.

By 2002, it seemed to me that I was no longer much needed within the LIGO Project. The students and postdocs I had trained, and other LSC theorists, could play the roles that I had been playing, and could do it at least as well I, if not better. So, with a sigh of relief (because by personality I did not really like working in a large project), I left day to day involvement with LIGO, and focused my attention largely on building at Caltech a research effort on computer simulations of colliding black holes and other sources of gravitational waves; see my Nobel Lecture for details.

One consequence of my departure from day-to-day LIGO work was my non-involvement in Advanced LIGO and its triumphant discovery of gravitational waves.

The credit for that ultimate success, and for all the rich insights about the universe that have begun to flow from it, belongs largely to the younger generation of LIGO/Virgo scientists and engineers, and also to my Nobel Prize co-laureates Rai Weiss and Barry Barish, who have continued to make major contributions in the Advanced-LIGO era.

I continue to help the LIGO Project whenever called on for help, but that is less and less often as time passes (and almost entirely on political issues and not technical issues).

MY STUDENTS AND POSTDOCS

Over the near-half-century of my career, my graduate students and postdocs have done much more important and impactful research, while in my group, than I myself. I take great pride in their accomplishments, some of which I describe in my Nobel Lecture.

In many cases they took research problems that I suggested, and with very little help from me, brought the problems into soluble form, solved them, and made major discoveries; an example is the work by Alessandra Buonanno and Yanbei Chen on quantum noise in Advanced LIGO interferometers (see my Nobel Lecture). In other cases, they identified important research problems themselves and, with little concrete input from me, brought the problems to fruition, with impactful results; examples are Carlton Caves' work on the origin of quantum noise and using squeezed vacuum to modify it, and Yuri Levin's work on thermal noise in interferometers (see my Nobel Lecture).

I patterned my style of working with students and postdocs after the styles of Wheeler, Dicke, and Zel'dovich (which I had observed up close) and of Robert Oppenheimer in Oppenheimer's Berkeley/Caltech years (the 1930s). I gave the students a lot of room and time and freedom to explore things on their own, flounder, and ultimately find themselves, with an occasional nudge from me. But I also gave them an intellectual environment in which to learn from each other, and from students and colleagues elsewhere – an environment that included weekly group meetings typically two hours long and sometimes far longer than that, with frequent participation by experimenters from the Drever/Whitcomb group and later the LIGO Laboratory, and by outside experts. It also included frequent trips to Santa Barbara to interact with the superb relativity group that James Hartle had created there, and frequent visits to Caltech by research leaders from around the world – for example, members of Zel'dovich's and Braginsky's groups, and Stephen Hawking and members of his Cambridge research group. We had an *Interaction Room,* with a huge blackboard, a refrigerator filled with drinks, and comfortable couches and chairs, in which we would gather for spontaneous discussions as well as organized discussions.

Over my 43 years of mentoring students and postdocs, roughly 2/3 of our time and effort went into gravitational-wave-related research, largely connected to LIGO or what would become LIGO, but also connected to LISA, Weber-type bar detectors, and sources of gravitational waves in all frequency bands. I describe some of this research in my Nobel Lecture.

The other 1/3 of our time has gone into a wide range of other issues in relativistic astrophysics, or relativity, including a highly enjoyable period of several years in which we asked ourselves whether the laws of physics permit an infinitely advanced civilization to build wormholes for rapid interstellar travel and machines for traveling backward in time. (Although such questions may seem weird or flaky, they are useful tools for probing the laws of physics in domains where experiment is not yet possible. For example, our research, and that of Hawking and his students, have convinced both Hawking and me that the poorly understood laws of quantum gravity control whether or not backward time travel is possible.)

A NEW CAREER AT THE INTERFACE OF SCIENCE AND THE ARTS

Since 2009 I have turned much of my effort in a very different direction: collaborations about science with artists, musicians and film makers. Christopher Nolan's movie *Interstellar* was one fruit of this, and with Stephen Hawking and my long-time Hollywood partner, Lynda Obst, I have a second science-inspired movie in the works. With the painter Lia Halloran, I am working on a book about the Warped Side of the Universe (objects and phenomena made largely or wholly from warped spacetime, most of them sources of gravitational waves). And I have been doing an occasional multimedia concert about the Warped Side of the Universe with composer Hans Zimmer and visual effects gurus Paul Franklin and Oliver James, using beautiful videos generated by numerical relativity physicists. I take great pleasure in these collaborations with brilliant and creative artists, who bring to our joint work talents and insights quite different from my own. These collaborations are my attempt to inspire non-scientists and especially young people about the beauty and power of science, in the same way as George Gamow's book *One, Two Three, ... Infinity* inspired me, 65 years ago.

MY FAMILY

This is a scientific biography, so I have chosen not to discuss my two marriages (to Linda Thorne, 1960–1975; and then to Carolee Winstein, 1984–...), nor Linda's and my children Kares Anne Thorne and Bret Carter Thorne (and his wife Regine Thorne), and granddaughter Larisa Anne Thorne. Suffice it to say that they all have been tremendously important in my life and have provided a balance to my scientific work that has helped make me more productive. They all went to Stockholm with me to share in the Nobel Week festivities.

Figure 3. My family in Stockholm, December 2017: Bret, Regine, Carolee, me, Linda, Larisa, and Kares. [Credit: The Nobel Foundation.]

Ligo and the Discovery of Gravitational Waves, III

Nobel Lecture, December 8, 2017 by Kip S. Thorne
California Institute of Technology, Pasadena, CA, USA.

INTRODUCTION AND OVERVIEW

The first observation of gravitational waves, by LIGO on September 14, 2015, was the culmination of a near half century effort by ~1200 scientists and engineers of the LIGO/Virgo Collaboration. It was also the remarkable beginning of a whole new way to observe the universe: gravitational astronomy.

The Nobel Prize for "decisive contributions" to this triumph was awarded to only three members of the Collaboration: Rainer Weiss, Barry Barish, and me. But, in fact, it is the entire collaboration that deserves the primary credit. For this reason, in accepting the Nobel Prize, I regard myself as an icon for the Collaboration.

Because this was a collaborative achievement, Rai, Barry and I have chosen to present a single, unified Nobel Lecture, in three parts. Although my third part may be somewhat comprehensible without the other two, readers can only fully understand our Collaboration's achievement, how it came to be, and where it is leading, by reading all three parts. Our three-part written lecture is a detailed expansion of the lecture we actually delivered in Stockholm on December 8, 2017.

In Part 1 of this written lecture, Rai describes Einstein's prediction of gravitational waves, and the experimental effort, from the 1960s to 1994, that underpins our discovery of gravitational waves. In Part 2, Barry describes the experimental effort from 1994 up to the present (including our first observation of the waves), and describes what we may expect as the current LIGO detectors reach their design sensitive in about 2020 and then are improved beyond that. In my Part 3, I describe the role of theorists and theory in LIGO's success, and where I expect gravitational-wave astronomy, in four different frequency bands, to take us over the next several decades. But first, I will make some personal remarks about the early history of our joint experimental/theoretical quest to open the first gravitational-wave window onto the universe.

SOME EARLY PERSONAL HISTORY: 1962–1976[1]

I fell in love with relativity when I was a teen age boy growing up in Logan Utah, so it was inevitable that I would go to Princeton University for graduate school and study under the great guru of relativity, John Archibald Wheeler. I arrived at Princeton in autumn 1962, completed my PhD in spring 1965, and stayed on for one postdoctoral year. At Princeton, Wheeler inspired me about black holes, neutron stars, and gravitational waves: relativistic concepts for which there was not yet any observational evidence; and Robert Dicke inspired and educated me about experimental physics, and especially experiments to test Einstein's relativity theory.

In the summer of 1963 I attended an eight-week summer school on general relativity at the *École d'Été de Physique Theorique* in Les Houches, France. There I was exposed to the elegant mathematical theory of gravitational waves in lectures by Ray Sachs, and to gravitational-wave experiment in lectures by Joe Weber. Those lectures and Wheeler's influence, together with conversations I had with Weber while hiking in the surrounding Alpine mountains, got me hooked on gravitational waves as a potential research direction. So it was inevitable that in 1966, when I moved from Princeton to Caltech and began building a research group of six graduate students and three postdocs, I focused my group on black holes, neutron stars, and gravitational waves.

My group's gravitational-wave research initially was quite theoretical. We focused on gravitational radiation reaction (whether and how gravitational waves kick back at their source, like a gun kicks back when firing a bullet). More importantly, we developed new ways of computing, accurately, the details of the gravitational waves emitted by astrophysical sources such as spinning, deformed neutron stars, pulsating neutron stars, and pulsating black holes. Most importantly (relying not only on our own group's work but also on the work of colleagues elsewhere) we began to develop a *vision for the future of gravitational wave astronomy:* What would be the frequency bands in which observations could be made, what might be the strongest sources of gravitational waves in each band, and what information might be extractable from the sources' waves.

Figure 1. John Wheeler, Robert Dicke, and Joseph Weber. Credit: Wheeler: AIP Emilio Segrè Visual Archives, Wheeler Collection. Dicke: Department of Physics, Princeton University. Weber: AIP Emilio Segrè Visual Archives.

We described this evolving vision in a series of review articles, beginning with one by my student Bill Press and me in 1972,[2] and continuing onward every few years until 2001,[3] when, with colleagues, I wrote the scientific case for the Advanced LIGO gravitational wave interferometers.[4]

Particularly important to our evolving vision was the extreme difference between the electromagnetic waves with which astronomers then studied the universe, and the expected astrophysical gravitational waves:

- Electromagnetic waves (light, radio waves, X-rays, gamma rays, ...) are oscillating electric and magnetic fields that propagate through spacetime. Gravitational waves, by contrast, are oscillations of the "fabric" or shape of spacetime itself. The physical character of the waves could not be more different!
- Electromagnetic waves from astrophysical sources are almost always incoherent superpositions of emission produced by individual charged particles, atoms, or molecules. Astrophysical gravitational waves, by contrast, are emitted coherently by the bulk motion of mass or energy. Again, the two could not be more different.
- Astrophysical electromagnetic waves are all too easily absorbed and scattered by matter between their source and Earth. Gravitational waves are never significantly absorbed or scattered by matter, even when emitted in the earliest moments of the Universe's life.

These huge differences implied, it seemed to me, that

- Many gravitational wave sources will not be seen electromagnetically.
- Just as each new electromagnetic frequency band (or "window") when opened — radio waves, X-rays, gamma rays, ...— had brought

great surprises due to the difference between that band and others, so gravitational waves with their far greater difference from electromagnetic waves, are likely to bring even greater surprises.
- Indeed, gravitational astronomy has the potential to revolutionize our understanding of the universe.

In 1972, while Bill Press and I were writing our first vision paper, Rai Weiss at MIT was writing one of the most remarkable and prescient papers I have ever read.[5] It proposed an L-shaped laser interferometer gravitational wave detector (*gravitational interferometer*) with free swinging mirrors, whose oscillating separations would be measured via laser interferometry. The bare-bones idea for such a device had been proposed earlier and independently by Michael Gertsenshtein and Vladislav Pustovoit in Moscow,[6] but Weiss and only Weiss identified the most serious noise sources that it would have to face, described ways to deal with each one, and estimated the resulting sensitivity to gravitational waves. Comparing with estimated wave strengths from astrophysical sources, Rai concluded that such an interferometer with kilometer-scale arm lengths had a real possibility to discover gravitational waves. (This is why I regard Rai as the primary inventor of gravitational interferometers.)

Rai, being Rai, did not publish his remarkable paper in a normal physics journal. He thought one should not publish until after building the interferometer and finding gravitational waves, so instead he put his paper in an internal MIT report series, but provided copies to colleagues.

I heard about Rai's concept for this gravitational interferometer soon after he wrote his paper and while John Wheeler, Charles Misner, and I were putting the finishing touches on our textbook *Gravitation*[7] and preparing to send it to our publisher. I had not yet studied Rai's paper nor

Figure 2. Rainer Weiss ca. 1970. Credit: Rainer Weiss.

discussed his concept with him, but it seemed very unlikely to me that his concept would ever succeed. After all, it required measuring motions of mirrors a trillion times smaller (10^{-12}) than the wavelength of the light used to measure the motions — that is, in technical language, splitting a fringe to one part in 10^{12}. This seemed ridiculous, so I inserted a few words about Rai's gravitational interferometer into our textbook, and labeled it "not promising".

Over the subsequent three years I learned more about Rai's concept, I discussed it in depth with him (most memorably in 1975, in an all-night-long conversation in a hotel room in Washington, D.C.), and I discussed it with others. And I became a convert. I came to understand that Rai's gravitational interferometer had a real possibility of discovering gravitational waves from astrophysical sources.

I was also convinced that, if gravitational waves could be observed, they would likely revolutionize our understanding of the universe; so I made the decision that I and my theoretical-physics research group should do everything possible to help Rai and his experimental colleagues discover gravitational waves. My major first step was to persuade Caltech to create an experimental gravitational-wave research group working in parallel with Rai's group at MIT.

Rai sketches the rest of this history, on the experimental side, in his Part I of our Nobel Lecture, and I recount some of it in my Nobel biography. I now sketch the theory side of the subsequent history.

SOURCES OF GRAVITATIONAL WAVES

When Bill Press and I wrote our 1972 vision paper, our understanding of gravitational wave sources was rather muddled, but by 1978 the relativistic astrophysics community had converged on a much better understanding. The convergence was accelerated by a two week *Workshop on Sources of Gravitational Waves* convened by Larry Smarr in Seattle, Washington in July–August 1978. The participants included almost all of the world's leading gravitational-wave theorists and experimenters, plus a number of graduate students and postdocs: Figure 3.

Some conclusions of the workshop were summarized in diagrams depicting the predicted gravitational-wave strain h as a function of frequency f for various conceivable sources:[8] three diagrams, one for short-duration ("burst") waves, one for long-duration, periodic waves (primarily from pulsars and other spinning, deformed neutron stars), and one for stochastic waves (primarily, we thought then, superpositions of emission from many discrete sources). Most relevant to this lecture is the segment of the burst-wave diagram that covers LIGO's frequency band: Figure 4.

Figure 3. Participants in the 1978 workshop on gravitational waves. Credit: Larry Smarr.

The waves here depicted are from:

- *Supernovae* (SN), that is, the implosion of the core of a normal star to form a neutron star, releasing enormous gravitational energy that blows off the normal star's outer layers.
- *Compact-binary destruction* (CBD), that is, the inspiral and merger of binaries consisting of two black holes, two neutron stars, or a black hole and a neutron star.

The supernova line in the figure was an estimated upper limit on the strengths of the waves from supernovae. More modern estimates predict waves much weaker. The box labeled CBD was the range in which the strongest compact-binary waves were expected.

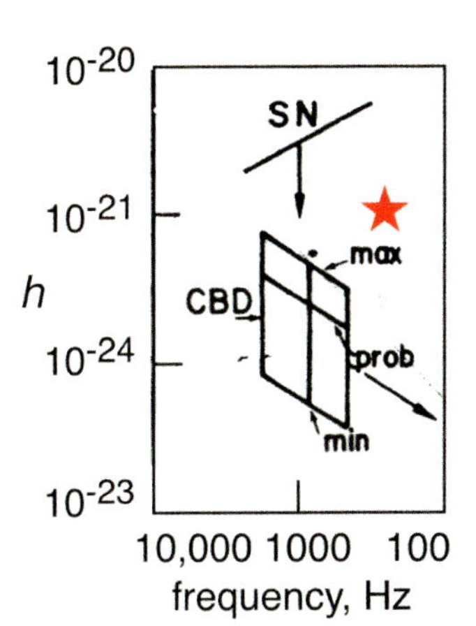

Figure 4. Segment of 1978 burst-source diagram.

Looking at this figure, we Workshop participants concluded that the strongest gravitational wave burst reaching Earth each year would have an amplitude of roughly $h \sim 10^{-21}$; and I (mis)remember that in our enthusiasm for this goal, we had T-shirts made up with the logo on them "10^{-21} or bust". However, colleagues with better memories than mine assure me we only discussed such T-shirts; but they were not made.

The first wave burst that LIGO finally detected, in 2015, was at the location of the red star, which I have added to this figure, and was from CBD: the inspiral and merger of two black holes (a "binary black hole" or BBH). Its amplitude was precisely 10^{-21} and its frequency was about 200 Hz — a bit stronger strain h and lower frequency than our 1978 estimates. This agreement of prediction and observation is partially luck. Our level of knowledge in 1978 was much lower than it suggests.

By 1984, when Weiss, Drever and I were co-founding the LIGO Project, I thought it likely that the strongest waves LIGO would detect would come from the merger of binary black holes (as did happen). My reasoning was simple:

- The amplitude of a compact binary's gravitational-wave strain h is proportional to the binary's mass (if its two objects have roughly the same mass).
- Therefore the distance to which LIGO can see it is also proportional to its mass (so long as the waves are in LIGO's frequency band, which means for binary masses between a few suns and a few hundred suns, i.e. "stellar-mass" compact binaries).
- Correspondingly, the volume within which LIGO can see such binaries is proportional to the cube of the binary's mass.
- The masses of then-known stellar-mass black holes were as much as ten times greater than those of neutron stars, so the volume searched would be 1000 times greater than for neutron stars.
- It seemed likely to me that this factor 1000 would outweigh the (very poorly understood) lower number of BBH in the universe than binary neutron stars, BNS.

Although this was just a guess, in planning for LIGO it led us to lay heavy emphasis on binary black holes, as well as on the much better understood binary neutron stars.

By 1989 when, under the leadership of Rochus (Robbie) Vogt, we wrote our construction proposal for LIGO[9] and submitted it to NSF, gravitational waves from compact binaries were central to our arguments for how sensitive our gravitational interferometers would have to be. The estimated event rates and strengths were so crucial to the scientific case for LIGO that we thought it essential to rely on rate estimates from astrophysicists who had no direct association with our project. For binary neutron stars (BNS), those estimates[10] (based on the statistics of observed binary pulsars in our own Milky Way galaxy) placed the nearest BNS merger each year somewhere in the range of 60 to 200 Mpc, with a most likely distance of 100 Mpc (320 million light years), and a signal strength as shown by the blue, arrowed line in Figure 5. (In 2017, when the first BNS was observed, its distance was about 40 Mpc — somewhat closer than expected — and its strength was as shown by the red, arrowed line in the figure.) For BBH merger rates, the uncertainties in 1989 remained so great that we did not quote estimates. (The first BBH seen, in 2015, was as shown by the red star.)

In the 1990s and 2000s, astrophysicists made more reliable estimates

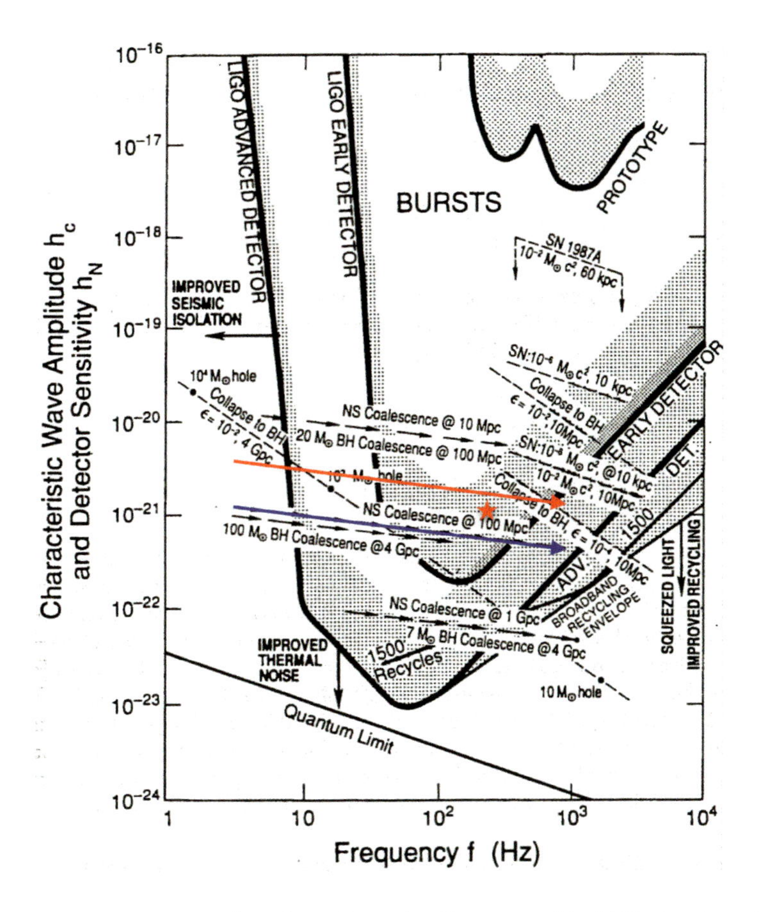

Figure 5. Figure A-4a from the 1989 construction proposal for LIGO, showing estimates of noise curves (solid) for Initial and Advanced LIGO interferometers, and the estimated strengths of waves from various sources. The tops of the stippled regions are the strength that a signal would need for confident detection with Gaussian noise and optimal signal processing. The quantum limit is for 1000 kg mirrors.

of BBH and BNS waves, with less than a factor 2 change in the BNS distances, and with the distance for the nearest BBH getting narrowed down to a factor ~10 uncertainty (~1000 uncertainty in the rate of bursts).[11]

INFORMATION CARRIED BY GRAVITATIONAL WAVES, AND COMPUTATION OF GRAVITATIONAL WAVEFORMS

Observables from a Compact Binary's Inspiral Waves

In 1986 Bernard Schutz[12] (one of the leaders of the British-German gravitational-wave effort) identified the *observables* (parameters) that can be extracted from the early inspiral phase of a compact binary's gravitational waves. From the gravitational-wave strain h as a function of time t, $h(t)$, measured at several locations on Earth, one can infer, he deduced:

- The direction to the binary.
- The inclination of its orbit to the line of sight.
- The direction the two objects move around their orbit.
- The *chirp mass*, $M_c = (M_1 M_2)^{3/5} / (M_1+M_2)^{1/5}$ (where M_1 and M_2 are the individual masses).
- The distance r from Earth to the binary (more precisely, in technical language, the binary's *luminosity distance*).

It is remarkable that gravitational astronomy gives us the binary's distance r but not its redshift z (fractional change in wavelengths due to motion away from Earth), whereas electromagnetic astronomy, looking at the same binary, can directly measure its redshift but not its distance. In this sense, gravitational and electromagnetic observations are complementary, not duplicative.

Figure 6. Bernard Schutz. Credit: Bernard F. Schutz.

The relationship between distance and redshift, $r(z)$, is crucial observational data for cosmology; for example, if the binary is not too far away, $r(z)$ determines the Hubble expansion rate of the universe today. Therefore, as Schutz emphasized, for binary neutron stars it should be possible to observe both the binary's gravitational waves (distance) and its electromagnetic waves (redshift) and thereby explore cosmology. That is precisely what happened in 2017 with LIGO's discovery of its first BNS, GW170817; see Barish's Part II of this lecture.

(In 1986, having identified the gravitational-wave observables for compact binaries, Schutz then started laying foundations for the analysis of data from gravitational interferometers.[13] He became the intellectual leader of this effort in the early years, before I or anyone else in LIGO began thinking seriously about data analysis. For some discussion of LIGO data analysis, see Weiss's and Barish's Parts I and II of this lecture.)

As a compact binary spirals inward due to radiation reaction, the strength of the mutual gravity of its two bodies grows larger, their speeds grow higher, and correspondingly, relativistic effects (deviations from Newton's laws of gravity) become stronger. This presents a *problem* (the need to compute relativistic corrections to the binary's waveforms), and an *opportunity* (the possibility that those corrections, when observed, will bring us additional information about the binary and can be used to test general relativity in new ways).

Post-Newtonian Approximation for Computing Inspiral Waveforms

The relativistic corrections are computed, in practice, using the *post-Newtonian approximation* to general relativity: a power-series expansion in powers of the bodies' orbital velocities v and their Newtonian gravitational potential $\Phi \sim v^2$. Motivated by the astronomical importance of these waveform corrections, several efforts were mounted to compute them beginning in the 1970s, and then the efforts accelerated in the 1980s, 1990s, and 2000s. I estimate that many more than 100 person years of intense work were put into this effort. The leading contributors included, among others, Luc Blanchet, Thibault Damour, Bala Iyer, and Clifford Will; and by now the computations have been carried up to order v^7 beyond Newton's theory of gravity.[14] As expected, at each higher order in the computation, there are new observables that can be extracted from the observed waves. These include, most importantly, the individual masses M_1 and M_2 of the binary's two bodies, and their vectorial spin angular momenta; and, if the binary's orbit is not circular, then its evolving ellipticity and elliptical orientation, and relativistic deviations from elliptical motion. And at each order, there are new opportunities to test, observationally, Einstein's general relativity theory — tests that are now being carried out with LIGO's observational data.[15]

Figure 7. Luc Blanchet, Thibault Damour, Bala Iyer, and Clifford Will. Credits: Blanchet: Luc Blanchet. Damour: Thibault Damour. Iyer: Bala Iyer. Will: Clifford M. Will.

Numerical Relativity for Computing Merger Waveforms

When the relative velocity of the binary's two bodies approaches 1/3 the speed of light and the bodies near collision, the post-Newtonian approximation breaks down. This, again, presents a *problem* (how to compute the waveforms) and an *opportunity* (new information carried by the waveforms).

The only reliable way to compute the waveforms in this collision epoch is by numerical simulations: solving Einstein's general relativistic field equations on a computer — *numerical relativity*. For this reason, in the 1980s I began urging my numerical relativity colleagues to push forward vigorously on such simulations.

Simulating BBHs was especially important, for several reasons:

- For neutron stars, with their small masses (about 1.4 suns each), the waves from the collision epoch are at such high frequencies that they will be difficult for LIGO to detect and monitor; almost all of the signal strength and extractable information will come from lower frequencies, where the post-Newtonian approximation is accurate.
- For black holes, by contrast, the collision epoch can produce waves at frequencies where LIGO is most sensitive. (That is precisely what happened with LIGO's first observed wave burst, GW150914; almost all of its signal strength came from the collision epoch, which could be analyzed only via numerical relativity).
- The waveforms from BBH collision and merger carry detailed information about *geometrodynamics*: the nonlinear dynamics of curved spacetime — about which we knew very little in the 1980s and 90s.

Figure 8. John Wheeler lecturing about geometrodynamics and related issues at Willy Fowler's 60th birthday conference in August 1971, in Cambridge England. Fowler is the Nobel Laureate with the shiny bald head in the front row. Credit: Kip Thorne.

In the late 1950s and early 1960s, John Wheeler identified geometrodyamics as tremendously important. It is the arena where Einstein's general relativity should be most rich, and deviations from Newton's laws of gravity should be the greatest. Black hole collisions, Wheeler argued, would be an ideal venue for studying geometrodynamics. Recognizing the near impossibility of exploring geometrodynamics analytically, with pencil and paper, Wheeler encouraged his students and colleagues to explore it via computer simulations.

With this motivation, Wheeler's students and colleagues began laying foundations for BBH simulations: In 1959–1961, Charles Misner, Richard Arnowitt and Stanley Deser[16] brought the mathematics of Einstein's equations into a form nearly ideal for numerical relativity, and Misner analytically solved the *initial-value or constraint* part of these equations to obtain a mathematical description of two black holes near each other and momentarily at rest.[17] Then in 1963, Susan Hahn and Richard Lindquist[18] solved the full Einstein equations numerically, on an IBM 7090 computer, and thereby watched the two black holes fall head-on toward each other and begin to distort each other. Sadly, Hahn and Lindquist could not compute long enough to see the holes' collision and merger, nor the gravitational waves that were emitted.

These calculations were picked up in the late 1960s, with some change in the detailed formulation, by Bryce DeWitt and DeWitt's student Larry Smarr, and were brought to fruition by Smarr and *his* student Kenneth Eppley in 1978.[19] In these simulations the two holes collided head on and merged to form a single, highly distorted black hole that vibrated a few times (rang like a damped bell), emitting a burst of gravitational waves, and then settled down into a quiescent state. Here we had, at last, our first example of geometrodynamics.

But head-on collisions should occur rarely, if ever, in Nature. When

Figure 9. Charles Misner, Richard Lindquist, Bryce DeWitt, Kenneth Eppley, and Larry Smarr. I have not been able to find a photo of Susan Hahn. Credits: Misner: Charles W. Misner. Lindquist: Wesleyan University Library, Special Collections & Archives. DeWitt: Kip Thorne. Eppley & Smarr: Larry Smarr.

two black holes or stars orbit each other, gravitational radiation reaction drives their orbit into a circular form rather quickly, so BBH collisions and mergers should almost always occur in circular, inspiraling orbits. The big challenge for the 1980s and 1990s, therefore, was to simulate BBHs with shrinking, circular orbits.

This was so difficult that by 1992 only modest progress had been made. To accelerate the progress, Richard Isaacson (the NSF program director who had nurtured the LIGO experimental effort with great skill, see Weiss's Part I of this lecture) urged all the world's numerical relativity groups to collaborate on this problem, at least loosely. Richard Matzner of the University of Texas at Austin led this *Binary Black Hole Grand Challenge Alliance,* and I chaired its advisory committee. To generate collegiality and speed things up, in 1995 I bet many of the Alliance's members that LIGO would observe gravitational waves from BBH mergers before numerical relativists could simulate the mergers; see Figure 10. I fervently hoped to lose, since the simulations would be crucial to extracting the information carried by the observed waves.

By early 2002, the Alliance had made much progress, but was still unable to simulate a full orbit of two black holes around each other. The computer codes would crash before an orbit was complete, and I was worried I might win the bet.

Alarmed, I left day to day involvement in the LIGO project and focused on helping push numerical relativity forward. Together with Lee Lindblom, I created a numerical relativity research group at Caltech, as an extension of the group I respected most: that of Saul Teukolsky at Cornell. With the help of private funding from the Sherman Fairchild Foundation, we grew our joint Cornell/Caltech *Program to Simulate eXtreme Spacetimes* (SXS) to the size we thought was needed for success: about 30 researchers.

The SXS program's first great triumph arose not, however, from the

Kip Thorne hereby wagers that LIGO will discover convincing gravitational waves from black hole coalescence before the numerical relativity community has a code capable of computing merger waveforms, to 10 per cent accuracy, as determined by internal computational consistency, for coalescences with random spin directions and magnitudes and random mass ratios in the range 1:1 to 10:1. The signatories below wager that Kip is wrong.

The loser(s) will supply a bottle or bottles of wine, value not less than \$100, to be consumed by the winner(s) and loser(s) together.

Agreed to this 17th day of July, 1995 in Austin, Texas by:

Kip S. Thorne

Richard Matzner
Wai-Mo Suen
Ed Seidel
Mark Scheel
Lawrence E. Kidder
Gregory B. Cook
Luciano Rezzolla
Mark Miller
Larry Shepley
Shyamal Mitra
Manoj Maharaj
Daniel Holz
Pablo Laguna
Roberto Gomez

Jörg Frauendiener
Dierdre Shoemaker
Bernd Brügmann
Béla Szilágyi
Nigel Bishop
Sascha Husa
Jeff Winicour
Mijan Huq
Luis Lehner
Robert Marsa
Scott Klasky
Marcus Berg
Juan F. Lara
Ethan Honda

Richard Matzner

Figure 10. My bet with Richard Matzner (photo) and members of his Binary Black Hole Grand Challenge Alliance. Credit: Matzner: Richard Matzner.

collaborative work of the SXS team. Rather it was a single-handed triumph by Franz Pretorius, an SXS postdoc. In June 2005, Franz cobbled together a set of computational techniques and tools into a single computer code that successfully simulated the orbital inspiral, collision, and merger of a BBH, one whose black holes were identical and not spinning.[20] Six months later, two other small research groups achieved the same thing, using rather different techniques and tools: a group led by Joan Centrella at NASA's Goddard Spaceflight Center, and another led by Manuela Campanelli at the University of Texas at Brownsville.[21] I heaved a sigh of relief; perhaps I would actually lose my bet!

But we were still a long way from meeting LIGO's needs: It was necessary to simulate BBHs whose two black holes have masses that differ by as much as a factor of 10, and spin at different rates and in different directions. And these simulations had to be carried out with a computer code that was highly stable and robust, and had a well calibrated accuracy that matched LIGO's needs. And it was necessary to carry out a large suite of simulations that covered the full range of parameters to be expected for LIGO's observed sources — seven non-trivial parameters: the ratio of the

Figure 11. Franz Pretorius, Manuela Campanelli, Joan Centrella, and Saul Teukolsky. Credits: Pretorius: New York Academy of Sciences. Campanelli: A. Sue Weisler / RIT. Centrella: Dwight Allen. Teukolsky: Saul A. Teukolsky.

holes' masses, and the three components of the vectorial spin of each black hole. We estimated that about a thousand simulations would be needed in preparation for LIGO's early BBH observations.

To achieve this goal, Teukolsky led the SXS team in constructing a code based on a formulation of Einstein's equations that is strongly hyperbolic and uses spectral methods — technical details that guarantee the code's accuracy will improve *exponentially fast* as the coordinate grid is refined. The resulting SXS code is called *SpEC for Spectral Einstein Code.*[22]

SpEC was far more difficult to write and perfect than the Pretorius, Centrella, and Campanelli codes, or codes created by several other numerical relativity groups (notably Bernd Brugman's group in Jena, Germany, and Pablo Laguna's Georgia Tech code, which grew out of Matzner's Texas effort). The other codes were perfected several years before SpEC and made major discoveries about geometrodynamics while SpEC was still being perfected. But SpEC did reach perfection a few years before LIGO's first BBH observation and then was used to begin building the large catalog of BBH waveforms to underpin LIGO data analysis;[23] and now that we are in the LIGO observational era, only SpEC has the speed and accuracy to fully meet LIGO's near-term needs.[24] And with great relief, I have conceded the bet to my numerical relativity colleagues.

Interfacing the output of the numerical relativity codes with LIGO data analysis was a major challenge. The interface was achieved by a quasi-analytic model of the BBH waveforms called the *Effective One Body* (EOB) Formalism, which was devised by Alessandra Buonanno and Thibault Damour;[25] and also achieved by the quasi-analytic *Phenomenological Formalism*, devised by Parameswaran Ajith and colleagues.[26] The numerical-relativity waveforms were used to tune parameters in these formalisms, which then were used to underpin the LIGO data analysis algorithms that discovered the BBH waves and did a first cut at extracting their information. The final extraction of information is most accurately done by direct comparison with the SpEC simulations.

Geometrodynamics in BBH Mergers

Just as I did not play a role in LIGO's experimental R&D, so also I did not play any role at all in formulating and perfecting the SXS computer code SpEC. My primary role in both cases was more that of a visionary. For SpEC a big part of that vision was inherited from Wheeler: Use SpEC simulations of BBHs to predict the geometrodynamic excitations of curved spacetime that are triggered when two black holes collide, and then use LIGO's observations to test those predictions.

By 2011, SpEC was mature enough to start exploring geometrodynamics. To assist in those explorations, we developed several visualization tools.

The first was a *pseudo-embedding diagram* (Figure 12), developed by

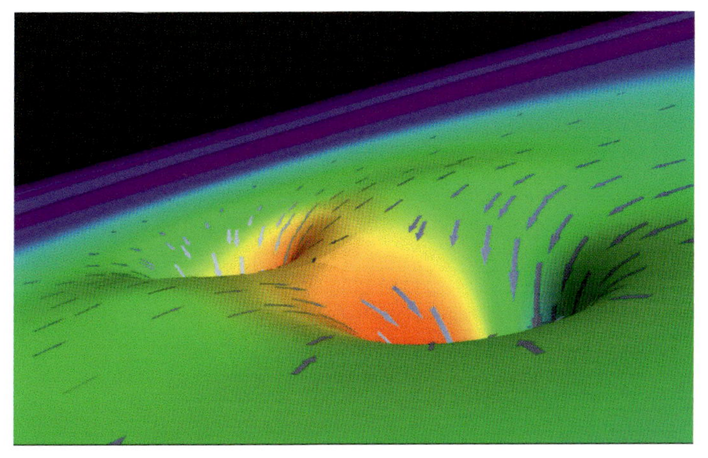

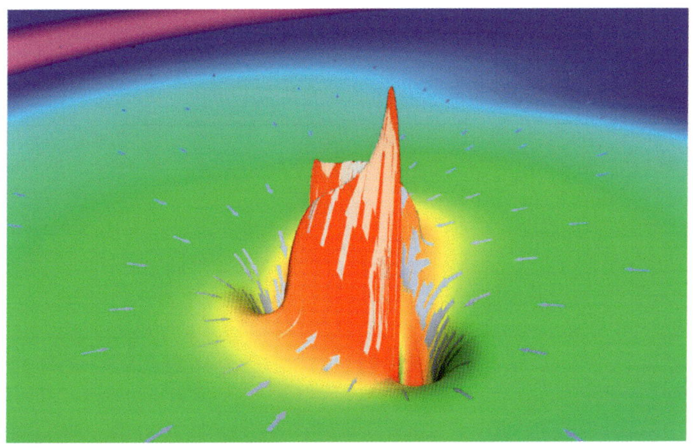

Figure 12. Snapshots (pseudo-embedding diagrams) from a movie depicting the geometry of spacetime around the GW150914 binary black hole 60 ms before collision, at the moment of collision, and 12 ms after the collision. Credits: SXS Collaboration.

SXS researcher Harald Pfeiffer. In this diagram, Pfeiffer takes the BBH's orbital "plane" (a two-dimensional warped surface), and visualizes its warpage (or, in physicists' language, its curvature) by depicting it embedded in a hypothetical, flat three dimensional space. The colors of the resulting warped surface depict the slowing of time: in the green regions, time flows at roughly the same rate as far away; in the red regions, the rate of flow of time is greatly slowed; the black regions (not often visible) are inside the black hole, where time flows downward. The silver arrows depict the motion of space.[27]

From a sequence of these diagrams (based on the output of an SXS simulation), Pfeiffer constructed a movie[28] of the BBH's evolving spacetime geometry. Figure 12 shows three snapshots from the movie for a BBH whose parameters are those of the first gravitational-wave burst that LIGO observed, GW150914:

- The first snapshot shows the BBH 60 milliseconds before collision. The space around each black hole dips downward like the water surface in a whirlpool, and the color shifts from green to red (time slows) as one moves down the tube.
- The second snapshot shows the BBH at the moment of collision. The collision has created a veritable *storm* in the shape of spacetime: Space is writhing like the surface of the ocean in a weather storm, and the rate of flow of time is changing rapidly.
- The third snapshot shows the BBH after the storm has subsided. It has produced a quiescent, single, merged black hole; and far from the hole, a burst of gravitational waves (depicted only heuristically as water-wave-type ripples) flows out into the universe.

These pseudo-embedding diagrams and movie have serious limitations. They depict only the BBH's equatorial plane and not the third dimension of our universe's space. The gravitational waves are not well depicted because they are essentially three dimensional. And some remarkable phenomena are completely missed, for example, two *vortices* of twisting space (one with a clockwise twist, the other counter-clockwise) that emerge from of each black hole, and also a set of stretching and squeezing warped-spacetime structures called *tendices*.[29]

The SXS simulations reveal the rich geometrodynamics of the BBH's spacetime geometry, and of its vortices and tendices. And the beautiful agreements between LIGO's observed gravitational waveforms and those predicted by the SXS simulations (e.g. Figure 6 of Barish's Part II of this lecture) convince us that geometrodynamic storms really do have the forms that the simulations predict — i.e. that Einstein's general relativity equations predict.

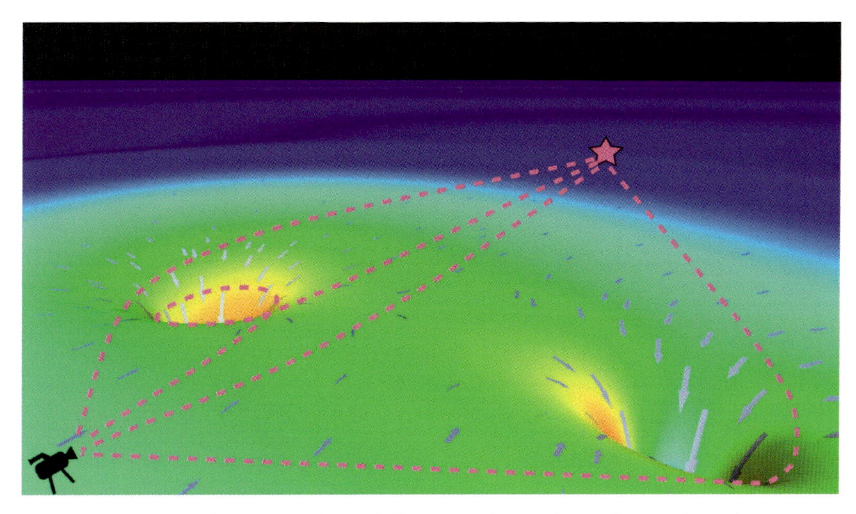

Figure 13. Light rays from a star, through the warped spacetime of GW150914, to a camera. Adapted from the movie[28] that underlies Fig. 12. Credit: SXS Collaboration.

If you and I were to watch two black holes spiral inward, collide and merge, with our own eyes or a camera, we would see something very different from the pseudo-embedding snapshots of Figure 12 and their underlying movie. Far behind the BBH would be a field of stars. The light from each star would follow several different paths to our eyes (Figure 13), some rather direct, others making loops around the black holes; so we would see several images of each star. (This is called *gravitational lensing.*) And as the holes orbit around each other, the images would move in a swirling pattern around the holes' two black shadows.

Teukolsky's graduate students Andy Bohn, Francois Hébert, and Will Throwe produced a movie[30] of these swirling stellar patterns from the SXS simulation of LIGO's first observed BBH, GW150914. Figure 14 is a snapshot from that movie.

Figures 12 and 14 and the geometrodynamic phenomena that I have described give a first taste of the exciting science that will be extracted from gravitational waves in the future. To that future science I will return below. But first I will dip back into the past, and describe briefly some contributions that theorists have made to the experimental side of LIGO.

THEORISTS' CONTRIBUTIONS TO UNDERSTANDING AND CONTROLLING NOISE IN THE LIGO INTERFEROMETERS

A major aspect of the LIGO experiment is understanding and controlling a huge range of phenomena that produce noise which can hide gravitational-wave signals. Theorists have contributed to scoping out some of

Figure 14. The BBH GW150914 as seen by eye, up close. Credit: SXS Collaboration.

these phenomena. This has been highly enjoyable, and it has broadened the education of theory students. I will give several interesting examples:

Scattered-Light Noise

In each arm of a LIGO interferometer the light beam bounces back and forth between mirrors. A tiny portion of the light scatters off one mirror, then scatters or reflects from the inner face of the vacuum tube that surrounds the beam, then travels to the other mirror, and there scatters back into the light beam (Figure 15, top). The tube face vibrates with an amplitude that is huge compared to the gravitational wave's influence, and those vibrations put a huge, oscillating phase shift onto the scattered light. That huge phase shift on a tiny fraction of the beam's light can produce a net phase shift in the light beam that is bigger than the influence of a gravitational wave.

This light-scattering noise can be controlled by placing baffles in the beam tube (dashed lines in Figure 15) to block the scattered light from reaching the far mirror. A bit of the scattered light, however, can still reach the far mirror by diffracting off the edges of the baffles.

Baffles and their diffraction of light are a standard issue in optical telescopes and other devices. But not standard, and unique to gravitational interferometers, is the danger that there might be coherent superposition of the oscillating phase shift for light that travels by different routes from one mirror to the other; such coherence could greatly increase the noise. In 1988 Rai Weiss recruited me and my theory students to look at this, determine how serious it is, and devise a way to mitigate it. Eanna Flanagan and I did so. To break the coherence, we gave the baffles deep saw teeth with random heights (Figure 15, bottom), and to minimize the noise

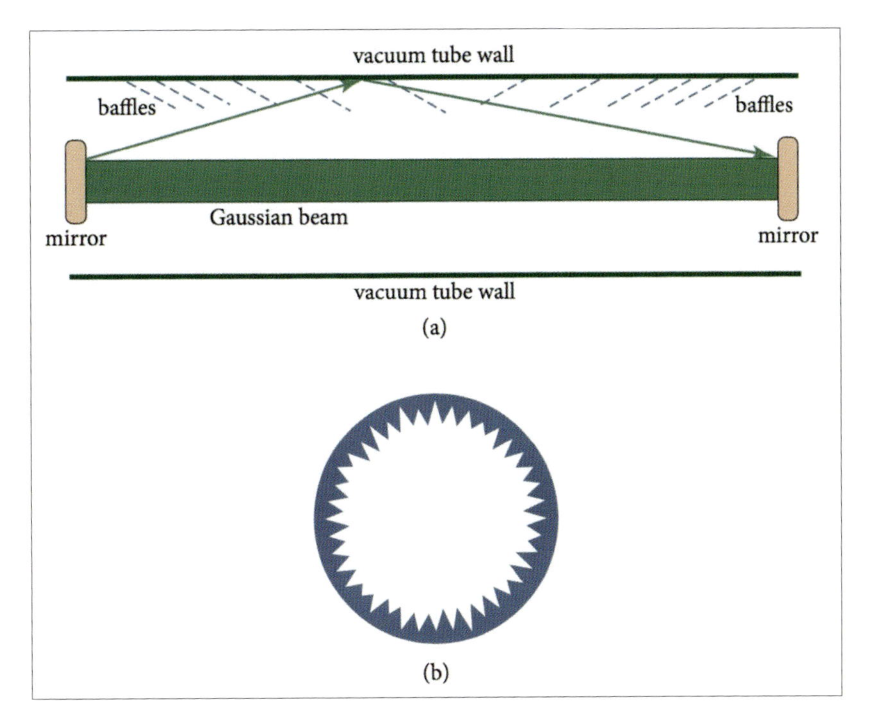

Figure 15. Top: A bit of beam light scatters off LIGO mirror, then scatters off vacuum tube wall, then travels to far mirror, and then scatters back into beam. Bottom: baffle to reduce noise and break coherence of scattered light. From K. S. Thorne and R. D. Blandford, *Modern Classical Physics* (Princeton University Press, 2017).

further we chose the teeth pattern optimally and optimized the locations of the baffles in the beam tube.[31] A segment of one of our random-saw-toothed baffles is my contribution to the Nobel Museum in Stockholm.

Gravitational noise

Humans working near a LIGO mirror create oscillating gravitational forces that might move the mirror more than does a gravitational wave. My wife, Carolee Winstein, is a biokinesiologist (expert on human motion). Using experimental data on human motion from her colleagues, we computed the size of this noise and concluded that, if humans are kept more than 10 meters from a LIGO mirror, the noise is acceptably small.[32] This was used as a specification for the layout of the buildings that house the LIGO mirrors. Theory students scoped out noise produced by the gravitational forces of seismic waves in the Earth,[33] and of airborne objects such as tumbleweeds.[34]

Thermal Noise

Thermal vibrations (vibrations caused by finite temperature) make LIGO's mirrors jiggle. These vibrations can arise in many different ways. Theory student Yuri Levin devised a new method to compute this thermal noise and to identify its many different origins.[35] Most importantly he used his method to discover that thermal vibrations in the coatings of LIGO's mirrors (which previously had been overlooked) might be especially serious. This has turned out to be true: In the Advanced LIGO interferometers, and likely in the next generation of gravitational interferometers, coating thermal noise is one of the two most serious noise sources; the other is quantum noise.

Quantum Noise and the Standard Quantum Limit for a Gravitational Interferometer

Quantum noise is noise due to the randomness of the photon distribution in an interferometer's light beams. In each Initial LIGO interferometer (Parts I and II of this lecture), the quantum noise had two parts: *photon shot noise,* caused by randomness in the arrival of photons at the photodetector (the interferometer's output); and *radiation pressure noise,* caused by randomness in the bouncing of photons off the interferometers' mirrors, which makes the mirrors jiggle.

Both forms of quantum noise must arise from light-beam *differences* in the interferometers' two arms, since the interferometer output is sensitive only to differences.

In the late 1970s, there was much debate among gravitational wave scientists over the physical origin of these differences. Theory postdoc Carlton Caves found the surprising answer[36]: Both the radiation pressure noise and the shot noise arise, he realized, from electromagnetic (*quantum electrodynamical*) vacuum fluctuations that enter the interferometer

Figure 16. Carlton M. Caves. Credit: Carlton M. Caves.

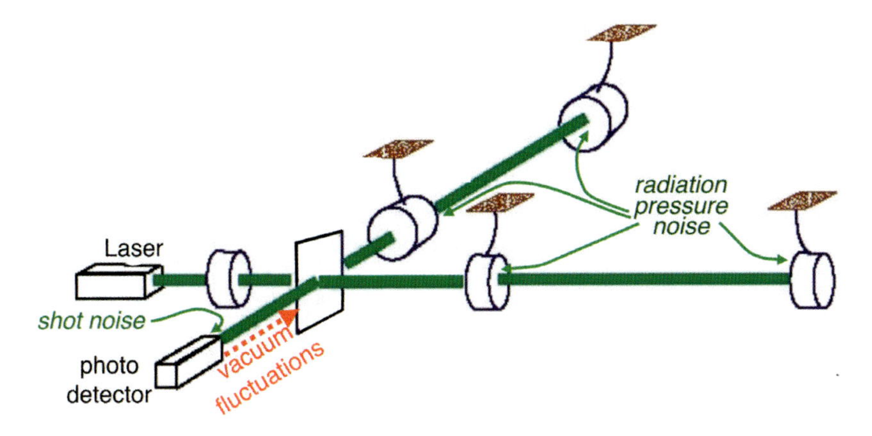

Figure 17. Vacuum fluctuations entering the output port of a gravitational interferometer beat against laser light to produce shot noise in the output photodetector and radiation pressure noise pounding on the mirrors.

backwards, from the direction of its output photodetector. These fluctuations beat against the laser light in the two arms to produce 1. radiation-pressure fluctuations (noise) that are opposite in the two arms, and 2. intensity fluctuations that also are also opposite and that therefore exit from the interferometer into the output photodetector as shot noise; Figure 17.

With this new understanding in hand, Caves noted the rather obvious fact that, when one increases the laser intensity I, the shot noise goes down proportionally to $1/\sqrt{I}$ and the radiation pressure goes up proportionally to $\sqrt{I}$; so the quantum noise curve (h as a function of frequency f) slides up and down a lower-limiting line as shown in Figure 18. That line is called the *standard quantum limit* (SQL) for an interferometer, and is given by Caves' simple formula

$$S_h{}^{1/2} = (8\hbar/mL^2\omega^2)^{1/2}. \quad (1)$$

Here S_h is the spectral density of the noise superposed on the gravitational wave signal, $\hbar$ is Planck's constant, m is the mass of each of the interferometer's mirrors, L is the length of the interferometer's two arms and ω is the gravitational wave's angular frequency.

In the late 1980s, Brian Meers at U. Glasgow (building on an idea of Ron Drever) proposed adding a signal recycling mirror to gravitational interferometers, in order to make them more versatile (see Weiss's and Barish's Parts I and II of this lecture), and by the late 1990s this new mirror was incorporated into the design for the future Advanced-LIGO interferometers. Strain and others used semiclassical (not fully quantum) theory to deduce the shot noise and radiation pressure noise in these Advanced-LIGO interferometers. This was worrisome because Advanced LIGO was expected to operate very near its standard quantum limit, SQL,

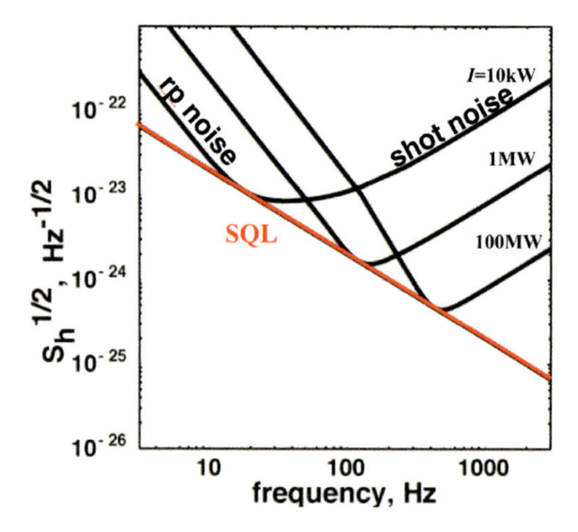

Figure 18. The shot noise and radiation pressure noise for various circulating powers *I* in the arms of the Initial LIGO interferometers.

where the semiclassical analysis might be flawed. So theory postdoc Alessandra Buonanno and graduate student Yanbei Chen carried out a full quantum mechanical analysis of the noise.

Their analysis revealed surprises:[37]

- The noise predictions of the semi-classical theory were wrong, so planning for Advanced LIGO would have to be modified, though not greatly.
- The interferometer's signal recycling mirror triggers the beam's light pressure in each arm to act as a frequency-dependent spring pushing against the mirrors, and so gives rise to an oscillatory, opto-mechanical behavior.
- The signal recycling mirror also creates quantum correlations between the shot noise and radiation-pressure noise. These correlations make it no longer viable to talk separately about shot noise and radiation pressure noise; instead, one must focus on a single, unified quantum noise.
- These correlations also enable the Advanced LIGO interferometer to beat Caves' SQL by as much as a factor 2 over a bandwidth of order the gravitational-wave frequency.

Quantum Fluctuations, Quantum Nondemolition, and Squeezed Vacuum

According to quantum theory everything fluctuates randomly, at least little bit.

A half century ago, the Russian physicist Vladimir Braginsky argued (in effect) that in gravitational wave detectors, when monitoring an object on which the waves act, one might have to measure motions so small that they could get hidden by quantum fluctuations of the object.[38] Later, in the mid-1970s,[39] Braginsky realized that it should be possible to create

Figure 19. Alessandra Buonanno and Yanbei Chen. Credits: Buonanno: S. Döring, Max Planck Society. Chen: Caltech.

Figure 20. Vladimir Braginsky. © Uspekhi Fizicheskikh Nauk 2012.

quantum nondemolition (QND) technology to circumvent these quantum fluctuations.

In 1980, Caves recognized that, although he derived his standard quantum limit [equation (1)] for an interferometer's sensitivity by analyzing its interaction with light, this SQL actually has a deeper origin: it is associated with the quantum fluctuations of the centers of mass of the interferometer's mirrors. The challenge, then, was to devise QND technology to circumvent those fluctuations and thereby beat their SQL.

Since the SQL is enforced by the electromagnetic vacuum fluctuations that enter the output port, Caves realized that a key QND tool might be to modify those vacuum fluctuations — and thereby, through their radiation-pressure influence on the mirrors, modify the mirrors' own quantum fluctuations.

More precisely, Caves[36] proposed to reduce the electromagnetic vacuum fluctuations in one quadrature of each fluctuational frequency (e.g. the $\cos \omega t$) quadrature) at the price of increasing the vacuum fluctuations in the other quadrature (e.g. $\sin \omega t$). (The uncertainty principle dictates that the product of the fluctuation strengths for the two quadratures cannot be reduced, so if one is reduced, the other must increase.)

One quadrature is responsible for shot noise, and the other for radiation pressure noise, Caves had shown; so by *squeezing the vacuum* in this way, one can reduce the shot noise at the price of increasing the radiation pressure noise—which is the same thing as one achieves by increasing the laser light intensity. (This use of squeezed vacuum has since become very important: The original plan for bringing Advanced LIGO to its design sensitivity entailed pushing up to 800 kW the light power bouncing back and forth between mirrors in each interferometer arm. However, such high light power produces exceedingly unpleasant side effects; the mir-

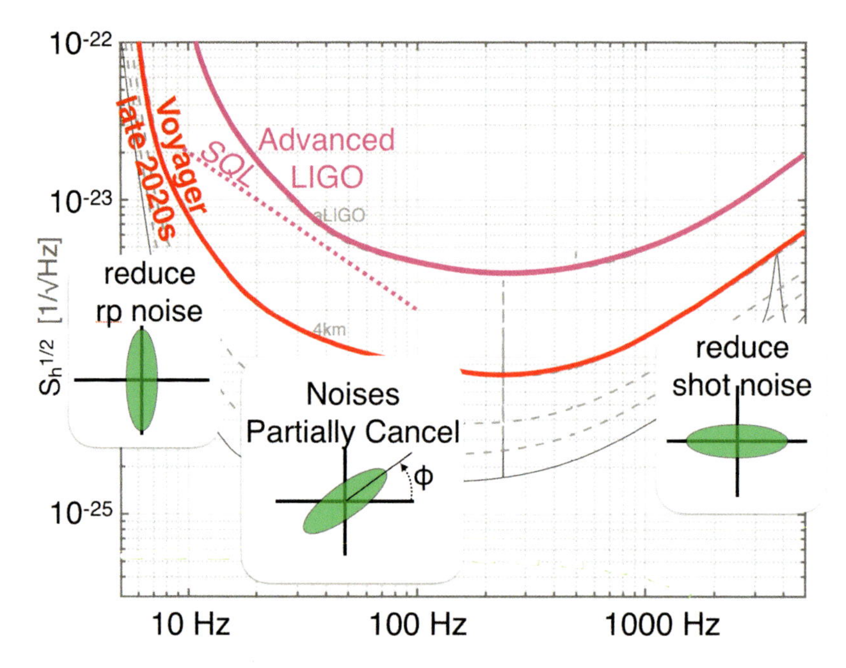

Figure 21. Noise curves for Advanced LIGO at design sensitivity and the proposed Voyager interferometer, and the SQL. The green ellipses are the input squeezed vacuum at high, intermediate, and low frequencies, which enable Voyager to beat the SQL.

rors have trouble handling it. Therefore, the new plan today, being implemented for LIGO's next observing run in early 2019, entails injecting squeezed vacuum into the output port in precisely the manner Caves envisioned, instead of a corresponding increase in light power.)

In Advanced LIGO, shot noise dominates at high gravitational-wave frequencies (well above 200 Hz), radiation-pressure noise dominates at lower frequencies (well below 200 Hz). Therefore, it is advantageous to inject vacuum that is squeezed at a frequency-dependent quadrature $\cos[\omega t - \varphi(\omega)]$, which produces a shot noise reduction ($\varphi=0$) at high frequencies, and a radiation-pressure reduction at low frequencies ($\varphi=\pi/2$). At intermediate frequencies an amazing thing happens — as was discovered by Bill Unruh[40] in 1981: the *two noises, shot and radiation-pressure, partially cancel each other out!* (See Figure 21.) As a result, the interferometer beats the SQL (it achieves quantum nondemolition), and with sufficient squeezing, it can do so by an arbitrarily large amount — in principle, but not in practice.

Although we have known this QND technique since 1983, in the 1980s and 1990s no practical method was known for producing the required frequency-dependent squeeze phase $\varphi(\omega)$.

In 1999, I discussed this problem in depth with my colleague Jeff Kimble (Caltech's leading experimenter in squeezing and other quantum-information-related techniques), and he devised a solution: Squeeze the vacuum at a frequency-independent phase, then send the squeezed vacuum through one or two carefully tuned Fabry-Perot cavities ("optical filters") before injecting it into the interferometer's output port.[41]

Among many different QND techniques that have been devised for LIGO interferometers,[42] this frequency-dependent squeezing, using *Kimble filter* cavities, is the one that currently looks most promising for future generations of gravitational interferometers: LIGO A+, Voyager, Cosmic Explorer, and Einstein Telescope (see Barish's Part II of this lecture). A small amount of QND will be required in LIGO A+, and a substantial amount in all subsequent interferometers.

THE FUTURE: FOUR GRAVITATIONAL FREQUENCY BANDS

Electromagnetic astronomy was confined to optical and infrared frequencies until the late 1930s, when cosmic radio waves were discovered by Karl Jansky. Later, other frequency bands were enabled by telescopes flown above the earth's atmosphere: ultraviolet astronomy in the 1950s, and X-ray and gamma-ray astronomy in the 1960s. Over the decades since then, ever wider frequency bands have been opened up. It is common to speak of electromagnetic "windows" onto the universe, with each window being a frequency band in which astronomers work: the optical, infrared, radio, ultraviolet, X-ray and gamma-ray windows.

Gravitational waves are similar. Within the next two decades, we expect three more gravitational windows to be opened, so we will have:

- The *high-frequency gravitational window* (HF; ~10 Hz to ~10,000 Hz; wave periods ~100 msec to ~0.1 msec), in which LIGO, VIRGO and other ground-based interferometers operate.
- The *low-frequency gravitational window* (LF: periods minutes to hours) in which will operate constellations of drag-free spacecraft that track each other with laser beams, most notably the European Space Agency's LISA (Laser Interferometer Space Antenna),[43] which is likely to be launched into space in 2030 or a bit later.
- The very-low frequency gravitational window (VLF; periods of a few years to a few tens of years), in which *pulsar-timing arrays (PTAs),*[44] are now operating and searching for gravitational waves.
- The *ultra-low frequency* window (ULF; periods of hundreds of millions of years), in which primordial gravitational waves are predicted to have placed peculiar, observable polarization patterns onto the comic microwave radiation.[45]

I will now describe LISA, PTAs, and CMB polarization in a bit more detail.

LISA: The Laser Interferometer Space Antenna

LISA will consist of three spacecraft that track each other with laser beams. The spacecraft reside at the corners of an equilateral triangle with separations of a few million kilometers. This triangular constellation travels around the Sun in the same orbit as the Earth, following the Earth by roughly 20 degrees. Each spacecraft shields, from external influence, a pair of *proof masses* (analog of a LIGO mirror), and uses thrusters to keep the spacecraft centered on the proof masses. The three proof-mass pairs, one in each spacecraft, move relative to each other in response to the tidal gravity of the Sun and the planets, and gravitational waves; and their relative motion is monitored by the laser beams using a technique called *heterodyne interferometry* (beating the incoming beam from a distant spacecraft against an outgoing beam). This is rather different from the type of interferometry used in LIGO.

The idea of a mission like LISA was discussed starting in 1974 by Peter Bender, Ronald Drever, Jim Faller, Rainer Weiss, and others. The presently planned orbital geometry (Fig. 22) was suggested by Faller and

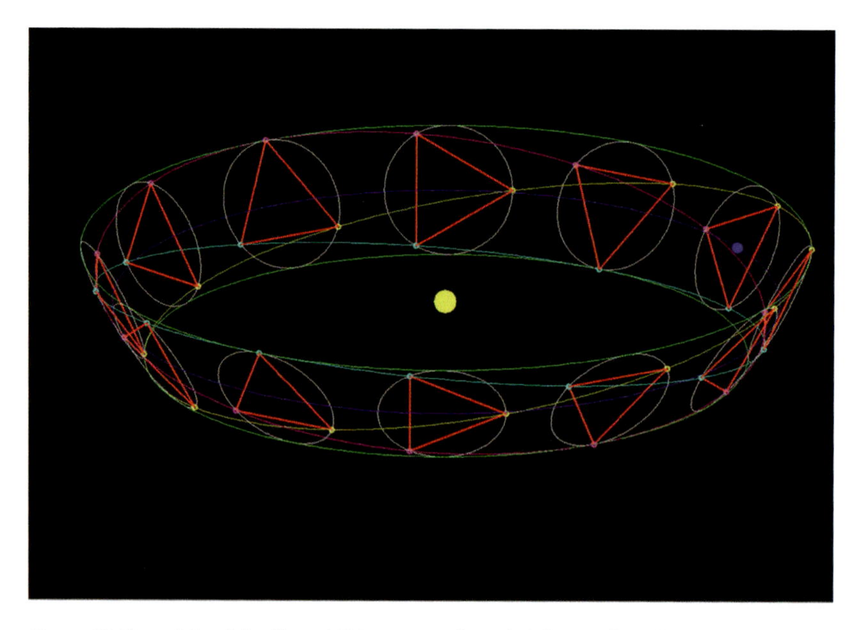

Figure 22. The orbits of the three LISA spacecraft. Each follows a free-fall (geodesic) orbit around the sun, and their configuration remains nearly an equilateral triangle. Credit: HEPL, Stanford University.

Figure 23. Peter Bender (right) discussing the LISA mission concept with Ronald Drever (left) and Stan Whitcomb (middle) in Padova, Italy, in 1983. Credit: Peter Bender.

Bender in talks in 1981 and 1984.[46] Bender then almost single handedly developed the LISA concept into a viable form through the 1980s and into the 1990s, leading NASA and ESA to develop a tentative plan for implementing it as a joint space mission. NASA dropped out in 2011 due mainly to cost overruns on the James Webb Space Telescope, leaving ESA to carry LISA studies forward alone, including a highly successful 2016 test of some of the most difficult technology, in the LISA Pathfinder Mission.[47] As of 2018 it appears that NASA may rejoin the LISA Mission as a junior partner to ESA and the launch might be as soon as 2030.

PTAs: Pulsar Timing Arrays

A Pulsar Timing Array (PTA) consists of an array of several pulsars whose pulse periods are monitored with very high precision by one or more radio telescopes (Figure 24). Heuristically speaking, when a gravitational wave sweeps over the Earth, it causes clocks on Earth to speed up and slow down in an oscillatory pattern; so when compared with Earth clocks, all the pulsars appear to slow down and speed up synchronously.

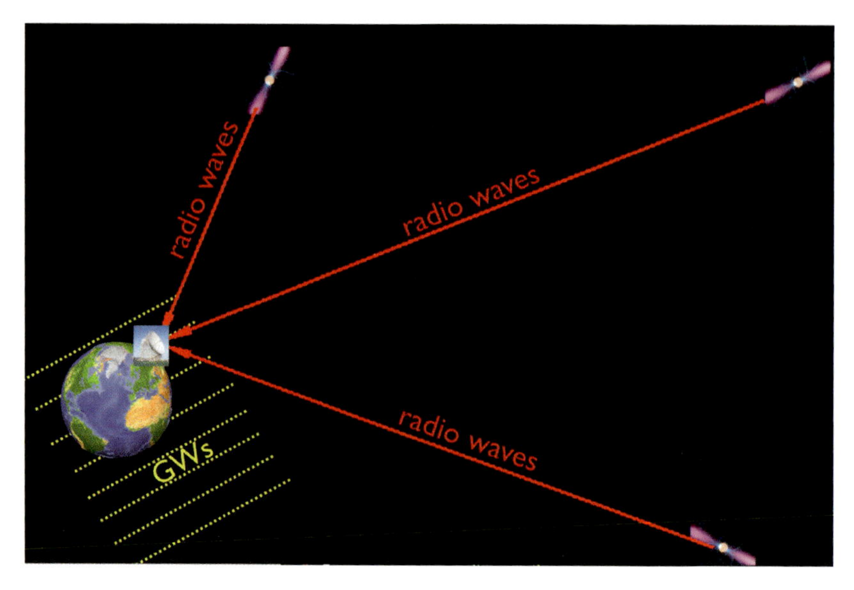

Figure 24. Pulsar Timing Array. An array of three pulsars sends radio-wave pulses to Earth, whose observed timings are synchronously modulated by gravitational waves sweeping over the Earth.

A more accurate description of how a PTA works is this:[48] The gravitational wave creates an effective anisotropic index of refraction for the space through which the pulsars' radio waves travels. This index of refraction makes the pulsars appear to speed up and slow down synchronously by amounts that depend on the angles between the direction to the pulsar and the direction to the gravitational-wave source and the wave's polarization axes.

The idea of using pulsar timing to detect gravitational waves was conceived independently in the late 1970s by Michael Sazhin and Steven Detweiler.[49] Currently three radio-astronomy collaborations are attempting to detect gravitational waves using PTAs: the NANOGrav collaboration in North America, the European PTA, and the Parkes PTA (Australia); and the three also work in a loose worldwide collaboration called the International PTA.

The primary target of these collaborations is gravitational waves from gigantic black hole binaries, weighing ~10^8 to ~10^{10} suns. Current PTA sensitivities are adequate to detect these waves at the level of optimistic estimates, and success may well come in the next decade.

CMB Polarization

The cosmic microwave background (CMB) radiation, studied intensely by astronomers, last scattered off matter in the era when the primordial plasma was recombining to form neutral hydrogen (at universe age ~380,000 years).

In the 1990s, several theoretical astrophysicists[50] realized that primordial gravitational waves (waves from our universe's earliest moments), interacting with the recombining plasma, should have created a so-called *B-mode* pattern of polarization in the CMB. Searching for that pattern on the sky has become a "holy grail" for CMB astronomers, as it may reveal details of the primordial gravitational waves. The pattern has been found, but it can also be produced by microwave emission from dust particles and by synchrotron emission from electrons spiraling in interstellar magnetic fields. So the challenge now is to separate those two foreground contributions to the B-mode polarization from the gravitational-wave contribution.[51] It is plausible that this may be achieved in the coming decade.

THE FUTURE: PROBING THE UNIVERSE WITH GRAVITATIONAL WAVES

I conclude this lecture with some remarks about the science that is likely to be extracted from gravitational waves in the coming few decades. I shall discuss sources that include matter (multi-messenger astronomy), then the gravitational wave exploration of black holes, and finally observations of the first one second of the life of our universe. For details on all the sources I discuss, I recommend a book by Michele Maggiore.[52]

Multi-Messenger Astronomy

LIGO/Virgo's first binary neutron star (BNS), GW170817 (see Barish's Part II of this lecture) is a remarkable foretaste of the discoveries that will be made in the high-frequency band via multi-messenger astronomy (astronomy using a variety of forms of radiation and waves). As ground-based interferometers improve:

- The event rate for BNSs will likely increase from ~ one per year now, to ~ one per month at LIGO design sensitivity (2020), to ~ one per day in Voyager (which could operate in the late 2020s; see Barish's Part II), to many per day in Cosmic Explorer and the Einstein Telescope (which could operate in the 2030s; see Barish's Part II); and the richness and detail extracted from multi-messenger observations will increase correspondingly.
- We will almost certainly also watch many black holes tear apart their neutron-star companions in black hole / neutron star binaries, from which we might be able more cleanly to extract neutron-star physics via multi-messenger observations, than from BNSs.
- We will very likely also see multi-messenger emission from a variety of types of spinning, deformed neutron stars, including pulsars, magnetars, and perhaps low-mass X-ray binaries.

- If we are lucky, we will see gravitational waves from the births of neutron stars in supernovae, and through combined gravitational, neutrino, and electromagnetic observations, discover the mechanisms that trigger supernova outbursts.
- And if we are lucky, we will see electromagnetic emission from some merging black-hole binaries, due to the black holes' interaction with matter in their vicinity, and we may thereby explore the black holes' near environments.

LISA and other low-frequency, space-based interferometers will participate in multi-messenger observations of a variety of astronomical objects and phenomena, including:

- White-dwarf binaries, and interactions between the two white-dwarf stars when they are very close together.
- AM CVn stars (a white dwarf that accretes matter from a low-mass helium-star companion).
- An enormous number of other binary star systems with gravitational-wave frequencies above about 0.1 mHz —with so very many between ~0.1 mHz and ~2 mHz that they will produce a stochastic background that dominates over LISA's instrumental noise.
- Possibly the implosion (collapse) of a few supermassive stars in galactic nuclei, to form supermassive black holes.

And of course, the most exciting prospect of all, is huge, unexpected surprises that entail multi-messenger emissions.

Exploring Black Holes and Geometrodynamics with Gravitational Waves
The high-, low-, and very low-frequency bands cover BBH inspirals over the entire range of known black-hole masses, from a few solar masses to $\sim 2 \times 10^{10}$ solar masses.[53]

In the high-frequency band of ground-based interferometers, BBHs with total mass up to about 1000 suns can be observed. As these interferometers improve, the rates of BBH events could increase from very roughly one per month in 2017 to a few per week at Advanced LIGO design sensitivity (~2020), to as much as one per hour in Voyager (late 2020s), to every black hole binary in the universe that emits in the high-frequency band, in Cosmic Explorer and Einstein Telescope (2030s). And with improving sensitivity, the maximum signal-to-noise ratio for BBH waves could increase from 24 today, to as much as 1000 in Cosmic Explorer and Einstein Telescope, with a corresponding increase in the accuracy with which the physics of black holes can be explored.

In the low-frequency band, LISA should see mergers of very massive

Figure 25. Embedding diagram showing the spacetime geometry of a small black hole orbiting a large black hole, in the large hole's equatorial plane. Credit: NASA/JPL-Caltech.

black holes (~ 10^3 to ~10^8 solar masses), with signal to noise as high as ~100,000, and corresponding exquisite accuracy for exploring geometrodynamics and testing general relativity.

LISA will likely also see many EMRIs: extreme mass-ratio inspirals, in which a small black hole or a neutron star or white dwarf travels around a very massive black hole on a complex orbit, gradually spiraling inward due to gravitational radiation reaction, and finally plunging into the massive hole. Figure 25 shows the spacetime geometry of the two black holes for the special case where the small hole is confined to the massive hole's equatorial plane; Figure 26 (from a simulation and movie by Drasco[54]) shows a segment of a generic orbit for the small hole, when the large hole spins rapidly.

The complexity of the generic orbit results from the combined influence of the massive hole's very strong gravitational pull (very large *relativistic periastron shift*), the curvature of space around it (not depicted in the figure), and the whirling of space (dragging of inertial frames) caused by its spin. Over many months, the orbit explores a large portion of the space of the massive black hole, and so the complicated gravitational waveform it emits carries encoded in itself a highly accurate map of the massive hole's spacetime geometry.[55] A major goal of the LISA mission is to monitor the waves from such EMRIs, and extract the maps that they

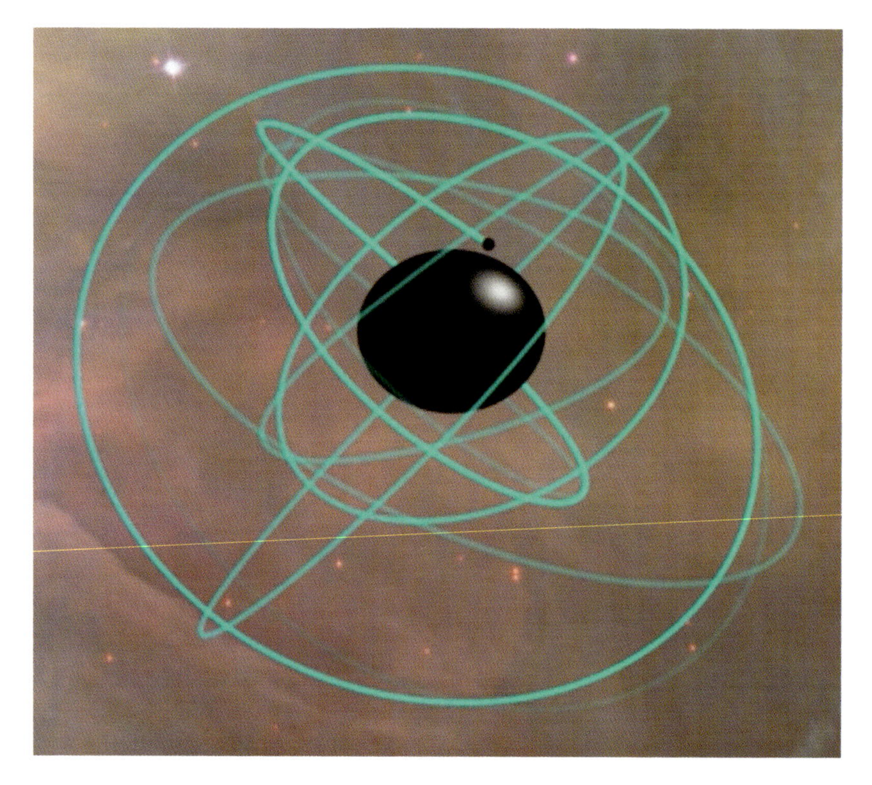

Figure 26. Segment of generic orbit for a small black hole orbiting a rapidly spinning large black hole. Credit: Steve Drasco.

carry, thereby determining with high precision whether the massive hole's spacetime geometry is the one predicted by general relativity: the *Kerr geometry.*

The struggle to understand the quantum mechanical phenomenon of information loss into black holes has led to speculations that instead of a horizon down which things can fall, a black hole has a *firewall*;[56] and also speculations that the firewall modifies the spacetime geometry from that of Kerr outside but near the firewall's location.[57] LISA's mapping project will search for any such modification. By this mapping project, LISA can also search for unexpected types of massive, compact objects, whose spacetime geometries differ from that of Kerr, for example naked singularities that are being orbited by much smaller bodies.

Exploring the First One Second of our Universe's Life

Every known type of particle or radiation, except gravitational waves, is predicted to be trapped by the universe's hot, dense plasma during the first one second of our universe's life. Therefore, gravitational waves are our only hope for directly observing what happened during that first one second.

Among the predictions that such observations might test is the origin of the electromagnetic force —one of the four fundamental forces of Nature. Theory predicts that, when the universe was very young and very hot, the electromagnetic force did not exist. In its place there was an *electroweak force.* As the universe expanded and cooled through an age of $\sim 10^{-11}$ seconds and a temperature of $\sim 10^{15}$ K, there was, according to theory, a *phase transition* in which the electroweak force came apart, giving rise to two new forces: the electromagnetic force, and the weak nuclear force.

If this was a so-called *first-order* phase transition (which it may well not have been), then it is predicted to be like the transition from water vapor to liquid water when the vapor is cooled through 100 C: the transition should have occurred in bubbles analogous to water droplets. Inside each bubble, the electromagnetic force existed; outside the bubbles, it did not exist. Theory predicts that these bubbles expanded at very high speeds, collided, and produced, in their collisions, stochastic gravitational waves. As the universe expanded, the wavelengths of these waves also expanded, until today, 13.8 billion years later, the wavelengths are expected to be in LISA's frequency band.[58] One of LISA's goals is to search for these stochastic gravitational waves produced by the birth of the electromagnetic force.

LIGO could see gravitational waves produced by a similar first-order phase transition when the universe was far younger, $\sim 10^{-22}$ seconds, and far hotter, $\sim 10^{21}$ K. In logarithmic terms, this time and temperature are roughly half way between the electroweak phase transition and the phase transition associated with grand unification of the fundamental forces. Unfortunately, this is an epoch at which no phase transition is predicted by our current understanding of the laws of physics.

Gravitational waves are so penetrating — so immune to absorption or scattering by matter — that they could have been generated in our universe's big-bang birth, and traveled to Earth today unscathed by matter, bringing us a picture of the big bang.

This picture, however, is predicted to have been distorted by *inflation,* the exponentially fast expansion of the universe that is thought (with some confidence) to have occurred between age $\sim 10^{-36}$ seconds and $\sim 10^{-33}$ seconds. More specifically, inflation should have *parametrically amplified* whatever gravitational waves came off the big bang. This amplification may well have made the primordial gravitational waves strong enough for detection, but the amplification will also have distorted the waves, so that the spectrum humans see is a *convolution* (combination) of what came off the big bang, and the influence of inflation.

Remarkably, we have the possibility, by the middle of this (twenty-first)

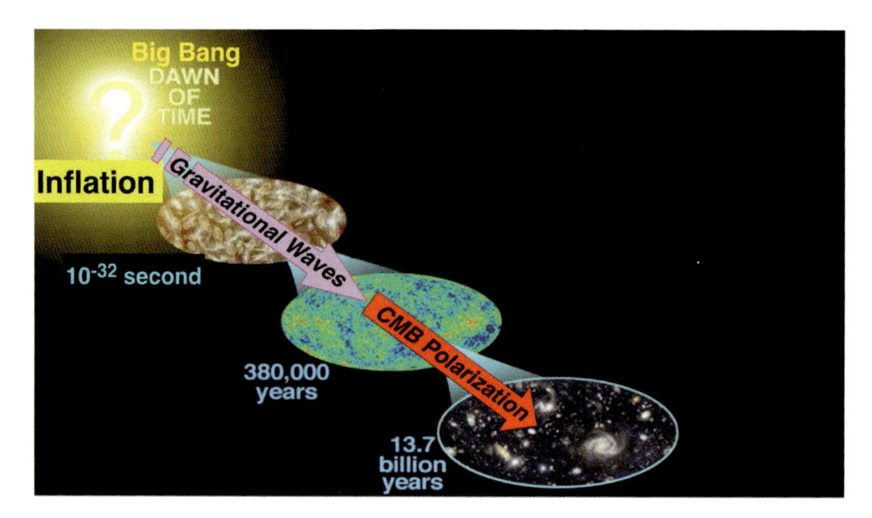

Figure 27. Primordial gravitational waves, amplified by inflation at universe age ~10^{-36} to ~10^{-33} sec, interact with primordial plasma at age 380,000 years, placing a polarization imprint on the CMB which is observed today at age 13.8 billion years. Credit: Adapted from WMAP # 020622 / NASA / WMAP Science Team.

century, to observe these primordial gravitational waves in two different frequency bands:

- In the extremely low-frequency band, by the B-mode polarization pattern that the waves place on the cosmic microwave background radiation, CMB; see above and Figure 27.
- At periods of seconds, between the high-frequency band and the low-frequency band, using a proposed successor to LISA: the Big Bang Observer,[59] which consists of several constellations of light-beam-linked spacecraft in interplanetary space (Figure 28).

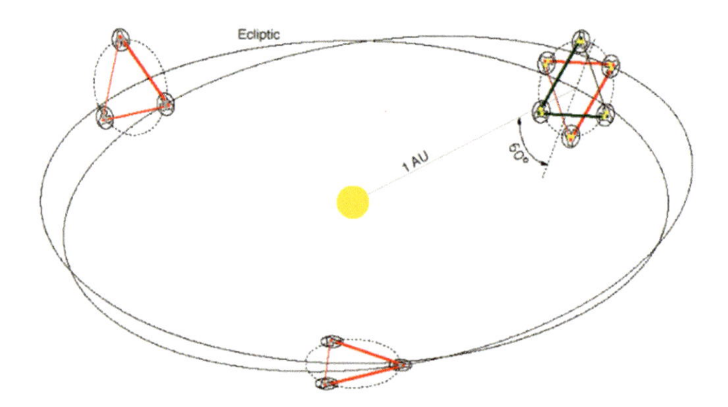

Figure 28. The Big Bang Observer's constellations of spacecraft, in the same orbit around the Sun as the Earth. Credit: Sterl Phinney.

Theorists' conventional wisdom dictates that what came off the big bang was the weakest gravitational waves allowed by the laws of Nature: vacuum fluctuations of the gravitational field. Inflation's parametric amplification was so strong, that even beginning with just vacuum fluctuations, the resulting primordial gravitational waves are likely to be strong enough for observation by both of these detectors, in both frequency bands—bands that differ in frequency and in wave period and wavelength by a factor of $\sim 10^{15}$.

I am skeptical of theoretical physicists' conventional wisdom, as I have seen it fail spectacularly in several ways during my career. I look forward to the possibility, indeed the likelihood, that the observations will differ from this conventional wisdom in one or both frequency bands, and that the observations will reveal enough about the birth of the universe to give crucial guidance to physicists who are trying to discover the laws of quantum gravity: the laws that governed the universe's big-bang birth.

CONCLUSION

Four hundred years ago, Galileo built a small optical telescope and, pointing it at Jupiter, discovered Jupiter's four largest moons; and pointing it at our moon, discovered the moon's craters. This was the birth of electromagnetic astronomy.

Two years ago, LIGO scientists turned on their Advanced LIGO detector and, with the data-analysis help of VIRGO scientists, discovered the gravitational waves from two colliding black holes 1.3 billion light years from Earth.

When we contemplate the enormous revolution in our understanding of the universe that has come from electromagnetic astronomy over the four centuries since Galileo, we are led to wonder what revolution will come from gravitational astronomy, and from its multi-messenger partnerships, over the coming four centuries.

ACKNOWLEDGMENTS

I gratefully acknowledge the US National Science Foundation (NSF) and the Sherman Fairchild Foundation for funding my theory group's gravitational-wave research.

REFERENCES

1 For greater personal detail that feeds into this lecture, see my Nobel Biography.
2 W.H. Press and K.S. Thorne, "Gravitational-Wave Astronomy," *Annual Reviews of Astronomy and Astrophysics*, **10**, 335–374 (1972).
3 C. Cutler and K. S. Thorne, "An Overview of Gravitational Wave Sources," *General Relativity and Gravitation, Proceedings of the 16th International Conference,* ed. N. Bishop and S. D. Maharaj (World Scientific, 2002), 72–111 [https://arxiv.org/pdf/gr-qc/0204090.pdf] .
4 K.S. Thorne et. al., "The Scientific Case for Advanced LIGO Interferometers," LIGO Technical Report LIGO-P000024-A-R, Caltech/MIT, January 2001 [https://dcc.ligo.org/LIGO-P000024/public].
5 R. Weiss, "Electromagnetically Coupled Broadband Gravitational Antenna", *Quarterly Progress Report* No. 105, MIT Research Laboratory of Electronics, 54–76 (April 15, 1972) [https://dcc-lho.ligo.org/public/0038/P720002/001/P720002-01.pdf].
6 M.E. Gertsenshtein and V. I. Pustovoit, "On the Detection of Low-Frequency Gravitational Waves", *Soviet Physics–JETP*, **16**, 433–435 (1963).
7 C. W. Misner, K. S. Thorne and J. A. Wheeler, *Gravitation* (W. H. Freeman, 1973).
8 R. Epstein and J. P. A. Clark, "Discussion Session II: Notes and Summary", in *Sources of Gravitational Radiation,* Ed. Larry Smarr (Cambridge University Press, 1979), pp. 477–497.
9 R. E. Vogt, R. W. P. Drever, F. J. Raab, K. S. Thorne, and R. Weiss, *Proposal to the National Science Foundation for the Construction, Operation, and Supporting Research and Development of a Laser Interferometer Gravitational-Wave Observatory* (California Institute of Technology, 1989) [https://dcc.ligo.org/public/0065/M890001].
10 Clark, J.P.A., Van den Heuvel, E.P.J. and Sutantyo, W., *Astronomy and Astrophysics,* **72**, 120 (1979).
11 LIGO Scientific Collaboration and Virgo Collaboration, "Predictions for the Rates of Compact Binary Coalescences by Ground-Based Gravitational-Wave Detectors", *Classical and Quantum Gravity*, **27**, 173001 (2010) [https://arxiv.org/pdf/1003.2480.pdf] and references therein.
12 B. F. Schutz, "Determining the Hubble Constant from Gravitational Wave Observations", *Nature*, **323**, 310 (1986).
13 Schutz, B.F. (ed.), *Gravitational Wave Data Analysis* (Proceedings of the NATO Advanced Research Workshop, held at St. Nicholas, Cardiff, Wales, July 6–9, 1987). Kluwer, Dordrecht (1989).
14 L. Blanchet, "Gravitational Radiation from Post-Newtonian Sources and Inspiralling Compact Binaries," *Living Reviews in Relativity*, **17**, 2 (2014); and references therein.
15 C. Cutler et. al., "The Last Three Minutes: Issues in Gravitational Wave Measurements of Coalescing Compact Binaries," *Physical Review Letters* **70**, 2984–2987 (1993). LIGO Scientific and Virgo Collaborations, "Tests of General Relativity with GW150914," *Physical Review Letters* **116**, 221101 (2016).
16 R. Arnowitt, S. Deser and C. W. Misner, "The Dynamics of General Relativity," Chap. 7 of Gravitation: An Introduction to Current Research, ed. L. Witten (Wiley, 1962), pp. 227–265 [https://arxiv.org/pdf/gr-qc/0405109.pdf]; and references cited therein.

17 C. W. Misner, "Wormhole Initial Conditions," *Physical Review* **118**, 1110–1111(1960).
18 S. G. Hahn and R. W. Lindquist, "The Two-Body Problem in Geometrodynamics," *Annals of Physics* **29**, 304–331 (1964).
19 L. Smarr, "Gauge Conditions, Radiation Formulae and the Two Black Hole Collision," in L. Smarr, ed., *Sources of Gravitational Waves* (Cambridge University Press, 1979), pp. 245–274, and references cited therein.
20 F. Pretorius, "Evolution of Binary Black-Hole Spacetimes," *Physical Review Letters* **95**, 121101 (2005) [https://arxiv.org/pdf/gr-qc/0507014.pdf].
21 M. Campanelli, C. O. Lousto, P. Marronetti and Y. Zlochower, "Accurate Evolutions of Orbiting Black-Hole Binaries Without Excision," *Physical Review Letters* **96**, 111101 (2006) [https://arxiv.org/pdf/gr-qc/0505055.pdf]; J. G. Baker, J. Centrella, D.-I. Choi, M. Koppitz and J. van Meter, "Gravitational-Wave Extraction from an Inspiraling Configuration of Merging Black Holes," *Physical Review Letters* **96**, 111102 (2006) [https://arxiv.org/pdf/gr-qc/0511103.pdf].
22 http://www.black-holes.org/SpEC.html .
23 https://www.black-holes.org/for-researchers/waveform-catalog .
24 I. Hinderer et. al., "Error-analysis and comparison to analytical models of numerical waveforms produced by the NRAR Collaboration," *Classical and Quantum Gravity,* 31, 025012 (2014).
25 A. Buonanno and T. Damour, "Effective One-Body Approach to General Relativistic Two-Body Dynamics," *Physical Review* **D59,** 084006 (1999) [https://arxiv.org/pdf/gr-qc/9811091.pdf].
26 P. Ajith et. al., "Phenomenological Template Family for Black-Hole Coalescence Waveforms", *Classical and Quantum Gravity* **24**, S689–S700 (2007) [https://arxiv.org/pdf/0704.3764.pdf].
27 In more technical language, the surface's shape, color, and arrows depict the 2-geometry of the orbital "plane", the lapse function, and the shift function.
28 https://www.youtube.com/watch?v=YsZFRkzLGew .
29 R. Owen et. al., "Frame-Dragging Vortexes and Tidal Tendexes Attached to Colliding Black Holes: Visualizing the Curvature of Spacetime," *Physical Review Letters,* **106**, 151101 (2011) [https://arxiv.org/pdf/1012.4869.pdf].
30 https://www.black-holes.org/gw150914 ; A. Bohn et. al., "What does a binary black hole merger look like?", *Classical and Quantum Gravity* **32**, 065002 (2015) [https://arxiv.org/pdf/1410.7775.pdf].
31 E. E. Flanagan and K. S. Thorne, "Light Scattering and Baffle Configuration for LIGO," LIGO Technical Report Number LIGO-T950101-00-R [https://dcc.ligo.org/DocDB/0028/T950101/000/T950101-00.pdf].
32 K.S. Thorne and C.J. Winstein, "Human Gravity-Gradient Noise in Interferometric Gravitational-Wave Detectors," *Physical Review D,* **60**, 082001 (1999) [https://arxiv.org/abs/gr-qc/9810016].
33 S.A. Hughes and K.S. Thorne, "Seismic Gravity-Gradient Noise in Interferometric Gravitational-Wave Detectors," *Physical Review D,* **58**, 122002 (1998) [https://arxiv.org/pdf/gr-qc/9806018.pdf].
34 T. Creighton, "Tumbleweeds and Airborne Gravitational Noise Sources for LIGO", *Classical and Quantum Gravity* **25**, 125011 (2008) [https://arxiv.org/pdf/gr-qc/0007050.pdf].
35 Y. Levin, "Internal Thermal Noise for LIGO Test Masses: A Direct Approach", *Physical Review D* **57**, 659 (1998) [https://arxiv.org/pdf/gr-qc/9707013.pdf].

36 C. M. Caves, "Quantum-Mechanical Noise in an Interferometer," *Physical Review D* **23**, 1693–1708 (1981).

37 A. Buonanno and Y. Chen, "Quantum Noise in Second Generation, Signal-Recycled Interferometric Gravitational-Wave Detectors," *Physical Review* **D 64**, 042006 (2001). Also "Scaling Law in Signal-Recycled Laser-Interferometer Gravitational-Wave Detectors," *Physical Review D* **67**, 062002 (2003).

38 V. B. Braginsky, "Classical and Quantum Restrictions on the Detection of Weak Disturbances of a Macroscopic Oscillator," *Soviet Physics JETP* **26**, 831–834 (1968).

39 V. B. Braginsky and Yu. I. Vorontsov, "Quantum-Mechanical Limitations in Macroscopic Experiments and Modern Experimental Technique," *Soviet Physics Uspekhi* **17**, 644–650 (1975). For Braginsky's own retrospective view of this work and subsequent developments up to 1996, see V. B. Braginsky and F. Ya. Khalili, "Quantum Nondemolition Measurements: The Route from Toys to Tools," *Reviews of Modern Physics* **68**, 1–11 (1996).

40 W. G. Unruh, "Quantum Noise in the Interferometer Detector," in *Quantum Optics, Experimental Gravity, and Quantum Measurement Theory,* eds P. Meystre & M.O. Scully (Plenum, NY, 1982), pp. 647–660.

41 H.J. Kimble et. al., "Conversion of Conventional Gravitational-Wave Interferometers into QND Interferometers by Modifying their Input and/or Output Optics," *Physical Review D* **65**, 022002 (2002) [https://arxiv.org/pdf/gr-qc/0008026.pdf].

42 For a review see S. L. Danilishin and F. Ya. Khalili, "Quantum measurement theory in gravitational-wave detectors." *Living Reviews in Relativity* **15**, 5 (2012).

43 http://sci.esa.int/lisa/ .

44 http://www.ipta4gw.org .

45 See, e.g., Sec. 20.4 of M. Maggiore, *Gravitational Waves, Volume 2: Astrophysics and Cosmology*, (Oxford University Press, 2018).

46 J. E. Faller and P. L. Bender, "A Possible Laser Gravitational Wave Antenna in Space", in *Precision Measurements and Fundamental Constants II,* B. N. Taylor and W. D. Phillips, eds., NBS Spec. Publ. 617 (1984), pp. 689–690; J. E. Faller et al., Space Antenna for Gravitational Wave Astronomy", *Proc. Colloq. Kilometric Optical Arrays in Space*, ESA SP-226, pp. 157–163 (April, 1985).

47 M. Armano et al. "Beyond the Required LISA Free-Fall Performance: New LISA Pathfinder Results down to 20 μHz," *Physical Review Letters* **120**, 061101 (2018).

48 This is one way of describing the derivation of the response of a PTA to a gravitational wave [which, for example, is sketched all too briefly in Exercise 27.20 of K. S. Thorne and R. D. Blandford, *Modern Classical Physics* (Princeton University Press, 2017)].

49 M. V. Sazhin, "Opportunities for Detecting Ultralong Gravitational Waves," *Soviet Astronomy* **22**, 36–38 (1978); S. L. Detweiler, "Pulsar Timing Measurements and the Search for Gravitational Waves," *Astrophysical Journal* **234**, 1100–1104 (1979).

50 U. Seljak and M. Zaldarriaga, "Signature of Gravity Waves in the Polarization of the Microwave Background," *Physical Review Letters* **78**, 2054–2058 (1997) [https://arxiv.org/pdf/astro-ph/9609169.pdf]; M. Kamionkowski, A. Kosowsky and A. Stebbins, "A Probe of Primordial Gravity Waves and Vorticity," *Physical Review Letters* **78** 2058–2061

[https://arxiv.org/pdf/astro-ph/9609169.pdf].

51 See, e.g., Sec. 20.4 of M. Maggiore, *Gravitational Waves, Volume 2: Astrophysics and Cosmology,* (Oxford University Press, 2018).

52 M. Maggiore, *Gravitational Waves, Volume 2: Astrophysics and Cosmology,* (Oxford University Press, 2018).

53 E. E. Flanagan and S. A. Hughes, "Measuring Gravitational Waves from Binary Black Hole Coalescences: I. Signal to Noise for Inspiral, Merger, and Ringdown," *Physical Review D* **57**, 4535–4565 (1998) [https://arxiv.org/pdf/gr-qc/9701039.pdf].

54 S. Drasco, "Binary Black Hole Inspiral at Natural Speed," https://www.youtube.com/watch?v=1VJU5odFhfc .

55 F. D. Ryan, "Gravitational Waves from the Inspiral of a Compact Object into a Massive, Axisymmetric Body with Arbitrary Multipole Moments," *Physical Review D* **52,** 5707–5718 (1995).

56 A. Almheiri, D. Marolf, J. Polchinski, and J. Sully, "Black Holes: Complementarity or Firewalls?" *Journal of High Energy Physics,* **2013**, 62 (2013) [https://arxiv.org/pdf/1207.3123.pdf].

57 See, e.g., S. B. Giddings, "Gravitational Wave Tests of Quantum Modifications to Black Hole Structure — with Post-GW150914 Update," *Classical and Quantum Gravity,* **33**, 235010 (2016) [https://arxiv.org/pdf/1602.03622.pdf].

58 See e.g., Sec. 22.4 of M. Maggiore, *Gravitational Waves, Volume 2: Astrophysics and Cosmology,* (Oxford University Press, 2018).

59 S. Phinney et al., *The Big Bang Observer: Direct Detection of Gravitational Waves from the Birth of the Universe to the Present,* NASA Mission Concept Study (2004).

Physics 2018

"for groundbreaking inventions in the field of laser physics"

one half to

Arthur Ashkin

"for the optical tweezers and their application to biological systems"

and the other half jointly to

Gérard Mourou and Donna Strickland

"for their method of generating high-intensity, ultra-short optical pulses"

The Nobel Prize in Physics, 2018

Presentation speech by Professor Anders Irbäck, Member of the Royal Swedish Academy of Sciences; Member of the Nobel Committee for Physics, 10 December 2018.

Your Majesties, Your Royal Highnesses, Honoured Laureates, Ladies and Gentlemen,

Sunlight is essential to life on Earth, and we are getting better and better at harnessing its energy. We also use light in many other ways, today often employing a laser as the light source. Two examples of lasers in everyday use are bar code readers and laser pointers. This year's Nobel Prize in Physics honours two inventions in laser physics that have led to pathbreaking new ways of using light. These methods have given us light-based tools with applications in medicine and other fields.

Arthur Ashkin is being rewarded for the invention of optical tweezers, and their application to biological systems. This invention is based on the ability of light to exert a force on matter, known as radiation pressure. The possibility of using light, via this force, to move physical objects may make one think of Star Trek and tractor beams, and sound like pure science fiction. Of course we can feel that sunbeams carry energy – they make us warm. But we cannot feel any small push – the force of this sunlight is too weak to do so. The starting point that led Ashkin to his optical tweezers was an experiment aimed at showing that the radiation pressure in an intense laser beam is actually strong enough to move microscopic particles. As it turned out, laser light could not only move such particles, but could also be made to grab them by focusing the beam with the help of a lens. This marked the birth of the optical tweezers, an elegant tool that has gained a broad range of applications and that lets us hold on to and move objects such as living cells without touching them. Ashkin's method has been successfully used to investigate various components of biological cells, among other things providing us with knowledge about the mechanics of tiny molecular motors that perform vital work inside these cells.

Gérard Mourou and Donna Strickland are being rewarded for an invention called chirped pulse amplification, or CPA. This is a method for creating extremely intense, short pulses of laser light. Efforts to create more intense laser pulses had been under way since the first laser was built in 1960. But by the mid-1980s these experiments had reached an impasse, since the intensity of the laser light destroyed the amplifying material itself. Using their CPA technique, Mourou and Strickland were able to get around this limitation. Their strategy was simple and elegant: First stretch out the laser pulse in time, thereby reducing its intensity and allowing it to be amplified. Finally, compress the pulse again to its original length of time, but now with much higher intensity. This method changed the landscape of research about high-intensity lasers, from something that had been carried out in a few large laboratories to something that could be done in many places around the world, leading to a powerful surge in development work. The pursuit of shorter and shorter pulses has enabled researchers to move closer to the attosecond level – which means one billionth of a billionth of a second. This opens the way for studying the movements of electrons in atoms and molecules. Numerous applications of laser pulses are made possible by the CPA technique. One example is eye surgery for correcting near-sightedness, in which laser pulses serve as ultra-precision surgical tools.

Dr Ashkin, Professor Mourou, Professor Strickland:

You have been awarded the 2018 Nobel Prize in Physics for your ground-breaking inventions in the field of laser physics. On behalf of the Royal Swedish Academy of Sciences it is my honour and great pleasure to convey to you our warmest congratulations. I now ask you to step forward to receive your Nobel Prizes from the hands of His Majesty the King.

Arthur Ashkin.

Arthur Ashkin

Biography

ARTHUR ASHKIN'S FATHER, Isadore Ashkenazi, was born in Tsarist Russia in 1891. Isadore was brought up with his older brother, Harry, both shown in Figure 1, in an orphan asylum and never told of the circumstances of how they ended up there. As it turned out, his brother made the mistake of not leaving Russia by the age of 18. That meant that he was subject to army service, which at that time was akin to a death sentence or prison. The solution to this problem was to use the passport of his younger brother and leave immediately, in 1910. After getting to America he could simply mail the passport back to Isadore for his use in leaving Russia. A serious problem arose when the passport never arrived. Isadore was now stateless, but he found a man who, in exchange of his accumulated savings, would arrange passage for him on a sealed train to Holland, and then steerage on a boat headed to New York. Despite his doubts, the arrangement worked out and Isadore found himself on Ellis Island, greeted by his brother, and accepted in the country. An agent changed his name to Ashkin to make it sound more American.

Since Isadore was a "good boy" in the orphanage, he was trained as a dental technician. Once in the United States he found employment and soon opened his own business, which flourished.

He was soon able to marry Anna Fishman, also an immigrant, born in 1895, who arrived at 2 years of age with her family from what is now Poland. They can be seen in Figure 2 as a young couple at Coney Island. They found lodgings in Brooklyn and soon had a family: a boy named Julius, a second son, two years younger, named Arthur, and a daughter named Ruth, five years younger. Anna and her three children can be seen in Figure 3, a photo taken in the late 1960s. Julius was a brilliant student, skipping several grades in elementary school and eventually earning a PhD in Physics from Columbia University at age 22. During World War II he worked on the atom bomb at Los Alamos and after the war taught at

Figure 1. Isadore – father of Arthur – and his older brother Harry in Kiev, Ukraine (circa 1908).

Figure 2. Father Isadore and mother Anna at Coney Island (circa 1925).

Rochester University and Carnegie Tech. He eventually became the head of the physics department at Carnegie Tech.

Their daughter, Ruth, Arthur's younger sister, majored in Greek and Latin, and a number of years later she earned a master's degree in Greek. However, she became a school teacher and devoted herself to teaching underprivileged children in a Brooklyn elementary school.

Ashkin is the product of Brooklyn's public school system during the Great Depression. He showed an early interest in science. As a ten year old he was fascinated by the Crookes radiometer. Seeing one in a pharmacist's window, he soon had one of his own. He understood the thermal effect of light causing molecules of air, under low pressure, to cause the movement of the vanes forward from the blackened side. A few years later, after exposing the radiometer outside for some time to make it hotter, he brought it back inside the house. It started to spin in the opposite direction, the same direction as if it was driven by light pressure.

In 1940 he enrolled at Columbia College with a major in physics, but his education was soon interrupted by the entry of the United States in World War II. He was drafted into the army but spent the war years as a technician at Columbia University's Radiation Lab, where he built magnetrons for radar. While at the Radiation Lab he was mentored by Sid Millman, who later became a department head at Bell Labs. After the war he completed his bachelor's degree and earned his Phi Beta Kappa key in his junior year. As a graduate student at Cornell University under the G. I. Bill, he earned his PhD doing electron-positron scattering under Professor Bill Woodward.

After getting his PhD, Arthur Ashkin was remembered by Millman at Bell Labs, who hired him. At Bell he initially worked on electron tubes for communication. He felt working in nuclear physics would put him in competition with his brother. He resented being called "Ashkin's brother Ashkin" by his brothers' colleagues. He switched to laser work in the early sixties. He had long been interested in how the momentum of light photons could be used to move matter. He formed the two-beam laser trap early on and in 1970 published his first paper on optical trapping.<1> The subject of the paper was the two-beam trap, but it also spoke of the possibility of trapping atoms, molecules and small particles. As a manuscript to be submitted to *Phys. Rev. Lett.*, it had to be reviewed by the Theoretic Physics Department in order not to besmirch the Lab's high reputation. It was rejected on three grounds:

1. There was nothing new;

Figure 3. From left to right: Arthur, his brother Julius, his mother Anna and his sister Ruth (late 1960s).

2. There was nothing wrong with the work – a comment reminiscent of the "Pauli insult"*;
3. It could be published somewhere, but not in *Phys, Rev. Lett.*

Ashkin's boss, Rudi Kompfner, was irate and simply said, "Damn it! Just send it in." It was immediately accepted. It was published and became the most referenced paper of that year. At over 5,000 citations it was one of the one hundred best atomic physics papers of the last century, as named in "The Physical Review the First Hundred Years, a Selection of Seminal Papers and Commentaries." In 1971, Ashkin achieved levitation of small glass spheres, which were used in many important experiments, such as the optical Millikan oil droplet experiment. Finally, he made his famous single beam tweezer trap <2> in 1983. He can be seen near the optical tweezer apparatus in Figure 4, a photo which was taken soon after the discovery for which he would be awarded the Nobel Prize.

Figure 4.
Arthur next to the optical tweezer apparatus, with the image of a scallion cell in the background in 1988.

* The comment for which theoretical physicist Wolfgang Pauli (1900–1958) is perhaps known combines utter contempt on the one hand with philosophical profundity on the other. "This isn't right," Pauli is supposed to have said of a student's physics paper. "It's not even wrong."

His work has been both widely used and widely recognized. He is a member of the National Academy of Engineering and the National Academy of Sciences, as well and the National Inventors Hall of Fame. He has been elected Fellow of the APS, the OSA, the IEEE, and the AAAS. He has received numerous awards, among them the Harvey Prize, the Rank Prize, the Keithly Prize, the Townes Award, and the Ives Prize. His most recent award is the 2018 Nobel Prize in Physics for Optical Tweezers for Application to Biological Systems. At 96 he is the oldest recipient of the Nobel Prize. It is generally recognized that he should have been included in the 1997 award, since optical tweezers were used to trap atoms in the prize-awarding experiments. After difficulties with atom trapping, Ashkin and Dziedzic trapped 20 Angstrom plastic spheres to demonstrate the ability of optical tweezers to trap particle of approximatively atomic size. Subsequently, successful atom trapping using the same technique was demonstrated.

Art Ashkin and his wife, Aline, have been happily married for almost 65 years. They met at Cornell, on the shore of Lake Cayuga. She was on a Cornell Outing Club excursion and he was occupied with his prospector's pick, splitting rocks of shale looking for fossils. She was curious about what he was doing. A conversation was struck up and they both realized that something special was happening. He asked for her phone number and she willingly gave it to him. She was a junior at Cornell, majoring in chemistry and he was finishing his PhD in nuclear physics. Two years later, after she graduated in 1954, they were married. They settle in Bernardsville, New Jersey and had three children. When Bell Labs opened a new laboratory in Holmdel, New Jersey, they moved to Rumson, New Jersey, where they have been living for the past 52 years. Once the children were all off to school, Aline started working as a high school chemistry teacher. She pursued that for fifteen years, obtaining a master's degree in science education along the way.

Their children did well. The oldest, Michael, and the youngest, Daniel, both went to the University of Pennsylvania. Michael's major was in Hebrew and Arabic; Daniel's was in Earth Science and Geology. Their daughter, Judith, went to Cornell to major in Spanish and Portuguese. Today, Michael is chair and professor of art in Cornell's School of Art, Architecture, and Planning. Judith teaches Tai-Chi and recently completed a Master's Degree in Acupuncture. Daniel, with a PhD in Ceramic Science from Rutgers University works on ceramics for CoorsTek, a company run by the same family who also make Coors beer.

Art and Aline are the sole survivors of their families. The parents of both are deceased as are their siblings. His brother and sister both died in their early sixties, as did her brother. Art and Aline are still alive, enjoying their five grandchildren and two great grandchildren. A recent photo of

Figure 5. From left to right: Arthur and his wife, Aline, Bob Wilson and his wife Betsy in October 2018 in Holmdel, New Jersey.

Aline and Art (left) is shown in Figure 5. It was taken on October 9, 2018, at the top of Crawford Hill in Holmdel, New Jersey, part of the laboratory of the same name at Bell Labs. They are in company of Bob Wilson and his wife Betsy (right). Bob who also spent his career at Bell Labs, did the work for which he was awarded the 1978 Nobel Prize in Physics on that same hill.

REFERENCES

1. "Acceleration and Trapping of Particles by Radiation Pressure," A. Ashkin, *Physical Review Letters*, **24**, No. 4, Jan. 26, 1970 (pp. 156–159).
2. "Observation of a single-beam gradient force optical trap for dielectric particles," A. Ashkin, et al. *Optic Letters*, **11**, May, 1986 (pp. 288–290).

Optical Tweezers and their Application to Biological Systems

Nobel Lecture, December 8, 2018 by
Arthur Ashkin*
Bell Laboratories, Holmdel, NJ, USA.

IT IS A PLEASURE to present this summary of my 2018 Physics Nobel Lecture [1] that was delivered in Stockholm, Sweden, on December 8, 2018 by my fellow Bell Labs scientist and friend René-Jean Essiambre. This summary is presented chronologically as in the Lecture. It starts with my fascination with light as a youngster and goes on to the invention of the optical tweezer and its applications to biological systems, the work that was recognized with a Nobel Prize. Along the way, I provide simple and intuitive explanations of how the optical tweezer can be understood to work. This document is a personal recollection of events that led me to invent the optical tweezer. It should not be interpreted as being complete in the historical attribution of the scientific discoveries discussed here.

CROOKES RADIOMETER: THERMAL EFFECTS

I have always been fascinated by the forces that light can exert on objects. I started to play with a Crookes radiometer [2] in my early teenage years and tried all kinds of experiments with it. I learned that thermal effects can explain the motion of the Crookes radiometer.

---

* Arthur Ashkin's Nobel Lecture was delivered by René-Jean Essiambre.

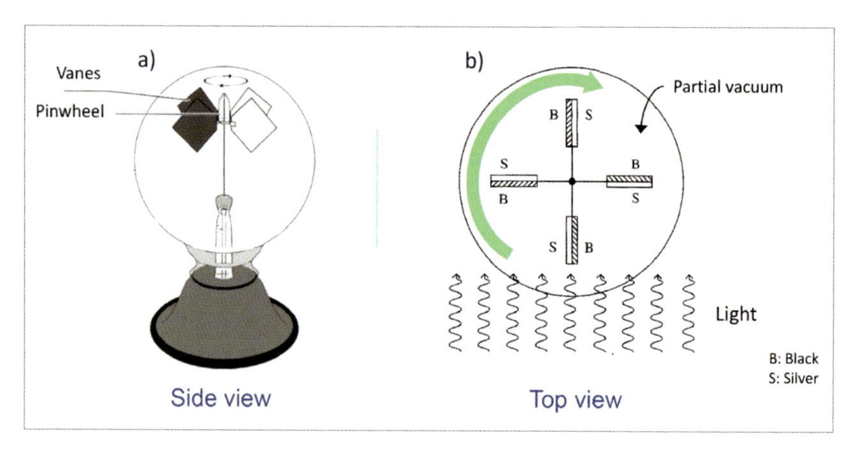

Figure 1. Crookes radiometer: a) side view b) top view showing the direction of rotation. Thermal effects are responsible for the rotation of the vanes.

Black surfaces absorb light which heats up the surface. In contrast, metallic surfaces reflect light with negligible heat generation. A Crookes radiometer, represented schematically in Fig. 1a, is a set of four vanes on a spindle with each vane having a black and a metallic side, everything being placed in a low-pressure glass bulb. They are disposed at 90-degree angles so as to always show the black side on the left and the metallic side on the right when looking at the radiometer from any side. Upon shining light on the radiometer, with light from the sun for instance, the vanes will turn in the direction where the black side moves away from the light. This direction of rotation originates from the heating of the black surface (thermal effect) by the absorption of the photons that in turn heat up the air around the black side of the vanes. The heated air moves around in the bulb resulting in a net force on the black surface that makes it move away from the light as represented in Fig. 1b.

LIGHT PRESSURE

The motion of the Crookes radiometer can be explained by thermal effects alone. During my teenage years I was acquainted with the experiment of Nichols and Hull [3], which demonstrated the effect of light pressure. This was accomplished with the development of high vacuum pumps that achieved lower pressure, therefore reducing the thermal effects until radiation pressure became the dominant effect. With a device like the Crookes radiometer, better vacuum results in the vanes rotating in the opposite direction as the one indicated on Fig. 1b.

HIGH-POWER MAGNETRONS

Shortly after the entry of the United States in World War II, I was drafted and asked to help with building a high-power magnetron in the Columbia Radiation Laboratory for radar application. I learned to make and solder high-power magnetrons for use with radar. I built a magnetron named the "rising-sun" magnetron because its geometry resembled the rays of the sun at sunrise. It operated at about 10 GHz, a frequency about 50,000 times smaller than visible light. The magnetron emitted high-intensity pulses at a frequency of 1000 cycles per seconds. If one shined it on a metallic vane of a phone earpiece and detected a thousand cycle frequency on an oscilloscope matching this frequency. I interpreted this as possibly the effect of light pressure on the receiver plate.

ARRIVING AT BELL LABS

I joined Bell Labs in New Jersey, in 1952, tasked to work on vacuum tube amplifiers. It was both interesting and time-consuming work. I joined Calvin Quate's sub-department, as it was called in those days, and was assigned to do an experiment on cancelling noise. This was a flawed idea, but it was not known at the time I started. Of course, it was doomed to failure, and I ended up being blamed for lack of positive results. After one year working fruitlessly on this topic, I was almost fired. But, fortunately, Quate saved me. This endeavor came to an end a few months later, with the arrival of Neville Robinson, a student of Rudolf Kompfner, who showed that the cancellation of noise was impossible. He showed that rather than being cancelled, the noise was transferred to other frequencies. I was barely saved, and thereafter started work of my own.

THE ADVENT OF THE LASER

Shortly after the first laser was demonstrated in 1960, I began to work with light pressure again, now with much more powerful and well-collimated light sources than previously available. The experiments performed in the early to mid-1960s were primarily exploring various linear and nonlinear effects of laser light on different materials and waveguides. This new tool that was the laser enabled a fast pace of discovery.

THE FIRST LIGHT PRESSURE EXPERIMENTS WITH A LASER

In 1966, I attended the International Quantum Electronics Conference (IQEC) in Phoenix, AZ. At the conference, Eric Rawson and his mentor Professor A.D. May from the University of Toronto presented a video showing particles behaving like "runners and bouncers" in the internal

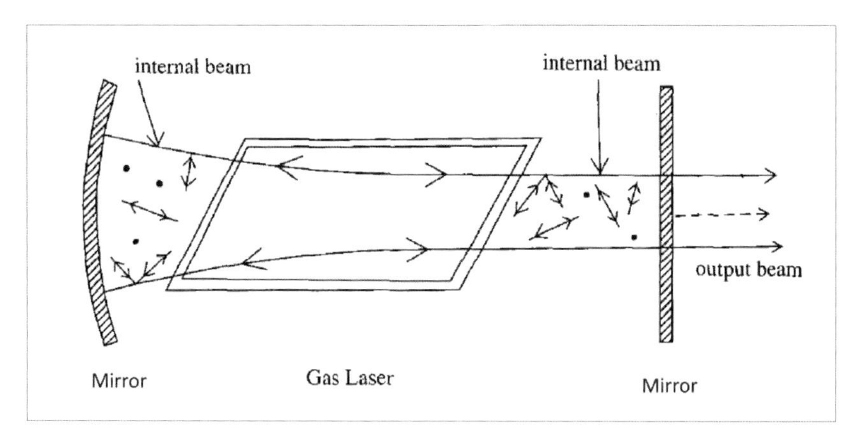

Figure 2. "Runners and Bouncers" in the internal beam of a laser cavity as observed by Rawson and May.

beam of a laser cavity [4], as depicted in Fig. 2. This curious behavior piqued my curiosity. One among various other possible explanations given for this phenomenon was light pressure. I figured out from a back-of-the-envelope calculation that light pressure could not be responsible for this behavior. Not long after the talk, all persons involved with the question agreed that the heating of the particles through absorption of light was responsible for the observed motion. The main impact of this experiment on me, however, was to reignite my desire to explore how light pressure from lasers could be observed and used to move small particles.

LIGHT PRESSURE ON A MIRROR AND ON A TRANSPARENT SPHERE

Let's consider the effect of light pressure on two different types of objects that do not absorb light: a highly reflective mirror and a highly transparent sphere. The mirror and the transparent sphere are assumed to be tiny (micrometer sized) and the effects of any other forces other than light pressure are considered negligible. First, what happens when a particle of light, a photon, hits a perfectly reflecting mirror at normal incidence as represented schematically in Fig. 3a? Of course, the photon is reflected, and its momentum is reversed. The mirror experiences a slight recoil, opposite to the change of momentum of the photon, so that the overall momentum is conserved. Now, let's consider a transparent object such as a small sphere made of glass or polystyrene as depicted in Fig. 3b. What follows is the core principle on which the operation of optical traps and optical tweezers is based. When a photon is incident on the sphere on its outer side, it generally goes through the sphere and is deflected towards the center of the sphere. This change of direction of the photon corre-

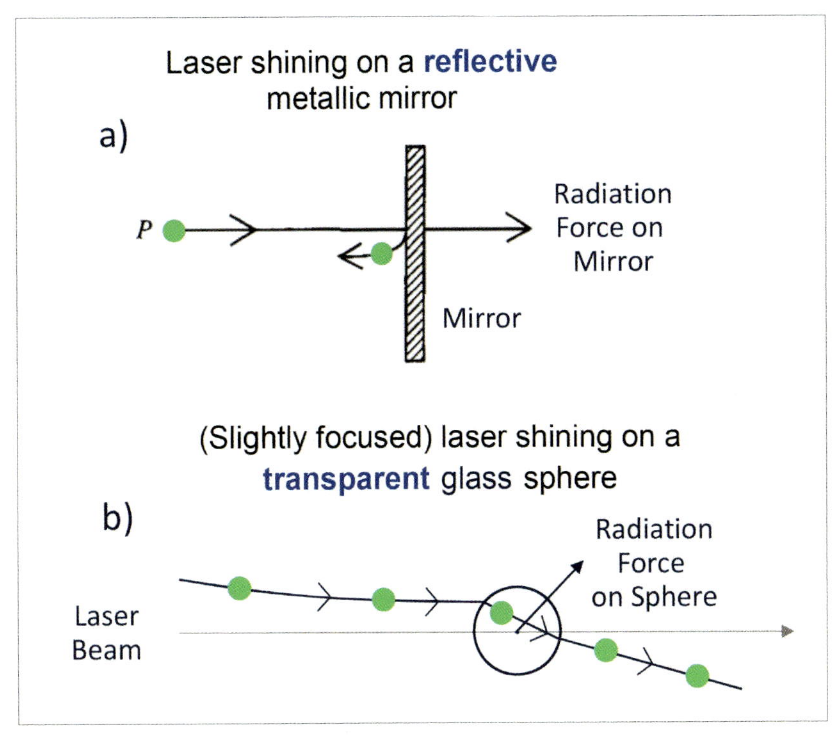

Figure 3. Two scenarios that show the impact of light pressure: a) a photon incident on an ideal reflective mirror; b) photons incident on a transparent sphere. When the mirror or the sphere are small (micrometers in size) and many photons are incident, the radiation pressure forces exerted on these objects can exceed gravity by several orders of magnitude.

sponds to a change of momentum. From conservation of momentum, the transparent sphere reacts and acquires a motion. Thus, a photon hitting the upper part of a sphere exits the sphere in the downward direction as shown in Fig. 3b. This change of direction of the photon induces a change of momentum that results in a force that makes the sphere move both forward and upward. On the other hand, when a photon hits the lower part of the sphere, the photon is deflected in the upward direction. The sphere then experiences a force that has a forward and a downward component. When two photons symmetrically located on each side of the sphere hit it simultaneously, the sphere moves only in the forward direction as the downward and upward components cancel. I gave the label "scattering force" to the component of the light pressure force on a transparent sphere as it arises from the scattering of light. As a consequence, the small sphere moves in the direction of the beam. An estimate of the scattering force exerted by a one-watt laser on a glass sphere of one micrometer in diameter gives a force that is six orders of magnitude larger than gravity. This is a tremendous force.

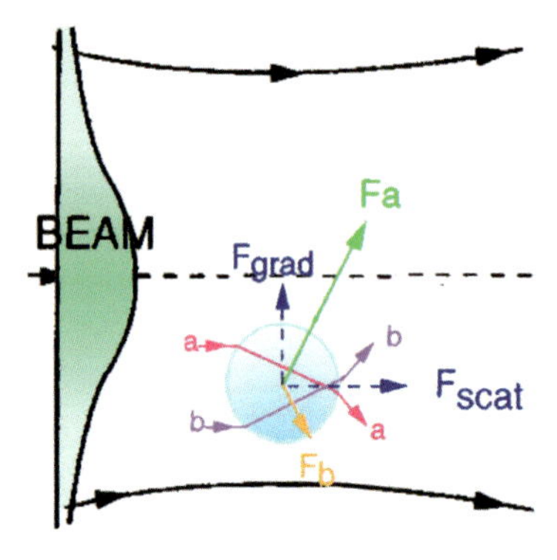

Figure 4. Forces exerted on a transparent sphere located off-axis in a laser beam. The force Fa resulting from photons hitting the upper part of the sphere is larger than the force Fb coming from the photons arriving on the lower part of the sphere due to a larger number of photons incident on the upper part of the sphere (higher laser intensity).

MOTION OF A SPHERE WITHIN A LASER BEAM

A laser beam is not uniform in the transverse direction as it follows the distribution of the spatial modes of the cavity. Let's consider what the net effect of light pressure is on a transparent sphere located off-center within a laser beam as depicted in Fig. 4. A typical laser beam has its highest intensity near its center, with the intensity decreasing gradually with the distance from the center. We refer to this difference in light intensity as light gradient, with a positive gradient being defined as the direction from low to high light intensity. In the case of the transparent sphere located off-axis as in Fig. 4, there is more light intensity hitting the upper part of the sphere (path a) closer to the beam center than the lower part (path b) further away from the beam center because it is closest to the region of the beam that has the highest intensity. Because there are more photons near the center of the beam, the net effect of the gradient radiation pressure is to push the sphere towards the center of the beam. I refer to this light pressure force as "gradient force" as the sphere is pulled towards the region having the highest gradient of intensity. As a result of light pressure, the sphere in Fig. 4 will move forward due to the scattering force and, at the same time, will move upward toward the center of the beam due to the gradient force. It will then stop when it hits the vertical microscope slide in Fig. 4 and be trapped at the center of the beam. It is trapped from the left by the laser beam and from the right by the microscope slide and can be referred to as an opto-mechanical trap.

TWO-BEAM OPTICAL TRAP IN LABORATORY NOTEBOOK

In 1969, I thought that placing a particle between two laser beams would create the first all-optical trap due to light pressure. Figure 5 shows two parts of an entry to my notebook. Figure 5a shows the schematic of the two-beam all-optical trap. In this figure, it uses a donut-shaped mode to trap a metallized sphere. Transparent spheres are discussed later in this

a)

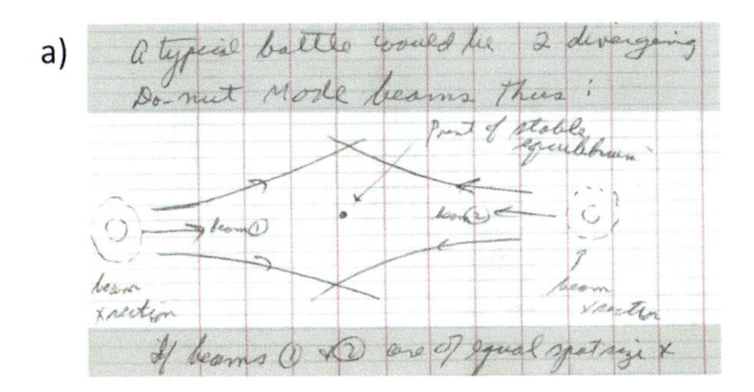
A typical bottle would be 2 diverging Do-nut Mode beams thus:

Point of stable equilibrium

beam ①

beam ②

beam xsection

beam xsection

If beams ① + ② are of equal spot size +

b)

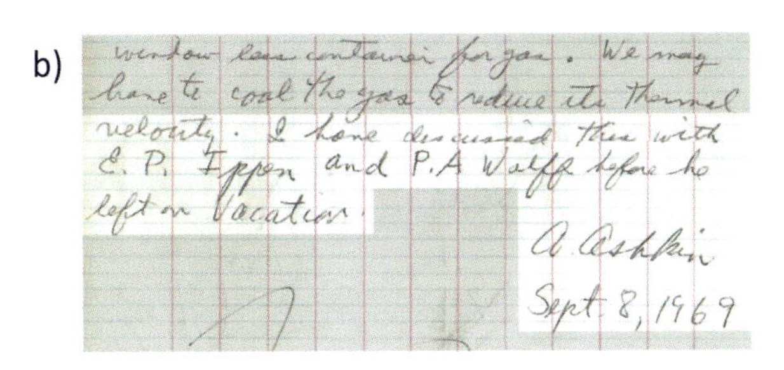
window less container for gas. We may have to cool the gas to reduce the thermal velocity. I have discussed this with E. P. Ippen and P.A Wolff before he left on vacation.

A. Ashkin
Sept 8, 1969

Figure 5. Two excerpts of an entry in my laboratory notebook at Bell Labs on September 8, 1969. It presents the idea of a two-beam all-optical trap. From Bell Labs archives.

entry. Figure 5b also shows the two witnesses of this entry, Erich P. Ippen, who became Professor at MIT and Peter A. Wolfe, who was my boss at the time. This led to the first paper on all-optical trapping in 1970 [5].

OPTICAL LEVITATION

In the two-beam trapping experiment, each laser provides the opposing force to trap the sphere. By using a single laser oriented upwards, one can compensate for gravity. As a result, the sphere is levitated in the air. Because light pressure can produce forces much larger than gravity, only low powers are necessary to achieve optical levitation. Figure 6 shows pictorially how the scattering force of a slightly diverging laser compensates for the gravitational force on the sphere. By changing the power of the laser one can change the height of the sphere.

The light pressure from the laser beam oriented upward can be used to levitate particles. Consider a glass ball resting on a microscope slide. If a low power laser beam goes through the ball from below, the ball experiences an upward

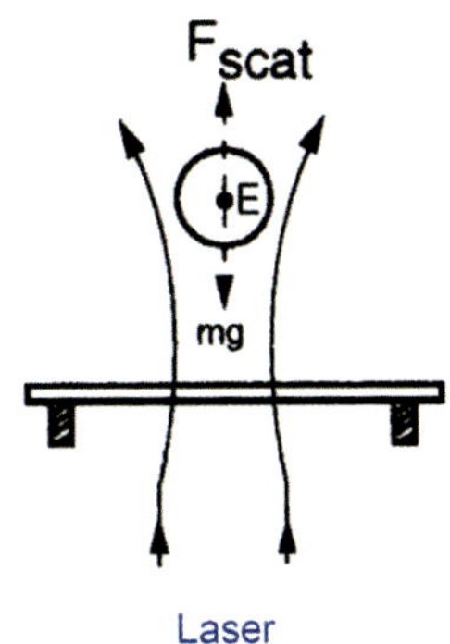

Figure 6. Transparent sphere levitating above a microscope slide as a result of light pressure from a laser pointing upwards.

force but does not move. As the power is increased above a certain threshold the scattering force exceeds gravity (neglecting van der Waals forces), and the ball will start to levitate. The final height reached depends on the laser power and its divergence.

OPTICAL TWEEZERS

Finally, in 1983, I discovered that a transparent sphere could be trapped by a single highly focused laser beam. I named this single-beam trap "optical tweezers". It is generated by using a large numerical aperture microscope objective. Such an objective produces a strong gradient of intensity along the direction of the laser beam near the laser focus as depicted in Fig. 7. This axial gradient generates a light pressure force in the backward direction relative to the beam and of a magnitude that can exceed the forward scattering force. At a certain distance downstream but close to the beam focus center, both the gradient and scattering light pressure forces are equal. This is the equilibrium point where a sphere is trapped. This invention of the optical tweezers is cited in the 2018 Nobel Prize for Physics.

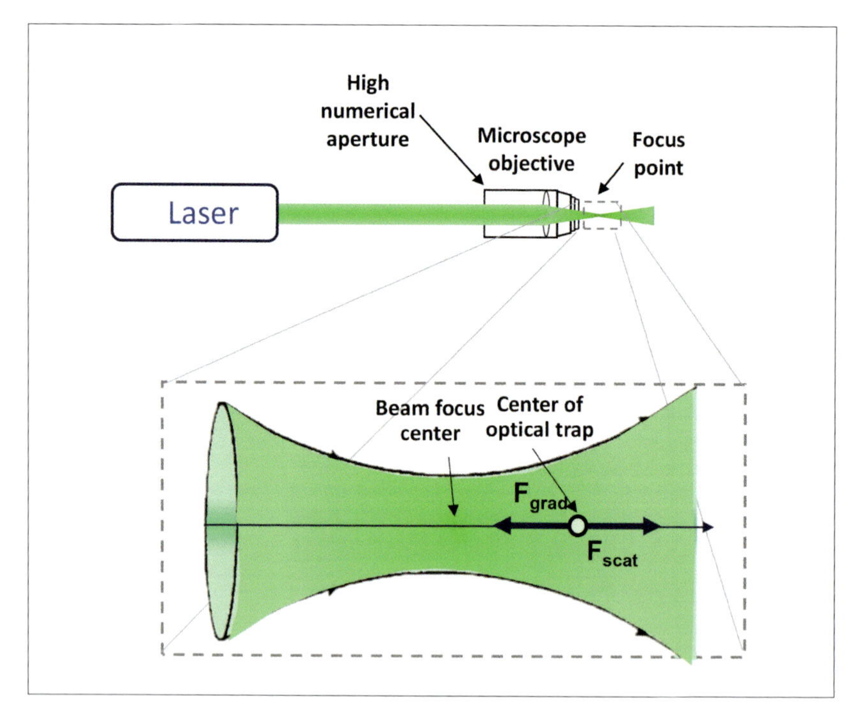

Figure 7. The layout of an optical tweezer (above) along with a magnification near the beam focus. A transparent sphere settles on the beam axis just beyond the beam focus where the gradient and scattering forces are equal.

OPTICAL MANIPULATION OF MICROORGANISMS

The understanding and development of optical tweezers were based on transparent spheres but were not limited to this ideal shape. They can also apply to other objects that can greatly depart from sphericity. It simply requires that the sum of the rays traversing a portion of an object is such that a net gradient force becomes comparable in magnitude to the scattering force. When this happens, the object can be "grabbed" by the optical tweezers and freely manipulated. The only requirement is that the object lets some light through. Note that most "small living things" are highly transparent at certain wavelengths where high-power lasers exist.

One of the first microorganisms manipulated by optical tweezers was the protozoan, *Paramecium*: a large, single-celled organism measuring between 50 to 330 μm in length. The internal components are called organelles and they range in size from a few μm to a few tens of μm. A picture of *Paramecium* with its organelles is shown in Fig. 8. The dimensions of these organelles are on the same order as the beam waist of a 1-micrometer laser, and therefore a natural subject to trap optically. Soon after we put together the set-up, we were able to manipulate its organelles. The circle, indicated by the arrow in Fig. 8, shows the organelle that could be moved around inside *Paramecium* or trapped in space. In the latter case, the trapped organelle eventually escapes the optical tweezer trap when it hits the walls of the moving cell.

In a period of over a year, we went on to use optical tweezers to manip-

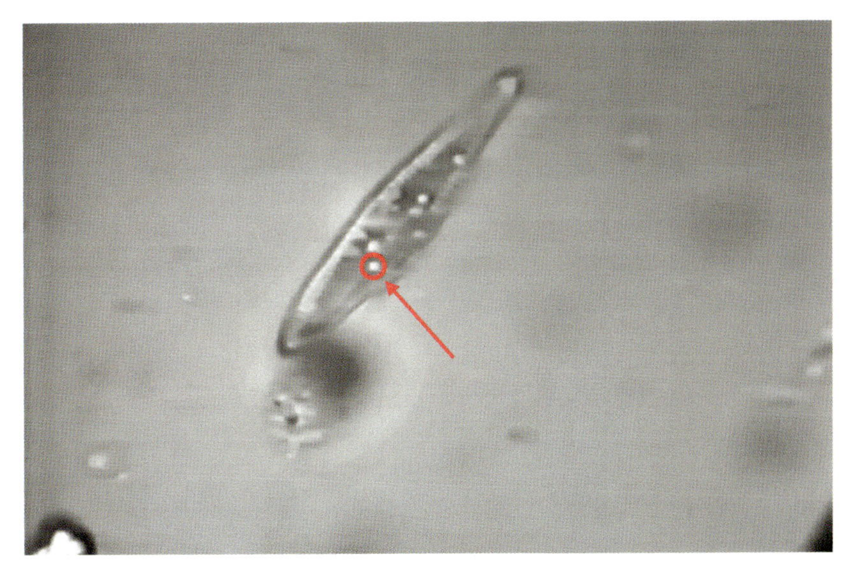

Figure 8. Picture of a paramecium where one can see the internal organelles. The arrow points to the specific organelle that we manipulated using an optical tweezer (see Nobel Lecture [1] for the video).

ulate "all sorts of living things". This included tobacco mosaic viruses (TMVs) that can be trapped easily by its ends. We also trapped sperm cells, manipulated chlorophyll in onion cells, and trapped various other microorganisms including bacteria.

OPTICAL TWEEZERS TO MEASURE THE MOTION OF KINESIN

Motor proteins are enzymes that move objects inside living cells. Optical tweezers are particularly well suited to measure the displacement of individual motor proteins, revealing their dynamics and properties.

Figure 9a shows a single motor protein, called kinesin, attached to a tiny plastic bead, about half a micrometer in diameter [6]. The bead is first captured by an optical trap and then positioned directly over a microtubule. The kinesin molecule binds to the microtubule and begins to walk out of the trap, carrying the bead with it. Its motion is fueled by molecules of adenosine triphosphate (ATP) in the solution. The displacement of the bead is measured using a position-sensitive detector that monitors tiny deflections of the laser beam produced by the bead moving inside the trap. Figure 9b shows three different records of displacement, all showing that kinesin takes discrete steps as it moves along the microtubule. Each step measures just 8 nm.

Steven Block is one of the pioneers who studied the properties of individual biomolecules using optical tweezers, and his lab was the first to measure directly the steps taken by kinesin motors. He started his research with optical traps within days of the publication of our study in *Nature*, which reported that living bacteria, such as *E. coli*, could be

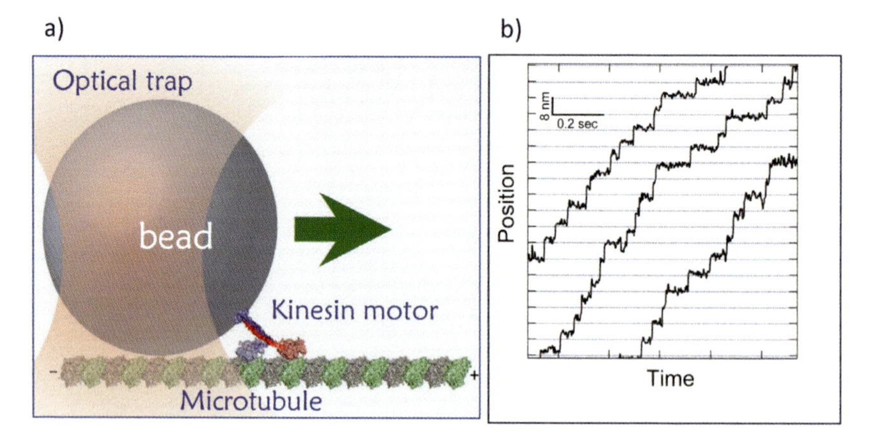

Figure 9. a) A schematic representation of the transport of a transparent bead by a kinesin molecule walking on a microtubule; b) three measurements of the position of the bead with time showing that the motion occurs by discrete steps of 8 nm. With permission from Steven Block.

trapped noninvasively by optical tweezers. Block first worked with his mentor, Howard Berg (at Harvard), to measure the nanomechanical properties of individual bacterial flagella, which propel swimming bacteria. He later built a series of sensitive optical trapping instruments that allowed him to study individual motor proteins, such as kinesin and myosin, nucleic acid enzymes, such as RNA polymerase, exonuclease, and helicase, and folded RNA molecules, like hairpins and riboswitches.

MEASUREMENTS OF GENE TRANSCRIPTION – DUAL OPTICAL TWEEZERS

Gene transcription by a single molecule of ribonucleic acid (RNA) polymerase (an enzyme) can be followed as it moves along a template of deoxyribonucleic acid (DNA), synthesizing a corresponding RNA [7]. In a technique represented in Fig. 10, one end of the DNA template is attached to one bead, while a single RNA polymerase enzyme is attached to the second bead. This forms a bead – DNA – bead chain, also called a single-molecule "dumbbell" assay. Each bead is trapped by a separate optical trap, with one trap being stronger than the other. In this configuration, the assay can be used to measure both the force and the displacement that the polymerase enzyme experiences. As the enzyme proceeds along the DNA template, transcribing the corresponding RNA, the distance between the two beads changes. The video shown in the Nobel Lecture [1] displays how the separation between the beads, and therefore the location of the RNA polymerase, changes over time. High-resolution records of the relative motion of the beads reveal interesting RNA polymerase behavior, including transcriptional pausing, stepping, backtracking, and termination.

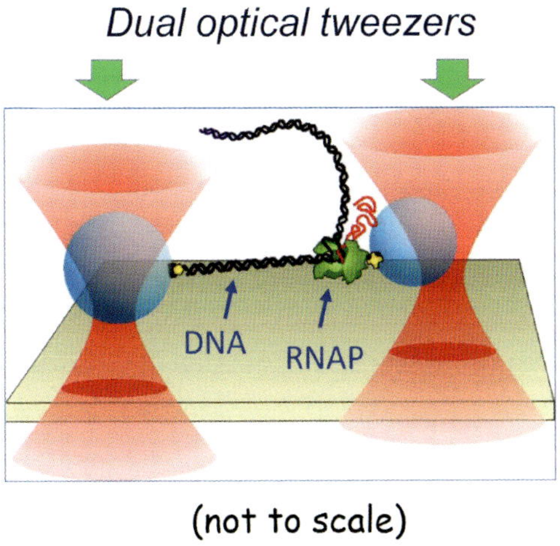

Figure 10. Two optical tweezers hold transparent spheres with one attached to a DNA template and the second attached to an RNA polymerase (RNAP) enzyme, actively transcribing RNA. With permission from Steven Block.

MEASUREMENT OF BASE-PAIR STEPPING BY RNA POLYMERASE

Figure 11 shows a measurement of the extension between the two beads of the single-molecule dumbbell assay as a function of time [7]. Note the transcriptional pauses, which occur reproducibly, at specific locations in the DNA template. This is an extremely sensitive apparatus that allows very precise measurements of displacement. At the large magnification shown in Fig. 11, we notice that the distance between the beads always changes in discrete increments of 3.4 Ångströms, which corresponds exactly to the distance between successive base pairs of the DNA double-helix. This was the highest-resolution measurement ever made directly on a single enzyme at that time. Optical tweezers, in combination with atomic force microscopy, is now being used to study these and other DNA-dependent processes, such as the work of Thomas Perkins at JILA/University of Colorado.

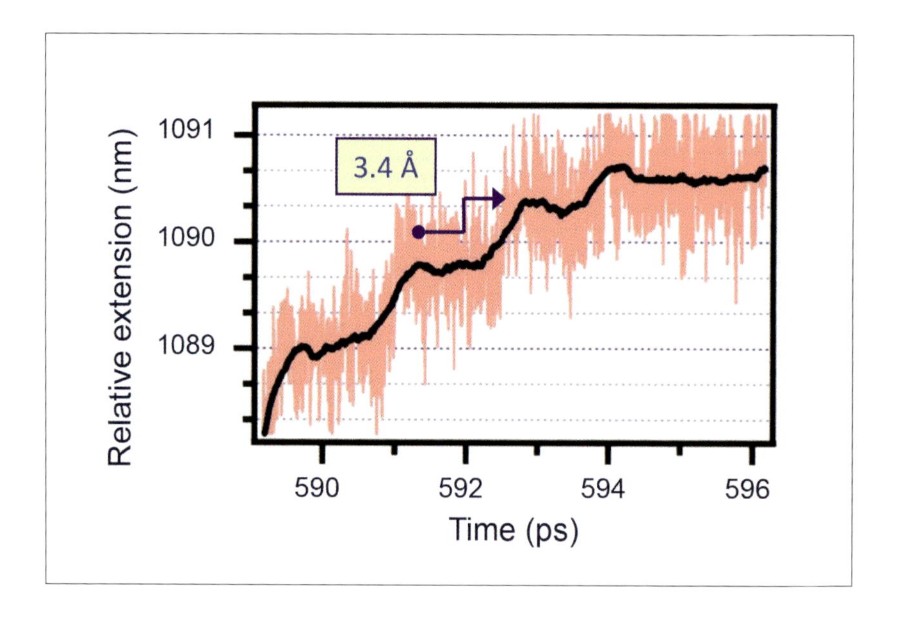

Figure 11. Measurements of the extension between the spheres as a result of the transcription of DNA by the RNA polymerase. A step size of 3.4 Ångströms was measured, which corresponds to the distance between the adjacent base pairs in the DNA double helix. With permission from Steven Block.

OTHER BIOLOGICAL APPLICATIONS OF OPTICAL TWEEZERS

There are many other applications of optical tweezers in biology such as

1) Measuring the properties of biopolymers produced in living organisms such as DNA and RNA
2) Measuring motion and forces of molecular motors, linear and rotary, that are agents of movement in living organisms
3) Studying the folding of proteins and structured nucleic acids
4) Studying the binding and assembly of biomolecular components
5) Micro-manipulation of small objects in general, including cells and organelles
6) ... and probably many other applications yet to come.

There are now several companies commercializing optical tweezers for diverse applications.

CONCLUSIONS

Since the invention of the optical tweezer in 1983, the technique has proven to be applicable to a wide range of particle sizes, from small numbers of atoms up to entire living cells. A key advantage has been the formidable power with which optical tweezers are able to manipulate and measure the motion of tiny objects. This has proven to be of particular value in studying the behavior of biomolecules. However, much work remains to be done in this area. Many of the nanomechanical properties of even the most essential proteins and nucleic acids remain unknown, and further investigation is needed to elucidate important processes involving these biomolecules. Advanced physical techniques, including optical tweezers, scanned-force microscopy, and the like, offer greatly improved capabilities to understand life's fundamental mechanisms.

ACKNOWLEDGEMENTS

I would like to acknowledge the contributions of my dear wife, Aline, my long-time research partner Joseph Dziedzic with whom many of the optical trapping experiments were performed, a close colleague, John Bjorkholm, and many other collaborators. I owe special thanks to René-Jean Essiambre for helping prepare and delivering the Nobel Lecture and for contributions to this summary. I am also indebted to Steven Block for graciously sharing materials, providing comments, and for a careful reading of the manuscript.

BOOK

I wrote a comprehensive book on optical trapping [8] with the help of my wife, Aline. It contains a description of the discovery and evolution of optical trapping from its inception until 2005. It also includes reprints of key articles on optical trapping. Of course, work continues as new discoveries are made. One of the motivating factors for writing the book were assertions made by certain researchers to the effect that magneto-optical traps (MOTs) were "much more important" than optical tweezers for trapping. I did not accede to this school of thought, and therefore decided to write a book demonstrating the power of using radiation pressure to trap tiny objects.

REFERENCES

1) Arthur Ashkin Nobel Lecture entitled "Optical Tweezers and their Application to Biological Systems", https://www.nobelprize.org/prizes/physics/2018/ashkin/lecture. Delivered in Stockholm, Sweden on December 8, 2018.
2) W. Crookes, "On Attraction and Repulsion Resulting from Radiation". *Phil. Trans. Roy. Soc.* London, Vol. **164**, pp. 501–527 (1874).
3) E. F. Nichols and G. F. Hull, "A preliminary communication on the pressure of heat and light radiation," *Phys Rev.* Series I, pp. 307–320 (1901). Also demonstrated by P. Lebedev, "Untersuchungen über die Druckkräfte des Lichtes," *Ann. Phys.* (Leipzig), Vol. **6**, pp. 433–458 (1901).
4) E. G. Rawson, A. H. Hara and A. D. May, "Orientation Stability and Propulsion of Dust Particles in a Laser Cavity," International Quantum Electronics Conference (IQEC), Phoenix, AZ, Paper 8C-3 (1966); E. G. Rawson and A. D. May, Propulsion and angular stabilization of dust particles in a laser cavity, *App. Phys. Lett.*, Vol. 8, pp. 93–95 (1966).
5) A. Ashkin, "Acceleration and trapping of particles by radiation pressure", *Phys. Rev. Lett.*, Vol. **24**, No 4, pp. 156–159 (1970).
6) K. Svoboda, C. F. Schmidt, B. J. Schnapp, and S. M. Block, "Direct observation of kinesin stepping by optical trapping interferometry," *Nature*, Vol. 365, pp. 721–727 (1993).
7) E. A. Abbondanzieri, W. J. Greenleaf, J. W. Shaevitz, R. Landick and Steven M. Block, "Direct observation of base-pair stepping by RNA polymerase," *Nature*, Vol. **438**, pp. 460–465 (2005).
8) A. Ashkin, "Optical Trapping and Manipulation of Neutral Particles using Lasers: A Reprint with Commentaries", edited by World Scientific Press (2006).

Gérard Mourou. © Nobel Prize Outreach AB. Photo: A. Mahmoud

Gérard Mourou

Biography

'YESTERDAY I HAD THE HONOUR of drinking coffee with Madam Joliot-Curie'. Those were the words of my father, who was an engineer in charge of the electricity network of The Three Valleys in Savoy. He went on to say that she was a very important personality who had received the Nobel Prize. I was about five years old and it was the first time that I had heard the Nobel mentioned. The rapid electrification of France after the war made it necessary for my father to travel. I was born in 1944 in Albertville, Savoy, with my two sisters, Jacqueline and Michèle. A few years later we moved to Moûtiers, the gateway to the Tarentaise and The Three Valleys, with its well-known skiing resorts, Courchevel, La Plagne and Val d'Isère, etc… I keep in mind an unforgettable memory of the town of Moûtiers, braced between those majestic mountains. Its cathedral and medieval bridge spanning over the tumultuous Isère were impressive and fascinated me. I learnt how to ski at Moûtiers on the slopes of the Champoulet hill. In springtime there were walks with the schoolmaster or priest on Thursdays in the mountains, so high up for the child that I was. In summertime my parents sent me to a boarding house with an authentic Savoyard family living in a chalet in which the main room was divided by a curtain into two parts separating the cows from the inhabitants. This was pure bliss, and those moments will always remain with me. I was nine years old when my father was transferred to La Voulte-sur-Rhône. The Rhône valley replaced my beautiful mountains. I had to leave my skis and my sledge behind. It was heartbreaking for me, reminiscent of the child Kane's Rosebud scene in the magnificent film "Citizen Kane". La Voulte, with its famous rugby team, was very sports-oriented and I replaced skiing with swimming, which I loved.

My father, being an electrical engineer, knew about electro-magnetism, which is a starting point for many fields in physics. He enjoyed sharing this knowledge with me in the way of small problems. I was fascinated by light and the speed of light, which travelled distances seven times the circumference of the Earth or the distance from the Earth to the Moon in one second. That is how he came to explain that we were nowhere near seeing Martians on Earth, contrary to what sensational newspapers were announcing. He taught me how to play chess, which we often did in the evening.

The summer holidays were either spent in Provence with my grandparents on my mother's side or in the Aude at Espéraza near Carcassonne. In Provence we spent our holidays with our cousins in the medieval village of St Martin-de-Pallières where the castle dominates the village and its park offered us a marvellous playground. The rhythm of our daily life was set by our grandmother, Mané, who we all loved and respected. It was a simple and modest way of life. We would go and retrieve drinking water from the fountain at the bottom of the village. The village school-teacher was my uncle, a voracious reader with an enormous library. Before the advent of television, he along with the school-teachers from the neighbouring villages had created a cinematographic club which regularly featured what today are the great classics. This period certainly awoke in me my deep interest in classic films. Sometimes we would have a visit of one of my mother's first cousins who had two daughters: Andrée and Marcelle. Marcelle, with whom I had much in common, was later to become my wife.

In the Aude at my grandparents' on my father's side, my grandfather was a fascinating character due to his joviality, his kindness, his talent for story-telling and his dexterity. My grandparents were very endearing, and it was a pleasure to spend my holidays with them. My grandmother was an excellent cook and as to my grandfather, he was a brilliant do-it-yourself man with golden hands. He had a workshop where we spent a lot of time together inventing and innovating. Although he had only received a primary education, he loved history and we spent many hours visiting Cathar castles or talking about how the locks along the marvellous Midi Canal worked. He was also an excellent gardener with two gardens at both ends of Espéraza, Fa and Couiza. As we left the table after one of those Sunday meals prepared by my grandmother, my grandfather would invariably ask which way we would prefer to go? To Fa or to Couiza? This was to amuse me a few years later when reading Proust and Madame Verdurin would also invariably ask her hosts the same question: a walk on the Swann side or on the Guermantes side? I was overwhelmed to stay with our relatives, hearing their typical and irresistible southern accent and expressions. It was a bliss to be soaked with my cousins during few weeks in a medieval town like Carcassonne or in a cathar city like Fanjeaux, where I had my first taste of cassoulet de Castelnaudary and blanquette de Limoux the ancestor of champagne.

My father wanted his son to have a technical education, and a very renowned school near Grenoble at Voiron offered advanced technical studies in preparation for the School of Arts et Métiers. It was the Voiron National Technical Lycée, in the Grenoble region, commonly known as the Nat. We sat a difficult exam in order to gain entry to the school in the fourth year. We studied general subjects: mathematics, French, English or Italian but also a number of technical subjects such as industrial design, mechanics and workshops. I was an average pupil, a dreamer and easily distracted. At that time these defects terrified my parents, especially my mother, but they would become precious assets later on in the course of my career as a researcher.

After a Technical High School Diploma and the baccalaureate, in 1963 I entered the University of Grenoble, where I passed a Master's degree in physics in 1967. Meanwhile Marcelle, who was an excellent gymnast at a national level, entered the École Normale Supérieure of Physical Education (ENSEP) which was then at Chatenay-Malabry near Paris. In the same year we married and went up to Paris, Marcelle in order to complete her training at the ENSEP, and I to start a postgraduate degree in physics (DEA) in order to do a Ph.D. We were living in Chatenay-Malabry near Sceaux. It was predestined. Our apartment belonged to the same landlord family that rented to Pierre and Marie Curie in Sceaux, south of Paris. The landlord was a charming talkative man who was proud to tell me that he had manuscript letters from Mme Curie. One day as we were in the garden, I asked him if it would be possible to see the letters. He went back to the house to bring me a note written by her, complaining of a leaking faucet that needed to be fixed urgently. The great couple was not working only on lofty problems. They had also their pedestrian moments. As I was in search of a post-graduate subject, I came across Jacques Ducuing, who had just come back from Harvard and who was starting a post-graduate degree in Non Linear Optics. That was in 1967 and it was the first time I had heard this term which, associated with the laser, immediately captivated me. I applied for this post-graduate degree straightaway.

Now I had to find a research laboratory. At the Optics Institute I came across an advertisement looking for a researcher in the laser field. It offered a position in research at the Ecole Polytechnique in Professor Vignal's laboratory directed by Alain Orszag. My career as a researcher was launched.

MY FIRST SCIENTIFIC WORKS AT THE ÉCOLE POLYTECHNIQUE

Of course, during the first months I was very impressed to find myself at the very spot where so many great scientists such as Ampère, Fresnel, Fourier and many others had distinguished themselves.

This was only 7 years after Theodore Maiman's laser demonstration.

The work I was assigned was in fact a prelude to the work that led to the Nobel Prize. It consisted in the analysis of the frequency of a Q-Switch laser. The latter was not constant and varied during the pulse time. It presented a frequency drift, which in English is called a 'chirp', during the pulse time. The idea of this project was to exploit this 'chirp' in order to reduce the duration of the pulse by compression. This experience in the nanosecond regime did not present much interest, the 'chirp' being too weak and unable to lead to spectacular results. However, it was very important for my experimental formation and would be demonstrated 10 years later with success and thus permit the production of the first femtosecond pulses corresponding to a number of optical cycles. Combined with amplification it would be the basis for CPA (Chirped Pulse Amplification), for which I was awarded the Nobel Prize in 2018.

After passing my DEA in 1968, in conditions complicated by the May demonstrations, in 1970 I passed my doctoral thesis of the 3ème cycle (equivalent at that time to a Masters thesis) on the drift frequency of Q-switched lasers. During this period we also had our first child, Julien.

THE CALL OF NORTH AMERICA, MY FIRST STEPS IN THE PICOSECOND FIELD

In 1970, I began my military service. At that time, it was possible to serve as a scientific cooperant in French-speaking countries. My thesis director Georges Bret, who founded the Quantel company, introduced me to Professor Marguerite-Marie Denariez-Roberge from Laval University. She was one of the first people in North America to have a picosecond laser. For three years I studied the kinetics of dye in the picosecond field. I was passionate about this field. I was still enrolled at the University of Paris and in contact with Professor Guy Mayer, with whom I prepared and passed my doctoral thesis (Doctorat d'État, which is equivalent to a PhD) in October 1973. During this period our second son, Vincent, was born.

After my thesis, I, together with my little family left for a post-doctoral year at the University of California in San Diego. Employed by Professor Michael Malley, an extraordinary man who had a picosecond laser but above all also had one of the first Optical Multichannel Analysers, OMA. The OMA allowed us to do away with photographic plates. It was far more sensitive while permitting the recording of light signals in real time.

Equipped with a detector well-suited to the task, I was the first to discover the manner in which to observe the movement of molecules or to measure the fluorescence time of the latter in the picosecond field. Personally, I consider this to be my first major discovery. I was experiencing epiphanic moments which attracted me to the field of ultrafast physics in a definitive and irreversible way.

RETURN TO FRANCE, INTRODUCTION TO ULTRAFAST PHYSICS ON THE PALAISEAU PLATEAU

Ever since I had left France, I remained on very friendly terms with Alain Orszag. We agreed that after my post-doctorate year in California, I would return to the Laboratory of Applied Optics, the LOA. We thus returned to France in 1974 and I introduced ultrafast optics to the LOA with the help of André Antonetti and also Gilbert Bourdet. At the same time the École Polytechnique ('X') was moved to the Palaiseau plateau and the LOA relocated to the ENSTA (Ecole Nationale Supérieure de Techniques Avancées) premises. We were living in this nice town of Dourdan in the south-west part of Paris. Our house was at the foot of the beautiful castle built by Philippe August in the XIV century. Julien and Vincent had a wonderful time.

At that period, I read David Auston's article about the use of the picosecond laser for electronic switching with just a gap on intrinsic silicon. The switching was realised by the creation of carriers, produced by the laser photons in the silicon gap. I marvelled at the simplicity and the elegance of the device which could produce perfectly synchronised high voltage pulses without any laser 'jitter'. Among the applications, I saw the switching of streak cameras which had a serious 'jitter' problem. The first experiments on switching were realised at the LOA with Antonetti but also Alain Migus who had joined us with financial help from the CEA (Commissariat for Atomic Energy). During this period, we had the pleasure of welcoming Michael Malley on a sabbatical year at the ENSTA. Michael was assembling the first femtosecond 'dye' laser with Alain Migus and Jean-Louis Martin, who had joined us, bringing with him information and advice given by Erich Ippen from the Bell Laboratories. Our switching experiments were presented in 1978 to the Conference on Lasers and Electro-Optics (CLEO) in Washington. My results attracted the attention of Wolf Seka from the Laboratory of Laser Energetics (LLE) at Rochester, NY. After an animated discussion, Wolf understood the interest of introducing femtosecond pulses in inertial fusion to diagnose implosion. A few months later, I received a phone call from Wolf telling me that the LLE was offering me a position as a scientist. I had thoroughly enjoyed my sabbatical year in the US and also had the feeling that I would evolve more easily in this field that appealed to me in the US. After talking it over with Marcelle, a few weeks later we decided to accept the offer and in September 1977 we left for Rochester along with our two children. However, both of them were sad to leave beautiful Dourdan.

ROCHESTER

The LLE had just been built. This laboratory was financed essentially by the Department of Energy (DOE) and dedicated to inertial laser fusion. My first impression was that I had made a big mistake. Laser fusion was not my field. I preferred smaller projects, less programmatic ones. However, a few months later my opinion was to change radically.

The University of Rochester in the 1980s played an important role in the development of the field of ultrafast science and technology. The Institute of Optics and the Laboratory for Laser Energetics (LLE) occupied centre stage. The Institute of Optics provided exceptional students and LLE a unique technical platform. Many of the techniques that the researchers in the field use today, like THz generation, picosecond electron diffraction (PED), electro-optic sampling (EOS), chirped pulse amplification (CPA), and jitter-free synchronization, were conceived and demonstrated by the ultrafast science group. The Institute of Optics students – Wayne Knox, Theodore Sizer, Irl Duling, Janis Valdmanis, James Kafka, Donna Strickland, Maurice Pessot, Jeffrey Squier and John Nees – formed the core. Their enthusiasm was infectious and contributed much to attract students from physics, such as Steve Williamson, Theodore Norris, and Kevin Meyer, and from electrical engineering, Daniel Blumenthal, John Whitaker, and Doug Dykaar, as well as faculty like C. W. Gabel, Robert Knox, Charles Stancampiano, Thomas Hsiang, Roman Sobolewski, Adrian Melissinos, Joseph Eberly, and David Meyerhofer.

In the early 1980s ultrafast science was dealing with eV energy-level phenomena. Our group extended its range into the meV on one side, with the introduction of THz beams and electro-optic sampling (EOS) techniques, and to the MeV-GeV on the other side, with chirped pulse amplification (CPA) and its ability to produce relativistic intensities. Work in this area mainly started at LLE in 1978 after my arrival. LLE was running a highly programmatic effort on inertial confinement fusion. At that time the director and founder, Moshe Lubin, and later Robert McCrory understood the importance of creating and supporting in parallel to the main research activity a group that would work on weakly related laser fusion projects, which could offer the flexibility and the type of environment that PhD research demands. I would be in charge of this group, known as the ultrafast science group.

I was impressed by the work of Dave Auston at AT&T Bell Laboratories, which demonstrated that electrical signals could be switched with picosecond precision. Here I had the opportunity to demonstrate that this simple technique could find some important applications in laser fusion because of the need for synchronised high voltage pulses for active pulse shaping or for jitter-free streak cameras.

An exceptional undergraduate student, Wayne Knox, shared an understanding of the importance of this line of research. At Wayne's high speed of progress, we extended Auston's work to very high voltage and applied it to the synchronisation of streak cameras. For the first time the streak camera could be used in accumulation mode. Weak luminescence signals could be accumulated, improving their signal/noise ratio. The jitter-free streak camera found immediate applications in photobiology with the group of Wayne's father, Professor Robert Knox. This technique is now routinely used in synchrotron-based femtosecond x-ray diffraction experiments. High voltage switching also has applications in active pulse shaping, as demonstrated in collaboration with John Agostinelli (student of C. W. Gabel), and in contrast improvement with Wolf Seka. This technique is still used today in high field science.

ELECTRON DIFFRACTION

A streak camera is a beautiful photon-electron transducer. It makes an electron replica of the photon pulses. The electrons are deflected across the phosphor screen, leaving a phosphorescent track. I was mesmerised by the thought that we could use this perfectly synchronised photoelectron pulse to perform time-resolved electron diffraction in the picosecond time scale by simply locating a sample under study in the camera drift region. We could study solid-liquid transformation simply by using a short optical pulse to produce the phase transition and the electron pulse to probe the structural change that would follow.

I asked a new student with great passion for research, Steve Williamson, if he would be interested in this project. This was an enormous challenge, as none of us had any kind of electron diffraction experience in steady state let alone in the transient regime. But Steve was a superb experimentalist, and in one year he built a complete "streak camera" and demonstrated the concept. We applied it by performing the first time-resolved structural transformation in the picosecond domain. It was the solid-liquid phase transformation of aluminium. Further work was conducted by Hani Elsayed-Ali, notably on surface melting. The activity was extended later to gas electron diffraction by Ahmed Zewail (Nobel Prize in Chemistry 1999). More recently – twenty years later – our picosecond electron diffraction experiment on aluminium was repeated by Dwayne Miller from the University of Toronto with a superior laser and shorter pulses. Note that Dwayne was at the University of Rochester in the chemistry department with a joint appointment in Optics when, in 1982, Steve did his seminal experiment. Today, time-resolved electron diffraction is becoming a very active field, rivalling time resolved x-ray diffraction.

FIRST STEP OF SINGLE CYCLE THZ GENERATION

We knew that the picosecond rise time produced by photoconductive switching could be used to produce THz transients, either from the gap itself by putting a switch in a coaxial waveguide transition or by exciting a microwave antenna. This simple experiment was performed by a dedicated undergraduate student, Daniel Blumenthal, from the electrical engineering department in collaboration with his adviser, Charles Stancampiano, and André Antonetti from the Ecole Nationale Supérieure de Techniques Avancées in France. The THz field became a very important domain once it was realised by Auston that the electric field could be time-resolved by the laser pulse itself. The field amplitude and phase could be measured, and a new THz spectroscopy technique was born that would replace infrared Fourier-transform spectroscopy. Besides spectroscopy, applications of these transients include THz imaging. Also, the methods of generation have been vastly advanced as demonstrated by X-C Zhang.

ELECTRO OPTICS SAMPLING: MEASURING ELECTRICAL SIGNALS WITH SUBPICOSECOND RESOLUTION

We could switch electrical signals with rise times in the subpicosecond domain, but it was difficult to measure them. Wide band sampling oscilloscopes could only go to 25 ps and the only way to measure the picosecond pulses was to use a second photoconductive gap with a fast photoconductive semiconductor. Of course, one solution was to try to use the electro-optic effect. The EO effect can have a purely electronic reaction with a sub-femtosecond response. But there is no free lunch, and this ultrafast response is paid for in terms of sensitivity. Kilovolts are usually necessary to detect a signal. So, it appeared the EO effect could not be a contender for fast measurements, as it was not sensitive enough. Janis Valdmanis, who had the idea to use lock-in detection in conjunction with the electro-optic effect, demonstrated this to be false. With his "golden hands," Janis showed that sub-millivolt, subpicosecond signals could be measured. The EOS technique became an indispensable tool to visualise THz electrical signals. For the first time, direct propagation of picosecond electrical pulses on transmission lines, both normal and superconducting (with low and high-Tc) could be investigated. EOS was also used in the measurement of the fastest transistor rise times and the switching of Josephson junctions. It was also used in the direct investigation of subpicosecond carrier dynamics in semiconductors, such as velocity overshoot. Most of the activity was coordinated by D. Dykaar and involved many students, like J. Whitaker, visiting scientist Roman Sobolewski and Professor Thomas Hsiang, from electrical engineering, as well as Kevin Meyer, a student from physics.

CHIRPED PULSE AMPLIFICATION

The generation and amplification of short pulses was, however, our main activity. Short pulses were used for everything. At that time, Ti:sapphire had not been invented, and dyes like rhodamine 6G were the main amplifier media. The leading laboratories were at AT&T Bell Laboratories with the group of Charles Shank, and with Erich Ippen and Hermann Haus at MIT. In our group, outstanding students were working on dye-based generation and amplification of ultrashort pulses. They were Theodore Sizer, Irl Duling, James Kafka and Theodore Norris. During one of our constant and endless discussions about novel ideas and concepts, we discussed in 1982 with Steve Williamson a possible way to get larger energy per pulse by using better energy storage media. Strangely enough, 1982 also marked the birth of our daughter Marie, which convinced me that I could do two very different things at the same time: generate a Nobel Prize-worthy idea and engender with Marcelle an adorable girl. From a bandwidth point of view, Nd:glass can in principle amplify subpicosecond pulses. However, unlike in dye, Nd:glass is almost too good of an energy storage medium. The major problem is that the pulse energy becomes too large, leading to high intensities and nonlinear effects. The nonlinear effects contribute to destroying the beam quality and ultimately lead to the "destruction" of the optical amplifier. Dyes, on the other hand, do not have this problem. They are mediocre energy storage media, due to their large amplifying cross-section. Therefore, the pulse energy stays below the critical intensity level where the nonlinear effects dominate. We were greatly influenced by the work of Dan Grischkowsky (IBM Yorktown Heights) and Anthony Johnson (AT&T Bell Laboratories) that demonstrated that by propagating a relatively long pulse in a fibre, the pulse will be the subject of broadening and stretching by a combination of self-phase modulation (SPM) and group velocity dispersion (GVD). As a result, the pulse is stretched with the spectral content of a much shorter pulse. It exhibits a linear chirp. At this point it can be compressed by using a Treacy grating pair, which exhibits a negative GVD to a value one hundred times the value of the input pulse. It looked to me that it would be simple to try to amplify the pulse in order to extract the amplifier energy and compress it later when the energy would be fully extracted. I asked a new student, Donna Strickland, if she would like to do this experiment. Donna was excited about it but also concerned that it might not be good enough for a Ph.D. thesis. She quickly demonstrated that this concept was working to the millijoule level.

It was at this time that Marcel Bouvier from Albertville joined the group. He was a shrew electrical engineer who made some impressive contributions. Notably by inventing a key device, called 1kHz Pockels cell that revolutionized the field. This laser component is now in the exhibit of the Nobel Museum. He also started the company MEDOX Inc. with Phillippe Bado a laser scientist in the group.

The key to CPA: the matched stretcher-compressor. The first approach to CPA was rudimentary and relied on an unmatched stretcher-compressor system. It was not perfect. After a certain amount of stretching, the compressor could not compress the pulse without causing significant wings on the pulse. The fibre-grating pair system was not matched over all orders. What we needed was a matched stretcher-compressor system so we could extract the energy better and compress it better. The matched stretcher compressor became our "Holy Grail." I was continuously thinking about it. One day I was skiing at Bristol Mountain with my wife Marcelle, and on the chairlift, I started to think about a paper I read the day before from Oscar Martinez. This paper was describing a compressor for communication applications at 1.5 μm. At this wavelength the GVD in fiber is negative and the pulses exhibit a negative chirp where the blue frequencies lead the red ones. To recompress the pulse at the fiber output, Martinez proposed a compressor with positive GVD that was a combination of a grating pair and a telescope of magnification unity. I realised that the Martinez compressor in the positive GVD region was in fact the matched stretcher of the Treacy compressor. This was exactly what we were looking for. I interrupted my day of skiing and went back to the laboratory, where I met Maurice Pessot, a new student in my group. I asked Maurice to drop what he was doing and show that the Martinez stretcher and the Treacy grating pair were matched. In a beautiful experiment, Maurice showed that an 80 fs pulse could be arbitrarily stretched 1000 times by the Martinez device and recompressed by the same factor to its initial value. A major hurdle in CPA was overcome. Fifteen years later, this stretching-compression system is still part of the standard CPA architecture.

En Route to the Petawatt. The stretcher-compressor was integrated in our first Joule level Nd:glass system by a visiting scientist, Patrick Maine, and a post-doctoral fellow, Philippe Bado. With Donna they demonstrated a pulse with one joule in 1 ps., i.e. 1 terawatt on a table top – called the "Maine event" since. It was at night and we were jubilant. Robert McCrory, the LLE director, was as usual working late and heard our noisy celebration. He poked his head in the laboratory curious to know what was going on. I told him that we had just demonstrated the generation of one TW with a new amplification technique. It was a thousand times improvement in power over standard techniques, and moreover, this technique could be scaled to a much higher energy than the kJ level using the glass development laser (GDL), a prototype chain at LLE. At that time, we paused and asked ourselves what the next scientific prefix after "tera" was. Nobody knew. We went to Bob's office and discovered that it was "peta." So, from now on, our next goal would be the petawatt. The first article on the possibility of producing petawatt level pulses was described in a French scientific journal, "En Route Vers Le Petawatt" and the first petawatt pulse was demonstrated by Michael Perry at Livermore ten years later. At that time, we decided with Patrick and Donna to call this new amplification technique chirped pulse amplification (CPA). Of course, Wayne, who was at AT&T Bell Laboratories by that time and always has something to say, called me to argue that people would get the acronym mixed up with "certified public accounting."

It was a great time with visits from bright people, like Michael Campbell and Michael Perry from LLNL who understood immediately the revolutionary nature of CPA. We had big plans to go together to the PW level. Also, we had See Leang Chin who came for a sabbatical and was the first to propose with Joseph Eberly to use T3 for the study of light matter interaction in the high intensity regime. With Henri Pepin and his group, Mohamed Chaker and Jean-Claude Kieffer, the contingent of Quebecois from INRS was growing. INRS would play an important role later in our decision to move to University of Michigan. Also, I don't want to forget the group of Adrien Mellinos who had the first the idea of using T3 on SLAC to demonstrate pair generation on SLAC.

We worked a lot to extend the technique to other materials, such as Alexandrite with Jeff Squier and Don Harter. That was before Ti:sapphire. Alexandrite was at that time the only broadband high-energy storage material available. A lot of the CPA work continued after our move to the University of Michigan with Ted Norris, Jeff Squier, François Salin and Gary Valliancourt producing the first kHz Ti:sapphire source – the workhorse of many ultrafast optical laboratories today. Let's also not forget Marcel Bouvier, our indispensable and reliable electrician.

MICHIGAN

However, by inventing CPA we created a new field with characteristics diametrically opposed to the fusion field, the LLE main mission. Our success was highly appreciated but it created some tension. One day I received a call from Duncan Steel and the dean of the College of Engineering from the University of Michigan, Charles Vest, inviting me to move all my group and their families to the University of Michigan Ann Arbor. After one month of negotiation, in August 1988, my group moved to Michigan. This coup was apparently perceived very positively by the MIT search committee looking for a new president. A few months later, Chuck Vest became President of MIT, a position that he held for 16 years.

With me, Henri Pépin's group followed with their equipment. They were the initiator of the ultra-high intensity field at Michigan.

In 1990, two years after our arrival we had been able to attract prominent scientists/professors like Janis Valdmanis, Donald Umstadter and Philip Bucksbaum from Bell Labs. We responded successfully to a call from the National Science Foundation to build an NSF Center. We named it the Center for Ultrafast Optical Science (CUOS) based on femtosecond optical pulses that can provide the shortest controlled bursts of energy, yet produce and enable the highest laboratory peak-power densities ever generated. These two characteristics have opened access to a number of new fields of research not previously available to basic science and applied technology. In the original establishment of the Center, it was pointed out that "ultrafast optical science is an inherently interdisciplinary effort implying scientists and technologists working on laser and optical physics, atomic and condensed-matter physics, chemistry, optical fibres, and electronics." The first important results on high field physics in gas and solid were obtained by the group of Don Umstadter and J.C. Kieffer with high energy electrons acceleration ... It was at CUOS in the early 1990s thanks to M. Bourier Pockels cell the kHz Ti: Sapphire was demonstrated by Jeff Squier and François Salin. This system became the workhorse of femtosecond research. As I was presenting in Bayreuth the kHz laser, Georg Korn introduced himself and expressed the desire to come to CUOS. I accepted and Georg stayed a few years with us where he participated in many important experiments. CUOS now includes researchers in all fields, as well as in plasma physics, accelerator physics, materials science, biophysics, and medicine, all working closely with scientists developing new ultrafast laser sources and measurement techniques – in short, in a "centre mode" of research.

In 1994, we spent 4 months on sabbatical at the University of Tokyo Roppongi at the laboratory of Shantaru Watanabe. We enjoyed immensely the time in Japan. We were hosted by Professor Hiroshi Takuma. It gave me also the opportunity to meet Professor Toshiki Tajima, the inventor of the wake-field accelerator, whom I did not know before. It was an epiphanic moment that started a fruitful collaboration between us that has had few discontinuities since. Marcelle started to take some Ikebana classes and later became a sensei of the SOGETSU Ikebana school.

FEMTO-MICROMACHINING AND EYE SURGERY

Ultrashort laser pulses offer both high laser intensity and a precise laser-induced breakdown threshold with reduced laser fluence. The ablation of materials with ultrashort pulses has a very limited heat-affected volume. The advantages of ultrashort laser pulses are applied in precision micromachining of various materials. Ultrashort-pulse laser micromachining have a wide range of applications where micrometer and submicrometer feature sizes are required.

With Ron Kurtz, Tibor Juhasz and students, we investigated refractive cornea surgery in vitro and in vivo by intrastromal photodisruption using a compact ultrafast femtosecond laser system. Two students, Detao Du and Xinbing Liu, demonstrated that in the femtosecond regime, photodisruption is associated with smaller and very deterministic threshold energy as well as reduced shock waves and smaller cavitation bubbles than with nanosecond or picosecond lasers. Our reliable all-solid-state laser system was specifically designed for real world medical applications. By scanning the 5 micron focus spot of the laser below the corneal surface, the overlapping small ablation volumes of single pulses resulted in contiguous tissue cutting and vaporisation. Pulse energies were typically in the order of a few microjoules. Combination of different scanning patterns enabled us to perform corneal flap cutting, femtosecond-LASIK, and femtosecond intrastromal keratectomy in porcine, rabbit and primate eyes. The cuts proved to be highly precise and possessed superior dissection and surface quality. Preliminary studies show consistent refractive changes in the in vivo studies. We conclude that the technology is capable of performing a variety of corneal refractive procedures at high precision, offering advantages over current mechanical and laser devices and enabling entirely new approaches for refractive surgery.

BACK TO FRANCE

In 2004 I was invited for the scientific evaluation of the Laboratory of Applied Optics (LOA). I was approached by the research director of the École Polytechnique, Maurice Robin, who asked me about the possibility of returning to France. Although I was very happy at Michigan with CUOS, it was an excellent opportunity for us to come back and almost 30 years later I returned as the director of the LOA in 2005.

The same year, ESFRI was in the process of updating its roadmap of large-scale research infrastructures. I took advantage of the opportunity to propose the Extreme Light Infrastructure (ELI) as a Pan-European facility. At the same time the Île-de-France region also had a call for major instruments and so we proposed the Apollon laser facility as well. These laser facilities, for me, were the extension of the ultra-high intensity at CUOS by seeking the tens of PW level almost 100 times more powerful than the Hercules laser at CUOS. It was a beautiful opportunity to fulfill our dream with Tajima. We succeeded in obtaining both projects. I was the initiator and PI of the ELI Preparatory Phase that started in 2008. After an agonising debate between the École Polytechnique, CEA, CNRS and the Institute d'Optics, it was decided to build the Apollon laser at the CEA site at L'Orme des Merisiers in an old accelerator facility that had been dismantled. It was decided also that the LULI would build and run the facility by 2019–2021.

The ELI project has as its goal to build an infrastructure of facilities providing the most advanced peak power laser systems in the world. This gargantuan power will be obtained by producing kJ of power over 10 fs. Focusing this power over a micrometer size spot, will bring forth the highest intensity. By producing, firstly, the highest electric field, secondly the shortest pulse of high energy radiations in the atto/zeptosecond regime and thirdly, electrons and particles with ultra-relativistic energy in the GeV regime, the laser signalled its entry into Nuclear Physics, High Energy Physics, Vacuum Physics and in the future Cosmology and Extradimension Physics. More precisely, ELI will be the first infrastructure dedicated to the fundamental study of laser-matter interaction in the ultra-relativistic regime ($I > 10^{24}$ W/cm^2). The infrastructure will serve to investigate a new generation of compact accelerators delivering energetic particle and radiation beams of femtosecond (10^{-15} s) to attosecond (10^{-18} s) duration. Relativistic compression offers the potential of intensities exceeding greater than 10^{25} W/cm^2, which will challenge the vacuum critical field as well as provide a new avenue to ultrafast attosecond to zeptosecond (10^{-21} s) studies of laser-matter interaction. After long debate it was decided that ELI will have three pillars located in three European emerging countries. Each countries will work on coordinated different topics: Czech Republic for the development of high energy particle radiation Beam Line, Hungary for Attosecond Source and Romania for Nuclear Physics.

In 2010, following the advice of Alexander Sergeev, I applied for a Russian Megagrant and was one of the winners. I had a joint appointment between the Institute of Applied Physics at Nizhny Novgorod and the University of Nizhny Novgorod. We were housed in a studio next to Sacha and Marina sharing our breakfast and meals together. We had long and friendly discussions between us on all subjects, sometimes exposing our French and Russian differences. Our collaboration started with the Russian Excel Laser and still continues with fine scientists like Efim Kazhanov and Sergey Mironov.

IZEST: ORBITAL DEBRIS, GOING BEYOND THE HORIZON

For ELI, 2011 was the end of the preparatory phase and the beginning of the construction phase. The respective countries managed the three facilities. I was 67 and had to become professor emeritus. With Toshiki Tajima, we proposed to create a unit to explore the prospective for

Extreme Light, IZEST for International Zeptosecond Exawatt, Science and Technology. IZEST is devoted to the investigation of Extreme Light beyond the Horizon set by the ELI infrastructure.

Among IZEST's achievements, we note: The International Coherent Amplification Network (ICAN) project, a laser system characterised by a novel architecture, based on the coherent combination of many CPA fibre lasers. ICAN, provides the laser with high peak power, high average power and good wall-plug efficiency. This is paramount to applications such as particle colliders, nuclear waste transmutators and space debris mitigation. The development of the working prototype is the XCAN program and is directed by Jean-Christophe Chanteloup.

After seeing the movie "Gravity", it occurred to me that the ICAN system could play a key role in space debris mitigation. The first conceptual demonstration was made by Rémi Soulard and Mark Quinn.

Finally, my future goal is dedicated to the increase of peak power towards the Schwinger regime. Here with Toshiki Tajima and Jonathan Wheeler, we are aiming to test a new paradigm: instead of producing high peak power by increasing the laser energy, we will increase the peak power by shortening the laser pulse to the atto and subattosecond regime. In this way we could produce high energy, attosecond single cycle pulses in the x-ray regime. The peak power could be exawatt, the wavelength in the x-ray and the intensity in the Schwinger regime, enough to produce PeV particles and vacuum materialisation.

Extreme Light Physics and Application

Nobel Lecture, December 8, 2018 by
Gérard Mourou
University of Michigan, Ann Arbor, MI, USA & École Polytechnique, Palaiseau, France.

I. INTRODUCTION

The advent of ultraintense laser pulses generated by the technique of chirped pulse amplification (CPA), along with the development of high-fluence laser materials has opened up entirely new fields of optics. A CPA laser exhibits stunning capabilities. It can generate the largest field, the largest pressure, the highest temperature and accelerating field, making it a universal source of high energy particles and radiation.

CPA technology produces a wide range of intensities extending from 10^{14} to 10^{25} W/cm^2. In the lower part of this range, the intensity regimes of 10^{14} to 10^{17} have applications that include micromachining, which can be performed on material regardless of its nature, i.e. ceramic, metal, biological tissue, cornea, etc. Extremely clean cuts of minimal roughness even at the atomic scale are produced. This attractive property led us to applications in ophthalmic procedures like refractive surgery, cataract surgery, corneal transplants and glaucoma treatment. Today, a million patients a year are benefiting from femtosecond interventions. In science, in the same intensity level, CPA makes possible to reach the attosecond frontier, offering a formidable tool to time resolved fundamental electronic processes.

For intensities >10^{18} W/cm^2 laser-matter interaction becomes strongly dominated by the relativistic character of the electron. In contrast to the nonrelativistic regime, the laser field moves matter more effectively, including motion in the direction of laser propagation, nonlinear modulation and harmonic generation, leading to high energy particle and radiation production. One of the hallmarks of this regime is Laser Wakefield Acceleration (LWA), where the electromagnetic energy from a laser pulse is transformed into kinetic energy producing accelerating gradients a thousand times higher than those applied in conventional accelerators. The electron beam can, in turn, produce a copious amount of keV radiation by betatron or Compton scattering.

For intensities at 10^{25} W/cm^2 the laser field becomes so large that protons and ions become relativist with GeV energies. The acceleration is directly produced by the light pressure. The source size is very small, and the large acceleration gradients combine to make this source brightness better than any existing.

The coupling of an intense laser field to matter also has implications for the study of the highest energies in astrophysics, such as ultrahigh-energy cosmic rays with energies in excess of 10^{20} eV. Intense laser fields can also produce an accelerating field sufficient to simulate general relativistic effects in the laboratory via the equivalence principle like the loss of information in Black Holes.

Many CPA applications offer great benefit to humankind. We discuss a few examples capitalising on the compactness of the CPA-based source. For instance, the generation of high energy protons and neutrons applied to the treatment of cancer, i.e proton therapy, or in nuclear pharmacology where short-lived radionuclides could now be created for therapy or diagnostics near the patient's bed.

In the environmental arena, owing to the efficient generation of high energy neutrons, it is one of our goals to use them to shorten the degree of radiotoxicity of the most dangerous elements, the minor actinides in nuclear waste.

Finally, looking onwards, particle production in "empty" space will remain one of the main objectives of the field. It is the historical path which guided the field in order to acquire an understanding of fundamental questions on the structure of vacuums, to give us a glimpse of the propagation of light in vacuums and how it defines the mass of all elementary particles.

A novel laser architecture to reach intensities at the Schwinger level is explored. Its paradigm is based on the transition of a single cycle visible light pulse into a high energy single cycle X-ray pulse. Successful transposition would give the field a formidable boost, equivalent to the one received when maser transitioned into laser, moving from GHz to PHz (Light) frequencies.

II. THE CPA TECHNIQUE

The key to high and ultrahigh peak power and intensity is the amplification of ultrashort pulses in the picosecond and femtosecond time scales. Over the past 40 years laser-pulse durations have continuously decreased from the microsecond domain with free running to the nanosecond regime with Q-switching, and finally to the picosecond and few-femtosecond regime with mode locking (Brabec and Krausz, 2000). With the advent of mode locking, the laser-pulse duration became so short that pulses could not be amplified without producing unwanted nonlinear effects. This caused a power and intensity plateau seen in Fig. 1. Before 1985, for reasonable sized systems, i.e., with a beam diameter of the order of 1 cm, the maximum obtainable power remained around 1GW, with focused intensities of about 10^{14} W/cm^2. Higher power could be obtained through the use of amplifying media with gain bandwidths that can accommodate the short pulse spectrum and high-energy storage media that have a small transition cross section σ_a. However, to be efficient, this approach requires the use of input pulses with a high laser fluence of a few J/cm^2. As we shall see later, good energy extraction from an amplifier calls for input pulses close to the saturation fluence, see Eq. 1. This level of fluence delivered over a short time leads to prohibitively large intensities, in excess of TW/cm^2. This is far above the limit of the GW/cm^2 imposed by the need to prevent nonlinear effects and optical damage in the amplifiers and optical components. As a result, the only alternative seemed to use low-energy storage materials such as dyes or excimers and increase the laser beam cross section. The latter is unattractive as it leads to large, low-repetition rate and costly laser systems. Therefore, high-intensity physics research was limited to a few large facilities such as the CO2 laser at Los Alamos National Laboratory (Carman et al., 1981), the Nd:glass laser at the Laboratory for Laser Energetics (Bunkenburg et al., 1981), and excimer lasers at the University of Illinois at Chicago and University of Tokyo (Luk et al., 1989; Endoh et al., 1989).

In 1985 laser physicists at the University of Rochester (Strickland and Mourou, 1985) and later (Maine and Mourou, 1988; Maine et al., 1988), demonstrated a way to simultaneously accommodate the very large beam fluence necessary for energy extraction in superior storage materials while keeping the intensity and nonlinear effects to an acceptable level. This technique was called Chirped Pulsed Amplification (CPA). CPA revolutionised the field in three ways. First, a TW system could fit on a table-top, thus delivering intensities 10^5–10^6 times higher utilising conventional technology. Second, the CPA architecture could be easily retrofitted to existing large laser fusion systems at relatively low cost. Today CPA is incorporated in all the major laser fusion systems around the world such as: Japan (Yamakawa et al., 1991), France (Rouyer et al., 1993), United

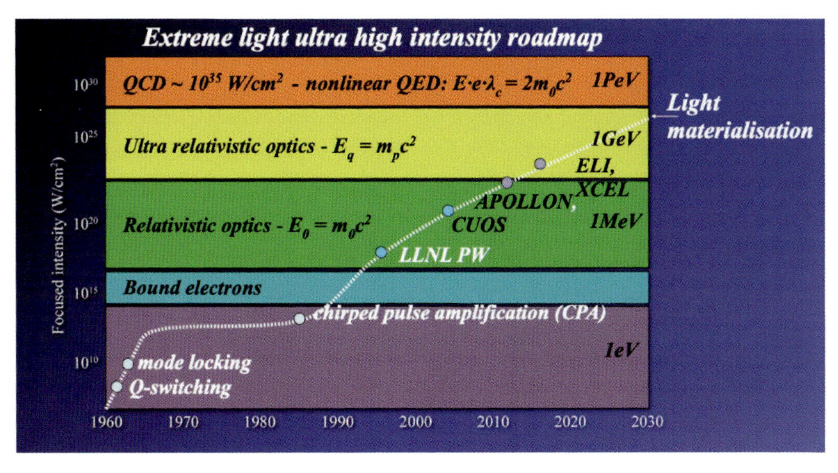

Figure 1. Laser intensity vs years.

Kingdom Vulcan Laser, United States Perry et al., 1999, etc. Third, because of their reduced size CPA lasers could be combined with large particle accelerators. In the case of synchrotrons (Wulff et al., 1997; Larsson et al., 1998; Schoenlein et al., 2000), they could be used to study time-resolved x-ray diffraction. With a linear collider such as SLAC one could produce fields higher than the critical field (Bula et al., 1996) and observe nonlinear QED effects such as pair generation from a vacuum. At present all the colliders are considering the incorporation of CPA technology to produce γ rays for photon-photon collisions to produce a γ-γ collider, (Telnov, 1990, 2000, 2001; Yokoya, 2000).

As we shall describe later, the availability of ultra-high intensity lasers has extended the horizon of laser physics from atomic and condensed-matter studies to plasma, nuclear, and high-energy physics, general relativity, cosmology, and physics beyond the standard model. It has also had a major effect in bringing back to university laboratories science that formerly could only be studied with large-scale facilities. Also at a relatively low peak power level, it made precision machining and ophthalmic procedures in the medical arena possible.

A. Amplification: The Energy Extraction Condition

Before 1985 all amplifier systems were based on direct amplification. As mentioned in the introduction, for extraction efficiency, a simple rule is that the energy per unit area be of the order of F_{sat}, the saturation fluence of the materials. This value is given by

$$F_{sat} = \frac{\hbar\omega}{\sigma_a} \quad (1)$$

where $\hbar$ is Planck's constant, ω is the angular laser frequency and σ_a is the amplifying transition cross section. F_{sat} is 0.9 J/cm^2 for Ti:sapphire and 4 J/cm^2 for Nd:glass and of the order of a mJ/cm^2 for dyes and excimers. It can be shown (Siegman, 1986) the output fluence F_{out} is given by

$$F_{out} = F_{sat} \ln\left[\frac{G_0 - 1}{G(t) - 1}\right] \quad (2)$$

where G_0 is the low signal gain and

$$G(t) = \exp\left[\sigma N_{tot}(t)\right] \quad (3)$$

$$\eta = \left[\frac{\ln G_0 - \ln G_f}{\ln G_0}\right] \quad (4)$$

The gain G_f at the end of the impulsion is given by

$$G_f = 1 + (G_0 - 1)\exp\left[-\frac{F_{pulse}}{F_{sat}}\right] \quad (5)$$

From Eqs. 4 and 5 we see that, to reach an efficiency close to unity, the laser input fluence F_{pulse} must correspond to few times F_{sat}. Fig. 2 illustrates this point for two different initial gains G_0 of 10 and 10^3.

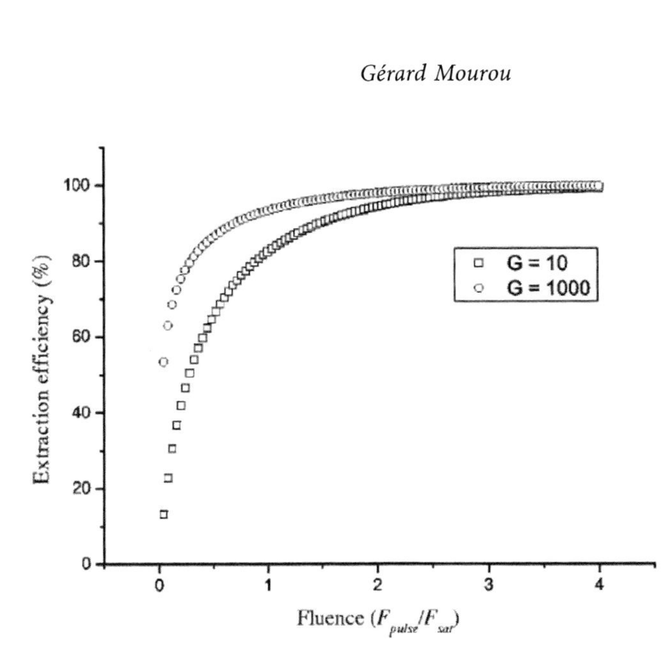

Figure 2. Amplifier efficiency. This illustrates the importance for the input pulse fluence Fpulse to be few times the saturation fluence F_{sat} to obtain a good extraction efficiency.

B. Amplification: The Propagation Condition

Prior to CPA the amplifying media were exclusively dyes, (Migus et al., 1982) and excimers (Endoh et al., 1989; Luk et al., 1989). Typical cross sections for these media are very large, in the range of 10^{-16} cm^2, implying a F_{sat} of only a few mJ/cm^2, or a power density of 1 GW/cm^2 for subpicosecond pulses. At this power density level, the index of refraction becomes intensity dependent according to the well-known expression

$$n = n_0 + n_2 I \qquad (6)$$

Due to the spatial variation of the laser beam intensity, this will modify the beam wavefront according to the B integral given by

$$B = \frac{2\pi}{\lambda}\int_0^L n_2 I(x)dx \qquad (7)$$

Here, B represents, in units of λ, the amount of wavefront distortion due to the intensity-dependent index of refraction, accumulated by the beam over a length L. For a perfectly Gaussian beam, B will cause the whole beam to self-focus above a critical power given by

$$P = \frac{\lambda_0^2}{2\pi n_0 n_2} \qquad (8)$$

For a nonlinear index $n_2=5\times10^{-16}cm^2/W$ for Ti:sapphire. When the laser beam exhibits spatial intensity modulations, n_2 will cause the beam to break up in filaments. In practice the small scale self-focusing represents the most severe problem in an amplifier system. The maximum growth rate g_m (Bespalov and Talanov, 1966) will occur for spatial frequencies K_m given by

$$K_m = \left(\frac{2\pi}{\lambda}\right)\left(\frac{2n_2 I}{n_0}\right)^{1/2} \qquad (9)$$

$$g_m = \left(\frac{2\pi}{\lambda}\right)\left(\frac{n_2 I}{n_0}\right) \qquad (10)$$

For intensities of the order of I = 1 GW/cm² in Ti:sapphire, $K_m=200\ cm^{-1}$, corresponding to 50 μm spatial wavelength. As seen in Fig. 3, these wave-front "irregularities" will grow at a rate of $g_m= 3\ cm^{-1}$ with an exponential growth rate G_m over the gain length L exactly equal to B,

$G_m = B$

For laser fusion, the beam is "cleaned" with spatial filters every time B reaches 3. For high-field experiments in which the spatial and temporal beam quality requirements are more stringent, B must be kept below 0.3 corresponding to a wavefront distortion of λ/20.

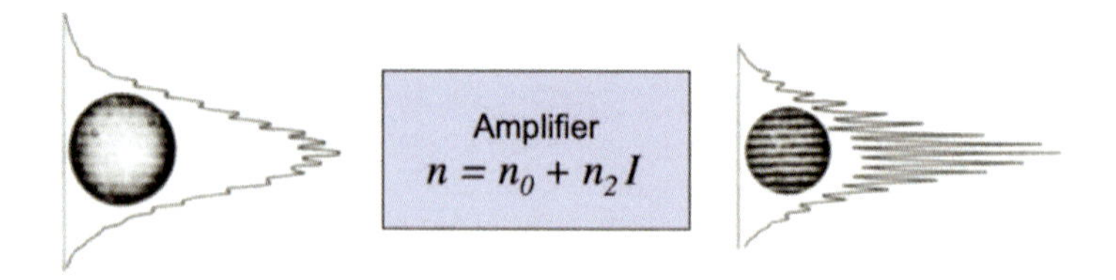

Figure 3. The intensity dependence of the Index of refraction, will create spatial variation of the laser beam intensity, producing undesirable filaments and hot spot.

C. The CPA Concept

We have seen that amplifying media with low cross sections offer the benefit of compactness. For instance, Nd:glass has a cross section of 10^{-21} cm^2, which means that we can store 10^3 to 10^4 times more atoms per unit volume and consequently, get 10^3 to 10^4 more energy before it self-oscillates, than with a dye or excimer with cross section 10^{16} cm^2. However, to extract all the energy in a picosecond pulse would require a beam with a fluence F_s of the order of 1 J/cm^2 or an intensity of 10^{12} W/cm^2 corresponding to a B of 10^3, or 10^3 times the acceptable value.

Therefore, in order to utilise superior energy storage materials, the laser scientist is confronted with the seemingly insoluble dilemma of increasing the input energy needed for energy extraction, while keeping the input intensity at an acceptable level. This problem was solved by the CPA concept. First the pulse is stretched by a factor of 10^3 to 10^4. This step does not change the input pulse energy and therefore does not affect the energy extraction capability, but it does lower the input intensity by the stretching ratio and hence keeps B to a reasonable level. Second, the pulse is amplified by 6 to 12 orders of magnitude, i.e. from the nJ to the millijoule or kilojoule level, before it is finally recompressed by the stretching ratio back to a duration close to its initial value, see Fig. 4.

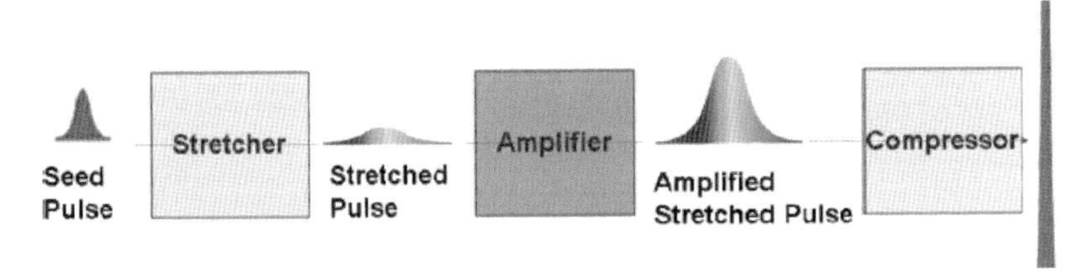

Figure 4. Chirped pulse amplification concept. To minimise nonlinear effects the pulse is first stretched several thousand times lowering the intensity accordingly without changing the input fluence J/cm^2. The pulse is next amplified by a factor of 10^6–10^{12} and is then recompressed by a factor of several thousand times closer to its initial value.

D. The Key Element: The Matched Stretcher-Compressor

In the first CPA set up (Strickland and Mourou, 1985) the laser pulse was stretched using an optical fibre with a positive group delay dispersion and was recompressed by a pair of parallel gratings (Treacy, 1969), with a negative group delay dispersion. Although this first embodiment led to a spectacular 100-fold improvement in peak power, the problem was that the stretcher and compressor were not matched over all orders. As a result, after recompression the pulse exhibited unacceptable pre-pulses and post-pulses.

This led the Rochester group to look for the ideal "matched stretcher-compressor." It came in 1987, when Martinez (Martinez, 1987) proposed a grating compressor with positive group delay dispersion for communication applications as shown in Fig. 5. In communication systems the wavelength of choice is 1.5 μm, a spectral region where the fibre exhibits negative group velocity dispersion. After propagation in a fibre the bits of information exhibit a negative chirp. It is therefore necessary to use a dispersive delay line after propagation with a positive group velocity dispersion to recompress the pulses. After examining this arrangement, the Rochester group came to the conclusion that the Martinez "compressor" was in fact the matched stretcher of the Treacy compressor that they were intently seeking. This can be easily shown by considering the arrangement shown in Fig. 5. When one uses a telescope of magnification 1, the input grating located at a distance f from the first lens will be imaged at the same distance.

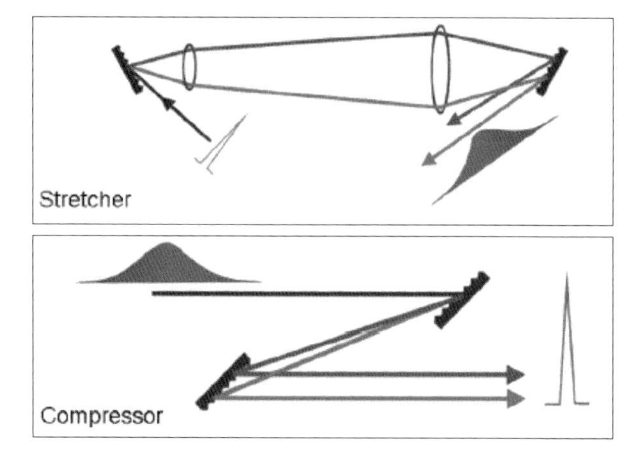

Figure 5. Treacy and Martinez grating arrangements. The Martinez grating pair used as a stretcher and Treacy grating pair used as a compressor. It was discovered and demonstrated by Pessot *et al.*, (1987) that these two grating arrangements are in fact matched over all orders. The pulse can be stretched and recompressed arbitrarily keeping the initial pulse unchanged. This grating arrangement is used in most CPA systems.

The phase conjugation properties of the "stretcher/compressor" were proposed and demonstrated in (Pessot et al., 1987) by stretching a pulse of 80 fs by a factor of 1000 using the Martinez arrangement and then compressing it back to exactly the same value using the Treacy compressor.

This demonstration represented a major step in chirped pulse amplification. This matched stretcher-compressor integrated into a CPA system was to produce the first terawatt pulse on a table top system, the so-called T^3 by the Rochester group. It was subsequently used for subpicosecond pulse amplification (Maine and Mourou, 1988; Maine et al., 1988) and for a pulse duration of 100 fs by Pessot et al. (1989). This arrangement stands today as the standard architecture used in CPA systems.

E. New Materials for CPA, OPCPA and Gain Narrowing

CPA was demonstrated initially with the two broadband-amplifying media that were available at the time, Nd:glass and alexandrite (Pessot et al., 1989). Shortly after, this initial work the concept was extended to Ti:sapphire (Vaillancourt et al., 1990; Kmetec et al., 1991; Squier et al., 1991; Sullivan et al., 1991) as well as Cr:LiSrAlF6 (Beaud et al., 1993; Ditmire and Perry, 1993) and Yb:glass (Nees et al., 1998). Among these materials Ti:sapphire has the advantage of the broadest bandwidth, with a high damage threshold and excellent thermal conductivity greatly enhanced at cryogenic temperatures (Backus et al., 1997).

Parametric amplifiers have also been proposed and demonstrated by Dubietis et al. (1992). It was used for large scale laser applications at the Rutherford Appleton Laboratory (Ross et al., 1997, 2000). This elegant technique, called OPCPA for Optical Parametric Chirped Pulse Amplification, is able, if the nonlinear propagation effects are kept under control, to provide an extremely large gain bandwidth. It offers the great advantage to be pumped by large-scale laser systems dedicated to laser fusion. OPCPA could be the natural companion of any large laser fusion system. A more detailed discussion of the OPCPA method is provided below.

In a conventional CPA system, one of the limitations in pulse duration comes from gain narrowing. Because of their wide spectrum, short pulses can be amplified only by materials with a gain bandwidth greater than the pulse spectrum. We note that materials with superior energy storage typically have a low transition cross section and broad gain bandwidth. However, large gain will lead to a reduction of the laser spectrum as it is amplified and therefore a longer output pulse. In the unsaturated regime – the linear regime – the laser spectrum will be subjected to a narrowing given by

$$G(\omega_a)\,\Delta\omega = \Delta\omega_a\sqrt{\frac{3}{G(\omega_a)-3}} \qquad (11)$$

where $\Delta\omega_a$ is the gain bandwidth and $G(\omega_a)$ the exponential gain. A gain of ten orders of magnitude will narrow the gain bandwidth by a factor of 3 to 4. A fraction of this gain, however, can be recovered in the saturated section of the amplifier.

F. "CAN" a Novel Architecture

Many applications involving high peak power, like high energy particle acceleration or transmutation of nuclear waste crucially need average power. They typically need peak power in the PW range with MW average power and excellent efficiency near 50%. Right now, a typical laser provides PW peak power but only with around tens watts average power and 0.1% efficiency. A far cry from what we would need for real applications.

To mitigate this problem, we proposed to build a novel amplifier based on a multitude (network) of phased fibres, called Coherent Amplifying Network (CAN). In a CAN fibre amplifier (Fig. 6) the active gain medium is confined within the core of the optical fibre. Fibres benefit from efficient heat removal due to the advantageous volume to lateral surface area. It can also be pumped efficiently by a laser diode that is well-tuned to the fibre. Hence the CAN laser could provide simultaneously, peak power, average power and efficiency. Continuous-wave fibre lasers with average powers in the kilowatt regime were first demonstrated in 2004.

Figure 6. CAN system is a fibre-laser based system built out of thousands of fibre lasers coherently added to increase the average power and pulse energy linearly. The fibre laser offers excellent efficiency (>30%) thanks to laser-diode pumping and provides a much larger surface cooling area, therefore making operation at high peak power and high average power possible with good efficiency and pristine beam quality.

Note also that CAN uses the CPA concept. An initial pulse from an oscillator laser is stretched and split into many fibre channels as many as 104. Each channel is amplified in several stages, with the final stages producing pulses of ~1 mJ at a high repetition rate. All the channels are combined coherently, compressed and focused to produce a pulse with a possible energy of >10 J at a repetition rate of 10 kHz. Of course, the cost of the system will be expensive but at this level any system will have a significant cost.

G. The Quest for PW Peak Power

As soon as the CPA concept was demonstrated at the millijoule and joule levels, it became clear that it could be extended to much higher energies using already built laser fusion systems to amplify nanosecond pulses in the 100–1000 J range. This means that with remarkably few alterations, that is, by chirping the pulse at the input and compressing it at the output, a laser chain built to produce TW with ns duration pulses could now produce petawatt PW with ps duration pulses (Maine et al., 1987).

The first experiment started at the CEA-Limeil on the oldest laser fusion system P102 in the early 1990s. The laser was capable of producing on the order of 100 joules in the nanosecond regime. A seed pulse provided by an oscillator was stretched amplified to several tens of joules and compressed to its initial value with the grating pair compressor. Of course, the main difficulty was the fabrication of the gratings with a decent damage threshold and efficiency that could handle the large pulse energy. The largest gratings at the time, 30 cm x 40 cm, were fabricated and integrated to the system. These first gratings were produced by Jobin-Yvon, now Horiba, with reflectivity of 90% and a damage threshold of 0.1 J/cm2 and proved sufficient for a convincing demonstration of high energy CPA.

With P102 retrofitted in CPA, we produced successfully 20 TW (Sauteret et al., 1991); and later 50 TW (Rouyer et al., 1993). Similar result few months later where obtained independently at ILE in Japan by Yamakawa et al. (1991). These results showed clearly that the CPA could work at any arbitrary energy level and would be capable to produce PW pulse. Indeed, as we predicted, the first Petawatt pulse was demonstrated by Perry et al. (1999), ten years after the first terawatt. One of the impressive hurdles overcome by Perry's group was the fabrication of meter-size diffraction gratings. They revolutionised the field.

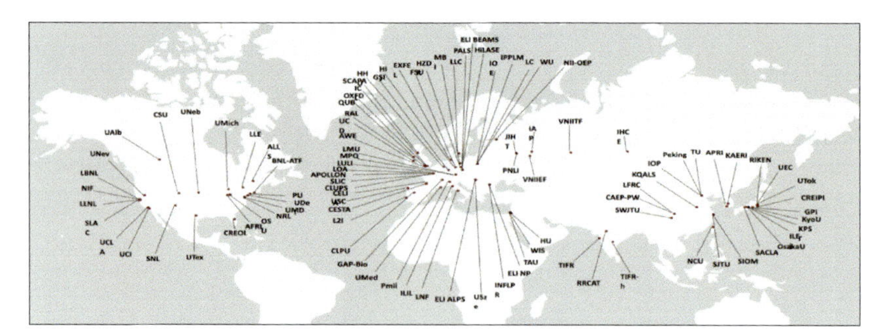

Figure 7. World map of lasers with peak power >100TW.

In parallel to the Nd:glass based petawatt systems, we have today a number of high-power Ti:sapphire-based systems. They exhibit much shorter pulses in the 20–30 fs range, and energies in the 5–10 J range, thus delivering peak power of 100 TW. A 100 TW class Ti:sapphire laser was first demonstrated at the University of California at San Diego (Barty et al., 1994). The leading laboratories at the present time in this area are CoReLS Research Center, in Korea, APRC in Japan with around 500 TW (Aoyama et al., 2002), Janus System at Lawrence Livermore, 200 TW, the Laboratoire d'Optique Appliquée (LOA) in France with 100 TW, the Max Born Institute in Germany with 100 TW, the Lund University in Sweden with 30 TW, and the Center for Ultrafast Optical Science University of Michigan with 200 TW. Today nearly one hundred 100 TW systems are operating (Fig. 7) with about 20 systems at the PW level (Fig. 8) existing or under construction. Groups in China (SIOM), Russia (APC), and the USA (LLE) are now aiming at the 100 PW level.

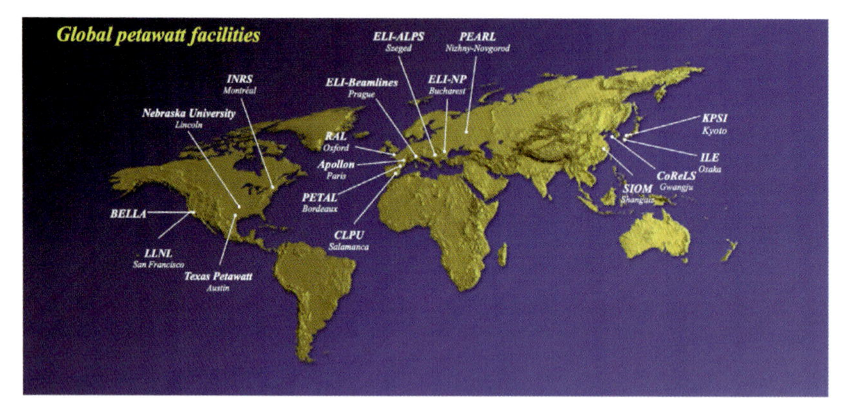

Figure 8. World map of lasers with peak power 1PW.

H. Optical Parametric Chirped Pulse Amplification (OPCPA)

In this section we discuss the differences between the CPA and OPCPA methods. Fig. 9 shows the conceptual layout of an Optical Parametric CPA, OPCPA system (Dubietis et al., 1992; Ross et al., 1997). Because we are reviewing only relativistic intensity laser we will not mention the large number of projects related to the sub-relativistic regime using OPCPA. As in CPA the objective in OPCPA is to stretch the pulse to a nanosecond duration and then amplify it to the joule or higher level by optical parametric amplification and recompress it back to close to its initial value. Note that the stretching is essential not only to keep the B integral low but also to pump and extract the energy efficiently. It is only during the stretched pulse that light can be transferred from the pump beam to the signal beam. Consequently, the pump pulse duration and that of the stretched pulse must be as equal as possible. The advantages of this technique are as follows:

1. Large bandwidth that could accommodate few-cycle pulses.
2. Ability to benefit from very large KDP crystals (100 × 100 cm2) developed for laser fusion.
3. Adaptability to existing laser fusion chains, which benefit from low-bandwidth well collimated nanosecond laser pulses at 532 nm.
4. No heat dissipation in the OPA crystal itself, this is important for high average power.
5. No transverse amplified stimulated emission, which is a major source of loss for large-aperture Ti:sapphire systems.
6. Ability to use an iodine laser as a pumping source.
7. Very simple amplification system.

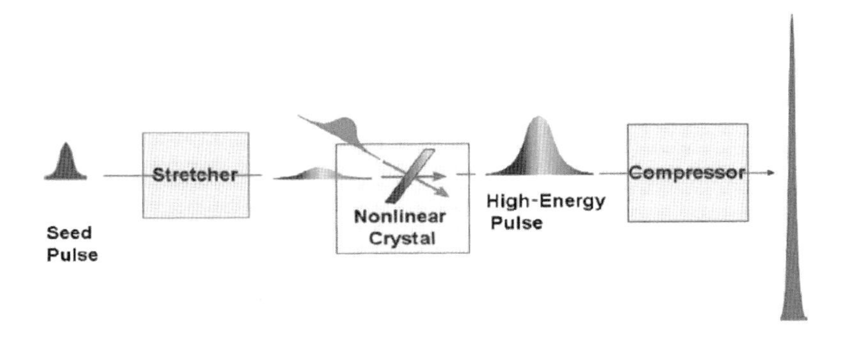

Figure 9. OPCPA concept. In the OPCPA the pulse is amplified by optical parametric amplification instead of regular optical amplification. Note that for efficiency the pump pulse and the stretched pulse must have approximately the same duration and the same spatial extend.

The disadvantages are as follows:

1. Lower efficiency than standard CPA. For a standard Ti:sapphire CPA, the efficiency can be as high as 50% from a long green pulse, say of 50 ns. The energy storage time of Ti:sapphire is 2 μs. So CPA is overall a more efficient system.
2. Very large stretching ratio, in the range of 10^6 to 1, i.e. 10 fs to 5 ns, necessary for energy extraction. This will make pulse compression down to the 10 fs regime difficult.
3. Gain is a significant function of the intensity. This means the pump-beam profile may affect the beam quality and needs a high level of control.

In both systems, the pulse duration will ultimately be limited by the grating bandwidth. At present no large gratings have the efficiency and the bandwidth necessary for efficient pulse compression much below 30 fs. Beam quality from CPA has been demonstrated to be excellent. The latest studies have shown that OPCPA can also provide good spatial beam quality (Collier et al., 1999). The potential of this technique has been demonstrated with the production of a 35 J, 85 fs pulse, equivalent to 0.4 PW, using a 10 cm-diameter beam (Collier et al., 2004). The possibility of reaching high energies seems to be more straight-forward with the OPCPA because it can benefit from kJ, ns fusion lasers that are already up and running. The pulse duration, however, will be limited by the grating bandwidth. CPA implementations must wait for large Ti:Sapphire crystals grown to 20×20 cm2 dimensions. These larger-scale crystals should become available as the demand for higher peak power increases.

III. BEYOND THE HORIZON: THE ZEPTOSECOND AND EXAWATT DOMAIN

In their pioneering experiment Grischkowsky and Ballant (1982) used a single mode optical fibre and were able to compress a picosecond pulse with nJ energy to the femtosecond level. This work triggered an enormous interest. To go higher in energy Orazio Svelto and his group (Nisoli et al., 1996) introduced a compression technique based on a fused silica hollow-core capillary, filled with noble gases to broaden the pulse spectrum before efficiently compressing to the 100 μJ level with chirped mirrors that introduce negative dispersion to the pulse phase. Refining this technique Orazio Svelto, Ferenc Krausz and collaborators (Nisoli et al., 1997) could compress a 20 fs into 5 fs or 2 cycles of light at 800 nm.

For higher energy, bulk compression was attempted by Rolland and Corkum (1988). In their implementation, the pulse is free-propagating in a solid and not guided anymore. The pulse was relatively long, around 50 fs, with an input energy of 500 μJ leading to an output pulse of 100 μJ in 20 fs. However this scheme is impaired by the beam bell shape intensity distribution. This leads to variation in the nonlinear response across the beam profile and a corresponding variation of the compression factor.

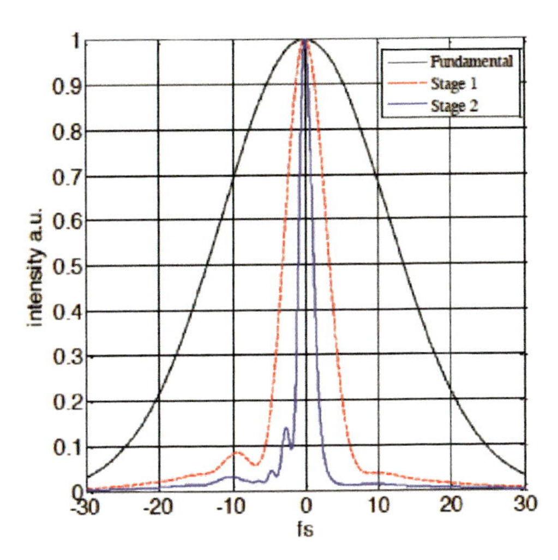

Figure 10. This element uses a uniform thickness plastic element of around .5mm. An incident flat top pulse will induce a uniform self-phase modulation across the element. After spectral broadening the pulse is compressed by 2 chirped mirrors to a near single cycle pulse into a 2.1 fs after two compression stages.

A. Large Energy Pulse Compression: Thin Film Compression (TFC)

Recently we have proposed a technique (Mourou et al., 2014) to compress 25 fs large energy pulses as high as 1 kJ to the 1–2 fs level. We call this approach the Thin Film Compressor, or TFC, and its principle is shown in Fig. 10. As shown in our simulation, this method is very efficient >50% and preserves the beam quality. Here the technique relies on the flat top nature of the high energy pulse and uses a thin "plastic" film of ~500 μm with a diameter of 20 cm. The element, that we call plastic for simplicity, will be transparent and exhibit a uniform thickness. The modelling of the pulse propagation leading to spectral broadening due to the nonlinear processes that occur within the film is described in Akhamov et al. (1992). After spectral broadening the pulse is compressed by chirped mirrors to a near single cycle pulse showing a pulse of 2.1 fs after two compression stages. This represents the shortest pulse duration that can be achieved based on the limitation defined by the laser wavelength, λ. Once the pulse is focused on λ^2 the pulse is in the desired λ^3 regime. The λ^3 regime is the focus volume that achieves the highest intensity for a given pulse energy. In order to increase the intensity, it is necessary to shift to shorter wavelengths that can support shorter pulse durations and tighter focusing.

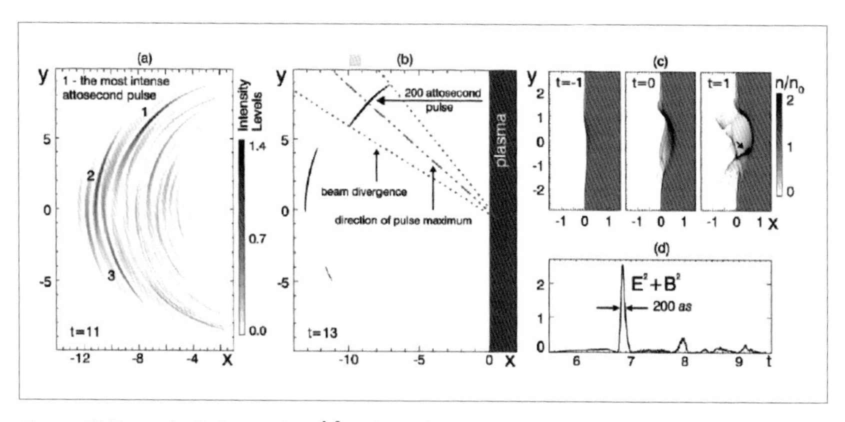

Figure 11. The pulse is focused on λ^2 and create a mirror with an indentation deforms due to (a). As it moves relativistically, it provides an isolated an individual pulse (b), (c).

The proposed method for up-converting to shorter wavelengths is through a relativistic plasma mirror, and especially in the laser conditions of the λ^3 regime (Mourou et al., 2006). The relativistic mirror is not planar and rather deforms due to the indentation created Fig. 11c by the focused Gaussian beam. As it moves relativistically in and out and sideways, the reflected beam is broadcast in specific directions and provide an isolated an individual pulse Fig. 11 a, b. In the relativistic regime (Naumova et al., 2004), predicts a pulse duration T compressed by the relativistic mirror-scaling like T = 600 [attosecond]/a_0. (Fig. 12) Here a_0 is the normalised vector potential, which is unity at 10^{18} W/cm^2 and scales as the square root of the intensity. For intensities on the order of 10^{22}–10^{24} W/cm^2 where a0 is on the order of 100 to 1 000, the compressed pulse could be only a few attoseconds, even zeptoseconds. Naumova et al. (2004) have simulated the generation of thin sheets of electrons of few nm thickness, much shorter than the laser period. It opens the prospect for X and gamma coherent scattering with good efficiency.

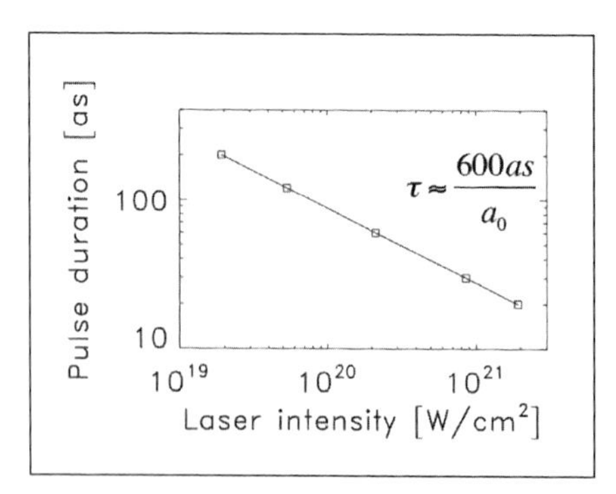

Figure 12. Naumova *et al.*, (2004) predicts a pulse duration scaling like *T*=600 (*attosecond*)/*a*$_0$.

IV. LASER PLASMA ELECTRON ACCELERATION

The concept of laser-plasma wake-field electron accelerator was proposed in 1979 by Toshiki Tajima and John M. Dawson (1979). In a laser-plasma accelerator (LPA), a plasma medium (e.g. fully stripped helium or hydrogen ions surrounded by free moving electrons) is used to transform electromagnetic energy from a laser pulse into kinetic energy of accelerated electrons by exciting high amplitude plasma density waves. See Fig. 13. An intense laser pulse (Esarey, 1996; Everett et. al., 1994; Clayton et. al., 1994; Umstadter et. al., 1996) causes the plasma electrons to move out of its path through the "photon pressure". The much heavier ions barely move and as a consequence are left unshielded. Some distance behind the laser pulse, the electrostatic force exerted by the ions on the electrons pulls them back to the axis, creating an electron density peak. The pattern of alternating positive and negative charges is referred to as a plasma wave or laser wake and supports an electric field. The wave oscillates at the plasma frequency which scales as the square root of the plasma density and has a wavelength typically around 10 to 100 μm. This is several orders of magnitude shorter than the typical RF period used in conventional accelerators. The amplitude of the plasma wave or strength of the electric field is proportional to the square root of the plasma density (number of free electrons per unit volume) and proportional to the laser intensity (for intensities $\geq 10^{18}$ W/cm^2). For typical densities (10^{18}–10^{19} electrons/cm^3) used in experiments, fields ranging from 10–100 GV/cm are produced, three orders of magnitude greater than with conventional technology. The wave's phase velocity is near the speed of light and electrons injected at the proper phase can be accelerated to high energies. To reach the same particle energy, plasma accelerators can then, in principle, be three orders of magnitude shorter than their conventional counterparts.

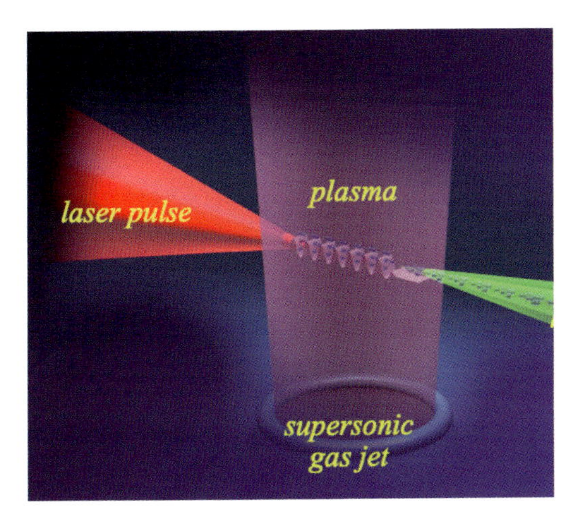

Figure 13. The Laser Wake Field Acceleration concept relies on the very large radiation pressure inducing a plasma wave. Electrons will be trapped in the wave and accelerated to energy of GeV/cm, a thousand time greater than conventional technology.

The wavelength of the plasma waves is also around three orders of magnitude smaller than the wavelength of the radiofrequency used in conventional accelerators. The generation of low energy dispersion electron bunches requires that the length of the bunch to be a small fraction of these wavelengths and/or the use of complex techniques only compatible with a large facility. In the case of plasma accelerators this condition implies the use of electron bunches shorter than 10 fs to get an energy dispersion below 10% typically. So far, these short pulses were produced by controlling the wavebreaking of the plasma waves or by using laser beam collision in the acceleration zone to produce a strongly localised injection. In a relatively short time the major hurdles, one after the other, have been surmounted.

Considering the beam monochromaticity, it has been simultaneously demonstrated (Mangles et. al., 2004; Geddes et. al., 2004; Faure et. al., 2004) to be of the order of 1%. In addition, high energy electrons have also been shown to the GeV level by several groups in the USA at the Lawrence Berkeley Laboratory, the Texas Petawatt Facility, University of Texas; in Korea at CoReLS; and in China at SIOM and Peking University.

That considered, the major stumbling blocks to laser-based acceleration (the repetition rate, average power and poor efficiency) due to the inherent low repetition rate of the driving laser are being addressed. Revolutionary laser infrastructures based on phased-array-optical fibre, like the Coherent Amplification Network (CAN) described in Section II. F. are being actively studied. (Mourou et al., 2013)

Accelerator technology has immense promise for innovation for various applications ranging from science with High Energy Particle Physics, to applications which include betatron and free-electron light sources for diagnostics or radiation therapy and proton source for hadron therapy as well as homeland security.

V. HIGH ENERGY PROTON BEAMS, PROTON THERAPY

It has been shown that laser thin-target interactions can produce plentiful MeV protons in a beam with superior transverse emittance. The proton generation is a direct consequence of electron acceleration. Electrons that are violently accelerated in the laser field can draw behind them protons that are on either the front or back surface of the target.

Highly energetic proton beams have been demonstrated at Livermore, LULI, CUOS, and Rutherford with intensities of 10^{18}–10^{20} W/cm^2. They could lead to important applications such as fast ignition for inertial confinement fusion as was pointed out by Roth et al. (2001) and proton therapy (Bulanov and Khoroshkov, 2002; Fourkal et al., 2002).

The proton used in radiotherapy and oncology provides several advantages. First, proton beam scattering on atomic electrons is weak and results in low irradiation of healthy tissues surrounding the tumour. Second, the stopping length for the proton with a given energy is fixed and avoids irradiation of the healthy tissues at the rear side of the tumour. Third, the Bragg peak of the energy losses provides substantial energy deposition in the vicinity of the proton stopping point.

Currently, proton beams with the required parameters are produced with conventional charged particle accelerators: synchrotron, cyclotron, and linear accelerators. The use of the laser accelerator is very attractive because its compactness is associated with additional possibilities for controlling proton beam parameters. The typical energy spectrum of laser-accelerated particles observed both in experiments and in computer simulations can be approximated by a quasi-thermal distribution with a cut-off at a maximum energy.

The effective temperature attributed to fast ion beams is within only a factor of a few from the maximum value of the particle energy. On the other hand, the above-mentioned applications require high-quality proton beams, i.e. beams with sufficiently small energy spread $\Delta E / E$. For example, for hadron therapy it is highly desirable to have a proton beam with $\Delta E / E=2\%$ in order to provide the conditions for a high irradiation dose being delivered to the tumour while sparing neighbouring tissues.

In the case of the ion injector, a high-quality beam is needed in order to inject the charged particles into the optimal accelerating phase. Bulanov and Khoroshkov (2002) have shown that such a beam of laser-accelerated ions can be obtained by using a double-layer target. Multilayer targets have been used for a long time in order to increase the efficiency of the laser energy conversion into plasma and fast particle kinetic energy; see, for example, Badziak et al. (2001, 2003). In contrast to the previously discussed configurations, the use of a double-layer target was proposed in order to produce fast proton beams with controlled quality. In this scheme the target is made of two layers with ions of different electric charge and mass.

A regime of ion acceleration that exhibits very favourable properties has been identified by Esirkepov et al. (2004). In this regime the radiation pressure of the electromagnetic wave plays a dominant role in the interaction of an ultra-intense laser pulse with a foil. In this radiation pressure dominant regime, ion acceleration appears due to the radiation pressure of the laser light on the electron component with momentum transferred to ions through the electric field arising from charge separation. In this regime, the proton component moves forward with almost the same velocity as the average longitudinal velocity of the electron component. Thus, the proton kinetic energy is well above that of the electron component. In addition, in the radiation pressure dominant regime the ion accel-

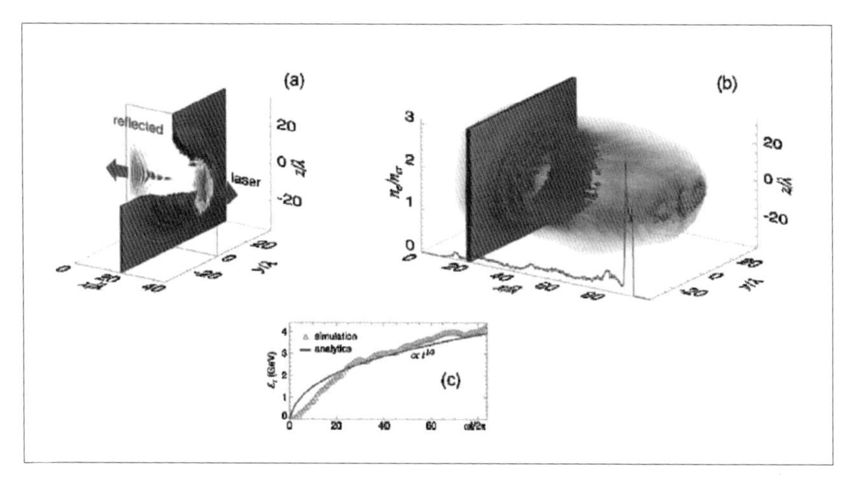

Figure 14. In this regime for intensity in the 10^{23} W/cm^2, the radiation pressure is dominant and scales with the laser-pulse energy.

eration mechanism is found to be highly efficient, and, as we shall explicitly show, the ion energy per nucleon is proportional to the laser-pulse energy. The main results of three-dimensional PIC simulations are shown in Fig. 14.

A. Proton acceleration in the single-cycle regime

Single-Cycled Laser Acceleration (SCLA) of ions (Zhou et al., 2016) relies on laser pulses which give access to a new ion acceleration regime. Historically, attempts in laser acceleration of ions relied on the multi-cycle, high-energy pulses available. The first experimentally realised laser ion acceleration was TNSA (Cousens et al., 2006). In this mechanism the target was thick, electrons penetrated through the thick target and ions were not adiabatically trapped and accelerated. Rather ions were accelerated on the surface of the fixed target over the sheath. For the foil thickness l we define the normalised electron areal density $\sigma = n_e\, l/n_c\, \lambda$ as the target parameter. One way to increase the adiabaticity and prolong the interaction for ion acceleration was to reduce the mass of the target. This reduced σ, as in the regime of Coherent Acceleration of Ions by Lasers (CAIL) (Yan et al., 2010) is far different from the TNSA regime. Radiation Pressure Acceleration (RPA) (Esirkepov, et al., 2004) increases a_0 and slightly decreases σ as compared with TNSA. SCLA, by the virtue of the decreased pulse length of the laser, also reduces σ and increases a_0. Thus, due to the reduction in σ and increase in a_0 for SCLA (and RPA) relative to TNSA, the coherence of the ion acceleration is enhanced. Two clear advantages then arise with SCLA over the longer pulse driven RPA: first,

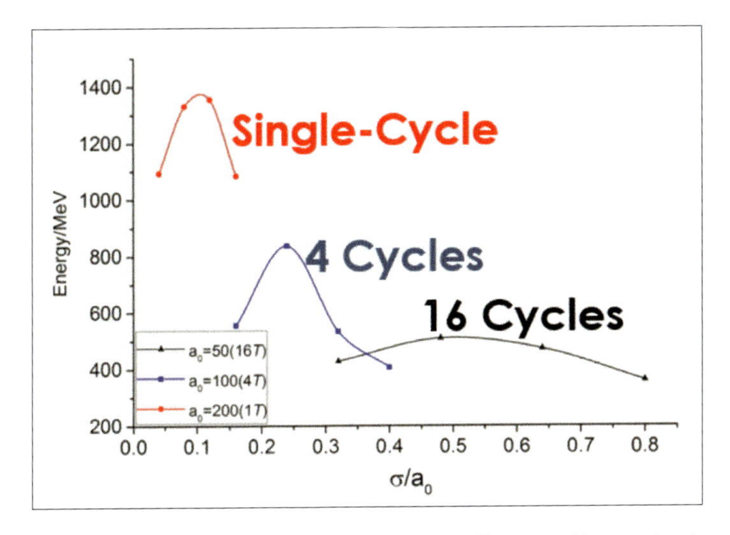

Figure 15. In the single-cycle regime, the acceleration efficiency of ions varies sharply. The fewer number of cycles translates into a larger laser vector potential and yields higher proton cut-off energy. The proton energy is increased by reducing the number of cycles from $\tau = 16T$ (black curve) where T is the period, to $\tau = 4T$ (blue curve). In particular, with the single-cycle pulse (red curve), the cut-off energy of the ions is increased by a significant amount and becomes quickly relativistic.

the pulse intensity is enhanced for a given laser energy as the pulse duration is reduced; second, the elimination of multi-cycle averaging over the oscillations of a longer pulse enhances the efficiency, coherence, and stability of the ponderomotive acceleration. See Fig. 15. With this combination, the ion acceleration in the limit of single-cycle laser pulses becomes far more robust, stable and intense due to the simple fact that the electron ponderomotive acceleration term $\langle v \times B \rangle$ no longer requires cycle-averaging as in the case of longer, multi-cycle pulses. The application of the single-cycle regime introduces a more coherent electron acceleration and sharper electron layer formation that creates the fields that will ultimately accelerate the trailing ions.

We know that in an ideal RPA light sail regime, the resultant maximum ion energy is inversely proportional to the total mass of the accelerated target. In a simple picture, the optimum thickness is achieved by decreasing it, namely, the lower the total mass, the higher the final maximum energy. However, other physical processes, such as transverse instabilities, will strongly affect the actual acceleration process and prevent it from reaching the optimum acceleration, particularly with current state-of-the art multi-cycle ultra-intense laser pulses. While for shorter pulse durations, especially for single-cycle pulses, the duration is too short for those instabilities to develop and the constraints caused by instabilities are strongly suppressed, which gives us more opportunity to approach to the ideal case. So compared to the traditional RPA, the optimal target thickness becomes smaller. Specifically, SCLA is located within the transparent area ($\sigma << a_0$) corresponding to smaller values for $\sigma_{opt.}$

VI. NEUTRON SOURCES

Neutron beamlines produced by CPA will enable a broad range of both fundamental and applied science. Thermal neutron sources (~meV–eV) are widely used for diffraction and spectroscopy experiments (Blomgren et al., 2006). and fast neutrons (~MeV) can be applied to radiography, medicine and material damage studies (Rynes et al., 1999). Laser generated sources of neutrons hold the advantages of a small source size and pulse duration and synchronicity with other secondary sources. Imaging using combined sources can yield information about the composition of materials. The intense source of γ-radiation planned at ELI-NP, for instance, also opens up the possibility of a revolutionary thermal neutron source generated through neutron halo isomers. This scheme avoids the need for moderation of the neutron energy and is predicted to generate neutron fluxes orders of magnitude higher than at existing spallation sources (Habs et. al., 2010).

A. Laser based sources

Laser-driven neutron sources are based on DD or DT fusion and (p,n) and (γ,n) reactions. There are two approaches: reactions between ions within a single laser-heated deuterium target or conversion in a secondary target of a laser driven ion or γ ray beam.

B. Single target scheme

Neutrons can be generated by irradiating thick solid deuterated plastic targets with intense lasers. A series of experiments with 1–100 TW lasers demonstrated yields up to 10^7–10^8 n sr^{-1} per shot (Pretzler et al., 1998; Izumi et al., 2002). Highly energetic deuterons heated directly by the laser at the front surface stream into the target and so initiate 'beam-target' fusion reactions with cold deuterons within the bulk of the solid. The angular distribution of the neutrons is closely related to that of the accelerated ions and is predominantly in the forward direction for short scalelength plasmas. Particle-in-cell code modelling of this mechanism is in good agreement with these experimental results and identified laser and plasma parameters, which lead to the production of an energy tuneable, quasi-mono-energetic forward-directed fast neutron beam. Fusion reactions can also be initiated in gases or a mist of sub-micron droplets produced using pulsed gas jets. The presence of solid-density clusters or droplets dramatically increases the absorption efficiency of short pulse intense irradiation, which can exceed 90% of a petawatt-class laser pulse (Gumbrell et al., 2008). Because the interaction is with an extended target several millimetres in length, 'beam–beam' reactions can occur between hot deuterons within the plasma as well as 'beam–target' reactions with the surrounding medium. If the emission is dominated by the thermonuclear process, then the duration of the neutron burst is limited to less than a nanosecond because the ions rapidly traverse the laser-heated volume. These experiments have been performed using D_2, CD_4 or D_2O and in principle can use a mixture of the tritiated forms of these materials. These sources are generally considered as isotropic, with their flux scaling strongly with the laser energy.

C. Double target scheme

In 'pitcher-catcher' experiments a primary target (pitcher) is irradiated with an intense laser pulse to generate an ion beam from the rear surface. The catcher is a slab of material which acts as a neutron converter in the same way as they are used on accelerator facility neutron sources. DD fusion in deuterated plastic catcher targets yielding ~10^4 neutrons per shot has been demonstrated using liquid droplets as the laser interaction target (20 μm diameter D_2O), enabling a high repetition rate.

The same arrangement can employ the large fluxes of γ rays produced in intense laser plasma interactions to initiate (γ,n) reactions in the catcher sample. These photo-neutrons are usually measured to diagnose the interaction rather than being optimised to investigate their potential as a source and so yields tend to be low (~100 neutrons per shot). In the next section we discuss a preferable option for generating neutrons using γ beams.

D. Neutron spallation

High intensity lasers can produce high energy protons in the several 100 MeV regime. These can be used to strike a target of a heavy metal, mercury or tantalum. Each impact can produce 20–30 high energy neutrons per spallation. This technique has the advantage of producing a beam of neutrons that can be modulated. This process may be linked to conventional nuclear reactor technology in an accelerator-driven system (ADS) to transmute long-lived radioisotopes in used nuclear fuel into shorter-lived fission products.

VII. X-RAYS

Since the dawn of the laser in 1960, scientists focused them to create non-linear effects and very quickly, as intensity increased, plasmas. Plasmas are ionised media known before the existence of lasers to be good emitters of X-rays. Naturally, laser-driven plasmas were studied for the generation of incoherent as well as coherent X-rays. The first proposal of a laser emitting X-rays was published in 1967 by Duguay and Rentzepis (1967), only seven years after the demonstration of the first ruby laser by Theodore Maiman. However, progress on the development of both incoherent and coherent X-rays was slow following delays in the evolution of the laser intensity. The demonstration and then the implementation of the CPA technique triggered a major change of paradigm for the field of laser-driven X-ray sources. This section gives a short overview of the different X-ray sources generated by laser that are available today with special attention on the impact of CPA technique.

A. Incoherent X-ray sources:

By focusing an intense laser on a gaseous, liquid or solid target, plasma is created that emits X-rays through Bremsstrahlung or after atomic inner shell excitation (K-α emission) of the cold solid part of the target. To be able to excite inner-shells, the electron kinetic energy has to be higher than the atomic level energy. Prior to CPA, acceleration of electrons inside the plasma was just sufficient to produce K-α emission from low atomic number elements like Aluminium (1.48 keV) and Copper (8 keV). The jump in intensity provided by CPA induced very strong electron acceleration in the plasma leading today to the production of K-α from elements as heavy as Gold (68.8 keV). Furthermore, in these conditions, Bremsstrahlung emission has been observed to extend up to several 100's of keV. Experiments have also used external laser-accelerated electron beams with energy up to several 100's of MeV leading to Bremsstrahlung emission extending also to 100's of MeV. All these developments are of high interest for medicine (radiotherapy or imaging) and for non-destructive imaging.

B. Coherent X-ray sources:

The interaction of an intense picosecond or femtosecond laser with a gas may generate the odd harmonics of the fundamental laser with a spectrum extending to a cut-off limit given by the law $E_c \approx 3.12\ U_p + I_p$ where I_p is the ionisation potential and $U_p = I\,\lambda^2$ and λ being the fundamental laser wavelength. It is apparent that by increasing the intensity, CPA allowed an extension the cut-off to much higher photon energies, although only up to a saturation point. Today multi-keV high harmonics have been demonstrated using a 4 μm CPA laser. Moreover, gas ionisation is detrimental for the phase-matching between the infrared laser and the high harmonics, thus limiting the energy transfer between them. Using ultrafast (the best being single-cycle) and intense laser pulses ensures strong high harmonic emission. High harmonics having their spectral phase locked demonstrate the emission of attosecond duration pulses. Last but not least, high harmonics maintain the optical properties of the fundamental laser: their wavefront might partially be controlled by the IR laser wavefront (Gauthier et al., 2008), their angle of polarisation is following that of the IR laser (Vodungbo et al., 2011) and they are very coherent. They are thus excellent X-ray sources for fundamental and societal applications.

High harmonics might be also produced by the interaction of a very intense laser (10^{19} W cm^{-2} and above) with solid target. In that case, the laser excites the electron from the target surface, pushing and pulling them away every half cycle. Therefore the electrons generate a train of attosecond X-ray bursts (Teubner and Gibbon, 2008). In some extreme cases, the laser can be back reflected by the relativistic electrons expelled from the surface. The laser is Doppler shifted to much higher photon energies, reaching the keV range. The pulse is also compressed down to 100's of attoseconds or shorter (Naumova et al., 2004; Tsakiris et al., 2006). The X-ray source is highly coherent and controllable by adjusting the laser and target parameters.

It is well-known that at high intensities, the interaction of a CPA laser with gas accelerates electrons up to 100's MeV and even at a few GeV over few millimetres of gas. Such high acceleration gradient is achieved in the so-called "bubble regime" where the laser pushed away the electrons from its path, leaving on its back a bubble with a deficit of electrons and thus an excess of ions. The expelled electrons can thus be accelerated inside this bubble due to the high electric field while for some conditions they also start to oscillate transversely producing betatron oscillations. This source emits radiation with a cone aperture varying as $1/\gamma^2$, γ being the Lorentz factor. The radiation is a "white spectrum" with a cut-off energy depending on γ^2. With a 100 TW-class laser, emission up to 20 keV has been demonstrated within a cone of a few degrees. The emission intrinsically has the duration of the laser (few fs) and the size of the focal spot (few μm) making for a very promising source for medical applications as well as for imaging.

As mentioned above, in the late 60s–early 70s research proposed different schemes for producing a real X-ray laser i.e. by creating a population inversion in plasma necessary for stimulated amplification of spontaneous emission (ASE). Although researchers were very active using non-

CPA lasers and saturated amplification was demonstrated in 1991, the emergence of CPA created a revolution. Thanks to CPA and the possibility of achieving ultra-high intensity, the energy of the pump laser was quickly reduced from kJ, to 10's of J and later joules with a subsequent jump in repetition rate from 10^{-4} Hz to 10 Hz. It is worth noting that the ASE regime without a cavity (normal X-ray mirrors are not suitable for a cavity) led to weakly coherent X-ray lasers. By seeding the amplifying plasma with fully coherent high harmonics produced by the interaction of a CPA laser with gas, the amplified beam is demonstrated to be fully coherent, polarised (either linear or circular depending of the incoming beam) (Zeitoun et al., 2004) and having an excellent wavefront (Godet et al., 2009).

C. X-ray CPA

Recently, it was proposed to use the CPA technique directly in the X-ray range for both free-electron lasers (FEL) and plasma-based X-ray lasers (PBXRL). Although the physics behind the X-ray amplification in FEL and PBXRL is different, the general concept is very similar: While amplifying an ultrashort X-ray pulse, typically femtosecond, most of the energy stored in the medium is left untouched because of a temporal mismatch (Wang et al., 2014; Oliva et al., 2012). By stretching the incoming femtosecond X-ray pulse, then amplifying and finally compressing the pulse, most of the stored energy may be transferred to the X-ray pulse. The concept was successfully demonstrated on an FEL (Gauthier et al., 2016).

Thanks to the CPA these sources are now mature and used for applications in various domains from biology, chemistry and physics.

VIII. ATTOSECOND SCIENCE

Attosecond science represents one of the frontiers in Ultrafast Optics, since it offers the possibility to initiate and control the motion of electronic wave packets inside atoms, molecules, nanostructures and solids. It is now a well-established research field, which offers formidable tools for the investigation of fundamental electronic processes (Calegari et al., 2016).

The production of attosecond pulses is based on high-order harmonic generation (HHG) in gases, which is a highly nonlinear process taking place when an intense (10^{13}–10^{15} W/cm^2) and short laser pulse is focused into a gas medium. The HHG process then results in the production of coherent extreme ultraviolet (XUV) radiation with pulse duration down to the attosecond regime. The physical processes giving rise to HHG can be understood in the framework of a quasi-classical three-step model (Schafer et al., 1993; Corkum et al., 1993). According to this model the most weakly bound electron tunnels through the energy barrier formed by the Coulomb field in the presence of the driving electric field. It then accelerates in the oscillating field as a free particle to eventually recombine with the parent ion, thus emitting a high energy photon. This process is periodically repeated every half optical cycle of the fundamental radiation, thus leading to a periodic emission of very short radiation bursts, with duration in the attosecond range, as first experimentally demonstrated by Paul et al. (2001).

For a number of important applications, such as pump–probe experiments, it is essential to isolate a single pulse of the train, which can then be synchronised with another optical pulse (Sansone et al., 2010a). The first experimental demonstration of isolated attosecond pulses was reported by Hentschel et al. (2001). A few developments in femtosecond laser technology were essential to achieve the generation of isolated attosecond pulses, in particular the stabilisation of the carrier-envelope phase (CEP) of the driving pulses (Jones et al., 2000) and the generation of high-peak-power sub-5-fs pulses (Baltuška et al., 2003). The most common way to generate high-peak power, few-optical-cycle pulses is to apply suitable post-compression schemes. The most common technique is based on spectral broadening in hollow fibres filled with noble gases, in combination with broadband dispersive delay-line with chirped mirrors (Nisoli et al., 1996, 1997).

Attosecond pulses were first employed for the investigation of ultrafast electron dynamics in atomic physics. Particularly interesting is the measurement of the delay in photoemission (Schultze et al., 2010; Klünder et al., 2011), the analysis of the process of tunnel ionization (Dudovich et al., 2006; Uiberacker et al., 2007), the investigation of electronic correlation in helium (Ossiander et al., 2017) to name but a few examples. The first application of attosecond pulses to molecular physics was reported in 2010, with the measurement of the electron localisation process in H2 and D2 molecules after ionisation induced by isolated attosecond pulses (Sansone et al., 2010b). By employing high harmonic spectroscopy, attosecond charge migration was measured and controlled in ionised iodoacetylene by analysing the harmonic light emitted after excitation of the neutral molecule with a strong NIR pulse (Kraus et al., 2015). Particularly interesting is the possibility to investigate the ultrafast electron dynamics in complex molecules where sudden ionisation by attosecond pulses may produce ultrafast charge migration along the molecular skeleton induce a nuclear rearrangement (Cederbaum et al., 1999; Remacle et al., 2006). The process of electron transfer in molecular complexes is of crucial importance in biochemistry since it triggers the first steps in a number of biochemical processes, such as photosynthesis and electron transport along DNA. Experimental evidence of charge migration initiated by iso-

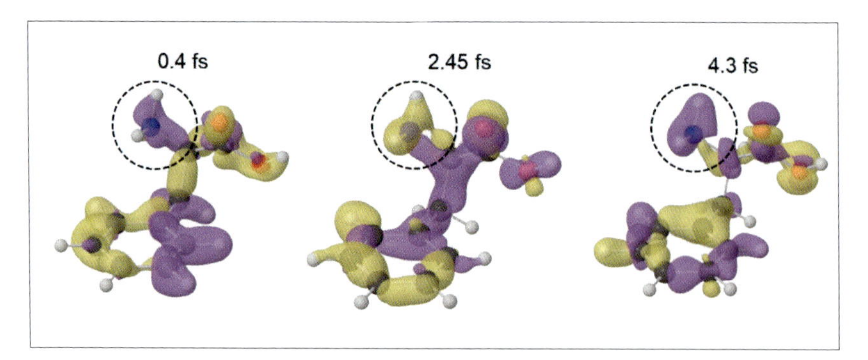

Figure 16. Electron density around the amine group of phenylalanine changes sign in ~2 fs.

lated attosecond pulses were first reported in 2014 in the amino acid phenylalanine- (Calegari et al., 2014). Charge migration was evidenced as an oscillatory evolution in the yield of a doubly-charged molecular fragment. Fig. 16 shows the calculated temporal evolution of the electronic wave packet generated by the attosecond pulses.

Attosecond technology has also been employed to study ultrafast processes in condensed matter. The first application of attosecond pulses to solids was reported in 2007, with the observation that photoelectrons from the 4f band of tungsten reach the surface ~100 attoseconds later than those from the Fermi-edge (Calegari et al., 2014). Another very interesting application was reported in 2013, with the observation of an insulator-to-conductor transition in fused silica (Schultze et al., 2013). It was demonstrated that the ac conductivity of fused silica can be increased by more than 18 orders of magnitude, within 1 fs, by using few-optical-cycle NIR pulses (Schiffrin et al., 2013). The process of tunnel ionisation has been investigated in solids by measuring the transmission of isolated attosecond pulses through a single-crystalline, free-standing Si membrane as a function of the delay with respect to a NIR pump pulse, which excites electrons from the valence band to the conduction band (Schultze et al., 2014). A photo-assisted tunnelling injection mechanism called the dynamical Franz-Keldysh effect has been observed in thin films of polycrystalline diamond (Lucchini et al., 2016). This effect is based on electronic intra-band motion, which competes with inter-band transitions. More recently, the role of inter- and intra-band dynamics in electron transfer between valence and conduction bands in GaAs monocrystalline samples has been investigated by employing attosecond pulses (Schlaepfer et al., 2018).

IX. ASTROPHYSICS

Over the last decade, ion acceleration driven by ultra-intense laser pulses has been emerging as a very exciting potential alternative to conventional acceleration technology. By offering radically new characteristics (Fuchs et al., 2006; Macchi et al., 2013), such as extremely compact acceleration (over less than 1 mm), ultra-short (less than 1 ps) and ultra-dense bunches (over 10^{13} particles/bunch in a single shot), laser ion acceleration offers very promising applications as in hadron therapy and for the production of medical radioisotopes. But already, laser-accelerated ions are used in ground-breaking applications in proton radiography (Chen et al., 2012), to study the concept of rapid ignition for Inertial Confinement Fusion (Roth et al., 2001), and in the so-called field of "laboratory astrophysics" (Remington et al., 1999; Albertazzi et al., 2014; Gregori et al., 2012).

A new emerging prospect in this laboratory astrophysics domain (Chen et al., 2018) is to investigate the question of the nucleosynthesis of heavy elements. At present, there is no complete physical scenario capable of explaining and reproducing the observed abundances of super-heavy elements in our Solar System as well as in other systems (Arnould et al., 2007; Thielemann et al., 2011; Reifarth et al., 2014). These super-heavy elements (i.e. roughly half of the elements heavier than Iron, and almost all of those beyond Bismuth) are postulated to be generated through the r-process (Arnould et al., 2007). This process proceeds through multiple neutron capture in a nucleus, rapidly increasing its mass number (A) until a β^--decay (emission of an electron and a anti-neutrino) takes place, which leads to an increase in atomic number (Z). It is generally accepted that this process can occur only under extremely high neutron flux of more than 10^{20} n/[cm^2.s] (Cowan et al., 1985) in order for multiple neutron captures to take place despite the small cross-sections involved. The main problem is that almost all of our knowledge relative to the dynamics of the r-process rests on theory and simulations, but their accuracy is hindered by large uncertainties (Mumpower et al., 2016; Panov, 2016) in the nuclear data for the involved heavy nuclei (e.g. neutron capture rates, β-decay and α-decay half-lives, masses). This is due to the fact that no facility based on conventional technology- can even come close to such an extreme neutron flux.

To reverse this, ultra-intense lasers offer a radically new prospect as they could be used to generate ultra-bright neutron beams in order to make direct measurements of neutron capture, i.e. (n,γ) nuclear reactions, as well as subsequent β-decay rates of the radioactive isotopes that are created by the neutron capture (Chen et al., 2018). For this, the required short duration and high neutron flux could be generated, from laser-accelerated protons, through spallation. Another advantage offered by ultra-intense laser facilities for such investigations is that they are equipped with auxiliary high-energy laser pulses, i.e. offering the possibility to perform nuclear measurements in a hot plasma environment that emulates astrophysical conditions.

To generate the required ultra-bright neutron beams through spallation, imminent multi-PW lasers offer the prospect of pushing the focused intensity of the laser beam on target by at least one order of magnitude compared to current lasers, i.e. to 10^{22} W.cm^{-2}, and even further, to 10^{23} W.cm^{-2}, with the help of refocusing plasma optics, allowing a reduction in the laser focal spot and a boost to the focused intensity (Nakatsutsumi et al. 2010, 2018). The immediate consequence should be to increase the maximum energy of the produced protons. Most notably, the domain above 200 MeV of maximum energy should be attainable (Chen et al., 2018), thus making it possible to significantly increase the throughput of neutron production by spallation. Spallation is a process that occurs when a light projectile (proton, neutron, or light ions) with a kinetic energy from several hundreds of MeV to several GeV, interacts with a heavy nucleus (e.g., lead) and causes the emission of a large number of hadrons (mostly neutrons) (Van der Meer et al., 2004).

Overall, multi-PW facilities like Apollon or ELI-NP can expect to produce $>10^{12}$ neutrons from the protons in the output of a Pb converter. Using a conservative estimate of 50% for the proton beam bandwidth, it will debunch over ~0.7 ns after 50 cm, i.e. by the time it reaches the Pb spallation converter. Taking this as the duration of the neutron bunch (as the individual spallation process takes place over a $\sim 10^{-22}$ s time-scale), and a source size imposed by the protons scattering in the Pb target (over 3 mm radius for protons at 250 MeV and 0.5 mm at 1 GeV), the conservatively estimated resulting peak flux will be $\sim 10^{22}$–5×10^{23} neutrons/[cm^2.s].

Also, because the repetition rate of the multi-PW lasers will be improved compared to present-day lasers, the time-averaged neutron flux should also become quite high. Using a repetition rate of 1 shot/min (for the largest facilities), the time-averaged flux neutron should reach ~1011–5×10^{12} neutrons/[cm^2.s]. This will greatly exceed those available on existing facilities and should permit multiple neutron capture (Couture and Reifarth, 2007).

We also note that such extreme brightness neutron sources will have a broad collateral effect aside from the nucleosynthesis application discussed here. Indeed, they could help satisfy the increasing demand for neutron sources (Hamm, 2010), e.g. for radioisotope production and even for more futuristic transmutation applications.

X. EXTREME LIGHT PROPAGATION IN QUANTUM VACUUM

Particle production in "empty" space is a historical path which has guided the field of Extreme Light with the ultimate goal of investigating laser-matter interactions in the new regime of ultra-relativistic optics reaching into the fundamental QED and possibly QCD regimes. The laser must necessarily be between the Petawatt and Exawatt regime to produce synchronised, high energy radiation and particle beams with extremely short time structures in the attosecond and zeptosecond time domain. These unique characteristics, unattainable by any other means, could be combined to offer a new paradigm to the exploration of the structure of a vacuum to respond to one of the most fundamental questions: how can light propagate in a vacuum; how can a vacuum define the speed of light; and how can it define the mass of all elementary particles?

A. Light propagation in a vacuum

The vacuum structure defines, for example, how light propagates, as first noted by Werner Heisenberg and his collaborators. The quantum fluctuations in the vacuum allow light-light scattering and the conversion of electromagnetic field energy into particle and antiparticle pairs. In particular, a photon traveling in the vacuum can fluctuate into a particle-antiparticle pair that is virtual because the energy of a single light photon is much smaller than that of a material particle pair. This transmutation of a photon into a virtual pair and back, is called vacuum polarisation since this process also alters the nature of the Coulomb law at short distance. If a second photon arrives just when the first photon exists in its electron-positron pair state, it can scatter from this virtual charged particle pair. In this way, light scatters from light. In direct extension of this argument, a strong electromagnetic field applied in the vacuum can deflect the virtual electrons and positrons. Therefore, there is an index of refraction of empty space filled with fields. In principle, light can be bent by applied electro-magnetic fields, just as gravity bends light through the deformation of space-time geometry. In this sense, empty space has a structure not all that different from that of a dilute gas. Because the Compton wavelength of an electron, $h/mc = 386$ fm (fermi = femtometer = one in 10^{15} part of a metre, the radius of a proton) is 3 million times shorter than a typical optical wavelength, vacuum structure does not massively obstruct the propagation of light. However, light propagating in the Universe over cosmological distances, in the presence of external magnetic fields experiences nonlinear vacuum effects such as photon splitting. It is important to recall that more than 50 years ago Julian Schwinger showed that a coherent ideal 'plane light wave' cannot scatter from itself, or be influenced by itself, no matter what the field intensity is. This is the only form of light known to which the vacuum is exactly transparent.

B. Electromagnetic field in a vacuum

The gap between the valence and conduction band of the best insulator, the vacuum, is twice the energy equivalent of the electron mass, $V_0 = 2mc^2/e = 10^6$ volts. Such high potential differences are commonly achieved in specialised nuclear accelerators (Tandems, Van der Graaff), however over a rather large distance. The vacuum does not begin to spark since the electron-positron pair must materialise on two ends of the potential well, and this is for laboratory devices a macroscopic distance apart. The electric field strength controls the speed of vacuum sparking. The field strength for which this vacuum decay occurs at the zeptosecond scale (light travels a distance of the Compton wavelength $\lambda_c = 10^{-12}$ m in 10 zeptoseconds) is the so called 'Schwinger' critical strength E_0 (named after Schwinger, though Heisenberg was well aware of the result) for which the potential step V_0 occurs over the electron's Compton wavelength, that is $E_0 = 1.3\times10^{18}$ V/m. Approach to this field strength is the intermediate term goal of ELI. This corresponds to an intensity $I_0 = 4.65\times10^{29}$ W.cm^{-2}. One speaks also of laser power when the intensity within the typical focal domain of 1 μm^2 is considered; this value is $P_0 = 4.65\times10^{21}$ W, that is 4650 EW (E = exa = 10^{18}) or 4.6 zettawatt.

For fields near the Schwinger value E0 we can observe massive materialisation of pairs for fields existing at time scales of fs to as. The materialisation of electrical fields into electron-positron pairs is a diagnostic tool allowing for understanding how well the energy is focused. The abundant formation of electron-positron pairs is the first of many vacuum effects to study while striving to focus the laser energy into smaller and smaller volumes.

The stability of the vacuum described above has been studied and evaluated at absolute zero temperature. At finite temperature the valence-conduction particle-hole excitation, here pair production, can be induced. The temperature has a significant effect on vacuum stability, which for the attosecond time scale corresponds to the mass of an electron, $k\,T_e = mc^2$. Still, already at a fraction of this value massive thermally assisted vacuum decay will be encountered. There are many other ways which are currently discussed to bridge the 1–2 orders of magnitude in the field strength associated with the ELI or IZEST project, such as the Lorentz boost experienced by the electromagnetic field colliding with an electron beam.

XI. EXTREME LIGHT AS A BLACK HOLE SIMULATOR

Extreme light offers the promise to create a "tabletop" black hole. It could help to prove whether information is truly lost when black holes evaporate. The idea that information could be lost this way has created a paradox in our current understanding of basic physics.

The debate over whether information is really lost during what is called Hawking evaporation (see Fig. 17) has persisted in the 40 years since Stephen Hawking combined quantum field theory with Einstein's theory of general relativity and discovered black hole evaporation. Almost all contemporary leading theoretical physicists have participated in this "black hole war". In quantum mechanics, the probability, or information, must be preserved before and after a physical process. The seeming loss of information as a result of the evaporation of a black hole therefore implies that general relativity and quantum mechanics, the two pillars of modern physics, may be in conflict.

So far investigations of this paradox have been mostly theoretical because of the difficulty of observing black holes in their later stages, when this potential contradiction is most acute. According to theory, a solar-size black hole would take 10^{67} to evaporate entirely, yet our universe is only about 10^{10} years old. Therefore, essentially all astrophysical black holes are too young to provide useful clues on the information loss paradox even if they are observed, such as that responsible for the gravitational waves observed by LIGO in 2016.

Recently a laboratory black hole to simulate this evaporation has been conceived (Chen and Mourou, 2017). Using state-of-the-art laser and nanofabrication technologies, it is projected to mimic black hole evolu-

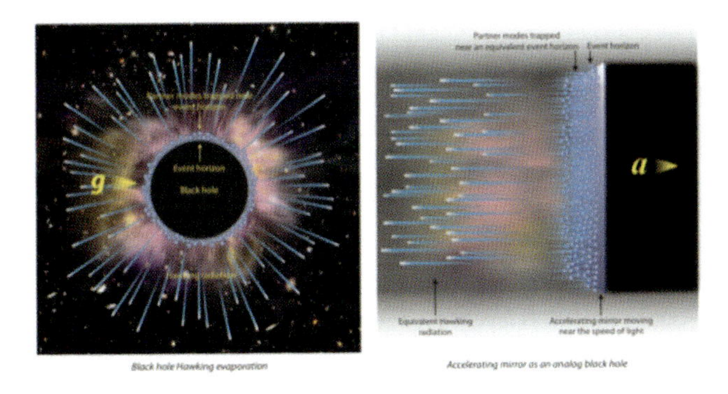

Figure 17. Accelerating mirror as an analog black hole. Left: Black hole Hawking evaporation and the trapping of the partner modes near the horizon. Right: An accelerating mirror also has a horizon and can also emit Hawking particles and trap their partner modes. The analogy between these two systems may be appreciated via Einstein's equivalence principle.

tions at their later stage and to reveal crucial details on how information may be preserved during black hole evaporation.

According to Einstein's equivalence principle, an accelerating mirror moving near the speed of light shares some common features with a true black hole (Einstein, 2002). In both cases, there exists an event horizon. Interacting with quantum fluctuations in vacuum near the horizon, both will emit Hawking particles and trap their partner modes (Fig. 17) until the black hole evaporates entirely or the accelerating mirror suddenly stops. By then the partner modes will be released. The purpose of this proposed experiment is to see whether and how the Hawking particles and their partners are entangled and therefore how the information would be preserved.

It is known that an intense laser traversing a plasma would push the intercepting plasma electrons to its back, the "plasma wakefields". Under extremely intense lasers, such density perturbations can be so concentrated that it can serve as a flying reflecting mirror. By properly tailoring the increase of the density of a thin-film target using nanofabrication technology, a relativistic plasma mirror accelerates as the driving laser continues to enter higher density regions. At the time when the laser leaves the thin-film target, the plasma mirror would abruptly stop its motion, which mimics the ending of the Hawking evaporation.

XII. TRANSMUTATION

Nuclear power wrestles with the problem of its spent fuel waste that still needs to be technically addressed (and a societal will needs to be forged). Among the three scenarios of nuclear waste disposal depicted in Fig. 18, i.e. (1) The no reprocessing approach (requiring 130,000 years for storage); (2) The spent fuel reprocessing (10,000 year storage requirement); and (3) The transmutation (which shortens the need for storage to 300 years and reduces the volume requirement by 100 times), one can find a laser-assisted path that allows for approach (3), which may be the most environmentally forward-looking.

The vision and its technology based on the combination of a CPA high power laser amplified within an OPCPA configuration by a pump created from efficient, high fluence fibre laser technology permits a new high repetition rate laser-driven neutron source for transmutation shown in Fig. 19. The OPCPA laser irradiates a nanometric foil ejecting deuterons onto a thin target to generate energy efficient neutrons by the fusion of D-T. The nascent investigation of laser ion acceleration through the physics of the Coherent Acceleration of Ions by Laser (CAIL) method (Tajima et al., 2009; Steinke et al., 2010) opens an efficient, compact, and economic path to the required neutron generation. Here CAIL introduces direct

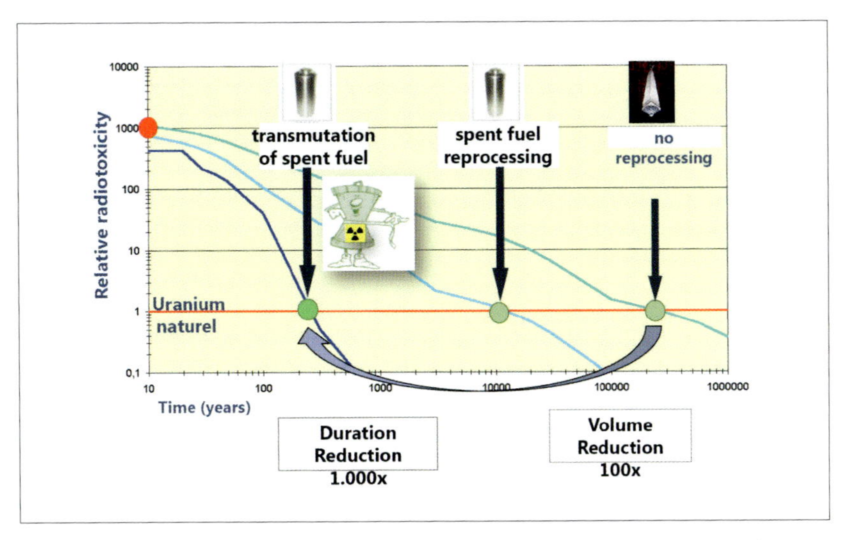

Figure 18. The radiotoxicity of spent nuclear fuel may be reduced from the level of no reprocessed case by about 1000 times if we transmute the spent fuel. This is why a substantial benefit may be gained by the transmutation. (Gales, 2018; Nifenecker et al., 2003).

ponderomotive acceleration of a thin sheet of electrons by the short-pulse laser, which in turn directly pulls ions behind the electrons. The condition of the resonant thinness of the target (and its limited mass) with the matched laser pulse intensity is the distinguishing feature of CAIL from the convention approach of Target Normal Sheath Acceleration, TNSA, (Snavely et al., 2000) in which electrons are heated by the laser absorption and no specific target mass relation exists to the laser intensity.

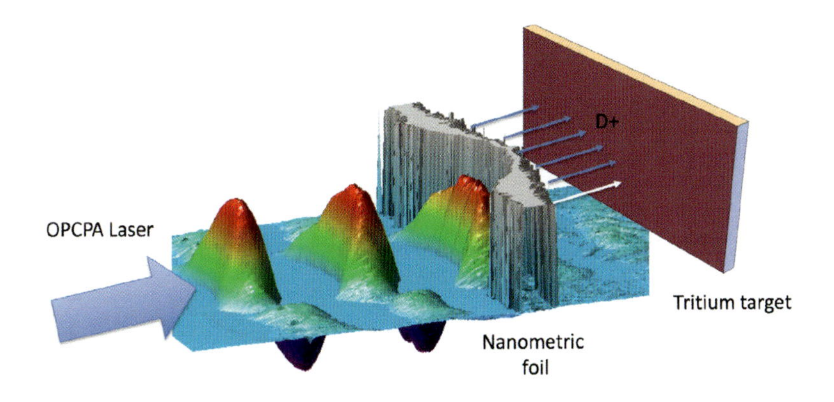

Figure 19. Neutrons are generated by an OPCPA laser irradiation of a nanometric deuteron foil, deuteron acceleration and interaction with tritiated solid target (Yan *et al.*, 2009).

The simulation of the CAIL process has been performed with the EPOCH PIC code (Brady et al., 2011). A linearly polarised laser of intensity 7.7x10^{17} W/cm^2 and wavelength 1 micron h been focused into a 3 micron beam footprint; this corresponds to $a_0 = 0.8$ where $a_0 = eE/(m_e\ c\ \omega_0) = 0.85\lambda_L(I/10^{18})^{1/2}$ where λ_L is the laser wavelength in μm and I is the laser intensity in W/cm^2. The foil is composed of deuterium with a target density of 1.0x10^{23} cm^{-3}. The efficiency of the energy conversion from the laser to ions of 20% has been shown numerically and it sensitively depends on the pulse length and target thickness. It has been shown previously (Steinke et al., 2010) that $a_0 = \sigma$ results in the highest efficiency where $\sigma = n_e/n_{cr} \cdot d/\ \lambda_L$ where ne is the electron density, n_{cr} the critical density (= $m_e\ \omega_L{}^2/4\pi e^2 = 1.1\text{x}10^{21}\ /\lambda^2$ μm where ω_L is the laser frequency and λ μm is the laser wavelength in μm), d is foil thickness and λ_L is laser wavelength. The theoretically maximum energy is given under the optimal condition of $\sigma = a_0$ as

$$\epsilon_{max} = (2\alpha + 1)Qmc^2(\sqrt{a_0^2 + 1} - 1) \qquad (12)$$

Such research opens the way to aspire for the future application of emerging laser technology toward the practical transmutation technology of nuclear waste.

XIII. FEMTOSECOND OPHTHALMOLOGY

CPA-based lasers have attracted significant interest as their potential for high-precision micromachining due to the low damage threshold and deterministic character of the interaction. Surgical applications, particularly in the transparent delicate cornea or lens in the case of a cataract, can take maximal advantage of these attributes. By improving existing procedures and enabling entirely new ones, femtosecond laser technology has the potential to become the preferred corneal laser scalpel in the 21st century.

Corneal laser surgery to correct myopia (short-sightedness), hyperopia (far-sightedness) and astigmatism is becoming the most commonly performed medical laser procedure. Most techniques rely on altering the curvature of the cornea by removing corneal tissue using direct photoablation with ultraviolet light from the excimer laser. Two techniques have demonstrated a high degree of clinical efficacy, photorefractive keratectomy (PRK) and laser-assisted in situ keratomileusis (LASIK).

LASIK has gained recent popularity due to its minimal effect on the corneal surface, which reduces pain and recovery time. However, LASIK requires the use of a mechanical blade (microkeratome) to give the excimer laser access to deeper corneal layers. In contrast to photoablative lasers, photodisruptive lasers operate in the near-infrared spectrum and are not absorbed (at least to the first order) in ocular media. Near-IR pulses can pass through transparent and limited thickness translucent material, affecting tissue only at the focus of the laser beam. In photodisruption, tissue effects are initiated by laser-induced optical breakdown (LIOB), which requires a small focal spot size to achieve a threshold fluence (energy/area) for plasma formation. Generation of a microplasma allows the target to absorb additional laser energy.

A. Optical Breakdown Energy Thresholds

An approximately square root dependence of the fluence threshold on the pulse duration is observed for pulses longer than 10 ps, below this value the dependence weakens significantly. Recent measurements down to 20-fs pulse durations have again confirmed these observations in corneal tissue (Loesel et al., 1999). These results suggest an optimal pulse duration for a corneal photo-disruptive laser in the few hundred femtosecond pulse duration range, where energy deposited in the tissue is significantly reduced. See Fig. 20. Further reduction of the laser pulse duration to a sub 100-fs level adds significant technical complexity and does not produce any further significant decrease of the threshold.

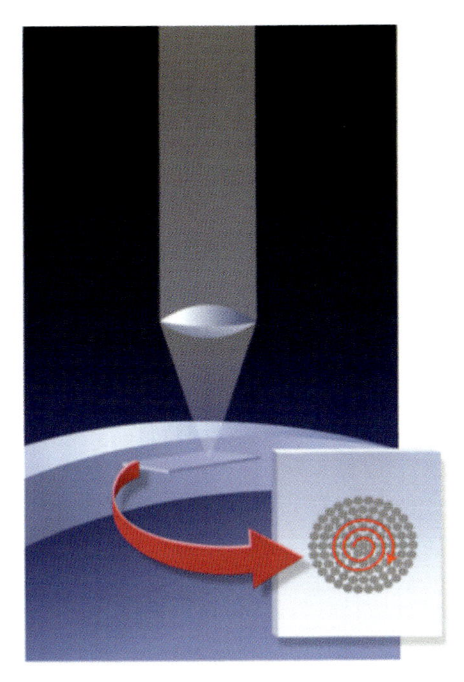

Figure 20. Intrastomal cut showing the importance of the pulse shortness: (a) 50 ps, (b) 100 fs. The picosecond laser cut is of poorer quality, manual dissection required to produce corneal cuts. The femtosecond laser produces a contiguous cut with surface quality similar to mechanical blades.

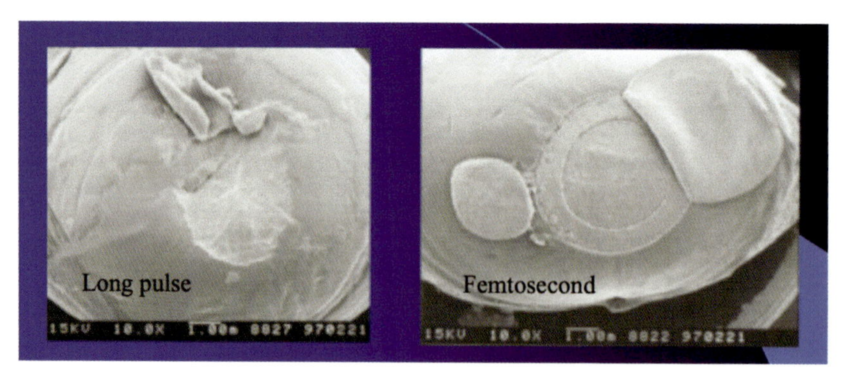

Figure 21. Femtosecond intrastromal scalpel.

B. Corneal Flap Cutting Procedure

(Squier et al., 1995; Ratkay-Traub et al., 2001)
Conventionally, a corneal flap is created with a mechanical microkeratome, a small motorised knife, to give the excimer laser access to deeper layers of the stroma during LASIK. For femtosecond laser-flap cutting (Fig. 21), the laser focus first is scanned along a spiral pattern inside the corneal stroma at a predetermined depth. The intrastromal cut is followed by several semi-circular shaped cuts at decreasing depth in the cornea to connect this intrastromal cut to the corneal surface, with a hinge present to maintain connection to the cornea. The depth is changed by movement of the focusing objective mediated by a computer-controlled galvo with accuracy of a few micrometers and speed that can accommodate a 10-kHz repetition rate. After the completion of the laser procedure, the contact glass is removed and the flap lifted, similar to what is done in the mechanical procedure. The technique offers some important advantage.

1. A decrease in LASIK flap complications.
2. Higher precision thickness of flap and residual tissue bed safe thin flaps.
3. Flexible hinge and bed parameters.
4. Planar flap, wavefront indicates less flap induced aberrations.
Immediate completion of interrupted procedures.

C. Extension of Corneal Flap Cutting to Corneal Transplants

The technique of flap cutting can readily be extended to corneal transplants procedures affecting 45,000 patients in the US. As shown in the Fig. 22, below, it can advantageously replace the trephines. It is not limited to full thickness transplants and can make easily partial thickness ones, 24a. Moreover, it can also create complex shapes, 24b, permitting self-locking that requires extensive suturing with a reduced healing time.

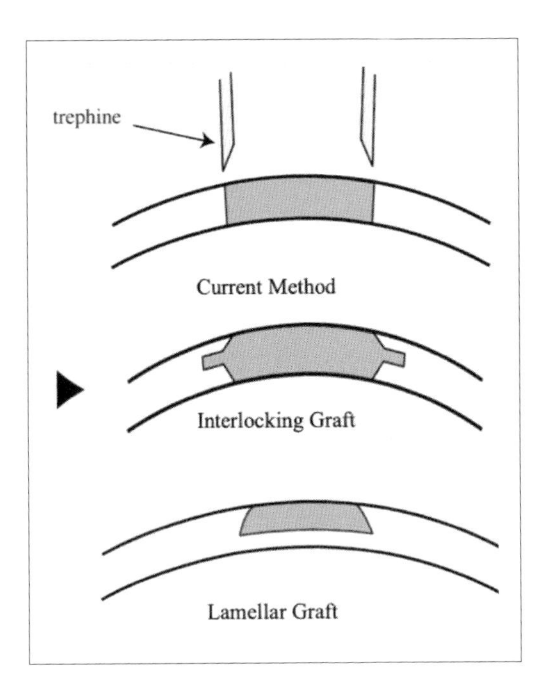

Figure 22. Today corneal transplant with trephine can only do a full thickness transplant. However, a femtosecond scalpel makes possible an interlocking graft. It shows also that a femtosecond laser can make a lamellar slice possible.

XIV. CONCLUSION

CPA has revolutionised the field of optics by providing the largest fields, the largest pressures, the largest temperatures and accelerations. The range of its applications has been considerable starting from the eV, the visible photon energy, and extending to the GeV, possibly the TeV. It is enlarging the field of optics from atomic to subatomic physics to include nuclear physics, high energy particle physics, astrophysics and cosmology. The applications are remarkably ubiquitous, going from the simple micromachining and eye surgery to the fundamental with TeV energy, like astrophysics, cosmology, gravitation or nonlinear QED. At intensities slightly above the solid or tissue damage threshold, it can be the perfect scalpel, cutting without any collateral damage. Through its interaction with gas, plasma, and solids, CPA can become a universal source of high energy radiations and particles. High harmonic generation in gases can yield attosecond X-UV pulses. The strength of its wake field in plasmas is such that it can accelerate particle with formidable gradients of GeV /cm, thousand times larger than conventional accelerating technique. The large energy acceleration provided by the laser through the Einstein equivalence principle, can be used to simulate a black hole and study, for instance, the loss of information paradox in a black hole. In addition, the electron being accelerated by a CPA laser in a gas can become by its lateral motion a powerful betatron source that emits radiation in the keV to

100 keV regime. Interacting with a solid target, high energy protons can be also efficiently produced. At very high intensity, the protons can become relativistic at an energy of 1 GeV. The protons in turn can produce neutrons. The latter can find a number of applications in the medical field like proton therapy or the production of nuclides for nuclear pharmacology. High energy protons can also be utilised for transmuting radiotoxic elements such as minor actinides which compose 60,000 metric tons of nuclear waste worldwide.

The CPA System provides ultra-high peak power in the PW level. However due to its low repetition rate in the Hz regime, its average power is in the 10–100 W range; which is far from enough for the applications such as particle collider or nuclear waste treatment. In addition, it is still plagued by a major flaw i.e, a very poor wall-plug efficiency of <1%. In this document we describe the project CAN which has the capability to provide both peak power, average power and efficiency.

Going to the Schwinger limit and beyond finally, current technology seems to limit the focused peak intensity to 10^{25} W/cm^2. To go higher, a novel compression technique is needed and presented. The intensity is obtained not by an augmentation of the energy but by the reduction of the laser pulse duration to the single cycle. By using relativistic mirrors, a gigantic ponderomotive force in the X-ray regime can be created with pulses as short as a zepto–attosecond. Once tightly focused it could easily produce intensity at or above the Schwinger limit. The ultra-high intensity could engender in a solid density a wake field with an acceleration gradient of TeV/cm, opening a fundamentally new vista beyond the Schwinger regime.

REFERENCES

1. Akhmanov, S. A., V. A. Vysloukh, and A. S. Chirkin, 1992, *Optics of Femtosecond Laser Pulses* (American Institute of Physics).
2. Albertazzi, B. et al., *Science* **346**, 325 (2014)
3. Arnould, M. et al., *Phys. Rept.* **450**, 97 (2007)
4. Aoyama, A., J. Ma, Y. Akahane, N. Inoue, H. Ueda, H. Kiriyama, K. Yamakawa, 2002, in *Technical Digest of Conference on Lasers and Electro-Optics* (CLEO) 2002 "Optical Society of America", p. 109.
5. Backus, S., C. Durfee, G. Mourou, H. C. Kapteyn, and M. M. Murnane, 1997, *Opt. Lett.* **22**, 1256.
6. Badziak, J., E. Woryna, P. Parys, K. Y. Platanov, S. Jablonski, L. Ryc, A. B. Vankov, and J. Wolowski, 2001, *Phys. Rev. Lett.* **87**, 215001.
7. Baetsle, L. H. et al., "Partitioning and transmutation in a strategic perspective," *Euradwaste*, **99**, 138 (2000).
8. Baltuška, A., T. Udem, M. Uiberacker, M. Hentschel et al., *Nature* **421**, 611 (2003).

9. Barty, C. P. J., C. L. Gordon III, and B. E. Lemoff, 1994, *Opt. Lett.* **19**, 1442.
10. Beaud, P., M. Richardson, E. Miesak, and B. T. Chai, 1993, *Opt. Lett.* **18**, 1550.
11. Bespalov, V. I., and V. I. Talanov, 1966, *JETP Lett.* **3**, 307.
12. Brabec, T., and F. Krausz, 2000, *Rev. Mod. Phys.* **72**, 545.
13. Brady, C. S. and T. D. Arber, "An ion acceleration mechanism in laser illuminated targets with internal electron density structure," *Plasma Phys. Control. Fusion*, 53, 074004 (2011).
14. Bula, C., K. T. McDonald, E. I. Prebys, et al., 1996, *Phys. Rev.Lett.* **76**, 3116.
15. Bulanov, S. V., and V. S. Khoroshkov, 2002, *Plasma Phys. Rep.* **28**, 453.
16. Bunkenburg, J., et al., 1985, *IEEE J. Quantum Electron.* QE-17, 1620.
17. Calegari, F., D. Ayuso, A. Trabattoni, L. Belshaw et al., *Science* 346, 336–339 (2014).
18. Calegari, F., G. Sansone, S. Stagira, C. Vozzi, M. Nisoli, J. Phys. B **49**, 062001 (2016).
19. Carman, R. L., R. F. Benjamin, and C. K. Rhodes, 1981, *Phys. Rev.* A **24**, 2649.
20. Cavalieri, A.L., N. Müller, Th. Uphues, V. S. Yakovlev et al., *Nature* **449**, 1029 (2007).
21. Cederbaum, L. S., J. Zobeley, Chem. *Phys. Lett.* **307**, 205 (1999).
22. Chen, P. and G. Mourou, 2017, "Accelerating Plasma Mirrors to Investigate the Black Hole Information Loss Paradox," *Phys. Rev. Lett.* **118**, 045001
23. Chen, S.N. et al., *Phys. Rev. Lett.* **108**, 055001 (2012)
24. Chen, S.N. et al., submitted to *Matter and Radiation at Extremes* (2018).
25. Clayton et al *Phys. Plasmas.* **1**, p1753 (1994).
26. Collier, J., C. Hernandez-Gomez, I. N. Ross, P. Matousek, C. N. Danson, and J. Walczak, 1999, *Appl. Opt.* **38**, 7486.
27. Collier, J. L., et al., 2004, unpublished.
28. Corkum, P.B., *Phys. Rev. Lett.* **71**, 1994 (1993).
29. Cousens, S., B. Reville, B. Dromey, and M. Zepf, "Temporal Structure of Attosecond Pulses from Laser-Driven Coherent Synchrotron Emission," *Phys. Rev. Lett.* **116**, 1–5 (2016).
30. Couture, A. and R. Reifarth, At. *Data Nucl. Data Tables* **93**, 807 (2007)
31. Cowan, J. J., et al. *Astrophys. J.* **294**, 656 (1985)
32. Ditmire, T., and M. D. Perry, 1993, *Opt. Lett.* **18**, 426.
33. Dubietis, A., G. Jonusauskas, and A. Piskarskas, 1992, *Opt. Commun.* **88**, 437.
34. Dudovich, N., O. Smirnova, J. Levesque, Y. Mairesse et al., *Nature Phys.* **2**, 781 (2006).
35. Duguay, M. A. and G. P. Rentzepis, *Appl. Phys. Lett.* 10, 350–352 (1967).
36. Einstein, A. and A. Engel. *The Collected Papers of Albert Einstein* Vol 7: The Berlin Years: Writings, 1918–1921. Princeton University Press, 2002.
37. Endoh, A., M. Watanabe, N. Sarukura, and S. Watanabe, 1989, *Opt. Lett.* **14**, 353.
38. Esarey, *IEEE Trans. Plasma Sci.* **24** p252 (1996).
39. Esirkepov, T., M. Borghesi, S. V. Bulanov, G. Mourou, and T. Tajima, 2004, *Phys. Rev. Lett.* **92**, 175003.
40. Everett et al *Nature* **368** p527 (1994).
41. Faure et al, *Nature* **431** p541 (2004).
42. Fourkal, E., B. Shahine, M. Ding, J. S. Li, T. Tajima, and C.-M. Ma, 2002, *Med. Phys.* **29**, 2788.
43. Fuchs, J. et al., *Nature Phys.* **2**, 48 (2006).
44. Gales, S., "Nuclear Energy and Waste Transmutation with High Power Accelerator and Laser Systems." [Online]. Available: https://indico.cern.ch/

event/617648/contributions/2517094/attachments/1442136/2220662/18_GALES_IZEST-Talk-Nuclear-Transmutation-040417.pdf. [Accessed: 22-Apr-2018].
45. Gauthier, J. et al, *Eur. Phys. Jour. D*, **48**, 3, 459-463 (2008).
46. Gauthier, D. et al, *Nature Communication*, **7**, 13688 (2016).
47. Geddes et al, *Nature* **431** p538 (2004).
48. Godet, J.P. et al, *Opt. Lett.*, **34, 16**, 2438-2440, 2009.
49. Gregori, G. et al., *Nature* **481**, 480–483 (2012).
50. Grischkowsky, D., A, C. Ballant, 1982, "Optical pulse compression based on enhanced frequency chirping," *Applied Physics Letters*, **41**.
51. Gumbrell, E. T. et al, New. *J. Phys.* **10**, 123011 (2008).
52. Hamm, R. (2010). Paper AP/IA-12, IAEA Proceedings Series, STI/PUB/1433, ISBN 978-92-0-150410-4.
53. Hentschel, M., R. Kienberger, Ch. Spielmann, G.A. Reider et al., *Nature* **414**, 509 (2001).
54. INTERNATIONAL ATOMIC ENERGY AGENCY, *Implications of Partitioning and Transmutation in Radioactive Waste Management*, no. 435. Vienna: INTERNATIONAL ATOMIC ENERGY AGENCY, 2004.
55. Jones, D.J., S.A. Diddams, J.K. Ranka, A. Stentz et al., *Science* **288**, 635 (2000).
56. Klünder, K., J. M. Dahlström, M. Gisselbrecht, T. Fordell et al., *Phys. Rev. Lett.* **106**, 143002 (2011).
57. Kmetec, J. D., J. J. Macklin, and J. F. Young, 1991, *Opt. Lett.* 16, 1001.
58. Kraus, P. M., B. Mignolet, D. Baykusheva, A. Rupenyan et al., *Science* **350**, 790 (2015).
59. Larsson, J., P. A. Heimann, A. M. Lindenberg, P. J. Schuk, P. H. Bucksbaum, R. W. Lee, H. A. Padmore, J. S. Wark, and R. W. Falcone, 1998, *Appl. Phys. A: Mater. Sci. Process.* **66**, 587.
60. Loesel, F.H., Tien, A.-C., Backus, S., et al., 1999, *Proc. SPIE*, 3565.
61. Lucchini, M., S. A. Sato, A. Ludwig, J. Herrmann et al., *Science* **353**, 916 (2016).
62. Luk, T. S., A. McPherson, G. Gibson, K. Boyer, and C. Rhodes, 1989, *Opt. Lett.* **14**, 1113.
63. Macchi, A., M. Borghesi, and M. Passoni, "Ion acceleration by superintense laser-plasma interaction," Rev. *Mod. Phys.* **85**, 751–793 (2013).
64. Maine, P., and G. Mourou, 1988, *Opt. Lett.* **13**, 467.
65. Maine, P., D. Strickland, P. Bado, M. Pessot, and G. Mourou, 1987, *Rev. Phys. Appl.* 22, 1657.
66. Maine, P., D. Strickland, P. Bado, M. Pessot, and G. Mourou, 1988, IEEE J. *Quantum Electron.* **24**, 398.
67. Mangles et al, *Nature* **431** p535 (2004).
68. Martinez, O. E., 1987, *IEEE J. Quantum Electron.* 23, 1385.
69. Migus, A., C. V. Shank, E. P. Ippen, and R. L. Fork, 1982, *IEEE J. Quantum Electron.* QE-18, 101.
70. Mourou, G., T. Tajima, and S. Bulanov, 2006, "Optics in the relativistic regime," *Rev. Mod. Phys.*, **78**, 2, p 309.
71. Mourou, G., B. Brocklesby, T. Tajima, and J. Limpert, 2013, "The future is fibre accelerators," *Nature Photonics* **7**, p 258.
72. Mourou, G., S. Mironov, E. Khazanov, and A. Sergeev, 2014, "Single cycle thin film compressor opening the door to Zeptosecond-Exawatt physics," *Eur. Phys. J. Spec. Top.*, **223**, 6, p1181.
73. Mumpower, M.R. et al., *Progr. in Part. And Nucl. Phys.* **86**, 86 (2016).
74. Nakatsutsumi, M. et al., *Optics Lett.* **35**, 2314–2316 (2010).

75. Nakatsutsumi, M. et al., *Nat. Comm.* **9**, 280 (2018).
76. Naumova, N. M., J. A. Nees, I. V. Sokolov, B. Hou, and G. A. Mourou, 2004, "Relativistic generation of isolated attosecond pulses in a lambda-cubed focal volume," *Phys. Rev. Lett.* **92**, 063902.
77. Naumova, N., I. Sokolov, J. Nees, A. Maksimchuk, V. Yanovsky, and G. Mourou, 2004, *Phys. Rev. Lett.* **93**, 195003.
78. Naumova, N. M., J. A. Nees, B. Hou, G. A. Mourou, and I. V. Sokolov, 2004, "Isolated attosecond pulses generated by relativistic effects in a wavelength-cubed focal volume," *Opt. Lett.* **29**, 778.
79. Nees, J., S. Biswal, F. Druon, J. Faure, M. Nantel, and G. Mourou, 1998, *IEEE J. Sel. Top. Quantum Electron.* **4**, 376.
80. Nifenecker, H., O. Meplan, and S. David, *Accelerator driven subcritical reactors.* CRC Press, 2003.
81. Nisoli, M., S. De Silvestri, O. Svelto, *Appl. Phys. Lett.* **68**, 2793 (1996).
82. Nisoli, M., De Silvestri, S., Svelto, O., Szipöcs, R., Ferencz, K., Spielmann, Ch., Sartania, S. & Krausz, F., 1997, "Compression of High-Energy Laser Pulses below 5fs," *Opt. Letters* **22**, 522.
83. Oliva, E. et al *Nature Photonics* **6**, 764–767 (2012).
84. Ossiander, M., F. Siegrist, V. Shirvanyan, R. Pazourek et al., *Nature Phys.* **13**, 280 (2017).
85. Panov, I.V., *Phys. of Atomic Nuclei* **79**, 159–198 (2016).
86. Paul, P.M., E.S. Toma, P. Breger, G. Mullot et al., *Science* **292**, 1689 (2001).
87. Perry, M., P. Pennington, B. C. Sutuart, et al., 1999, *Opt. Lett.* **24**, 160.
88. Pessot, M., J. Squier, and G. Mourou, 1989, *Opt. Lett.* **14**, 797.
89. Pessot, M., et al., 1987, *Opt. Commun.* **62**, 419.
90. Potemkin, A. K., M. A. Martyanov, M. S. Kochetkova, and E. A. Khazanov, 2009, "Compact 300 J/ 300 GW frequency doubled neodimium glass laser. Part I: Limiting power by self-focusing," *IEEE Journal of Quantum Electronics* **45**, 336.
91. Pretzler, G. et al., *Phys. Rev.* E **58**, 1165 (1998).
92. Ratkay-Traub I1, Juhasz T, Horvath C, Suarez C, Kiss K, Ferincz I, Kurtz R., Ultra-short pulse (femtosecond) laser surgery: initial use in LASIK flap creation. Ophthalmol Clin North Am. 2001 Jun;14 (2):347–55, viii–ix.
93. Reifarth, R. et al., *J. Phys. G: Nucl. Part. Phys.* **41**, 053101 (2014).
94. Remacle, F., R. D. Levine, *Proc. Natl. Acad. Sci. U.S.A.* **103**, 6793 (2006).
95. Remington, B. A., et al. *Science* **284**, 1488 (1999)
96. Ross, I. N., J. L. Collier, P. Matousek, et al., 2000, *Appl. Opt.* **39**, 2422.
97. Ross, I. N., et al., 1997, *Opt. Commun.* **144**, 125.
98. Roth, M., et al., 2001, *Phys. Rev. Lett.* **86**, 436.
99. Rouyer, C., et al., 1993, *Opt. Lett.* **18**, 214.
100. Rynes, J. et al., *Nucl. Instrum. Meth.* A **422**, 895 (1999).
101. Sansone, G., L. Poletto, M. Nisoli, *Nature Photon.* **5**, 655 (2010).
102. Sansone G., F. Kelkensberg, J. F. Pérez-Torres, F. Morales et al., *Nature* **465**, 763-766 (2010).
103. Schafer, K.J., B. Yang, L.F. DiMauro, K.C. Kulander, *Phys. Rev. Lett.* **70**, 1599 (1993).
104. Schiffrin, A., T. Paasch-Colberg, N. Karpowicz, V. Apalkov et al., *Nature* **493**, 70 (2013).
105. Schlaepfer, F., M. Lucchini, S. A. Sato, M. Volkov et al., *Nature Phys.* **14**, 560 (2018).

106. Schoenlein, R. W., S. Chattopadhyay, H. H. W. Chong, et al., 2000, *Science* **287**, 2237.
107. Schultze, M., M. Fieß, N. Karpowicz, J. Gagnon et al., *Science* **328**, 1658 (2010).
108. Schultze, M., E.M. Bothschafter, A. Sommer, S. Holzner et al., *Nature* **493**, 75 (2013).
109. Schultze, M., K. Ramasesha, C.D. Pemmaraju, S.A. Sato et al., *Science* **346**, 1348 (2014).
110. Siegman, A. E., 1986, *Lasers* University Science Books, Mill Valley, CA", p. 362.
111. Snavely, R. et al., "Intense high energy proton beams from petawatt-laser irradiation of lasers", *Phys. Rev. Lett.* 85, 2945 (2000).
112. Squier, J., F. Salin, G. Mourou, and D. Harter, 1991, *Opt. Lett.* 16, 324.
113. Squier, Du, J., Kurtz, R.M., Elner, V., et al., 1995, in *Ultrafast Phenomena IX*, Barbara, P.F. et al., Ed. (New York: Springer), 254.
114. Steinke, S. et al., "Efficient ion acceleration by collective laser-driven electron dynamics with ultra-thin foil targets," *Laser Part. Beams*, **28**, 215 (2010).
115. Strickland, A. D., and G. Mourou, 1985, *Opt. Commun.* **56**, 212.
116. Sullivan, A., et al., 1991, *Opt. Lett.* **16**, 1406.
117. Tajima T and Dawson J. M., *Phys. Rev. Lett.* **43**, 267 (1979)
118. Tajima, T., D. Habs, and X. Yan, "Laser Acceleration of Ions for Radiation Therapy," *Rev. Accel. Sci. Technol.*, **2**, 201 (2009).
119. Telnov, V. I., 1990, *Nucl. Instrum. Methods Phys. Res.* A **294**, 72.
120. Telnov, V. I., 2000, *Int. J. Mod. Phys.* A **15**, 2577.
121. Telnov, V. I., 2001, *Nucl. Instrum. Methods Phys. Res.* A **472**, 43.
122. Teubner, U. and P. Gibbon, *Reviews of Modern Physics* (2008).
123. Thielemann, F.-K. et al., Progr. in *Part. And Nucl. Phys.* **66**, 346–353 (2011)
124. Treacy, E. B., 1969, *IEEE J. Quantum Electron.* **5**, 454.
125. Tsakiris, G. D. et al, *New Journal of Physics,* **8**, 19 (2006).
126. Uiberacker, M., Th. Uphues, M. Schultze, A.J. Verhoef et al., *Nature* **446**, 627 (2007).
127. Vaillancourt, T. G., B. Norris, J. S. Coe, and G. A. Mourou, 1990, *Opt. Lett.* **15**, 317.
128. Van der Meer, K., et al., *Nucl. Instrum. Methods Phys. Res. Sect.* B **217** 202–220 (2004).
129. Vodungbo, B. et al, *Optics Express,* **19**, 5, 4346 (2011).
130. Wang, Y. et al, *Nature Photonics,* **8**, 381 (2014).
131. Wulff, M., D. Bourgeois, T. Ursby, L. Goir, and G. Mourou, 1997, in *Time Resolved Diffraction*, edited by J. R. Helliwell and P. M. Rentzepis (Clarendon Press, Oxford), pp. 195–228.
132. Yamakawa, K., H. Shiraga, Y. Kato, and C. P. J. Barty, 1991, "Prepulse-free 30-TW, 1-ps Nd:glass laser," *Optics Letters* **16**, 1593.
133. Yan, X. Q., T. Tajima, M. Hegelich, L. Yin, and D. Habs, "Theory of laser ion acceleration from a foil target of nanometer thickness," *Appl. Phys. B Lasers Opt.* **98**, 711–721 (2010).
134. Yokoya, K., 2000, *Nucl. Instrum. Methods Phys. Res.* A **455**, 25.
135. Zeitoun, P. et al, *Nature,* **431**, 426 (2004).
136. Zhou, M. L., X. Q. Yan, G. Mourou, J. A. Wheeler, J. H. Bin, J. Schreiber, and T. Tajima, "Proton acceleration by single-cycle laser pulses offers a novel monoenergetic and stable operating regime," *Phys. Plasmas* **23**, 43112 (2016).

Donna Strickland. © Nobel Prize Outreach AB. Photo: A. Mahmoud

Donna Strickland

Biography

LIFE

Donna Theo Strickland was born on May 27, 1959, in Guelph, Ontario, Canada. She studied engineering physics at McMaster University in Hamilton, Canada, and optics at the University of Rochester in the United States. She earned her doctorate in optics in 1989. Her PhD supervisor was Gérard Mourou, future Nobel Laureate. She worked at the National Research Council, and then lived in the United States where she worked at Lawrence Livermore National Laboratory and Princeton University. She has been affiliated with the University of Waterloo in Canada since 1997. Donna Strickland is married with two children.

WORK

Professor Strickland developed chirped pulse amplification (CPA) with Gérard Mourou, her doctoral supervisor while at the University of Rochester. CPA enables the most intense laser pulses ever and the research has led to tools with applications in medicine, industry, science, the military and security.

AUTOBIOGRAPHICAL

On the day I was born, my father marked the day by buying a copy of a newspaper. He performed this same ritual to mark the birth of each of his three children. My mother was a very organized person and kept all of my mementos. Near the end of her life, when she sold the family home, she gave me her collection of my childhood memorabilia. This newspaper was preserved in a plastic bag. Along with it was a note from my mother, pointing out an article that she thought I would now find particularly interesting.

It was a piece on the first female engineering graduate from the University of Toronto. She was the only woman in a graduating class of 450. The accompanying photo shows her seated, holding a bouquet of roses, and surrounded by several of her male classmates. It looks a lot like Marilyn Monroe's number *Diamonds are a Girl's Best Friend* from the movie *Gentlemen Prefer Blondes.* The men are all looking at her adoringly and each is holding up a glass. According to the article, they were toasting their queen. The story describes her as a girl, a maid and, as I say, a queen. Not once was she called a woman.

Back then, my home town had a population of less than 40,000 people. Guelph's most notable former resident is John McCrae, the field surgeon in World War I who penned the poem *In Flanders Fields.* It's also known as the Royal City, Guelph being the surname of several royals, including George IV, the king at the time the city got its name.

My mother trained as a teacher but was a homemaker when I was a child. My father was an electrical engineer. I'm the second of their three children, appearing between my older sister Anne and my younger brother Rob. I would say that we had the stereotypical good, middle-class life. I was a daddy's girl. When I was young, I was always climbing up on my dad's lap, like in the family photo. I was always asking him to read my favourite story, *Molly Whuppie.* It is an English fairy tale about three sisters having to fend for themselves. Molly was the youngest, but she used her wits and her gumption to save all three of them from the horrible giant. My dad always asked, "Again? Can't you pick a different story?" But he always read me *Molly Whuppie.* He started calling me Molly Whuppie, which got shortened to Whup and that is what he always called me from then on. My dad was an avid newspaper reader, reading it from cover to cover, including the comics. He liked the comic *Cathy*, because she reminded him of me. I have several of the *Cathy* strips that he cut out of the papers. They are about the dad just wanting Cathy to be happy and Cathy mostly wanting to be an independent woman, more interested in a career than settling down and starting a family.

I have always loved going to school. I was one of those rare kids who was happy to have summer vacation over so I could go back to school. I started this love affair with school at Victory Public School. I met my new best friend, Susan, in Grade 1. We went all through school together, including skipping Grade 3 together. We went onto Willow Road School for Grades 7 and 8 and then Guelph Collegiate Vocational Institute, GCVI, for Grades 9 through 13. In the earliest grades, I excelled at reading, but as I got older, math and science became my favourite subjects. I would say writing was what I found hardest to do, so I shied away from subjects like history. In high school, when I made a comment to my friends about how I was glad I was done having to take any more history and geography courses, the teacher overheard and asked what courses I wanted to take. When I responded that I wanted to take math and physics, she told me

Figure 1. Strickland family November 1964.

that they were boys' subjects. I could not believe she would say such a thing. I had never once thought there were girls' and boys' subjects. Most of the top math students in my year were girls and none of my classmates thought that was odd. There were boys in the class who were excellent at writing essays and stories and I certainly didn't think that was strange. This was the 1970s and women's lib was all the rage. We girls were told we could do anything we wanted, and I believed it.

Education was important to both of my parents. My father's father had grown up in a small fishing village in Newfoundland and had no formal education. He learned to read and do math after he retired. Both of his sons went to university, which was rare coming from a small town on Cape Breton Island in Nova Scotia. My mother grew up on a farm outside a small village in Ontario. Her brother took over the family farm, while my aunt and my mother went off to university. Again, this was rare for girls from small villages to go off to university. Growing up, our parents never said to any of their children, "if you go to university." It was always, "when you go to university." So, the three of us understood from a very young age that we would be well educated. We went on many family car trips, mostly down east to see the Strickland family, in Cape Breton. My mother would bring along a large binder to share facts and history about the places we would see. We went to museums, historical sites and toured mines. My mother liked to tell the story of one family trip to the science centre about an hour from our home. As my mom told the story, when my dad was looking at one of the displays, he called me over to him and said I am going to want to see this because this is the future. It was a laser. So according to Strickland family lore, it was my father who introduced me to what would eventually become my life's work.

I was incredibly shy in high school. I had a few very close friends, but I wasn't very well known in the high school. I remember one day when I was asked if I was Edith's daughter, followed by someone else asking if I was Rob's sister and finally someone else wondered if I was Anne's sister. I didn't seem to be known by anyone as me. Even though I enjoyed math and physics I always felt it made me seem very nerdy to the other students. I remember when I won the school's prize for the highest mark in Grade 11 physics, I dreaded going up on stage to receive the award. I thought it would label me a supreme nerd. It turned out to be a learning moment for me. The other kids were wonderful about it and said things like it must be nice to be so smart. While at high school I enjoyed being in the school band, even though I really couldn't play the clarinet. I also enjoyed the outers club, which went on camping trips. The rest of my family didn't really like to rough it, but I enjoyed camping. Winter camping was my favourite because you didn't have to deal with the millions of mosquitoes that were present in the spring, or spiders in the tent.

While I was in high school, my dad received a diagnosis of terminal spinal cancer and was told he only had a year to live. I was spared from knowing this devastating news at the time. My mother realized that she would have to be the breadwinner for the family. She felt very fortunate when a teaching vacancy opened up at the high school where she had taught before getting married. This high school was my high school, so we were both at GCVI when I was in Grades 12 and 13. My father was not a quitter. He researched his own disease and asked his doctor about a new radiation treatment he'd read about. The physician wasn't optimistic, but my dad wanted to try it. He was one of the lucky ones and the radiation therapy worked. My mom kept her teaching job, since it would be a few years before they would know if the remission was permanent. We had my dad for another two decades.

With my mother's paycheque, my parents were able to afford a cottage on Lake Huron after my father sold the family home that he had inherited in Cape Breton. The family has enjoyed many summer weekends together up at the cottage ever since. You can see the family photo taken there in 1985, the year I published the paper that earned me a Nobel Prize.

So many times in my life I heard my mother talk about how she wished she had gone into science or mathematics. She was always sure that she would have found university easier taking the subjects that she had been very good at while in high school. People discouraged her when she was young because they felt women just didn't go into those disciplines. I think hearing her story of regret made me determined to pick a career in a field I was good at and enjoyed, regardless of what others thought or said. I was very clear in my own head about where my strengths were and what I wanted to do.

I would like to acknowledge my homeroom teacher, Jim Forsyth, who was also my physics teacher in Grade 13. When I returned to Canada as a faculty member at the University of Waterloo, he read that I had developed chirped pulse amplification. He contacted me through my mother asking if I would be willing to be placed on GCVI's wall of fame. I wasn't sure that I belonged on this wall that included John McCrae. Jim said that he wanted to have a female scientist on the wall as a role model for the female students. I agreed to his request and he made it happen. I have been on GCVI's wall of fame for two decades for the development of CPA. They recently have rewritten the citation to say that I have received the Nobel Prize for CPA. Now it doesn't seem so strange for me to be on GCVI's wall of fame.

I decided that when it came time to pick a university, I would go where my closest friends were not going. My best friend was going to take engineering at the University of Waterloo, where I teach physics today. My sister was already there. I knew if I went there, I'd stick with them and continue as I had been. I wanted to get over my shyness, and going away would force me to meet people and stand on my own.

While deciding what to take at university I couldn't make up my mind whether I wanted to pursue engineering like my father and my sister or if I should take physics, which I thought would be more fun. I looked over the course calendars of all the Ontario universities and I found that McMaster University had an engineering physics program. I decided that program would let me walk the line between engineering and physics. When I saw that one of the engineering physics programs was about lasers and electro-optics, I just thought what fun it must be to study lasers. That clinched it for me and off to McMaster I went. I made wonderful new friends at MAC. My first-year roommate, along with another pair of roommates from down the hall in the first-year residence, recently took a trip to Nashville together to celebrate turning 60. I was actually only turning 59, but I didn't let that stop me from joining the birthday bash. It was this same group of girlfriends who helped me shop for the gowns and dresses I needed for Nobel Week. I graduated from McMaster in 1981 with a Bachelor of Engineering degree. Unlike the woman who had graduated in 1959, I was not the only woman in engineering, and in fact there were three women who graduated in the discipline of engineering physics that year.

Even as a child, I knew I belonged in school and decided I would get a PhD, since I was told that was the ultimate in education. Towards this goal of going to graduate school, I took a research job at McMaster the summer after second year. I worked in the laser group supervised by Brian Garside. I had two different research projects in two different labs. The labs in the subbasement of Burke Sciences Building were undergoing renovations and the air conditioning had been turned off. This was a very hot summer and the men were working with their shirts off. I didn't see that I wanted to do this too, so I decided that I would spend most of my time on the other project in another building. I was to try to melt germanium onto gallium arsenide to make a p-n junction to be a fast optical detector. I had to clean out an old evaporator, including cleaning and fixing a diffusion pump. It took me most of the summer, but I did get to try to make one sample. To test it I had to silver paint some electrical connections to it. I have shaky hands and the small test sample flew out of my tweezers onto the floor littered with solder drops that all looked the same as my device. I was down on my hands and knees looking for it when one grad student came in and asked what I was doing. He got down on the floor with me and started looking. One by one as the men came in and asked what we were doing they each got down on the floor with us and looked. We found it! I wondered if the guys would do this for any summer student or did it help my cause that I was the only woman in the group. I remain very good friends with the grad students I worked with that summer.

Unfortunately, when I finally tested my device, it was not a diode and my summer project was a bust. Luckily, I had impressed my supervisor anyway and he was willing to write me good references for my next summer job and for grad school. The summer after third year, I worked at the National Research Council in Ottawa with John Rolfe. That year my project was investigating polishing fiber bundles. The group needed a way to get light down a cryostat that did not have windows. I don't remember being much more successful that summer than I had been the previous summer. I made many fiber bundles, but I don't remember ever getting to shine light through any of them to see if they worked.

I knew I wanted to keep working with lasers and I asked the grad students where I should go. I was advised by one to consider the two optics schools in the United States: The Institute of Optics at the University of

Rochester, or the University of Arizona Optical Sciences Center. I didn't get through the foreign student admissions at Arizona, but Rochester accepted me and off I went. During my first week at the Institute of Optics, I met a Canadian graduate student who offered to show me around the campus. When he learned that I was interested in working with lasers, he introduced me to Gérard Mourou at the Laboratory for Laser Energetics. I went into his ultrafast laser lab that had a red dye laser pumped by a green laser beam and I thought what fun it would be to work on such lasers. They reminded me of a Christmas tree and Christmas is my favourite time of year. Gérard was not a professor at the time and so he became my research supervisor. By the time I was in third year, the institute realized that so many students wanted to work with him, that he should be a professor and so I became one of Gerard's first PhD students. He is also the person with whom I now share half of the 2018 Nobel Prize in Physics.

While working in Gérard's group I met my future husband Doug Dykaar. He was a graduate student in electrical engineering. The group he worked with used the ultrafast lasers in Gérard's lab. As Doug describes our first meeting – the lights were off, except for the flashlamps that pumped the infrared lasers – anyone else would think it was romantic. Optics labs are often dark to help us see the laser beams. Doug and I became good friends one week when we worked together to the wee hours of the night on a project about electro-optic sampling with infrared light. We used my laser with Doug's electro-optic sampling apparatus. Doug and I share a fondness for dessert and a great new dessert place had opened up near the lab. We quite often took a break in the afternoon and went and ate cake together. We didn't actually start dating until five years after that first meeting, but we think the group thought we were dating long before that.

Gérard gave me a theoretical paper by Stephen Harris of Stanford University. It was about high order harmonic generation that could yield coherent radiation out in the extreme ultraviolet. Lasers do not work at these high frequencies. I decided that it looked like a good PhD project for me to work on. To make it happen, we would need a more intense laser than Gérard's group had at the time. At first, I tried pulse compression of a high energy laser, but it didn't work. At the time, it wasn't possible to increase the intensity of the beams without damaging the laser itself. Gérard came up with a new idea that we now call chirped pulse amplification, CPA for short. The simple idea that Gérard came up with was to stretch the pulse by chirping it. That means the frequency changes through our laser pulse. In a bird's chirp, the sound frequency changes in time through the note. It was the stretched, chirped pulse that we would amplify.

Figure 2. Strickland family 1985.

It took me about a year to build the amplification system. To get a lot of energy into a short pulse, we needed to first stretch it to make it a long pulse, amplify the long pulse and then compress it. I want to thank Marcel Bouvier for helping me with this project. Gérard had managed to get an old laser amplifier that I could use for the project. The electronics needed to be fixed and Marcel was the electronics engineer for the group. Marcel had also developed a new type of Pockel cell driver for the lasers in the group. I used one of these Pockel cells in the regenerative amplifier I built for the CPA system. Marcel went on to start a company to sell these Pockel cells, and I still use one of his Medox Pockel cells to this day in my lab at Waterloo. I also want to thank my friend and colleague Steve Williamson. I did not have any way to measure the duration of the amplified compressed pulses. Steve had a streak camera that would work. He brought his streak camera into the lab one night and together we measured the pulse duration. CPA worked. The amplification process did not distort the pulse chirp and we had a short intense laser pulse.

To answer a commonly asked question, I never wondered at the time if I would be awarded a Nobel Prize for this. I was just trying to do one of the world's best PhDs. That was my goal. We were quick to publish our results. We wanted to be sure we were first, so submitted to *Optics Communications,* a journal with a relatively quick turnaround. When I finished writing, the article came to only three pages in length. I simply wanted to state the idea, explain the laser details and give the data. *Compression of Amplified Chirped Optical Pulses* by Strickland and Mourou came out December 1, 1985. It was my first published paper. It was awarded the Nobel Prize in Physics 33 years later. At the time we received the award, the paper had more than 4,000 Google citations, with about 200 new ones per year. I would like to note that during the first few years after publication, it wasn't that highly cited. Large laser systems take time to develop and were done mostly at the large labs. CPA didn't really become mainstream until titanium doped sapphire was developed. This allowed academic researchers with small labs to build CPA systems. Then the field took off.

CPA formed the basis of my PhD, but I had to do a scientific study with it for my PhD. The high harmonic generation was going to be too difficult to finish in time. For a six-month sabbatical, See Leang Chin from Université Laval in Quebec came to Rochester to study multi-photon ionization with the CPA laser. Gérard agreed that I could work with See Leang on the ionization studies for my thesis. In the end, there was a group of us working on this project that included a new student, Steve Augst, a new professor David Meyerhofer and a theorist Joe Eberly, who was a professor in both physics and optics.

Long before finishing my PhD, I knew that I wanted to be Paul Corkum's second post-doctoral fellow. His first post-doc had been a grad student at McMaster when I was an undergrad with the group. Paul worked in the laser and plasma section of the physics division of the National Research Council in Ottawa. Paul was already considered to be Canada's leading expert in ultrafast optics. Now he is renowned for his development of the new field of attosecond science, but he didn't get to that until after I worked with him. I took so long to get my PhD that I almost lost the opportunity to work with Paul. He called me in January 1988 and told me that if I could promise to be done by the end of the summer, he would hold the job for me. I was so happy when See Leang showed up and suggested that we all work together on multi-photon ionization. That helped me keep my promise and I was able to join Paul's group in September 1988. Technically I did not graduate with my PhD from Rochester until February 1989. With Paul, I worked on the link between continuum generation and self-focusing. I still find continuum generation to be like magic seeing all the colours appear out of nothing, even though I helped discover why all the colours were generated. After that we worked on Coulomb imaging of iodine. It was a wonderful three years. I tell everyone that doing a post-doc is the best. You get to do research full time. While at grad school, you are busy doing classes and worried about things like qualifying exams. As a professor, you are the one responsible for finding the money to do the research, while also teaching and doing committee work. As a post-doc, you have only one responsibility and that is to have fun doing research.

Doug and I were married in 1991 right after I finished my post-doc. After leaving Rochester, Doug was lucky enough to have found a job he loved working at Bell Labs in New Jersey. I wanted to find a science job in New Jersey so I could live with my husband, but I didn't manage to find one. I took a job at Lawrence Livermore National Laboratory in California, where I again worked on developing new CPA lasers and also using the 10TW CPA laser that they had in the group to finally study high harmonic generation. My supervisor at Livermore was Mike Perry. I worked on developing CPA in chromium doped LiSAF with Todd Ditmire, a fantastic young grad student. Todd went on to build the second Petawatt CPA laser as a young faculty member at the University of Texas. I worked with John Crane on the harmonic generation.

I had a fun year at Livermore, but I wanted to live with my husband and working at a lab in the United States as a non-citizen is difficult. Since Doug had a dream job, I kept looking for a new job in New Jersey. I took a job as a member of technical staff at Princeton University's Photonics and Electro-Optic Materials Center. I worked with Warren S. Warren there and Stephen Forrest, who was the director of the center. It turned out to be a blessing that I had this position. I went through both of my pregnancies while on this job. Pregnancy made me nauseous morning, afternoon and night. I would have found it hard to teach as a faculty member during this time. I somehow managed to get my job done even though I did spend quite a bit of time in the sick room.

Doug and I have two children, Adam and Hannah. They are both adults now and the best of friends, despite being quite different in their temperament and interests. They have always been close, ever since they were tiny children. Doug and I always encouraged them to follow their interests so long as they are doing what makes them happy and that plays to their strengths. Hannah seems to be following in her parents' footsteps. She is

Figure 3. Strickland-Dykaar wedding August 21, 1991.

currently a grad student studying astrophysics. Adam has taken his own route and is studying comedy writing and performance.

Doug and I continued to both look for academic jobs. The University of Waterloo was the first university that offered me a job. I started there in 1997. Since I had followed Doug to New Jersey, he agreed to follow me to Waterloo. He took a job in industry. I set up an ultrafast laser group. Teaching is rewarding and I enjoy sharing my excitement about physics with budding scientists. In addition to teaching and conducting research, I believe strongly in serving the scientific community. Among my professional activities, I was president of the Optical Society of America (OSA), and was on the board of the Canadian Association of Physicists as the director of Academic Affairs.

The early-morning call from Sweden on October 2, 2018, changed my life. I am grateful that the Royal Swedish Academy of Sciences recognized my work and I feel so honoured to be among the many eminent scientists I join as a Nobel Laureate. With one phone call my work is out in the world more than ever before, my name is mentioned in the same breath as trailblazers Marie Curie and Maria Goeppert Mayer, and I have had dinner with royalty. I will use the platform the Nobel Prize affords me to continue to advocate on behalf of science and the many scientists who devote their lives to probing the fundamental questions.

Generating High-Intensity Ultrashort Optical Pulses

Nobel Lecture, December 8, 2018 by
Donna Strickland
University of Waterloo, Waterloo, Canada.

THIS LECTURE is about how light interacts with matter and how the interaction changes when the power of the light gets very high. When the power of the light is low, the interactions are linear. That is, the response of the interaction varies linearly with the power of the light. If you double the power of the light impinging on the material, you will absorb twice as much power. At the time of my PhD thesis, scientists were investigating nonlinear interactions with media. With nonlinear absorption, when the incident light power is doubled the absorption rate is more than doubled. My PhD research project was to study a highly nonlinear experiment and to do the experiment I needed first to develop a very intense laser. This is the reason that I developed Chirped Pulse Amplification (CPA) along with my PhD supervisor, Prof. Gérard Mourou. With the new intense CPA laser system, we were able to show that the interactions were not the nonlinear interactions that we were expecting. We had to rethink how intense light interacted with matter, and the development of CPA helped usher in a new field of study that was emerging called high intensity laser physics.

The interaction of light with matter has been studied over the centuries. Scientists wondered if light was made up of particles or whether it could be a wave. Experimental observations showed evidence of both explanations of light. The fact that a tree casts a well-defined shadow led people to think light had to be made of particles travelling in straight lines. If the path of a light particle went through the tree, it was simply blocked by the tree, resulting in the shadow. On the other hand, wave-like phenomena were also observed. When light passes through a small aperture there are fringes on the edge of the resulting light pattern that can only be explained using wave theory.

In the mid-19th century, theorist James Clerk Maxwell combined the equations resulting from various experimental studies on time varying electric and magnetic fields. There were four equations in total coupling electric and magnetic fields. To make the four equations mathematically consistent, Maxwell had to add one more term concerning time varying electric fields inducing a magnetic field. This last piece of the puzzle could not be experimentally observed with mid-19th century technology. By combining the four complete mathematical equations, Maxwell was able to show theoretically that time varying electric and magnetic fields travelled as waves. These electro-magnetic waves travelled at the speed of light. He then theorized that since light travels at the speed of light, it must be an electro-magnetic wave. By the end of the 19th century Heinrich Hertz had experimentally confirmed that time varying electric fields did indeed travel as waves. Because of the success of Maxwell's equations leading to the wave equation describing the time varying electro-magnetic waves, scientists were led to believe that light must be a wave.

By the end of the 19th century, scientists such as Hertz were carrying out experiments on ejecting electrons from material by shining light on the material. If light was indeed a wave, the brighter the light, the faster the electrons should be travelling when they leave the material. These experimental observations would be analogous to watching stones be being thrown up on a beach by a water wave. A small ripple in the water will not move a stone very much, but when the waves get big, the stones can move quickly and travel further away from the water. The power of a wave is related to the amplitude of the wave.

The experiments with light ejecting electrons used different colours of light. The colour of light is given by the wavelength of the light, which is the distance between the crests of the wave. Of the colours that we can see with our eyes, red has the longest wavelength of about 0.7 micrometers (one micrometer is one millionth of a meter) and the shortest wavelengths of light that we can see are violet light with wavelength around 0.4 micrometers. When the scientists shone red light on the material, no electrons were ejected no matter how powerful the red light was made. With green light, electrons came off the material but at low speed. When the power was increased, more electrons were ejected but always at the low speed. With violet light, the electron speed was higher than with green light, but again the speed did not increase with increased light power. The number of electrons ejected increased with light power. These experiments flew in the face of the theory that light was a wave.

In 1905, Albert Einstein was able to explain this effect. Of all the physics theories that Albert Einstein discovered, it is his work on this phenomenon known as the photo-electric effect that Einstein was awarded his Nobel Prize for. From these experimental observations of light causing electrons to be ejected from materials, Einstein figured out that light is quantized in its energy. There is a minimum energy unit of light that we now call the photon. A photon is a wave-like particle. Einstein realized that the energy of a photon is given by the wavelength of the light. The total energy in a light pulse is then given by the energy of a single photon multiplied by the number of photons.

To understand why the photon picture of light explains the experimental observations, I will use an analogy with gravitational energy because it's harder for us to feel the energy of light. We know from everyday experience that if we drop a ball to the ground, the ball picks up speed as it falls. The ball would be moving faster when it hits the ground if dropped from a higher position above the ground. We will imagine playing basketball with a child's basketball net. The photons will be the basketball players trying to drop the ball through the net and we will determine the speed the ball has when it enters the net. The speed of the ball through the net is analogous to the speed of the electron as it is ejected from the material by the photons.

As depicted in Figure 1, a red photon has the longest wavelength and so has the smallest energy. It is like a child sized basketball player. A red photon playing basketball, no matter how they try, even on their tippy-toes cannot reach the net to drop their electron through. And no matter how many of these child-sized photons there are, electrons are never going to get through the basketball net. This is equivalent to no electrons coming off the material when irradiated by red light at any power.

Figure 1. Energy of photons is depicted as height. Of the visible colours, red photons have the least energy, violet photons the greatest energy, with green photons in between. The height of the ball held up by the photon, compared to the height of the basketball net is equivalent to the comparison of photon energy with the energy that holds the electrons to the atoms.

Now if you have a green light, that's like an adult sized photon, playing with a child's basketball net and they can dunk the electron through the net, but only barely. They're only standing slightly taller than the basketball net and so when they drop their ball through it has only picked up a small speed before reaching the net.

But the violet photon is like a pro basketball player. They're the very tall photons. Their height is well above the net and when they drop their electron through the net, the electron has a lot more speed. But it wouldn't matter how many of these violet photons there were, they would all be dropping electrons at the same speed through the net. There would be more electrons coming through the net.

Usually when people study the photoelectric effect, it is about the study of quantum mechanics, but that's not what I want to concentrate on for this lecture. What the experiment and theory of the photo-electric effect tells us is about how light interacts with matter. It is always one photon interacting with one atom at a time. If that photon has enough energy, more than the energy that the atom is holding onto its electron, it can send that electron on its way. The kinetic energy of the electron is given by the difference in the energy of the photon and the energy the atom was using to hold onto the electron. As the power of the light is increased, it means that there are more photons in the pulse and so more photons are available to interact with more atoms, but always one photon meeting one atom at a time.

And that's how we understood how light interacted with matter through the beginning of the 20th century. And then along came Maria Goeppert Mayer, the second woman to be awarded a Nobel Prize in physics. Her Nobel Prize work was about nuclear shell structure, but I will discuss the work she did for her PhD in 1930 and published in 1931[1]. I cited that 1931 paper in my own PhD thesis.

Maria Goeppert Mayer started a whole new area of physics known now as multiphoton physics. She worked out the quantum mechanical theory of an atom simultaneously absorbing two photons to leave the atom in an excited electronic state. In order for light to be linearly absorbed, the energy of the photon must match the energy of the atomic energy between the ground state and the excited state. For the case of two photon absorption that Goeppert Mayer solved, the addition of the energies of the two photons must equal the energy difference of the atomic states. Of all possible two photon processes, absorption is the most likely because it uses a resonant interaction. This is the same idea as pushing a child on a swing. To make the swing go higher you only push one time each period or in other words only when the swing has reached its maximum height. If you pushed at other times, it would disrupt the swinging motion rather than increase the swinging motion to greater heights.

If instead of absorption, we consider the process of multiphoton ionization, two little red photons would be simultaneously absorbed and the combined energy of the two red photons would be the same energy as a violet photon of twice the energy. The two red photons would cause an electron to be ejected as if it was irradiated by a single violet photon.

Goeppert Mayer had theoretically predicted multi-photon absorption, but no one had observed the effect experimentally. In fact, no one would see a multi-photon effect for another 30 years. It was Peter Franken's group at the University of Michigan that were the first to see a multiphoton effect. They were not investigating multiphoton absorption but rather a nonlinear optical process known as second harmonic generation [2]. To clarify, multiphoton physics refers to the study of the medium undergoing a multiphoton process and nonlinear optics is the study of the resulting light from a multiphoton process. Harmonic generation is not a resonant process, but one where an atom will simultaneously absorb the energy of two photons but then quickly release the energy. The energy will be released as a single photon having twice the energy. The photon having twice the energy of the input photons is known as the second harmonic. Since the photon energy is doubled, the wavelength is halved. Using an optical spectrometer that could measure both the red optical signal as well as the generated violet light, the Franken group measured a very small signal in the violet when they had a very powerful red light.

Second harmonic generation was then first observed in 1961, which begs the question of why it took 30 years to see any type of multiphoton effect similar to what Maria Goeppert Meyer had predicted in 1931. What was special about 1961 was that in 1960, the laser was first demonstrated.

Because this is a Nobel lecture, I want to honour all of the people that have been awarded Nobel Prizes for developments that led to the invention of the laser. Nicolay Basov, Alexander Prokhorov and Charles Townes were honoured for developing the maser. The maser was the precursor of the laser and the m is for microwave rather than light in lasers. Technologically it was easier to make a maser than a laser. Masers were first demonstrated in the 1950s. Art Schawlow was then awarded a Nobel Prize for laser spectroscopy, but he had done a lot of the pioneering work on converting maser technology to the optical wavelengths.

But I want to give credit to Theodore Maiman. There was a race on at the end of the 1950s and into 1960 to see who would be the first person to demonstrate the laser. Ted Maiman won the race [3]. He was working at Hughes Aircraft.

The laser was born in 1960 and that's why Peter Franken's group could see a nonlinear optical effect. Why did laser light lead to observations of nonlinear optics? Regular light sources such as the sun or a light bulb emit photons of every colour, which is why the light appears white. The photons go off in all directions. They also don't communicate with each other. They emit at random times. Because the photons are emitted at random times, the crests of some waves overlap troughs of other waves and the waves cancel each other out. The overall amplitude of the combined waves then is not that high. In other words, the density of photons is low. On the other hand, a laser emits a beam of light where all the photons travel in one direction. The light from a laser will also only have one colour. The photons in the laser communicate with each other so that the photons are emitted in a way to have all the crests of the waves of each photon line up together. The waves of each of the photons add together, making themselves into a giant wave and a giant wave means the density of photons is very high.

So now we need to discuss why the higher density of photons leads to the observation of multiphoton interactions. The density of the photons is given by the total number of photons in a given volume. For a light beam, two dimensions of the volume are given by the area of the beam. The beam area can be reduced by focusing the beam with a lens. The shorter the focal length of the lens, the smaller the beam diameter. The smallest beam diameter is limited to the dimension of the wavelength. For light, these wavelengths are on the order of 1 micron. The third dimension of the volume of light is given by the length of the pulse. The pulse length can be given by a spatial dimension, l, or the temporal length of the pulse τ. The two lengths are related by the speed of light, c such that, $l = c\tau$. The shortest pulses to date are on the order of two wavelengths although pulses this short have yet to be amplified in a laser amplifier.

In a linear interaction, one photon interacts with one atom at time. The rate of interaction is given by the probability of finding a photon in the interaction volume of the atom. The interaction volume differs from the actual size of the atom. The interaction volume is given by what is known as the cross-sectional area of the atom. The easier it is for the atom to interact with the light, the larger is this area, but it is about the area of the atom, which has dimensions of 0.1nm squared. This area is more than one hundred million times smaller than the focused spot area of the light. The third dimension of the interaction volume is given by the interaction time. The interaction time increases as the photon energy approaches the energy of the electronic state of the atom. If the photon is absorbed in the interaction, the process is a resonant process and the interaction time can be quite long, such as microseconds. This time scale corresponds to a length, l, of 300 m so that the interaction volume can be large. With linear interactions, the probability of finding a single photon in the interaction volume increases linearly with photon density.

To see a second order nonlinear process such as the Franken group observed, there must be a non-negligible probability of finding two photons in the interaction volume. Unlike two-photon absorption, second harmonic generation is a non-resonant process and so the time scale of the process is quite short and the interaction volume is the cross-sectional area of the atom and a length of a few microns. With regular light, the photon density is too small to have an observable chance to see the effect. The Franken group used a laser that had 3 J of energy with a 1 ms time duration pulse. With this 300W of laser power focused into the medium, the efficiency of the frequency doubling process was 10^{-7} or one violet photon for every 10 million red photons [3].

By the time I was working on my PhD in the 1980s, scientists were working on higher order nonlinear harmonic generation. The goal of the work was to generate coherent radiation equivalent to laser radiation, but out in the ultraviolet and possibly beyond that to the extreme ultraviolet. My supervisor Gérard Mourou gave me a paper [4] written by Stephen Harris of Stanford University to read and consider whether I could experimentally show the type of effects that Harris had considered theoretically. Harris had determined various transitions in atoms and ions that would have resonances for high order even harmonics. There are reasons of symmetry that do not allow even numbers of photons to be absorbed in isotropic, homogeneous media such as any gas medium so the even number of photons could not be absorbed with this multi-photon resonance. The even harmonic would also not be generated. By finding resonances for high order even harmonics, this would allow the next higher odd harmonic to have a better chance of being generated than the lower orders, which would not experience any resonant enhancement.

I determined that twice ionized nickel would have a resonance with the 8th harmonic of the radiation from a particular type of laser that we had in our research group, a Nd:YAG laser. My original thesis project was then to generate the 9th harmonic of the 1-micron radiation from this laser. There are many reasons that I won't go into here, but I never did accomplish these experiments. To try and do a ninth order nonlinear optical experiment, I didn't just need a pulsed laser. I needed a high intensity laser. This is why I was the student who worked with Gérard to develop Chirped Pulse Amplification.

As discussed previously, nonlinear interactions are dependent not on the total energy in the laser pulse, but rather the energy density. To generate the highest order nonlinearities, you want the highest energy density, which is energy per unit volume. Because we are discussing light waves, we more often refer to the intensity, which is the energy per unit area per unit time. The area of the beam is fixed by the focusing element and the wavelength of the light. The energy per unit time is the power of the light. To increase this power, you have two choices, increase the energy or decrease the time duration. Ideally, to maximize the intensity, you want both high energy and short pulses.

At the time of my PhD project, inside the Laboratory for Laser Energetics (LLE) at the University of Rochester, where I was working on the experiments for my thesis, both high energy lasers and short pulse lasers existed. Within Gérard's research group there were several short pulse laser systems. During the 1980s, the shortest pulses were created with dye lasers. Dye lasers had very large gain bandwidths that could support short pulses. I will discuss later the connection between short pulses and large spectral bandwidths. The dye lasers at LLE operated with pulses as short as 100 femtoseconds (fs). The 100 fs pulses stretch over a distance of only 30 microns. Compare this length to a 1 second pulse of light, which would stretch over two thirds of the distance from the earth to the moon, 300,000 km. Dye lasers have high gain, which means the lasing medium can be quite short to achieve the maximum power stored in the lasing medium. It is this characteristic of high gain, that makes dye lasers poor energy storage lasers. To store the maximum energy, the gain medium must stay in the excited state for a long time. To be a high gain medium, the opposite is true. The atoms must want to give up the energy in the excited state to allow for gain by stimulated emission. Dye lasers are therefore always low energy lasers with a maximum energy of about 1mJ. The maximum power of the dye laser was then 1 mJ per 100 fs or 10 GW. On the other hand, there was a high energy laser system known as the Omega laser at LLE. The main research goal at LLE is laser fusion, which is a process based on the total laser energy that can be applied to the fuel target. The Omega laser could deliver one kJ of energy, but the laser pulses had to be longer than 1 ns to not damage the laser rods. The gain medium of the Omega system was Nd doped glass. The energy of the glass laser was a million times greater than the dye system, but the pulse duration was 10,000 times longer. The power in the pulses was higher and reached a TW or a million MW.

The gain bandwidth of the Nd:glass laser is large enough to support amplifying pulses as short as 1 ps. The pulse duration was limited by the nonlinear optical effect of self-focussing. Self-focusing is a different nonlinearity than already discussed. At low power, the light is transmitted through the medium at a lower speed than light travels in vacuum. The difference in speed is determined by the index of refraction of the medium. At low power, the refractive index is given just by the material itself. When the intensity gets sufficiently high, the interaction changes such that the index of refraction changes instantaneously with light intensity. A laser beam is more intense in the centre of the beam compared to the edges. At high intensities then, the light in the centre of the beam travels slower than the light at the edges. This causes the light beam to start to focus in on itself, which causes the beam to get smaller and more intense. This is then a runaway process where the beam finally collapses to a small enough size that the intensity is large enough to damage the material.

Scientists want the high peak power at the output of the laser system so they can study the various nonlinear processes, but the lasers themselves are damaged by the nonlinear processes if the power gets too high inside the gain medium. This was the problem that had to be overcome. High intensity laser pulses had to be generated without destroying the laser medium itself. The solution was CPA.

The CPA idea is beautiful in its simplicity as depicted in Figure 2. Start with a short pulse from an oscillator and then stretch it to be long enough to not allow nonlinear interactions in the lasing medium. Amplify the stretched pulses. After amplification the long, high energy pulse can be compressed back to its short pulse duration creating the high-power pulse at the output of the system.

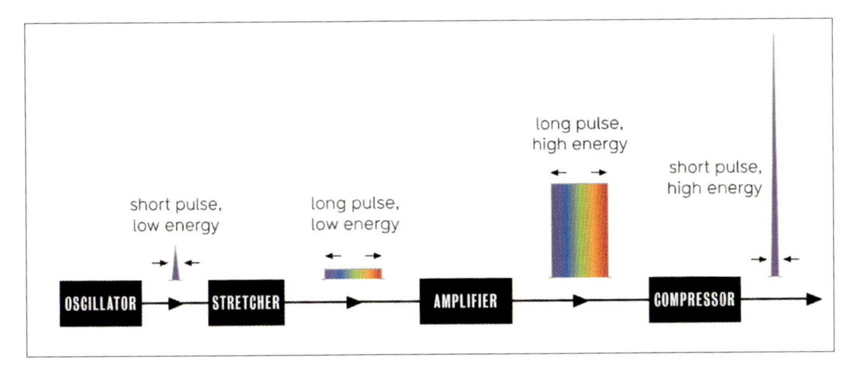

Figure 2. Schematic of chirped pulse amplification, showing that the power remains low in the amplifier because of the stretcher, but the power is very high at the output of the pulse compressor.

The CPA concept works best for lasing medium that can support short pulses because of a very large gain bandwidth and also support high energy because they are a good energy storage medium. The best compromise between these two criteria is titanium doped sapphire, which was being discovered at the same time as CPA was being developed [5]. Most CPA lasers today use Ti:sapphire as the lasing medium both for the oscillator and the amplifiers. However, in the mid 1980s the short laser pulses came from dye lasers that had wavelengths of 0.6 μm and the large energy storage lasers were Nd:glass operating at wavelengths of 1.0 μm, so it was not possible to simply stretch the short dye laser pulses and amplify them in the high energy glass amplifiers.

To demonstrate the CPA concept we used a mode-locked Nd:YAG laser, which could generate pulses with durations of about 100 ps at the same 1μm wavelength as Nd:glass amplifiers. As we wanted pulse durations shorter than 100 ps, we first had to use optical fiber pulse compression techniques that were being developed at that time [6]. In order to generate short pulses, you need large spectral bandwidths. If you have a single wavelength, the wave goes on forever. If you add a wave that has a slightly different wavelength to the first wave, such that at one time, which we will call t = 0, both waves crest, then at a later time the two waves will add up to zero because one will be at a peak when the other is at a trough. As shown in Figure 3, if you keep adding waves with different wavelengths so that they all peak at t = 0, then when the waves all add together, only at t = 0 do they all add constructively. At all other times there is some amount of destructive interference, that is, there are peaks and troughs adding together to make zero. The more wavelengths you can add, the shorter the pulse becomes. This is why there is an inverse relationship between spectral bandwidth and minimum possible temporal duration. The laser process that ensures all the wavelengths are adding so that all the crests add together at one time in order to produce a short pulse is known as mode-locking. The Nd:YAG laser used mode-locking to produce the ~ 100ps pulses.

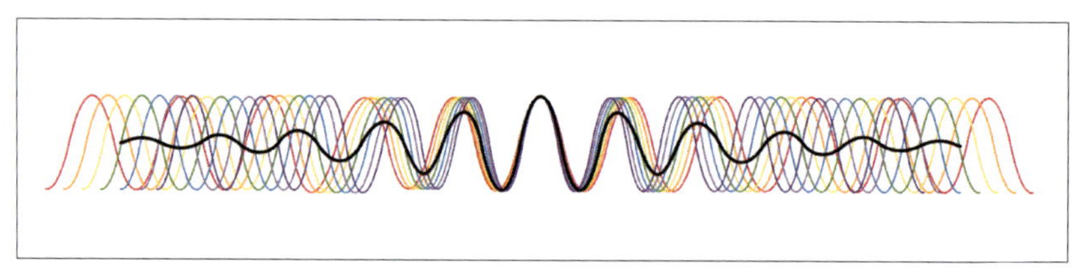

Figure 3. Demonstration of adding different wavelengths in a controlled manner leading to shorter pulses.

Fiber pulse compression uses a nonlinear optical technique known as self-phase modulation (SPM) to generate more spectral bandwidth. It is the same nonlinearity that leads to self-focusing in the bulk media, but rather than the intensity varying spatially across the beam, the intensity varies temporally over the pulse duration. In an optical fiber, the beam is guided and so the beam continues to travel as a plane wave down the fiber. The beam area is very small in the fiber, leading to the intensity of the light being high and remaining high over the long lengths of the fiber. The laser pulse has a time varying intensity. The index of refraction changes with intensity at these high intensities, causing the peak of the pulse to move slower than the leading and trailing edges of the pulse.

Because the light at the peak of the pulse moves slower than the leading edge, as the pulse propagates, the distance between the leading edge and the peak lengthens. The number of phase fronts remains constant and so the phase fronts must have greater distance between them. This longer distance between phase fronts is a longer wavelength than the original wavelength. At the same time, the peak is getting closer to the trailing edge of the pulse with propagation, and so the crests of the waves are being pushed together to create shorter wavelengths. The pulse duration does not change because of SPM, as the beginning and end of the pulse both continue to propagate at the low intensity speed. If the newly generated colours can be timed together rather than the red at the front and the blue at the back, the extra bandwidth would allow a shorter pulse duration. This retiming of the colours is known as pulse compression.

Propagation in the fiber not only creates new colours by the nonlinear interaction SPM, but these colours disperse along the fiber because the ordinary refractive index is wavelength dependent. The reddest of the colours travel fastest, leaving the bluest colours to trail at the rear of the pulse. The dispersion leads to the pulse durations becoming longer with propagation. By 1984, scientists had shown that the best pulse compression occurred when the fiber length was sufficient to not only create the needed extra spectral bandwidth, but that the pulse had stretched sufficiently so that the colours spread out in such a way that the frequency of the light changed almost linearly with time throughout the pulse [6]. The frequency sweep through the pulse is known as a chirp, in the same way that a bird's chirp has its sound frequency change in time. We could have called the technique stretched pulse amplification, but chirped pulse amplification sounded better.

It was already known by 1969 that a pair of parallel gratings could compress a linearly chirped pulse back down to its minimum pulse duration by completely eliminating the chirp [7]. The work on fiber optic pulse compression then gave us the path forward to demonstrate CPA. We would generate the needed spectrum and stretch the pulse in an optical fiber. This stretched pulse was then amplified in a Nd:glass amplifier and a pair of parallel gratings compressed the pulses to deliver short, energetic pulses at the output [8].

The Mourou group had a mode-locked Nd:YAG laser to pump a short pulse dye laser. Dye lasers needed green colour pumps, which was achieved by generating the second harmonic of the 1μm beam from the Nd:YAG laser. The beam intensity was sufficient to convert about 10% of the power to green light, but that left 90% of the infrared beam wasted. It was this wasted beam that I used for the input to the original CPA laser. It is lucky that we needed to use a long fiber for the spectral enhancement and pulse stretching as there was no room in the dye laser lab for me to build the CPA system. We were able to string the fiber in the ceiling down to another lab at the other end of the building where we had the space to build the system.

We used 1.4 kilometers of specialty optical fiber from Corning Inc. This fiber had a 9 μm core and so was single mode for the 1 μm light from the Nd:YAG laser. Corning had donated 2.5 km of fiber to us for this project. Unfortunately, only one end was available on the spool and so I had to unwind and rewind the fiber onto a different spool and I damaged the fiber during this process and was left with two pieces of optical fiber. I used the longer piece which was 1.4 km in length. The 2 W, Nd:YAG laser beam was focused into the fiber. At this power, the spectral bandwidth was increased to 4 nm Full Width Half Maximum (FWHM). The pulse duration increased to 300 ps at the output of the fiber.

To amplify the pulses a regenerative amplifier was built. This type of amplifier uses a cavity similar to an oscillator, but the focusing is weaker in the amplifier compared to an oscillator allowing a larger beam diameter in the gain medium. The oscillator must start with one spontaneously emitted photon. This photon then stimulates another and the then the two become four. This is exponential gain and how all lasers start. However, the gain must reach the saturation level where the gain becomes linear. Otherwise the amplification is very inefficient. As shown in Figure 4, if you consider one single pass through an amplifier with low input signal, then at the input side of the gain medium, there are insufficient photons to de-excite the atoms, leaving most of the energy behind in the gain medium. In order to extract that energy, the input to each amplifier stage must be at the saturation level. Each type of laser gain medium has its own saturation level and it is related to the energy storage. Nd:glass amplifiers have a large saturation energy of ~ 5 J/cm2. In oscillators and regenerative amplifiers, the pulses make multiple passes through the gain

Figure 4. Demonstration of low signal gain being energy inefficient. The figure shows the output photon number of 8 having started with one photon at the input. The stimulated emission has left only 7 of the atoms de-excited and almost all the atoms still in the excited state. This energy would be wasted.

medium such that they reach the saturation level by the final pass through the gain medium. To keep increasing the energy of the pulses, the size of the beam must keep increasing so that at the output of each amplification stage, the energy per unit area remains the same. In our case the pulse energy was increased almost a million times, from a few nJ to about 2 mJ. The repetition rate was decreased from 80 MHz to 10 Hz.

The self-focusing nonlinearity limits the maximum pulse intensity or energy density. The saturation energy of the gain medium determines the maximum energy per unit area. From these two limits, you can determine the minimum pulse duration that can be amplified without causing self-focusing. To achieve the maximum possible energy from a Nd:glass laser, the pulse should be stretched to 1 ns. As we only achieved 300 ps with our 1.4 km fiber, we did not extract the maximum energy from the regenerative amplifier.

The final step in CPA is the pulse compression. We used a pair of parallel gratings and a roof top prism. Gratings cause the colours to diffract at different angles as depicted in Figure 5. If you look at the path of the red beam, you will see that it has the longest path to the output of the compressor, whereas the blue wavelength beam has the shortest path. After diffracting off the pair of gratings, the colours will be along a line perpendicular to the beam path. Because we want to have a circular beam at the output of the compressor, this line of colours is reflected by the roof prism back through the pair of gratings. In this way all the colours come back together into one circular beam, but the path length difference between the reddest and bluest colours has been doubled.

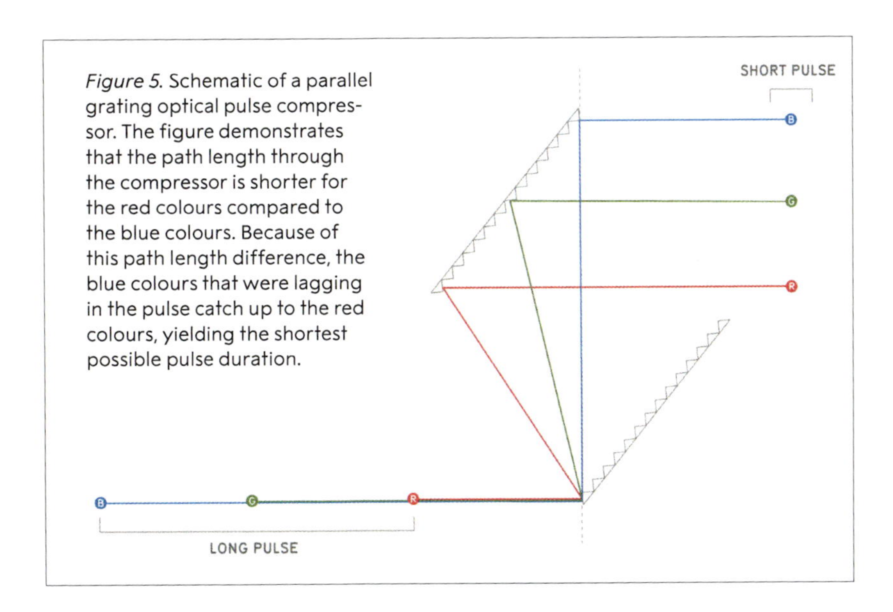

Figure 5. Schematic of a parallel grating optical pulse compressor. The figure demonstrates that the path length through the compressor is shorter for the red colours compared to the blue colours. Because of this path length difference, the blue colours that were lagging in the pulse catch up to the red colours, yielding the shortest possible pulse duration.

By placing the two gratings a certain distance apart, you can have the opposite dispersion than the fiber dispersion. That is, the colours that were stretched out over 300 ps can now be compressed to the minimum duration allowed by the spectral bandwidth. As I pointed out, even with the best fiber pulse compression, the chirp after the fiber is not perfectly linear. The opposite dispersion of the pair of gratings is also not perfectly linear, and so the pulse duration does not compress completely to the bandwidth-imposed limit. The compressor was set up with the stretched oscillator pulses, as we had the type of pulse duration measurement device, known as an autocorrelator, that could operate at the high repetition rate of the oscillator. We compressed the pulses to 1.5 ps.

To measure the amplified and compressed pulses at the 10 Hz repetition rate, I needed to use a streak camera. My colleague Steve Williamson had such an instrument and together we measured the pulse duration to be 2 ps. This was the limit of temporal resolution of the streak camera, but it showed that the amplification process had not changed the chirp, as the compressed pulse duration was ~ 2 ps before and after amplification.

This first CPA laser system delivered 1 GW power pulses, with 2 mJ of energy in 2 ps duration. At Rochester, along with Patrick Maine, we continued to develop CPA lasers to deliver TW laser power from a laser system that could sit on one optical table, which we referred to as a table top terawatt or T-cubed laser [9]. It was with the TW system that I completed my PhD research.

Rather than study high order harmonic generation for my PhD, we decided to study multi-photon ionization. This is an extension of the perturbative treatment worked out by Goeppert Mayer for the case of two photon absorption. Again, using gravitational energy as an analogy for the electromagnetic energy, an electron in an atom sits down in a well. The height difference to the top of the well from where the electron sits is known as the ionization potential. In order to ionize by a multiphoton process, the atom needs to absorb the energy of enough photons so the total absorbed energy is greater than the ionization energy. With a multi-photon process, the ionization rate would vary with the light intensity to the power given by the number of photons required to exceed the ionization energy threshold. We used various noble gases, with Xenon being the easiest to ionize. With xenon, it would take 11 photons to ionize the atoms with 1 μm photons so we expected to see the ionization rate vary to the eleventh power of intensity. This was not the case. With our infrared photons and ultrahigh laser intensity of 10^{15} W/cm^2, we were no longer in the multi-photon regime. We had such a large photon density that the light energy could once again be thought of as a giant wave.

At high intensities, rather than the electron getting enough kinetic energy to jump out of the potential well, the potential well is distorted by the interaction with the light wave as demonstrated in Figure 6. The well is bent over by the wave at the frequency of the light. The well is tipped one direction for half the period of light and then the other direction for the second half of the period of the light. The electron then is no longer contained by the well, for a brief moment of time. The longer the wavelength, the longer the time duration that the potential well is bent allowing a longer time for the electron to escape. Quantum mechanics also allows the electron to tunnel through a barrier and so the potential does not need to be completely tipped, but our work with long wavelengths in the infrared and the very high intensity showed that the electrons did simply ionize over the barrier [10]. These ionization studies [11] helped usher in a new field of study, high intensity laser physics.

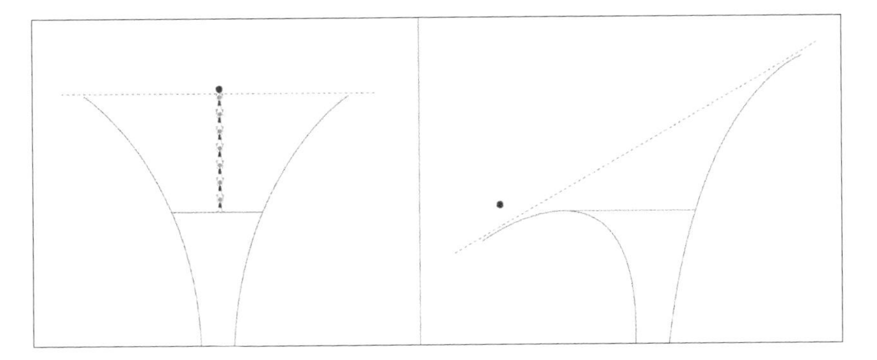

Figure 6. On the left panel, the perturbative effect of multi-photon ionization is depicted. The electron gets sufficient energy from simultaneous absorption of multiple photons to escape from the potential well. On the right panel, the process of over-the-barrier ionization is shown. The interaction with the high intensity light wave bends the potential energy wall over. If the light power is sufficiently strong, the well is bent as low as the electron energy level and the electron is free to escape for the duration of the half period of the wave.

Whereas some physical processes are a result of the total energy that can be applied, other processes occur because of the force applied per unit time. A mechanical example is the use of a hammer. If you simply push on a nail, it is difficult to move the nail, but if you hit the nail quickly with the hammer, the nail moves easily. This is why I like to say that I built a laser hammer when I built the CPA. The laser hammer has led to new types of laser machining. Unlike the thermal machining processes carried out by long pulse or continuous wave lasers, which depend on the total energy deposited to the material, short pulse laser machining leads to very precise cutting and drilling. This is because there is very little heat delivered to the material. Any heat spreads through the material, melting and deforming the material. If you can have a very short pulse, material can be cut with very low energy and so very little heat. This short pulse, low energy machining results then in very clean cuts and the drilling of very small holes. The machining is a result of the material being ablated or in other words ripped apart. The ablation is a result of the electrons being ripped off the atoms by the high intensity laser ionization process. Because this type of machining does not depend on depositing the energy in the material, it can machine transparent material such as glass or the cornea in the eye. The light is then not absorbed by the material and the laser ablation of material occurs only at the focal point. The machining can therefore be done inside the transparent material. It is because of this machining precision, that CPA lasers have found applications in machining and in particular machining the corneal flap for laser eye surgery.

The laser ushered in a new field of experimental study in nonlinear optics and multi-photon atomic processes. The development of CPA furthered our understanding of the interaction processes, which not only include exciting the electrons' kinetic energy but also the distortion of the potential energy. The new field of high intensity laser physics has pushed new technologies to be developed that lead to shorter pulse durations or higher powers or both. These new technologies continue to lead us to a new understanding of how light and matter interact.

REFERENCES

[1] Goeppert-Mayer M. (1931). "Über Elementarakte mit zwei Quantensprüngen," *Annals of Physics* **9** (3): 273–295. Bibcode:1931AnP...401..273G. doi:10.1002/andp.19314010303.
[2] Franken, P. A., Hill, A. E., Peters, C. W. and Weinreich, G. (1960). "Generation of Optical Harmonics," Phys.Rev.Lett. 7 (4): 118–119. Bibcode:1961PhRvL...7..118F. doi:10.1103/*PhysRevLett*.**7**.118
[3] Maiman, T. (1960). "Stimulated Optical Radiation in Ruby," *Nature* **187** (4736): 493–94. *Bibcode:1960Natur.187..493M. doi:10.1038/187493a0.*
[4] Harris, S. E. (1973). "Generation of Vacuum-Ultraviolet and Soft-X-Ray Radiation Using High-Order Nonlinear Optical Polarizabilities," Phys.Rev.Lett. 31 (6): 341–344. doi:10.1103/*PhysRevLett*.**31**.341
[5] Moulton, P. F., (1986). "Spectroscopic and laser characteristics of Ti:Al_2O_3," *J.Opt.Soc.Am.* **B 3**(1): 125–133. doi:10.1364/JOSAB.3.000125
[6] Tomlinson, W. J., Stolen, R. H. and Shank, C. V., (1984). "Compression of optical pulses chirped by self-phase modulation in fibers," *J.Opt.Soc.Am.* **B 1**(2): 139–149. doi:10.1364/JOSAB.1.000139

[7] Treacy, E. (1969). "Optical pulse compression with diffraction gratings," *IEEE J. Quantum Electron.* **5**(9): 454–458.
doi: 10.1109/JQE.1969.1076303

[8] Strickland, D. and Mourou, G. (1985). "Compression of amplified chirped optical pulses," *Opt. Commun.* **56**(3): 219–221.
doi: 10.1016/0030-4018(85)90120-8

[9] Maine P., Strickland D., Bado P., Pessot M., and Mourou G (1988). "Generation of ultrahigh peak power pulses by chirped pulse amplification," *IEEE J. Quantum Electron.* **5**(9): 454–458.
doi: 10.1109/3.137

[10] Keldysh L.V. (1965). *Sov.Phys.JETP* **20**(5):1307.

[11] Augst, S., Strickland, D., Meyerhofer, D. D., Chin, S. L. and Eberly, J. H. (1989). "Tunneling ionization of noble gases in a high-intensity laser field," Phys.Rev.Lett. 63 (20): 2212–2215. doi:10.1103/*PhysRevLett.***63**.2212

Physics 2019

"for contributions to our understanding of the evolution of the universe and Earth's place in the cosmos"

one half to

James Peebles

"for theoretical discoveries in physical cosmology"

and the other half jointly to

Michel Mayor and Didier Queloz

"for the discovery of an exoplanet orbiting a solar-type star"

The Nobel Prize in Physics, 2019

Presentation speech by Professor Mats Larsson, Member of the Royal Swedish Academy of Sciences; Chairman of the Nobel Committee for Physics, 10 December 2019.

Your Majesties, Your Royal Highnesses, Esteemed Nobel Laureates, Ladies and Gentlemen,

This year's Laureates in Physics have taken us along on a journey that began when the Universe was very young – less than 400,000 years – and which continues to this day.

The young Universe can be described entirely according to the laws of physics. The formation of the first atoms marked the first step towards the Universe we know today. The temperature was about 3,000 degrees Celsius, and the radiation that so far had been confined in a primordial soup of electrons and protons now began to break free, as atoms were formed by a fusion of protons and electrons. Since it was discovered by humans in the mid-1960s, this cosmic radiation has provided the oldest and purest information we have about the early Universe. By that time, James Peebles already realised that cosmic background radiation plays a crucial role in how stars and galaxies are formed. His theoretical discoveries in physical cosmology have subsequently laid a foundation for interpreting the measurements of background radiation that are being carried out in satellite-based missions, using more and more refined technology. Fluctuations in the background radiation that was frozen out during the transition between the impenetrable primordial soup and a transparent early Universe consisting of hydrogen and helium could be given a theoretical interpretation and meaning. Peebles' introduction of cold dark matter and his re-introduction of Albert Einstein's cosmological constant – also known as dark energy – were the final pieces of the puzzle in the cosmological standard model, which describes the Universe at an extraordinarily detailed level. We know that the Universe is entirely dominated by dark matter and dark energy, but its physical origins remain shrouded in mystery.

The sun, the moon and the brightly shining planets in our own solar system, as well as the stars that are visible with the naked eye, have been known to humanity since prehistoric times. But are there planets that orbit stars similar to our own sun? Is our solar system unique, or are there other planetary systems? Until quite recently, in a historical perspective, these questions remained unanswered. The reason is simple; planets orbiting other stars cannot be directly observed, since the light they emit is too faint. Instead we must look for the slightly rocking motion a star makes if a planet, for example one the size of Jupiter, is rotating around it. Michel Mayor and Didier Queloz built an instrument – a spectrograph – that can measure such movement by utilising the Doppler effect. Many people are aware of how the Doppler effect influences sound. We hear a higher-pitched sound from an emergency vehicle that is approaching us, but a lower-pitched sound as it moves away from us.

In October 1995, Mayor and Queloz announced the discovery of a Jupiter-like planet orbiting the star 51 Pegasi in the constellation known as Pegasus, about 50 light years away from Earth. It moves around its star at very high speed; a Pegasi year takes just over four days, compared to Earth's one year and Jupiter's 12 years. Other astronomers were quickly able to confirm this discovery, and since then the new field of "exoplanets" has literally exploded. Today more than 4,000 exoplanets within a few thousand light years of earth have been observed, enabling researchers to draw the conclusion that in our own Milky Way galaxy alone, there are perhaps 100 billion planetary systems. Technological development is progressing rapidly, and the question of whether there is life elsewhere in the Universe than in our own solar system will engage a new generation of astronomers.

Professors Peebles, Mayor and Queloz:

You have been awarded the 2019 Nobel Prize in Physics for your outstanding contributions to our understanding of the development of the Universe from its early childhood to the present day, and our Earth's place in the cosmos. On behalf of the Royal Swedish Academy of Sciences, it is my honour and great pleasure to convey to you our warmest congratulations. I now ask you to step forward to receive your Nobel Prizes from the hands of His Majesty the King.

James Peebles. Photo: A. Mahmoud

James Peebles

Biography

MY IMPRESSION IS that each of us is born with distinct personal characteristics. It seems quite observable in our three daughters, and I think I see it in myself. I believe I was born to be a physicist of the kind who may be a little weak on the mathematics but has some sort of intuitive grasp of the science. I remember at a very early age pestering my mother to be allowed to put together the parts of the coffee percolator after she washed it. I enjoyed taking other things apart, though I did not always do so well in reassembling them. And I remember coming upon an explanation of compound pulleys in a schoolbook of one of my older sisters. I thought that was neat, and still do. To me physics is compound pulleys, all the way down. I also inherited or somehow acquired the tendency to dream, sometimes about physics. That may have been a little detrimental to my career, because dreaming can help postpone action. But on the whole, it was seriously beneficial.

I was born in what was then the city of St. Boniface. It was meant to be the francophone sister city of anglophone Winnipeg, in the province of Manitoba in the center of Canada. But the arrangement was unstable: as Winnipeg grew St. Boniface shrank to a charming neighborhood in greater Winnipeg. I began education in grade one at King George V school in Norwood, an anglophone suburb of St. Boniface. We moved to St. Vital, then a rural municipality just south of Winnipeg, where I attended Windsor junior high school and then Glenlawn Collegiate for grades 10 to 12. It was a small high school then; I guess about twenty-five graduated in my class. Glenlawn has grown a lot larger.

I realize now that I was not a satisfactory student in high school because I did not pay much attention to my teachers. I was not rebellious, just a dreamer, little motivated to do more than what was required to pass the examinations. I cannot remember whether I had given any thought at all to what I wanted to do with the rest of my life when I graduated from high school. My father had wanted to go to university, but the great depression and his overbearing father prevented that. So, my father took a job at the Winnipeg grain exchange, where several companies looked after the shipping and sales of the harvest from the great prairies to the west. It was not a very inspiring job, but important to have during the depression, and he stuck with it to the end of his life. He found solace in alcohol. My mother found solace in St. Mark's Anglican Church. On the rare occasions when she was able to persuade me to accompany her, I was deeply bored. My father would have nothing to do with religion. He was a handy person, and I enjoyed working with him.

My older sister Audrey went to normal school, which in those days was preparation for teaching in public schools. Before me that had been the highest degree of education in my immediate family. While attending university I lived at home and took the bus in. My summer jobs generated enough money for books and tuition.

As I said, I had not given thought to what to study at the university. But I knew I liked to build things such as model airplanes, and I had the impression that engineers built things, so I enrolled in engineering at the University of Manitoba (which I will term the U of M for short). I liked many of the classes, and learned a lot, some of lasting value. I was exposed to calculus in what might be termed an engineering point of view, which suited me, for as I said I don't have a strong intuitive feeling for mathematics. I remember with pleasure engineering drawing with india ink on linen fabric. We don't do that sort of thing anymore, at least not in my field, but the manner of visualization it required was valuable and has stayed with me. And the courses allowed intervals of free time that I filled by playing hearts, the card game. I particularly liked the physics courses. I remember, in my second year in engineering, complaining to a friend from Glenlawn Collegiate, Dale Loveridge, that I was running out of physics courses to take. He replied that I could transfer to physics. To the best of my recollection that thought had not occurred to me. So why, if I was born to be a physicist, did it take someone else to make me realize that I ought to transfer to physics? I can only say that I tend to be vague about such things. Anyway, I made the change in my third year at the U of M and felt at home. I guess I could have made my way through life as a mediocre engineer, but Dale directed me to something for which I am far better suited.

The courses in physics were fascinating and the students compatible with my inclinations, intellectually and socially. We spent a good deal of our spare time playing another card game, bridge. But we also spent a lot of time arguing about mathematics, which was OK, and physics, which I loved. I learned a lot from those discussions, and even more from the lectures. Our courses did not get very far into the 20th century, but that was fine for me. When I arrived at Princeton University as a graduate student,

I found I had to work hard to catch up with what the other students knew about modern physics. But I think I had a better than average education in the foundations, including good old classical physics.

I remember the day my closest friend among the students in physics at the U of M, John Moore, came to me saying, Jimmy you have to meet Alison. That was because her surname is the same as mine. Maybe we're related, but her line of Peebles came from Ireland, mine from England, so the relation looks kind of distant. We liked each other, and our physics friends saw us married and shipped off to Princeton in 1958. Al has been my best friend since we met.

Graduate study at Princeton was Ken Standing's idea. He was a professor of physics at the U of M and had been a graduate student in nuclear physics at Princeton ten years earlier. He formed the opinion that Princeton is the only place for me. I don't imagine he could have seen how right he was. It was a real pleasure to talk to him the last time we met, in the Spring of 2016. But he had started to exhibit the symptoms of Parkinson's Disease, which soon took him. Ken had a productive career in precision measurements of masses of macromolecules, which biophysicists value, and he loved to spend time in his cabin in the beautiful woods toward the eastern edge of Manitoba, in the Precambrian Shield. I owe a lot to Ken.

I entered Princeton with the intention of doing something fancy in particle physics. I wrote one paper on that subject, which I see has gathered five citations, one of them mine. I was saved from a dismal future in that direction through the help of two fellow graduate students, both also from the U of M. Bob Pollock was a year ahead of me, we were friends while both of us were there, and he and Jean, and Al and I, remained good friends. Pollock was a gifted experimentalist. Soon after I arrived at Princeton I was approached by Professor Donald Hamilton, who wanted to discover whether I was another Pollock, and if so whether he could persuade me to join his experimental atomic beams group. A short conversation revealed that I am no Pollock, and we parted as friends. I did not know Bob Moore while at the U of M; he was a few years earlier than us. But Moore led me to Professor Robert Henry Dicke's Gravity Research Group.

After war research on radar and other electronics, Bob Dicke spent a decade at the laboratory bench in Princeton on what might be termed quantum optics. But then he decided that the study of the physics of gravity was seriously neglected, and that the great advances in electronics during the war would allow many of the classical experiments in gravity physics to be done better and would allow new experimental probes into the nature of gravity. He quite abruptly changed his direction of research to the empirical study of gravity physics. The first twelve PhD dissertations he guided had nothing to do with gravity. The last of these is dated

1959. The next, dated 1961, is mine. The twenty-six dated after 1960 include only one that has nothing to do with gravity.

The abrupt switch of direction may seem bold. But at about the same time another member of the faculty, John Archibald Wheeler, decided to turn the direction of his research to the theoretical study of gravity. This cannot have been entirely coincidental, but the two had quite different philosophies. Wheeler accepted Einstein's general theory of relativity and explored to great effect its consequences and ways to reconcile it with quantum physics. Dicke seemed to be almost personally offended by the scant empirical support for general relativity, and he enthusiastically explored questions that many had considered settled without empirical support at the precision possible then. Is the period of a mechanical oscillator, measured at rest relative to the oscillator, really independent of its motion relative to distant matter? Is the period defined by a spectral line quite independent of the atom's motion? Are parameters of physics such as the strengths of the gravitational and electromagnetic interactions really independent of motion? Might these parameters be evolving as the universe expands?

I was impressed by what Wheeler was doing and enjoyed interacting with his many graduate students and postdocs, but I was not inclined to join his group. Bob Moore took me to the weekly evening meetings of Dicke's Gravity Research Group. The group cannot have been much more than a year old when I arrived, in the autumn of 1958, but graduate students and postdocs had already started ambitious experiments, while others were looking into such arcane things to me as the dating of historical eclipses, for the purpose of checking the orbits of the moon around the earth and the earth around the sun. Dicke had some of his graduate students and postdocs working with him on a repetition of the Eötvös experiment that demonstrates that the gravitational acceleration of a free test particle depends very little if at all on its composition. Eötvös had to observe his balance from a distance, using a telescope. Dicke buried his balance and used his elegant feed-back techniques to monitor the electrostatic force needed to hold the balance fixed. The measurements have since been done even better, but Dicke showed the way. We heard progress reports and discussions of these projects, and thoughts about what other things might be investigated. Some of Wheeler's students, who were looking into the theoretical side of general relativity and quantum physics, sat in on the Gravity Group meetings. And Dicke brought occasional visitors. It was a fascinating tour of physics. The Gravity Group meetings showed me what I wanted to do and taught me a lot about how to do it.

Dicke directed me to the issue of whether the strength of the electromagnetic interaction, represented by the fine-structure constant (in the old-fashioned units I still use)

$$\alpha = e^2/\hbar c \quad (1)$$

might be evolving as the universe expands. This led me to learn a lot of nuclear physics, because if α evolves then the decay rates of long-lived isotopes change, increasing or decreasing according to how a change in α changes relative energy levels. And that could mean the radioactive dating of minerals and meteorites based on the assumption of constant decay rates would produce inconsistent results from different isotopes. So I read a lot of geology, and learned fascinating things such as the great extinctions. And I cooked up a relativistic classical field theory that allowed α to evolve without serious violation of the Eötvös experiment. All of this went into my doctoral dissertation. My bounds on the possible rate of change of the value of α are modest compared to what has been done since by observations of the spectra of galaxies and quasars at redshifts well above unity. And I have never reexamined my theory of how α might evolve, to see if it truly makes sense. But this was an excellent learning experience.

I learned the standard thinking about the expanding universe from the book, *The Classical Theory of Fields.* It is part of the marvelous series on theoretical physics by Landau and Lifshitz. The books in this series do not deal much with phenomenology. The closest I find in *The Classical Theory of Fields* is in a footnote (on page 332 in my edition, the 1951 translation from the 1948 Russian edition). It cautions that the validity of the assumption that the universe is close to homogeneous and isotropic in the large-scale average remains an open question. That was a very sensible remark. My other reference was Tolman's *Relativity, Thermodynamics and Cosmology,* published in 1934. It too is thin on phenomenology. I was left with the early impression that the subject of cosmology was pretty much free of the empirical physics I enjoy applying, and I saw lots of room to go about creating some physics.

Dicke had suggested that the universe may have expanded from a hot dense early condition, leaving a remnant sea of thermal radiation that was cooled by the expansion. I saw that this would imply interesting thermonuclear production of light isotopes. Toward the end of 1964 I learned that I had been reinventing the wheel; George Gamow published most of my ideas in 1948. But he left room for more detailed analyses of the evolution of the isotope abundances. And I did hit on new ideas about how the sea of thermal radiation would affect the gravitational assembly of matter into galaxies and groups and clusters of galaxies. I discuss all that in my Nobel Lecture.

I gave a one-term graduate course on these ideas about physical cosmology in the fall of 1969, and John Wheeler insisted that I turn my lectures into a book. To that end he took notes that he gave me at the end of each lecture. The sight of that great physicist taking notes in his elegant hand so unnerved me that I promised to produce a book. I meant its title, *Physical Cosmology,* to indicate that I did not intend to get into the subtleties of what might be termed astronomical cosmology: evidence from stellar evolution ages and the extragalactic distance scale. I don't think I thought of it at the time, but the title also helps distinguish this book from the bloodless approach in Tolman's *Relativity, Thermodynamics and Cosmology* and Landau and Lifshitz's *The Classical Theory of Fields.* I meant to explore the physical processes that are observed to operate, or we might imagine operate, in an expanding universe. At about the time of publication of my book, in 1971, Steve Weinberg published his book, *Gravitation and Cosmology.* It presents more complete theoretical considerations. Mine is more complete on the phenomenology and the physics that might show how the phenomenology all hangs together. The two books mark the start of the growth of physical cosmology from its near dormant condition in the early 1960s to a productive branch of physical science by the end of the 1960s. But my role in how that happened is discussed in my Nobel Lecture.

How Physical Cosmology Grew

Nobel Lecture, December 8, 2019 by
P. James E. Peebles
Princeton University, Princeton, NJ, USA.

I BEGAN STUDYING the large-scale nature of the universe in 1964, on the advice of Professor Robert Henry Dicke at Princeton University. Bob guided my doctoral dissertation and from then on, I counted on him as my professor of continuing education.

The usual thinking at the time was that the universe is homogeneous in the large-scale average, and that it is expanding and evolving as predicted by Einstein's general theory of relativity. The schematic nature of this cosmology, and its scant observational support, worried me. But I saw a few interesting things to look into, the results suggested more, and that continued through my career. I review my story at length in the book *Cosmology's Century* (Peebles 2020). Here I recall a few of the steps along the path to the present standard and accepted cosmology that is so much better established than what I encountered in the early 1960s.

Cosmology became more interesting with the discovery that the universe is filled with a near uniform sea of microwave radiation with a thermal spectrum at a temperature of a few degrees Kelvin. This CMB (for cosmic microwave background radiation) proves to be a remnant from the hot early stages of expansion of the universe. Theory and observations in this great advance converged in a complicated way.

In 1964 Bob Dicke explained to three junior members of his Gravity Research Group, Peter Roll, David Wilkinson, and me, why he thought the universe might have expanded from a hot dense early condition. In this hot big bang picture space would be filled with a near uniform sea of thermal radiation, left from the hot early conditions and cooled by the expansion of the universe. Bob suggested that Peter and David build a microwave radiometer that would detect the radiation, if it's there, and he suggested that I think about the theoretical implications of the result. We knew there may be nothing to detect. But we were young, the project did not seem likely to take too much time, and it called for interesting experimental and theoretical methods. I expected I soon would return to something less speculative. That did not happen because the sea of radiation was discovered and gave employment to David and me for the rest of our careers.

Peter Roll went on to a career in education, putting computers into teaching laboratories. Figure 1 shows David and me with Bob Dicke, in a photograph taken about a decade after identification of the presence of the sea of microwave radiation. A balloon carried the instrument in front of us above most of the atmosphere, and a radiometer detected the difference of responses to a pair of horn antennae separated by 90°, so each is tilted 45° from the vertical. As the instrument rotated around its vertical axis this difference of responses made a precision map of variations of the

Figure 1. Left to right David Wilkinson, Jim Peebles, and Bob Dicke, in the late 1970s.

radiation intensity across the sky. You see four horns: two pairs of antennae that operate at two radiation frequencies. This is one of a series of experiments by David and colleagues, along with groups at a few other places, that placed increasingly tight bounds on the departure from exact isotropy. That was leading to the critical developments in the early 1980s to be discussed.

The evidence I know is that the sea of microwave radiation was first detected in the late 1950s as unexpected excess noise in experiments in microwave communication at the Bell Telephone Laboratories. To account for this excess the engineers assumed that radiation from the environment entering through the side and back lobes of their antenna contributes about 2 K to the total noise received (DeGrasse et al. 1959). But this was a fudge; their antenna rejects ground radiation better than that. The unexplained excess consistently appeared in later experiments. It remained a "dirty little secret" at Bell Labs until 1964, when Arno Penzias and Robert Wilson, both new to the Bell Radio Research Laboratory at Crawford Hill, New Jersey, resolved to look into the problem. They carefully searched for the explanation of this puzzling excess microwave noise, whether originating in the instrument or somehow entering from the surroundings. News of the Princeton search for radiation from a hot early universe showed them a possible solution: maybe the Bell excess noise is from a sea of radiation.

Bell Laboratories showed us in Princeton credible evidence that we are in a sea of microwave radiation, and that the radiation is close to uniform because the excess noise is close to the same wherever in the sky the antenna points. It proves to be what Dicke had suggested we look for, a fossil from the hot early stages of expansion of the universe. How did the Princeton group react to being scooped by Penzias and Wilson? My recollection is excitement at the realization that there actually is a sea of microwave radiation to measure and analyze. Why did the Nobel committee not name Dicke with Penzias and Wilson for the identification of this radiation? Naming Penzias and Wilson was right and proper, because they refused to give up the search for the source of the excess noise and, equally important, they complained about it until someone heard and directed them to Bob Dicke. Bob directed the search for the radiation that explains the Bell Labs anomaly that so puzzled Penzias and Wilson.

At Bob Dicke's suggestion I had been thinking about the significance of finding or not finding a sea of radiation. A negative result, a tight upper bound on the radiation temperature, would have suggested an interesting problem. The great density of matter in the early stages of expansion of an initially cool universe could have made the electron degeneracy energy large enough to have forced conversion of electrons and protons to neutrons. The problem with this is that neutrons and their decay protons would have readily combined to heavier elements, contrary to the known large cosmic abundance of hydrogen. So, I proposed a way out: postulate a sea of neutrinos with degeneracy energy large enough to have prevented electrons from combining with protons. In the Soviet Union Yakov Zel'dovich saw the same problem with a cold big bang and he offered the same solution, lepton degeneracy. Since Zel'dovich was an excellent physicist it is no surprise that he reached the same conclusion, given the problem. The interesting thing is that we saw the problem at essentially the same time, independently. The consideration somehow was "in the air." I think any experienced physicist can offer other examples of apparently independent discoveries. It seems to have taken a sociologist, Robert Merton (1961), to recognize that this is a phenomenon that deserves to be named. He termed it "multiples in scientific discovery." He also named the phenomenon "singletons in scientific discovery," which he argued may be less common.

I saw that a universe hot enough to have left a detectable sea of thermal radiation would have tended to leave the abundances of the elements in a mix characteristic of the rapid expansion and cooling of the early universe. In an unpublished preprint in late 1964, I estimated that a reasonable upper bound on the primeval helium abundance requires a lower bound on the CMB temperature, $T_o \gtrsim 10$ K, in the absence of degeneracy. My estimate of the foreground radiation from observed stars and radio-loud galaxies indicated that a sea of thermal radiation at this temperature would be readily detected above the foreground.

We might pause to review why I had a lower bound on T_o. During the course of expansion of the early universe, when the temperature fell through the critical value $T_c \gtrsim 10^9$ K, detailed balance would have switched from suppression of deuterons by photodissociation to accumulation by radiative capture. When deuterons accumulate, they can merge to heavier isotopes by the more rapid particle exchange reactions. The smaller the present temperature T_o, the further back in time the temperature passed through T_c, hence the greater the baryon density at T_c, thus the more complete the incorporation of neutrons in deuterons before the neutrons can have decayed, which means the greater the helium production. The amount of element production is determined by the combination ρ_b/T^3, where ρ_b is the present baryon mass density. My upper bound on the primeval helium abundance, $Y = 0.25$ by mass, is reasonably close to the present standard value, and my lower bound on the mass density, $\rho_b = 7 \times 10^{-31}$g cm^{-3}, which I of course took to be all baryons, is not much above the established value of the present baryon density. So, my 1964 bound on the CMB temperature is a factor of three high. I have not attempted to discover why.

After I had worked out these considerations I learned that George

Gamow already presented the physics of element buildup in a hot big bang in two memorable papers published in 1948 (Gamow 1948a,b). Gamow had earlier proposed that the chemical elements were produced in the hot early stages of expansion of the universe by successive neutron captures, beta decays keeping the atomic nuclei in the valley of stability. His graduate student, Ralph Alpher, computed the element abundances to be expected in this picture, and he and his colleague Robert Herman (1948) found the first estimate of the CMB temperature based on Gamow's picture. Their value is closer than mine, $T_o \simeq 5$ K. Their story is complicated, however, because they used a smooth fit to the measurements of the neutron capture cross section as a function of atomic weight, and they extrapolated this smooth fit to lower atomic weight through atomic mass 5. Alpher knew there is not a reasonably long-lived isotope at mass 5, so he made the sensible working assumption that nuclear reactions to be discovered bridge the gap. Eliminating this and the other gaps allowed a computation of the buildup of the heavy elements. The Alpher and Herman normalization of ρ_b/T_o^3 is based on their fit to measured abundances of the heavy elements. The detective work establishing this is in Peebles (2014).

Following up an idea with a detailed computation was not Gamow's style. But Fermi and Terkevich at the University of Chicago soon worked the first computation of the buildup of element abundances in a hot big bang using realistic nuclear reaction rates. They established that there would be little element buildup beyond helium, a result of Alpher's mass-5 gap. Gamow (1949) reported their result. I could compute in more detail and show evidence that the predicted light isotope abundances coming out of a hot big bang could match the observations. I first analyzed this in 1964, unpublished because I realized I had reinvented the wheel. Soon after that we realized there is a sea of microwave radiation, and after that I published a better computation in Peebles (1966).

Meanwhile, in the Soviet Union, Zel'dovich knew Gamow's ideas but thought they must be wrong because the theory predicts an unacceptably large primeval helium abundance. To check the prediction, he asked Yuri Smirnov (1964) to compute element production in the hot big bang model, along the same lines I was taking in the USA.

In the UK, Hoyle and Tayler (1964) knew the evidence that the helium abundance in old stars is large, and not inconsistent with Gamow's (1948a,b) ideas. Hoyle asked John Faulkner to check Gamow's estimate of deuterium buildup. That was followed by more detailed computations by Wagoner, Fowler, and Hoyle (1967). Tayler (1990) recalls that in 1964 he and Hoyle realized that Gamow's theory predicts the presence of a sea of thermal radiation, a fossil from the early hot conditions, but they supposed it would be obscured by all the radiation produced since then.

So, consider the situation in 1964. In the USSR, Zel'dovich thought Gamow's hot big bang theory is wrong because it overpredicts the helium abundance. In the UK, Hoyle knew the evidence that the prestellar helium abundance is large, and maybe consistent with Gamow's theory. But Hoyle expected the fossil radiation that would accompany it would be uninterestingly small. In the USA, I did not know about Gamow yet, but I knew there was a chance of detecting fossil radiation from a hot big bang that made helium because the foreground at microwave frequencies looked likely to be small. Also, in the USA, 30 miles from Princeton, Penzias and Wilson had a clear case of detection of microwave radiation of unknown origin. All of this was tied together the following year. It is a charming example of a Merton multiple.

Yet another multiple was the recognition of the role of the sea of thermal radiation in the gravitational growth of the galaxies, independently by Gamow, Zel'dovich and his group, and me. I hit on what might be a singleton, the analysis of the effect of the dynamical interaction of matter and radiation in a hot big bang cosmology (in Peebles 1965 and many later papers). The early universe would have been hot enough to have thermally ionized matter, and the Thomson scattering of the CMB by free electrons and the Coulomb interaction of electrons and ions would have caused plasma and radiation to act as a viscous fluid. That meant small departures from exact homogeneity in the early universe would tend to oscillate as acoustic waves. Oscillation would be terminated when the plasma cooled to the point that it combined to neutral atoms, freeing the radiation and allowing gravity to draw the baryonic matter into clumps. The termination of acoustic oscillations is a boundary condition that favors discrete wavelengths. That imprints distinctive patterns on the distributions of matter and radiation. The effects became known as BAO, for baryon acoustic oscillations. By the late 1960s the hot big bang cosmology community had grown large enough that several of us, particularly Joe Silk (1967), more or less independently worked out the viscous fluid description of the evolution of departures from homogeneity. I developed the basic ideas of the modern approach to the growth of cosmic structure that describes the radiation by its distribution in phase space. My first graduate student, Jer-tsang Yu, and I applied this theory in the numerical solutions of the effects of BAO on the distributions of matter and radiation published in Peebles and Yu (1970).

It took some time to connect BAO theory to observations of the effect in the distributions of matter and radiation. The BAO effect in the angular distribution of the CMB was discovered and well measured at the turn of the century, and at the time there was a hint of detection in the galaxy space distribution (as reviewed in Peebles 2020). The BAO signature in the galaxy distribution was particularly well seen in the galaxy two-point position correlation function. The matter power spectrum shown in Peebles and Yu has a series of roughly equally spaced bumps, at the wavelengths of the modes favored by the boundary condition, the decoupling of matter and radiation. The correlation function is the Fourier transform of the power spectrum. The Fourier transform of a sine wave is a delta function. The Fourier transform of a series of bumps, which is an approximation to a sine wave, produces an approximation to a delta function, a bump in the correlation function. I presented the prediction of this bump in Peebles (1981, Fig. 5). Tom Shanks (1995) set out to find it, but the available data were not adequate. Daniel Eisenstein rediscovered the bump through a consideration of the Greens's function for the matter. It is a different argument on the face of it but physically equivalent to mine. Daniel and colleagues demonstrated the bump in data from the Sloan Digital Sky Survey (Eisenstein, Zehavi, Hogg, et al. 2005). It is a sign of the times that my 1981 paper is single-author and the 2005 paper lists 50 authors. But they were needed for the data to detect this subtle effect. And we might note that the bump is much weaker than what Yu and I had considered because the nonbaryonic dark matter weakens the BAO effect. Anyway, I consider the connection of theory and observation of the effect of BAO to be a multiple.

For most of the time between BAO theory and observation it was not at all clear to me that there would be a detection. The BAO theory assumes standard physics, including the general theory of relativity. That is an extrapolation from the tests in the solar system and smaller, on scales $\gtrsim 10^{13}$ cm, to the scales of cosmology, $\sim 10^{28}$ cm. Would you be inclined to trust an extrapolation of fifteen orders of magnitude? The theory also assumes cosmic structure grew out of departures from homogeneity associated with small near scale-invariant spacetime curvature fluctuations. There were other possibilities. What is more, there was a hint that some of these assumptions fail because they predict that the sea of microwave radiation, the CMB, has a close to thermal spectrum. Prior to 1990 the measurements suggested a significant excess over thermal at wavelengths shorter than the theoretical Wien peak. That might mean violent events in the early universe released a lot of energy, contributing some of it to the CMB and some to rearranging the matter. Or maybe the universe is not very close to homogeneous; maybe we observe a mix of radiation temperatures from different regions. Either might be expected to have spoiled the BAO signatures computed in linear perturbation theory.

This uncertain situation was resolved in 1990 by two brilliant experiments, one carried by the USA NASA satellite COBE, the other by the Canadian University of British Columbia rocket COBRA. Both established that the spectrum is very close to thermal (Mather, Cheng, Eplee, *et al.*

1990; Gush, Halpern, and Wishnow 1990). That demonstration, a clear multiple, eliminated a serious challenge to the BAO theory. John Mather rightly was named a Nobel Laureate for his leadership in the spectrum measurement. Herb Gush was equally deserving; awards can be capricious.

Prior to the demonstration that the CMB spectrum is wonderfully close to thermal I had to consider the possibility that there is a real and substantial departure from that equilibrium condition. The interpretation would be messy. I didn't want to think about it, so while awaiting clarification of the spectrum measurements I turned to another program, statistical measures of the galaxy distribution and motions relative to the mean homogeneous expansion of the universe. There were several catalogs of galaxy positions ready and waiting for analyses. Most important was the catalog assembled by Donald Shane and his collaborators, mainly Carl Wirtanen, at the Lick Observatory of the University of California. They counted galaxies in small cells in the sky, logging some one million galaxies by scanning photographic plates with a traveling microscope. This heroic effort took them ten years. Converting to data suitable for computation of statistical measures was a considerable effort too. Graduate students in physics seem to have a sense of where interesting things are happening and gather around. Graduate students Jim Fry, Mike Seldner, Bernie Siebers, and Raymond Soneira did much of the heavy lifting, along with my colleague on the faculty, Ed Groth.

Since I like images, I was pleased with the map we made of the large-scale galaxy distribution. And I was delighted to have the chance show the map to Donald Shane and ask whether it looks like what he saw. He laughed and said, "I was looking at this one galaxy at a time."

The Lick and other catalogs are compilations of galaxy angular positions with approximate distances. The statistical measures I used are N-point position correlation functions and their Fourier or spherical harmonic transforms. These statistics allow convenient translations from angular to the wanted spatial functions. And the N-point functions scale in a predictable way with the characteristic distances of the galaxy samples, assuming the universe is a stationary random process. That was particularly important because it allowed a test for systematic errors by checking the scaling of the angular correlation functions with depth. Another singleton in my career is the successful demonstration of scaling published in Groth and Peebles (1977). It showed that we had reliable measurements of the low order galaxy position correlation functions at separations from a few tens of kiloparsecs to a few megaparsecs. Methods and results for this program are assembled in my book, *The Large-Scale Structure of the Universe* (Peebles 1980).

Why did I devote so much effort to this program? I enjoy this kind of analysis. And I had the vague feeling that the results might offer a hint to how the galaxies and their clumpy space distribution got to be the way they are. That happened, more or less, as follows.

By 1980 it had become clear that the sea of microwave radiation is far smoother than the space distribution of the galaxies. But the mass concentrations in galaxies and groups and clusters of galaxies were supposed to have grown by gravity out of the initially close to homogeneous early universe of the hot big bang theory. How could this growth of mass concentrations have so little disturbed the CMB? Surely the gravitational gathering of mass concentrations in the early universe would have drawn the radiation with it, dragged by the coupling of plasma and radiation. That would have seriously rearranged the radiation. Such a disturbance to the radiation was not observed in the measurements by David Wilkinson and his students and by others in the growing community of empirical cosmologists. Bruce Partridge (1980), who had moved on from the Princeton Gravity Group to Haverford College, presents a considerable list of the increasingly tight bounds on the CMB anisotropy we had in the years around 1980. So why is the CMB so smooth? In yet another of Merton's multiples, I and Zel'dovich's group in the USSR independently guessed the answer: Suppose the baryonic matter that stars and planets and people are made of is only a trace element, and that most matter is dark and interacts weakly if at all with radiation and our type of baryonic matter (Doroshkevich, Khlopov, Sunyaev, Szalay, and Zel'dovich 1981; Peebles 1982). The CMB would slip freely through this nonbaryonic dark matter, allowing mass concentrations to grow while disturbing the CMB only by the weak effect of gravity and by the interaction with a modest amount of our baryonic matter. In pursuing this line of thought I had some advantages over Zel'dovich and colleagues, the other main group active on the theoretical side of empirically based cosmology in those days. They assumed the dark matter is one of the known neutrino families with a rest mass of a few tens of electron volts (the mass allowed by the condition that the mass density of the neutrinos thermally produced along with the CMB not exceed what cosmology would allow, and the mass indicated by a laboratory experiment later falsified). The rapid motions of these neutrinos in the early stages of expansion of the universe would have smoothed the primeval mass distribution to a mass scale typical of rich clusters of galaxies. That would mean the first generations of bound mass concentrations were much larger than galaxies. These concentrations would have to have fragmented to form the galaxies. But I knew rich clusters are rare, and most galaxies are not near any of the clusters. And we all know that gravity tends to gather together, not cast away. Thus, it was pretty clear to me that the USSR scenario is not viable. Wanted instead was nonbaryonic dark matter that had been effectively cold in the early universe, meaning its pressure had not suppressed the early gravitational formation of small clumps of matter that would have merged to form the hierarchy of clumps we observe around us. I knew that elementary particle physicists had been speculating about forms of nonbaryonic matter that would have this wanted property. I also knew the relativistic prediction of the gravitational disturbance to the radiation produced by the departure from a homogeneous mass distribution. Rainer Sachs and Arthur Wolfe (1967) had worked that out. And I had a well-checked statistical measure of the space distribution of the galaxies, which I took to be the wanted measure of the mass distribution needed to normalize the model.

The model I put together from these pieces predicts that the disturbance to the CMB caused by the formation of the observed matter distribution would cause the CMB temperature to vary across the sky by a few parts per million. That is much less than the upper bounds from the CMB anisotropy measurements we had when I published this prediction in Peebles (1982). The CMB anisotropy was detected some 15 years later and found to agree with my computation within the modest uncertainties. This is no surprise because I guessed at the right physical situation, the computation is not complicated, and I had a reliable calibration from the galaxy space distribution.

The new form of matter in my 1982 proposal became known cold dark matter, or CDM, the "cold" meaning the dark matter pressure in the early universes was small enough not to have excessively smoothed the primeval mass distribution. I added the assumption that general relativity survives the immense extrapolation to the scales of cosmology, and that mass concentrations grew out of primeval spacetime curvature fluctuations. The introduction of this CDM cosmological model might be counted as a singleton, because I don't know that anyone else independently put all these pieces together. I just assembled pieces I already had, to be sure, but that's not unusual; we build on what came before.

There was a remarkable multiple in 1977. Five groups, independently as far as I can tell, introduced the idea of a new class of neutrinos with rest mass ~ 3 GeV. They became known as WIMPs, for weakly interacting massive particles. WIMPs have the properties I needed, though the particle physicists who proposed WIMPs in 1977 certainly couldn't have foreseen that. And they were at best vaguely aware of the astronomers' evidence of subluminal mass around galaxies. Yet the WIMP idea appeared not long after the astronomers had good evidence of subluminal mass around galaxies, and not long before I needed nonbaryonic cold dark matter to account for the smoothness of the CMB.

My 1982 CDM cosmology was greeted with more enthusiasm than I felt it warranted because I could think of other models that would equally well fit what we knew then. The CDM model is particularly simple, to be sure, but does that mean it is the best approximation to the real world? In particular, my 1982 paper assumed for simplicity that the universe is expanding at escape velocity, but by that time I already knew what I considered to be reasonably good evidence that the expansion is faster than that.

Expansion at escape velocity, in the relativistic Einstein-de Sitter cosmological model that assumes space curvature and Einstein's cosmological constant may be ignored, would mean that whenever we happened to flourish and take an interest in the expanding universe, we would find that the rate of expansion is at escape velocity. That is, we would not have flourished at any special time in the course of expansion of the universe. This seems comforting somehow. I liked the thought, prior to 1982, but it proves to be wrong. The early indication came from Marc Davis, who had been a graduate student in Dicke's Gravity Research Group and moved on to Harvard and the Smithsonian Center for Astrophysics. Marc had worked with me in analyses of the theory of evolution of the galaxy distribution. He knew my hunger for measurements of galaxy redshifts that would improve the statistical measures, and he found that his new position had the resources for a systematic galaxy redshift survey. That was something new then. Marc invited me to join him in the data analysis. The results in Davis and Peebles (1983) surprised me by suggesting that we do flourish at a special epoch.

These redshift data yielded a probe of the relative motions of the galaxies. That gave a measure of galaxy masses, which indicated that the mean mass density is less than required for escape velocity. The community opinion was that this seems quite unlikely. One way out supposes most of the mass is not in the galaxies, but is more broadly spread, which would reduce the gravitational attraction of neighboring galaxies, reducing their relative velocities, as wanted. But that didn't seem right to me. Davis and I found consistent galaxy mass estimates from the relative motions of galaxies over a range of a factor of ten in separation. If mass were more broadly distributed than galaxies, shouldn't we see that more of the mass is detected as we increase the scale of the measurement? Also, the popular idea then was that mass is more broadly distributed than galaxies because galaxy formation had been suppressed in regions with lower mass density. It would have made galaxies more tightly clustered than mass, as wanted. But if galaxy formation were suppressed in low density regions then galaxies that did manage to form there ought to show signs of a deprived youth: irregulars or dwarfs. This was not seen in the Center for Astrophysics data.

From the early 1980s through the mid-1990s I played the role of Cassandra, emphasizing the growing evidence that the universe is expanding faster than escape velocity to people who for the most part would rather not think about it. I remember a younger colleague saying I only did it to annoy. I knew it teases, but I meant it, and I regret nothing. The evidence was reasonably good then, and it is well established now, that we flourish at a special time in the course of evolution of the universe, as the rate of expansion is becoming significantly more rapid than escape.

In 1984 I introduced the accommodation to the low mass density that proves to work: add Einstein's cosmological constant, Λ (Peebles 1984), in what became known as the ΛCDM theory. At the time others were starting to pay attention to my arguments for low mass density and were thinking about the benefits of adding Λ. Turner, Steigman, and Krauss (1984) proposed it, for example. The largest part of their paper is a discussion of the idea that the mass of the universe is dominated by relativistic products of the recent decay of a postulated sea of massive unstable particles. Their last three paragraphs are considerations of the benefits of adding Λ. From the choice of emphasis, I take it that they considered the hypothetical particle species with its relativistic decay products to be less adventurous than the addition of Λ. And Λ is odd indeed. Anyway, I think I was the first to present actual computations of the effect of adding Λ. Einstein wrote his constant as λ. I don't know who introduced the change to Λ. I believe Michael Turner, the University of Chicago, introduced the change of name to dark energy. But whatever the name we don't understand the physical interpretation, though it's clear now that we need something that acts like Λ.

In the years around the mid-1990s I again acted in my self-appointed role of Cassandra, because I was not at all confident that the ΛCDM theory is a good approximation. The tests were not yet all that tight, and I could think of other models that fit the data about as well. In the late 1990s I was finishing my latest and maybe most elegant alternative to ΛCDM when I learned that the CMB anisotropy measurements revealed features characteristic of the theory Jer Ju and I had worked out a quarter century earlier. So, I abandoned the search for alternatives.

I remain surprised and impressed at how well ΛCDM passes ever more demanding tests. But I continue to hope that challenges to ΛCDM will be found and help guide us to a still better more complete theory.

I have written four books on the state of research in cosmology. I meant the title of the first, *Physical Cosmology*, to indicate that I did not intend to get into the subtleties of what might be termed astronomical cosmology: evidence from stellar evolution ages and the extragalactic distance scale. I don't think I thought of it at the time, but the title also helps distinguish my book from the earlier bloodless treatises on cosmology. I meant to explore the physical processes that are observed to have operated, or might be expected to have operated, in an expanding universe, and to explore how theory might be shaped to observations. At about the time of publication, in Peebles (1971), Steve Weinberg (1972) published his book, *Gravitation and Cosmology*. It is more complete in the mathematical considerations. Mine is more complete in the considerations of phenomenology and of how the phenomenology might be related to physical processes. The two books signal the change of physical cosmology from its near dormant state in the early 1960s to the start of a productive branch of research in physical science by the late 1960s.

My second book on cosmology, *The Large-Scale Structure of the Universe*, published in 1980, is a sort of catalog of the statistical measures I had devised and applied, the methods of analyses of how these measures might be expected to have evolved in an expanding universe, and the observational consequences of the evolution. I did not aim to arrive at a standard model for cosmology. Ideas about that were much too confused, a result of the still quite limited evidence. I meant this book to be a working guide to how we might proceed in research in physical cosmology. As it happened, thoughts about a standard model were seriously disrupted a few years later by my argument for dark matter that is not baryonic. Writing this book helped me introduce what came to be known as the Cold Dark Matter cosmological model, in 1982. I still consult *The Large-Scale Structure of the Universe* for reminders of methods.

My third book, *Principles of Physical Cosmology*, is much larger than the second, which in turn is much larger than the first. This one was published in 1993, at about the end of the time when it was practical to aim to present in one volume a reasonably complete assessment of the state of research in the physical science of cosmology. One certainly would not consider aiming for that now. Research in cosmology in the mid-1990s was an active turmoil of multiple ideas and promising-looking but confusing results from model fits to measurements in progress. That situation quite abruptly changed at the end of the decade, when research converged on a well-tested standard model, the ΛCDM cosmology.

The convergence was driven by three great observational programs. One is the tight measurement of the redshift-magnitude relation that reveals the departure from the linear low redshift limit. That feat generated a Nobel Prize. Second is the precision measurement of the cosmic microwave radiation anisotropy spectrum. That was a comparably important accomplishment that certainly merits a Nobel Prize. The third, the measurement of the cosmic mean mass density, was the main focus of empirical research in cosmology from the early 1980s through the mid-1990s. Its story is more complicated, and not as well recognized and understood as it ought to be. The three made the case for a cosmology that is hard to resist. I have once again given in to the impulse to write a book. This one, *Cosmology's Century*, describes how these three programs, with other results from brilliant ideas and elegant experiments, along with the wrong turns taken and opportunities missed, got us to a well-tested cosmology (Peebles 2020).

The establishment of cosmology is a considerable extension of the reach of well-tested physical science, and the story is simple enough that it offers a good illustration of the ways of physical science. In particular, I am impressed by the many examples of Merton's multiples in scientific discovery. I have mentioned examples from the history of cosmology, and this story has quite a few more. We all can think of examples in other branches of physical science. Some multiples may be coincidences, pure and simple. Some may be artifacts of our tendency to present the history of science in a linear fashion that makes unrelated developments appear related. I can imagine some multiples grew out hints communicated by gestures or thoughts not completed that suggest meaning within our shared culture of physical science. It happens in everyday life, why not in science? And I picture the broad general advance of physical science as a spreading wave that touches many and might be expected to trigger any particular idea more than once, apparently independently. As we sometimes say, thoughts may be "in the air." But I must leave a firmer assessment to those better informed about the ways we interact.

Meanwhile, let us not forget the great lesson that the established social constructions of science are buttressed by rich and deep webs of evidence. Surely there is a better more complete cosmology than ΛCDM. But we may be confident that the better theory will predict a universe that is a lot like ΛCDM, with something analogous to its cosmological constant and dark matter, because the universe has been examined from many sides now and found to look a lot like ΛCDM.

I confess to having been unhappy with the Nobel Prize Committee for not recognizing Bob Dicke's deep influence in the development of gravity physics and cosmology. The committee had their reasons, of course; their considerations can be complicated. But I am satisfied now because my Nobel Prize is closure of what Bob set in motion, his great goal of establishing an empirically based gravity physics, by the establishment of the empirically-based relativistic cosmology.

REFERENCES

Davis, M. and Peebles, P. J. E. 1983, *The Astrophysical Journal*, **267**, 465.
DeGrasse, R. W., Hogg, D. C., Ohm, E. A., and Scovil, H. E. D. 1959, *Journal of Applied Physics*, **30**, 2013.
Doroshkevich, A. G., Khlopov, M. I., Sunyaev, R. A., Szalay, A. S., and Zel'dovich, Ya. B. 1981, *Annals of the New York Academy of Sciences* **375**, 32.
Eisenstein, D. J., Zehavi, I., Hogg, D. W., et al. 2005, The Astrophysical Journal, **633**, *560.*
Gamow, G. 1948a, *Physical Review*, **74**, 505. Gamow, G. 1948b, *Nature*, **162**, 680.
Gamow, G. 1949, *Reviews of Modern Physics*, **21**, 367.
Groth, E. J., and Peebles, P. J. E. 1977, *The Astrophysical Journal*, **217**, 385.
Gush, H. P., Halpern, M., and Wishnow, E. H. 1990, *Phys. Rev. Lett.* **65**, 537.
Hoyle, F. & Tayler, R. J. 1964, *Nature*, **203**, 1108.
Mather, J. C., Cheng, E. S., Eplee, R. E., Jr., et al. 1990, *The Astrophysical Journal*, **354**, L37.
Merton, R. 1961, *Proceedings of the American Philosophical Society*, **105**, 470.
Partridge, R. B. 1980, *Physica Scripta*, **21**, 624.
Peebles, P. J. E. 1965, *The Astrophysical Journal*, **142**, 1317.
Peebles, P. J. 1966, *Phys. Rev. Lett.*, **16**, 410.
Peebles, P. J. E. 1971, *Physical Cosmology*. Princeton: Princeton University Press.
Peebles, P. J. E. 1980, *The Large-Scale Structure of the Universe*. Princeton: Princeton University Press.
Peebles, P. J. E. 1981, *The Astrophysical Journal*, **248**, 885.
Peebles, P. J. E. 1982, *The Astrophysical Journal*, **263**, L1.
Peebles, P. J. E. 1984, *The Astrophysical Journal*, **284**, 439.
Peebles, P. J. E. 1993, *Principles of Physical Cosmology*. Princeton: Princeton University Press.
Peebles, P. J. E. 2014, *European Physical Journal H*, **39**, 205.
Peebles, P. J. E. 2020, *Cosmology's Century*. Princeton: Princeton University Press.
Peebles, P. J. E. and Yu, J. T. 1970, *The Astrophysical Journal*, **162**, 815.
Sachs, R. K., and Wolfe, A. M. 1967, *The Astrophysical Journal*, **147**, 73.
Shanks, T. 1985, *Vistas in Astronomy*, **28**, 595.
Silk, J. 1967, *Nature*, **215**, 1155.
Smirnov, Y. N. 1964, *Astronomicheskii Zhurnal* **41**, 1084; English translation in *Soviet Astronomy* **8**, 864, 1965.
Tayler, R. J. 1990, *Quarterly Journal of the Royal Astronomical Society*, **31**, 371.
Turner, M. S., Steigman, G., and Krauss, L. M. 1984, *Phys. Rev. Lett.*, **52**, 2090.
Wagoner, R. V., Fowler, W. A., and Hoyle, F. 1967, *The Astrophysical Journal*, **148**, 3.
Weinberg, S. 1972, *Gravitation and Cosmology: Principles and Applications of the General Theory of Relativity*. New York: Wiley.

Michel Mayor. © Nobel Prize Outreach AB. Photo: A. Mahmoud

Michel Mayor

Biography*

MY FAMILY, CHILDHOOD AND EDUCATION

I was born in 1942 in Lausanne, Switzerland, a small town along Lake Leman where I started elementary school. My father was a police commissioner with positions in different cities. When I was six years old, my family moved to Cully, a superb village in the vineyards along the same lake and four years later to Aigle, a small town of less than ten thousand inhabitants in the upper Rhone Valley, in the midst of the mountains.

During my 5 years at school in Aigle, when I was aged 11–16, I had an exceptional teacher of science, Edmond Altherr. He was the Director of the College and a teacher, but nevertheless he continued to do research at home, on the nematodes, during his whole life. Nematodes are an animal phylum that includes some worms that are parasites of the human body. Although that does not look like a charming subject, he was able to convey a lot of enthusiasm about the study of nematodes and so he was able to stimulate our interest in science. A great pedagogue, really.

Apart from school, I was an active member of the scouts: hiking, ski-

Figure 1. @Sian Prosser, Royal Astronomical Society.

Figure 2. 1948, Lausanne, end of my first year at school!

* This biography is an updated version of Mayor, M and Cenadelli, D, 2018 in *The European Physical Journal* H, **43**, pp. 1–43.

Figure 3. 1960, Science high school in Lausanne, but also some arts. (MM second from left).

Figure 4. My family in 1977, Françoise and our three children.

ing, camping in high altitude mountains and every kind of outdoor activity. As a teenager, I began doing one of my most exciting sports: climbing. With Aigle so close to mountains, living there allowed me to regularly practice these sports. Climbing provides me with immeasurable pleasure but... at one stage I was very happy to be rescued from a deep crevasse without too much damage!

I was obliged to move back to Lausanne to go to high school, as there was none in Aigle. After high school, I went to the University of Lausanne to study theoretical physics. When I was a child, I was always interested in science, but did not dream of being an astronomer. I was fascinated by every domain of sciences: wildlife, geophysics, geology, the Alps ... But I loved mathematics, so I chose to study physics and mathematics at the University of Lausanne. After the first year I had to choose between them, and I chose physics. Professor Karl Gerhard Stuekelberg was my most important professor for theoretical physics. Students were fascinated by his very profound vision on physics and I chose to do my masters degree in theoretical physics (in 1966). My thesis was on the interaction of particles with large spin and was based on formalism of Steven Weinberg. In mathematics I had the privilege to have Georges De Rham as one of my professors. He is known for his major contributions in differential topology, but also as a great alpinist!

In 1966 when I finished my studies at the university, I got married to Françoise. She was then a student in the natural sciences at the University of Lausanne. We have been blessed with three children, Anne, Claire and Julien and today five grandchildren.

For our children, Haute-Provence Observatory was a second home as the whole family came with me for numerous observing runs (at least during school holidays). It seems that our children have acquired the virus of scientific curiosity by listening to the discussions with friends at home or at Haute-Provence Observatory. Today, all of them share the same passion for research but in quite different domains.

Figure 5. ... and our family in December 2019!

FIRST STEPS IN SCIENCE

After graduation, I got a position as a PhD student in the field of galactic dynamics at the University of Geneva.

The origin of galactic spiral structure had been a long-standing problem: if it was a result of a strong differential rotation, as appeared, then such galaxies would not keep their shape, but would wind themselves up. Professors C.C. Lin and Frank H. Shu proposed a theoretical explanation in 1963. I wanted to test the consequences of their idea on local stellar velocity fields to see how the spiral structure perturbed the mean flow of stars. This was the start of my interest in stellar kinematics, and I started my PhD study in this domain.

Figure 6. A galaxy similar to our Milky Way with a magnificent two spiral arm structure.

By the end of my PhD I was looking for observations to test these ideas, but the stellar velocities in catalogues were not adapted to my question. I needed new data. This was my critical motive for moving from theory to instrumentation, with a specific focus on the determination of stellar kinematics. By chance, at an N-body Colloquium in Cambridge, I met Roger Griffin and we talked about a new spectrographic method to measure stellar velocities by cross-correlation (Fellgett 1953, Griffin 1967). I was really impressed by this technique. It was evident we could make much progress in this domain, and I realized that this was what I needed to get my data. Back in Geneva, I discussed it with the director of our institute, who was in favor of this as a way to develop the observatory. But I think he wondered how I, a theorist, would dare build an instrument.

I needed advice, and André Baranne – professor of optics at Marseille Observatory in France – provided it. In fact, he found the problem interesting as he designed the optics... Sometimes it is easy to collaborate with people like this case, in science. This was my first cross-correlation spectrograph with more technical possibilities than Roger had envisaged and with a computer that made it very efficient.

A first CORAVEL instrument was installed in 1977 on our 1-meter Swiss telescope at the Haute-Provence Observatory and a second one on the 1.5-meter Danish at the ESO La Silla Observatory in Chile to have access to the southern sky. The huge efficiency gain of this new kind, fully optimized instruments opened for me the domain of the stellar kinematics. During more than 15 years, I have visited so many different domains: dynamics of globular clusters, cepheids and supergiants in the Magellan Clouds, stellar rotation and with Antoine Duquennoy double stars of the solar vicinity.

(Duquennoy and Mayor, 1991). The statistical properties of binary stars are fossil traces of stellar formation processes. The mass-ratio distribution of double stars was one of the observed results. The precision of the CORAVEL was just sensitive enough to detect companions at the very bottom of the main sequence... close to the domain of brown dwarfs and giant planets. The opportunity offered by the Haute-Provence Observatory to develop ELODIE at the very beginning of the nineties opened the path to the detection of brown dwarfs AND giant planets. Based on our long radial velocity monitoring of stars of the solar vicinity we have selected 142 single G and K stars as the stellar sample for our ELODIE program.

What conditions played a significant role in that discovery? The impressive efficiency of the cross-correlation spectroscopy, the on-line reduction of the measurements with its impact on the observing strategy, the large size of the stellar sample, the number of observing nights allocated to our program by the OHP observing committee and obviously the precision of the spectrograph. One hot Jupiter among 142 solar-type stars corresponds to the frequency of this kind of giant planets... we have not been especially lucky!

The discovery of 51 Pegasi b in 1995 completely changed the priorities of my research.

I was happy to have been associated with that first epoch of exoplanet discoveries.

After this first detection, not only have we continued our search in the northern hemisphere but in 1998 we have initiated a systematic search for exoplanets in the southern sky.

CORALIE, a spectrograph almost identical to our northern instrument, was installed on our quite new 1.2 m EULER telescope at La Silla observatory. Small telescope ... but today having discovered more than 150 exoplanets.

At the very end of the nineties, in an answer to a call for proposals issued by ESO, I took the lead of the development of a new spectrograph to achieve a radial velocity precision of 1 m/s. Only after 3 years, in March 2003, we got the first light and the contractual goal of a precision of 1 m/s. For our consortium having built that instrument, the reward as measured by the number of 500 observing nights allocated for five years was at the level of the challenge.

Five hundred nights on a 3.6 m telescope devoted exclusively to a comprehensive program to detect and characterize exoplanets was a superb adventure for my colleagues and me.

Many of my PhD students are actively working in the field of exoplanets with outstanding results, for example Willy Benz, Didier Queloz, Nuno Santos, Christophe Lovis and Pedro Figueira.

The Kepler space mission has provided an exceptional harvest of transiting planets.

The comparative planetology requires having exoplanet radii (via transits) and masses (via Doppler spectroscopy). As the Kepler field is in the northern hemisphere, we have been obliged to develop a northern copy of our HARPS instrument (an instrument developed with Francesco Pepe as principal investigator). This program of that spectrograph installed on the 3.5 m Galileo telescope at La Palma Island, Spain, was focused on the physics of very low mass planets.

In 2007, I became emeritus professor at the University of Geneva with the privilege, still today, to continue contributing to transforming the old dream of Greek philosophers in the very active domain of present-day astrophysics.

Plurality of Worlds in the Cosmos: A Dream of Antiquity, a Modern Reality of Astrophysics

Nobel Lecture, December 8, 2019 by
Michel Mayor
University of Geneva, Geneva, Switzerland.

IT IS AMAZING TO CONSIDER that the question of the plurality of worlds in the universe was already discussed in antiquity by Greek philosophers. In a very famous letter of Epicurus (341–270 BC) we can read "Worlds are in an infinite numbers some of them similar to our own one, some others being different... living species, plants and all the other visible things could exist in some worlds and could not in others."

The question of the plurality of worlds in the universe has been continuously present during the last two millenia. We can, for example, quote this sentence by the philosopher and theologian Albertus Magnus (Circa 1200–1280) "Do there exist many worlds, or is there but a single world? This is one of the most noble and exalted questions in the study of Nature."

In 1277, Etienne Tempier, Bishop of Paris, with the agreement of Pope John XXI, asked that the question of plurality of worlds be taught at the Sorbonne. We can also mention the two major contributions of Immanuel Kant (1755) in his *Universal Natural History and Theory of Heaven* and Pierre-Simon Laplace in his *Exposé du système du Monde.* Both contribu-

Figure 1. 200 billion stars, but how many planetary systems are there in the Milky Way? This photo illustrates the huge number of stars seen in a very small fraction of the disk of our Galaxy. How do we detect planetary systems hosted by these stars?

tions introduce the notion of protoplanetary nebula, having noticed that all planets are moving in the same plane and sense of rotation.

PARADIGM SHIFT DURING THE SECOND HALF OF THE 20TH CENTURY

How many planets are there in the Milky Way? How many planets are similar to Earth?

It is interesting to look at the astronomical literature of the twentieth century for estimations of other planetary systems in the Milky Way. Before 1943, the estimations were between zero and at most a few. It was supposed that the formation of the protoplanetary nebulae results from the close encounter of two stars. The very low probability of such an event (close to zero!) is at the origin of these pessimistic estimates. In the early 1940s, claims of planet discoveries around some of the closest stars to the solar system (claims which were later found to be erroneous) produced a complete paradigm shift, with estimates of planetary systems in our galaxy as large as hundreds of billions (see Dick, S.J. 1991) It is interesting to note that this paradigm shift was actually the result of spurious detections of planetary systems in the twentieth century.

During the past three decades, improvements in astronomical instrumentation and the development of new observational techniques made it possible to transform the old philosophical concept of "plurality of worlds" in the universe into an active field of modern astrophysics.

Today, more than 4,000 exoplanetary systems have been detected, and we are beginning to discover planets in the so-called habitable zones of host stars. These Earth-like exoplanets have physical conditions suitable for the development of the complex chemistry of life. In the last 25 years these discoveries have completely transformed our understanding of planetary populations and the process of planetary system formations.

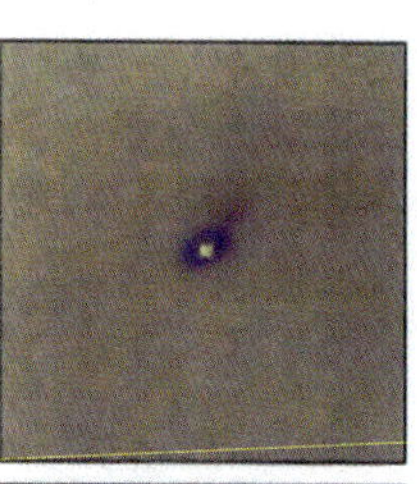

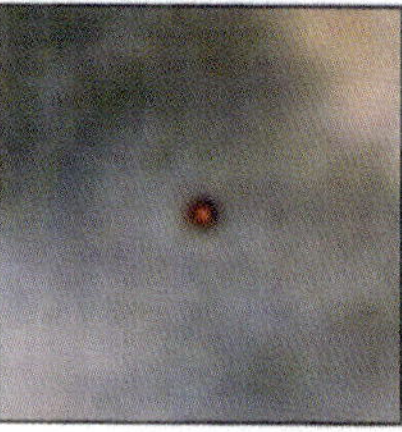

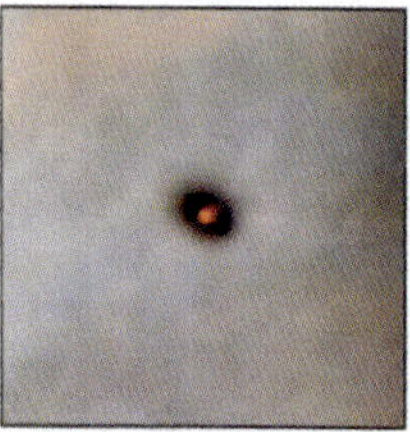

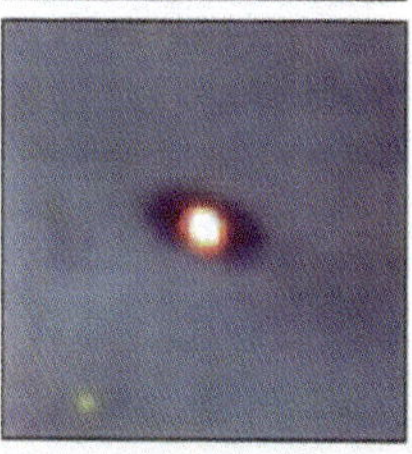

Figure 2. 1995, Images of protoplanetary disks: The Hubble Space Telescope reveals protoplanetary disks around very young stars of the Orion Nebula. (McCaughrean & O'Dell 1996).

Young stars formed by gravitational collapse of turbulent giant molecular clouds should have extremely large rotational velocities. However, the observed rotational velocities of stars at the bottom of the main sequence (stellar masses less than about 1.2 times the solar mass) are extremely small. Otto Struve suggests that the excess of angular momentum, if not present in the stars themselves, should be present in the protoplanetary nebulae. Consequently, protoplanetary disks are byproducts of the stellar formation itself and we can anticipate that most of stars (if not all of them) should host planetary systems.

In the 1970s, an excess of infrared luminosity in the spectra of very young stars finally revealed the presence of protoplanetary disks. Then in 1995, direct imaging of very young stars moving out of the Orion Nebula showed that most of them are surrounded by disks of dust and gas (see M.J. McCaughrean and C.R. O'Dell 1996).

No doubt – most (if not all) stars should host planetary systems.

How can we detect these systems?

Before discussing our contribution to the detection of planets, I would like to mention an extraordinary discovery made by Alex Wolcsczan and Dale Frail in 1992. By measuring the anomalous arrival times from the neutron star PSR B1257+12, they deduced the presence of two planets with masses only a few times the mass of our Earth. More recently, an additional planet was discovered orbiting the same pulsar. It may be possible that these

Figure 3. Alex Wolszczan and MM in front of the secondary mirror of the Arecibo radio-telescope in 2002 (20th anniversary celebration of the discovery of the planetary system hosted by the neutron stars PSR B1257 +12).

planets, whose orbits are almost circular – like most low-mass planets around normal stars – were formed from the debris of the destruction of a small stellar companion, although other formation scenarios are possible.

DOPPLER SPECTROSCOPY AS A PATH TO THE DETECTION OF EARTH-LIKE PLANETS

The possibility of detecting the gravitational influence of orbiting planets on the radial velocities of stars was suggested long before the Doppler technique was precise enough to allow such measurements (Belorizki, D. 1938, Struve, O. 1952).

In the eighties, several teams explored the possibility of developing spectrographs with the goal of achieving a precision better than about 15 m/s, a precision requested for the detection of gaseous giant planets. Among these different approaches, only a few were used in a systematic search to detect gaseous giant planets: in 1979, Bruce Campbell and Gordon W. Walker introduced an HF absorption cell in front of the spectrograph in order to achieve internal precise wavelength calibration, while in 1992, Geoffrey Marcy & R. Paul Butler designed an iodine cell for the same purpose.

THE PERMANENT QUEST FOR HIGHER AND HIGHER PRECISION: THE FIRST STEP WITH CORAVEL

We started to build instruments at the Haute-Provence Observatory in the South of France. Our first cross-correlation spectrometer CORAVEL, installed in 1977 on our 1m-telescope, achieved a precision of 300 m/s. It was a very exciting period of my life. The efficiency of CORAVEL was amazing, about 4000 times the efficiency of the ancient spectroscopic technique using photographic plates. (Baranne, Mayor, Poncet 1979). With such efficiency, it was easy to revisit many areas of astrophysics. CORAVEL was used for a diverse range of studies, including the dynamics of globular clusters, for example Omega Centauri and the pulsation of Cepheids in the Magellanic Clouds (small galaxies some 150,000 light years from the Earth).

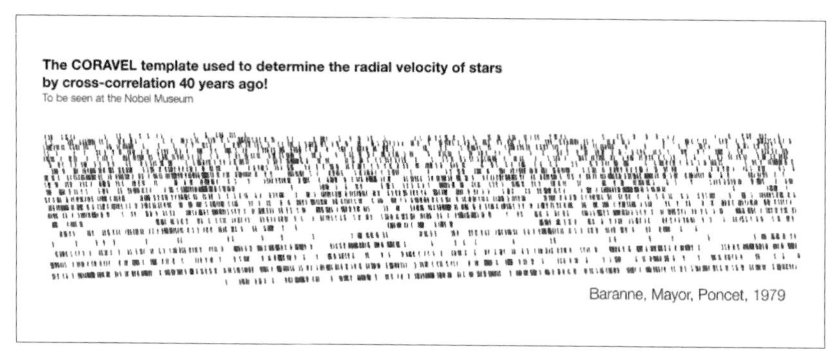

Figure 4. The CORAVEL template used to determine the radial velocity of stars by cross-correlation 40 years ago! To be seen at the Nobel Prize Museum.

The majority (about 2/3) of solar type stars have a stellar companion. Together, Antoine Duquennoy and I in 1991 made a 15 year-survey of several hundred stars relatively close to the solar system to determine the statistical properties of double stars: the distributions of their characteristics are seen as fossil tracers of stellar formation mechanisms. While CORAVEL was not designed to search for exoplanets, by 1989 we had discovered an m sin $i = 11$ Jupiter mass companion after combining our measurements with similar ones by David Lathan (Latham et al. 1989).

Recent astrometric measurements made by the GAIA satellite (Kiefer 2019) reveal a very small inclination of this system to the line of sight, in turn revealing the true mass of the companion to be that of a very low mass M star. Nevertheless this early detection demonstrates that spectrograph precision was approaching the level needed to detect real planets.

Figure 5. Roger Griffin was the first to demonstrate, already in the early sixties, that radial velocity measurements can be efficiently made by a cross-correlation spectrograph. (Credit: Roger Griffin).

A SMALL TECHNICAL NOTE

What is the principle of a cross-correlation spectrograph? I would like to illustrate the key point of this technique. Stars at the lowest part of the main sequence have thousands of atomic absorption lines in their spectra. When you disperse stellar luminosity into its different wavelengths, these absorption features appear as narrow lines in their spectra. Stars are moving in the sky relative to our solar system. The component of the velocity along the line of sight is called the *radial velocity*. Due to the motion of a star, you will see a small change in the position of lines resulting from the Doppler effect. If you measure the wavelength shift of an atomic transition, you will have the possibility of getting the radial velocity of that star. At the level of precision needed, this is a difficult task, although the basic idea of the instrument is quite simple.

To precisely measure the positions of the absorption lines, we need a lot of photons... but stars are faint. The central feature of the cross-correlation technique is an instrumental design allowing one to use thousands of atomic lines simultaneously. The first demonstration of its feasibility appeared in 1967 by Roger Griffin, while the first proposal had been made by Peter Fellgett in 1953. To detect planets, we have to measure extremely small changes in stellar wavelengths. Our spectrograph HARPS (Mayor *et al.* 2003) is able to detect changes of velocities of 0.3 m/s. This velocity corresponds to a Doppler shift of only a billionth of the wavelength... a shift of only a few silicon atoms on our detector. A planet hosted by a star will induce a small wobble in its velocity. For example, the velocity of our Sun is affected by the gravitational influence of Jupiter and as a result moves at 12 m/s around the centre of gravity of the solar system. The Earth also induces a wobble in the Sun's velocity, but of only 8 cm/s – the discovery of Earth-type planets is a real challenge.

It is easy to understand the cross-correlation technique and its capability to *concentrate all the Doppler information* when we have a look at the first CORAVEL spectrometer. The stellar spectra obtained by a cross-dispersed optics (echelle grating and grism) is projected on a template (see figure). This template is a glass plate coated with chromium, except on the position of atomic absorption lines. When the stellar lines match the holes in the template, the transmitted light is minimal. On the other hand, if the stellar spectrum is Doppler shifted, matching will not be perfect and the amount of transmitted light will be larger. An optical device allows one to determine how much the stellar spectrum must be shifted in order to minimize the transmitted therefore the mean radial velocity from several thousands of atomic lines. With CORAVEL, the cross-correlation is made optically. In our subsequent instruments (ELODIE, CORALIE, HARPS and ESPRESSO), the stellar spectra are registered with a low noise: large CCD and the cross-correlation is done numerically with a digital template. However, the principle is exactly the same.

It is interesting to note that several other processes which have a global effect on atomic lines can benefit from the cross-correlation technique and its capability to concentrate diluted physical information. For example, the stellar rotation velocity is easily determined as the Doppler broadening affects all the stellar atmospheric lines (Benz, Mayor 1980) while the measurement of *Fe/H* gives the mean stellar metallicity (Mayor 1980). As a result of the huge efficiency of cross-correlation spectroscopy, the technique is frequently used today to determine stellar radial velocities, rotational velocities as well as stellar metallicities.

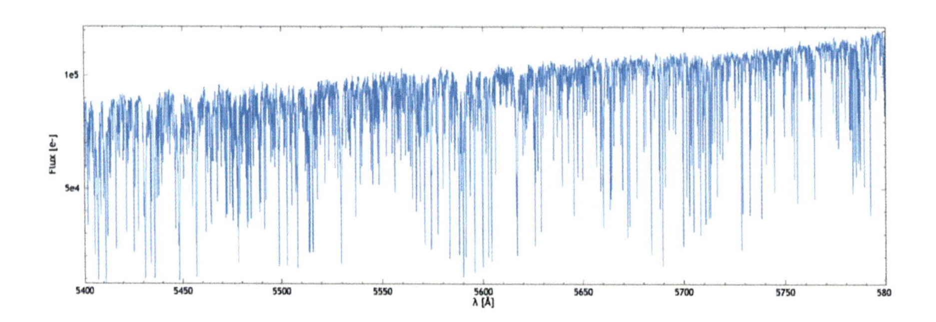

Figure 6. Spectrum of HD 85512 obtained with the ESPRESSO spectrograph installed at the ESO Paranal Observatory (Chile). (Pepe et al. 2020) Only 400 Ångström appear in the figure while the true spectral window of ESPRESSO is about 10 times larger. With ESPRESSO we can measure the important Doppler information contained in the spectra of solar-type stars or colder. The cross-correlation technique allows one to concentrate Doppler information from several thousand absorption lines to measure precise stellar radial velocities. (Courtesy of the ESPRESSO Consortium/dace.unige.ch.)

THE QUEST FOR HIGHER PRECISION: THE SECOND STEP WITH ELODIE

By the end of the 1980s, the evolution of technology allowed for the development of a new spectrograph. In 1988, the director of the Haute-Provence Observatory asked André Baranne and me to design a cross-correlation spectrograph adapted to the 1.93-meter telescope at that observatory. Two significant technological developments were critical to improve the precision of the spectograph: the possibility of having a large CCD detector and the existence of optical fibers of high quality. Here is not the place to present the technical details.

I would just mention that we needed to have a very, very stable illumination of the optics of our instrument to achieve the desired precision and to maintain it over several years. Optical fibers offer that possibility, guiding the stellar light from the telescope to the spectrograph in a stable environment in a thermally controlled enclosure below the telescope.

One crucial aspect provided by our computer-controlled spectrograph CORAVEL was the possibility of immediately having the fully reduced stellar radial velocity in a few seconds after the end of the measurement. We wanted to conserve that unique characteristic with the new spectrograph. The situation was not straightforward with the new instrument and at the end of his graduate studies, in 1990, Didier Queloz took charge of this important part of the software.

In science, it is only in exceptional cases that you can do things by yourself – at least in the field of modern astronomical instrument. I have to thank all the technicians and engineers of Haute-Provence and Geneva for their contribution to the success of the ELODIE instrument. Special thanks are especially due to André Baranne, our Chief Optician. ELODIE was a big success from 1993 onwards, immediately resolving velocity variations down to 10–15 m/s (a factor of 20 to 30 better than CORAVEL) and providing a precision which allowed for the detection of exoplanets.

Figure 7. André Baranne, the father of the optics of CORAVEL, ELODIE and CORALIE.

Figure 8. The ELODIE instrument installed on the 1.93-meter telescope at the Haute-Provence Observatory (OHP) was built by the technical staffs from OHP and Geneva Observatory. A special mention is due to André Baranne (Optical Engineer at Marseille Observatory). André is on the second row, just below the word OHP. He discovered the white pupil mounting, broadly used today in many astronomical spectrographs. (Alain Vin is missing in this photo!)

SEARCHING FOR EXOPLANETS

How could we detect a planet? A planet does not produce any luminosity. It just reflects a small part of the luminosity of a star it received. Let us look at our solar system. Jupiter reflects one billionth of the luminosity of the Sun. In 1995, it was not possible to directly get images of exoplanets due to this large luminosity contrast between the host star and planets. Therefore, we were obliged to use an indirect technique. As a result of the gravitational influence of a planet, the host star moves around the center of gravity of the system. We could then measure shifts of wavelength caused by the Doppler effect.

During the spring of 1994, my two young collaborators Antoine Duquennoy and Didier Queloz and I began a program with the new ELODIE instrument to search for possible brown dwarfs or gaseous giant planets orbiting solar-type stars. Antoine and I wanted to extend our study concerning double stars (Duquennoy & Mayor 1991) to explore the domain of very small mass ratios. But I had no a priori expectation of what we would find. At that time, brown dwarfs, stars which are not massive enough to have nuclear reactions in their core, were still undetected.

Figure 9. Antoine Duquennoy.

The lower limit for their masses was estimated to be only a few times the mass of Jupiter, overlapping the domain of gaseous giant planets.

Observing time on a telescope is only given on a competitive basis. We got seven observing nights every second months. Unfortunately, in June 1994, Duquennoy died in a car accident and then we were only two to do the observations. That search, among a sample of 142 solar-type stars, began during the spring of 1994, and already at the end of our first season of observations, we noted that the velocity of the star 51 Pegasi showed a periodic variation, which could be interpreted as being caused by the influence of a planet: a planet with a smaller mass than that of Jupiter. We observed an orbital period of 4.2 days, which disagrees with theoretical predictions. We had found a gaseous giant planet with an orbital period of four days rather than the 10 years (or more) than everyone expected – a factor of 1,000 out!

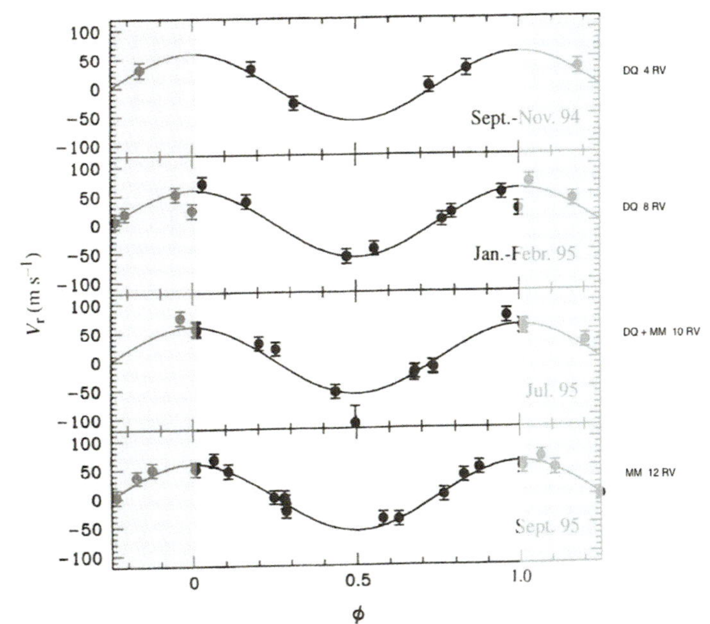

Figure 10. 51 Peg, Four seasons.

There had been many claims of discoveries of planets in the past, which were later found to be wrong. That is one of the reasons why we decided to postpone the publication of our finding for an additional season. We were certain of the quality of all of our measurements, but there was a risk of bad interpretations. We could be misled by other physical processes, such as those related to stellar magnetic activity. A very interesting example of the confusing effect of magnetic activity is given by the star HD 166435. A periodic variation (with a period of 3.8 days) was observed during several observing periods in 1998. However, a photometric variation of the luminosity and color indicates an intrinsic cause of that variability resulting from a very large magnetic spot with a rather long lifetime. (Queloz *et al.* 1998) We concluded that the radial-velocity variations were not due to gravitational interactions with an orbiting planet but, instead, originated from line-profile changes stemming from spots on the surface of the star. The quasi-coherence of the radial-velocity signal over more than two years, which allowed a fair fit with a binary model, makes the stability of this star unusual among other active stars. It suggests a stable magnetic field orientation where spots are always generated at about the same location on the surface of the star.

Another concern came from the existing scenario for the formation of giant planets in the nineties. As the quantity of dust is limited in an accretion disc, the formation of gaseous giant planets requires the agglomeration of ice particles. Ice particles only exist at sufficiently large distances from solar-type stars... and the formation of gaseous giant planets could only exist at distances larger than about five astronomical units and have orbital periods larger than 10 years! (Boss 1995)

The period of the companion of 51 Pegasi, 4.2 days, was much too short. We did not understand how it was possible to produce a planet with such a short period, but by July 1995, our data was so consistent that we ventured to announce the discovery of the first extrasolar planet orbiting a Sun-like star (Mayor, Queloz 1995).

The discovery of this first planet with its very short orbital period made it necessary to take into account the orbital migration of planets during the formation period in an accretion disc. This mechanism had already been studied 15 years before the discovery of 51 Pegasi b by Peter Goldreich and Scott Tremaine (1980), (see also: Papaloizou & Lin 1984, Ward 1986, Lin & Papaloizou 1986). However, the prediction of the migration of exoplanets had never been used to build observing strategies!

Soon after the discovery of 51 Pegasi b, Lin *et al.* (1996) showed that short-period gas giant could results from the gravitational interaction of the young planet with the accretion disk.

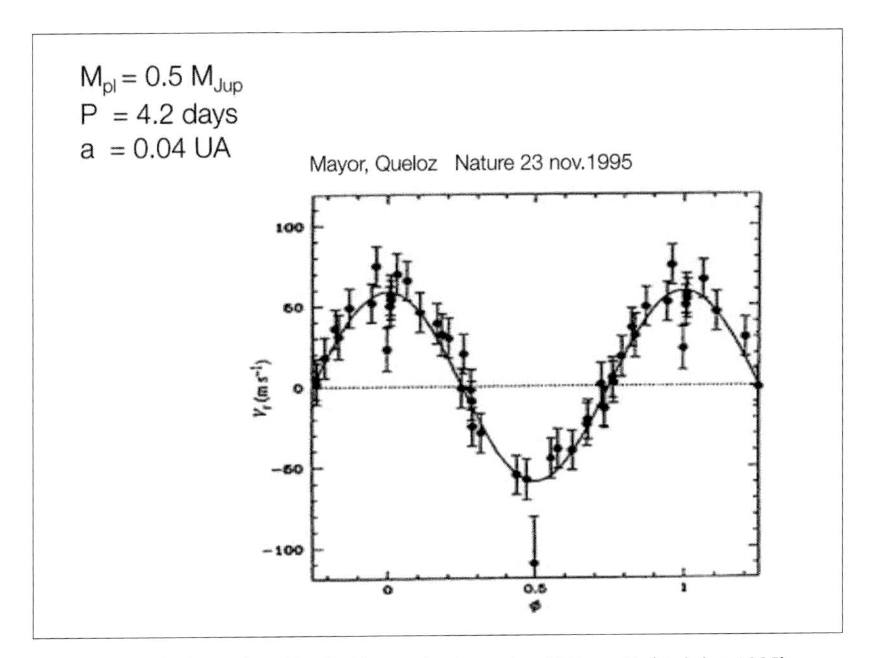

Figure 11. The first exoplanet hosted by a solar-type star: 51 Pegasi b (October 1995).

Since 1995, the observational evidence for orbital migration has deeply changed every scenario of planetary formation.

A few months after the discovery of 51 Peg b, the detection of several short period planets was announced by the California team (Marcy & Butler 1996, Butler et al. 1996,1997). Clearly, 51 Peg b, the first « Hot Jupiter » is not a unique object with exceptional characteristics.

We continued the search for planets in the northern sky. We moved to the southern sky and started observations at La Silla Observatory (ESO) located in Chile. Firstly with CORALIE, a slightly improved copy of the ELODIE spectrograph. Then with the HARPS spectrograph on the ESO 3–6-meter telescope at La Silla.

We may wonder why we continue to search for planets when we have already found more than 4000 of them. In fact, the goal is not simply to detect an additional planet. It was probably the case at the beginning, but today we want to have a global view of planetary systems and to understand their formation and evolution.

The formation and evolution of planetary systems involves a very broad spectrum of physical process: orbital migration, lifetime of accretion discs, detailed mechanisms of planetary formation, interaction between planets, chemical composition of host stars ...

Figure 12. Exoplanet pioneers at the Wyoming conference in 2011. From left: Alex Wolczscan, Michel Mayor, Nathalie Bathalia, William Borucki, David Charbonneau and Geoff Marcy. (Credit: Geoff Marcy)

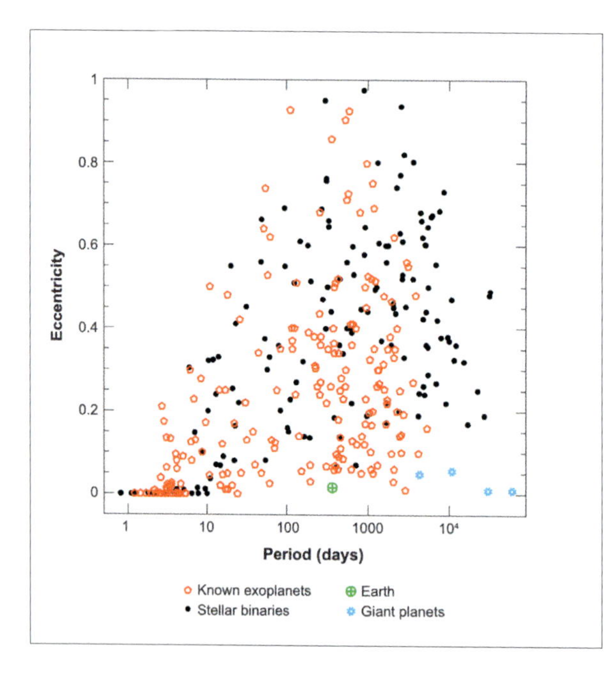

Figure 13. The observed diversity of planetary systems:

- orbital periods as short as a few hours.
- orbits with a large range of eccentricities.

Period-eccentricity diagram for the sample of known exoplanets (in 2007!) in comparison with stellar binaries. The Earth and giant planets of the Solar System are indicated as well. (Udry & Santos 2007.)

The observation of the diversity of planetary systems has been used to constrain theoretical models of their formation. Several teams have explored the relative importance of these different processes. Our understanding of planetary formation results from the dialogue between theory and observations. (Benz, W. *et al.* 2014)

Another very nice possibility exists for planet detection. If a planet passes between a star and an observer's line of sight, it blocks out a tiny part of the star's light. As a result, we can observe a periodic diminishing of the stellar luminosity due to the transit of the planet. The depth of the depression is directly proportional to the relative size of the planet compared to its host star. Before the announcement of the discovery of 51 Pegasi b, we immediately tried to detect possible planetary transits... but the inclination of the orbital plane was not adequate.

Upon detecting another short period planet in the summer 1999 (P = 3.5 days), we were able to predict the exact time when it might transit in front of its star HD 209458. At the predicted time, on September 9 and 16 of that year, the first planetary transit was observed, which proved that indeed, we were observing gas giant planets such as Jupiter or Saturn (Charbonneau *et al.* 2000). The bulk density of that giant planet is as low as 0.3 grams per cubic centimeter. This planetary transit was independently measured on November 6, 1999 by the California team (Henry et al. 2000).

Figure 14. M. Mayor and D. Queloz at La Silla Observatory (ESO, Chile) in front of the 1.2-meter EULER telescope and in the distance, the 3.6-meter telescope. These two telescopes have made significant contributions to the detections of exoplanets since 1998 and 2003 respectively.

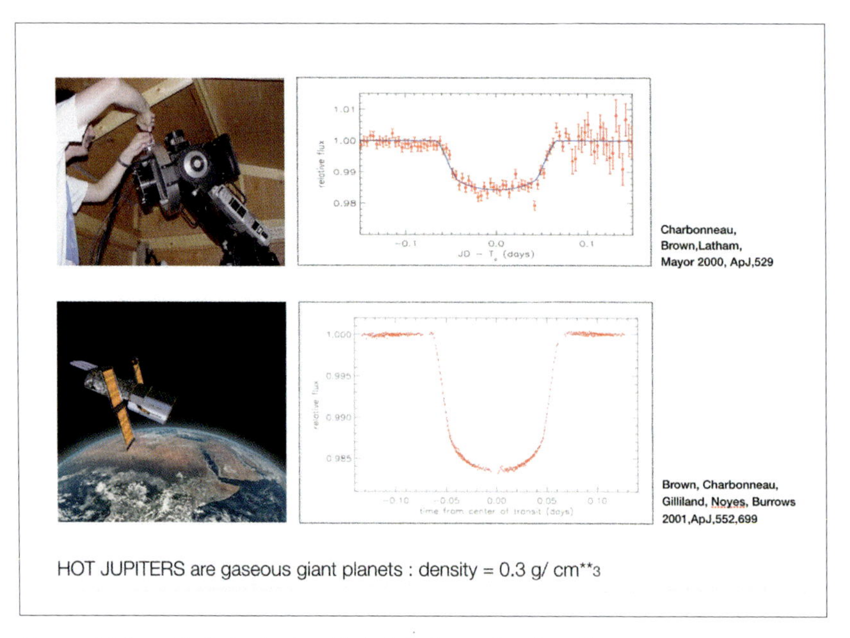

Figure 15. 9th and 16th September 1999: A first planetary transit.

Soon after we were able to measure the Rossiter-MacLaughlin effect for a planet: a spectroscopic transit, which allows the measurements of the projected angle between the stellar spin axis and the planet's orbital axis (Queloz *et al.*2011). New results show a large variety of angles, with occasionally very inclined orbits and in few cases even retrograde orbits. These cannot be explained solely by planetary migration. The evolution of planetary systems becomes even more complicated with the possible dynamic influence of distant stellar companions (via the Kozai effect).

About twenty years after the discovery, we finally succeeded in detecting the reflected light from 51 Pegasi b (Martins *et al.* 2015) thereby obtaining a direct estimate of 0.46 (+0.06, −0.01) M Jup for the mass of the planet.

The observation of an exoplanetary transit opened the door to the study of the internal composition of planets, therefore creating a new field of astronomy: Exoplanetology. This first detection of a planetary transit also played a crucial role in the decision to built space missions devoted to detecting exoplanetary transits.

CHEMICAL CLUES FOR STARS WITH PLANETS

The chemical composition of a planet, including both its interior and atmosphere, is likely to be related to the chemical composition of the protostellar cloud, and this will be reflected in the composition of the stellar atmosphere. The precise determination of the stellar chemical abundances provides important constraints on the mechanisms of planetary formation.

The very first detections of exoplanets immediately led to the suspicion that there should be a relation between stellar metallicity and the occurrence of giant planets. Systematic surveys of the metallicity of large stellar samples have confirmed the strong positive correlation between the frequency of giant planets and the chemical composition of the host stars (Santos et al. 2004, Fischer, Valenti 2005, Sousa *et al.* 2011).

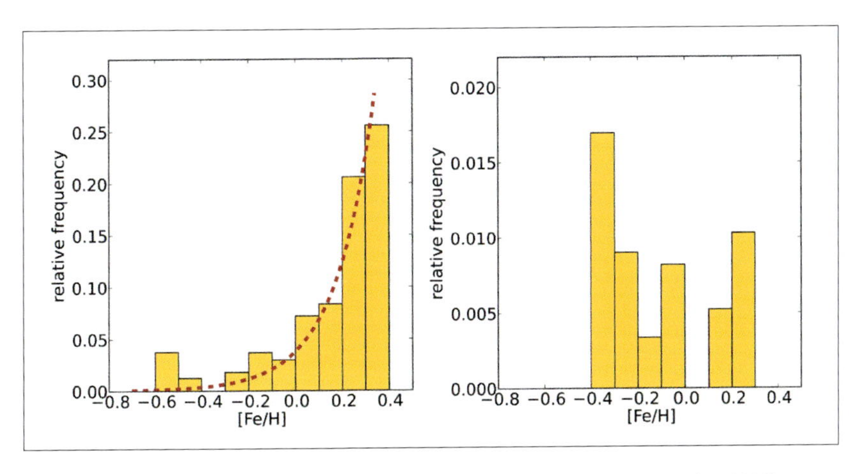

Figure 16. Metallicity distribution of planet hosting stars (Mayor, Lovis, Santos 2014) In the left panel, the frequency of giant planets as a function of stellar metallicity is shown based on results from the HARPS planet search program. In the right panel, we illustrate the same plot for stars that host only Neptune- or Super-Earth-like planets. These plots show a clear correlation between the presence of giant planets and the metallicity of the star. This trend is not seen for stars hosting lower-mass planet.

Figure 17. Garik Israelian and Nuno Santos: two colleagues who have contributed so much to the study of the chemical composition of stars with or without planets. (Copyright [right-hand photo] Susana Neves)

HARPS, THE THIRD STEP TOWARDS HIGHER PRECISION AND THE PATH TO THE DETECTION OF ROCKY PLANETS

The sensitivity of the HARPS spectrograph has improved to the point where it now allows us to detect much lower mass planets. This can be considered as part of the quests for rocky planets. Recall that while Jupiter induces a change of velocity of the Sun at the level of 12 m/s, the Earth induces a change of only 8 cm/s. High precision is required in order to detect rocky planets.

We were able to design a new, much more sophisticated instrument that works in a vacuum with temperature controlled at the level of a few milli-Kelvin degrees during the night. In 2000, I took the lead role in the construction of a new spectrograph called HARPS, which was fully optimized to search for very low mass planets (Mayor *et al.* 2003). That new spectrograph, installed at La Silla in Chile in 2003, was sensitive enough to detect velocity changes smaller than 1 m/s and therefore to discover even lighter planets, right down to the mass of the Earth. Francesco Pepe played a major role in the development of that instrument as project engineer.

Our obsessional search for higher and higher velocity precision has been rewarding. With the HARPS spectrograph, we have detected a new population of Super-Earth and Neptune mass planets: a population of extremely common planets orbiting solar-type stars (planets with between 1 and 20 Earth-masses). That rich sub-population has been beautifully confirmed by the Kepler space mission.

Figure 18. The third step: HARPS.

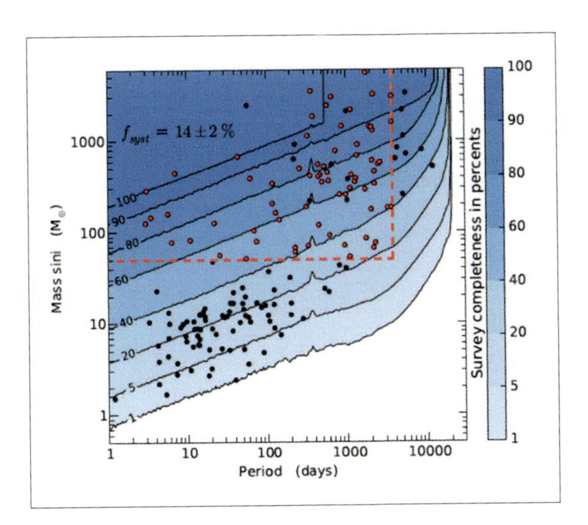

Figure 19. After 16 years we are continuing to conduct a large and systematic survey of stars in the southern hemisphere with the CORALIE spectrograph on the EULER telescope and the HARPS spectrograph on the 3.6-meter telescope. We try to design surveys with controlled detection bias in order to obtain distributions of planetary systems as functions of mass, orbital period, host star metallicity, etc. These distributions provide constraints on planet formation scenarios and tell us (for example) that planets more massive than 50 Earth-masses are hosted by about 14% of solar-type stars. We also note that "hot-Jupiters," while being the first planets detected are rare (about 1%); most giant planets have longer orbital periods from several months to several years (M. Mayor & M. Marmier). We can also note the extreme abundance of planets with masses between a few Earth-masses and 20 Earth-masses. (Super-Earth).

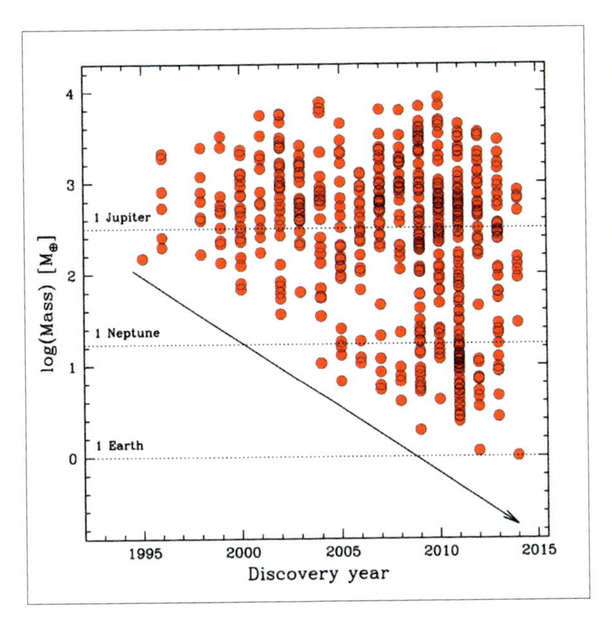

Figure 20. After the discovery of 51 Pegasi b, we can observe the amazing number of planet discoveries. We can also note the result of the improvement in the sensitivity of Doppler spectrographs. today allowing the detection of planets with masses as small as the Earth (at least for relatively tight orbits!).

The Kepler space mission, with its harvest of several thousand planetary transits, has provided planetary radii for a large number of Earth-type planets. We need to know the mass of these planets to constrain the bulk density of their composition. As Kepler candidates are in the northern sky, we have been obliged to develop a copy of HARPS, currently installed at La Palma Observatory in Spain on the Galileo 3.5-meter telescope. We devote an extremely large number of observing nights to studying the inner composition of planets having only a few times the mass of our Earth.

The combined data from HARPS radial velocities and planet diameters derived from planetary transits is of a special interest for planets with masses less than 20 Earth masses. These measurements, for example, allow the study of the transition from rocky to Neptune-like planets (see a recent Radius-Mass diagram from Frustagli et al. 2020, derived from combined radial velocities and diameters for transiting planets and reproduced below).

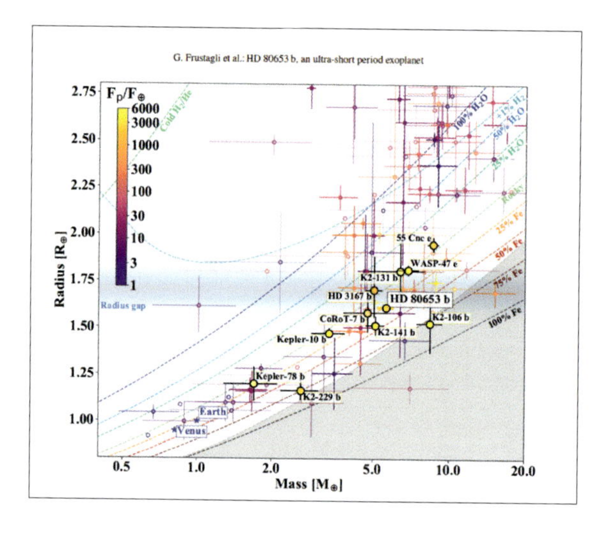

Figure 21. Mass-radius diagram of planets smaller than 2.8 Earth radii. (Frustagli *et al.*, 2020). The dashed lines show planetary interior models for different compositions as labelled (Zeng *et al.*, 2019). Planets are color-coded according to the incident flux Fp, relative to the solar constant received on the Earth (for the full description see Frustagli *et al.* 2020).

The huge harvests of detections made by space missions like CoRoT, Kepler, TESS as well as ground-based experiments like SuperWASP have demonstrated the potential of the transit technique. The present focus is on detecting Earth twins. We know that we have a huge number of rocky planets in the galaxy. The problem is to detect planets as close as possible to us for follow-up studies and especially planets which are located in the so-called habitable zone of the star, that is, the zone at a distance from the star where the complex chemistry for life development has had the chance to emerge. We already have the ability to detect Earth twins with our present instrumentation.

Increasing the precision

Radial velocity via cross-correlation spectroscopy:
A path to the detection of Earth-type planets

SPECTRO	year	precision	Telescope	
CORAVEL	1977	300 m/s	1 m	OHP
ELODIE	1994	13 m/s	1.9 m	OHP
CORALIE	1998	6 m/s	1 m	ESO Chile
HARPS	2003	1 m/s	3.6 m	ESO Chile
HARPS-N	2013	1 m/s	3.5 m	IAC La Palma
ESPRESSO	2018	0.1 m/s	8.2 m (x4)	ESO Chile

Figure 22. Over the past 40 years, the precision of the different generations of cross-correlation spectrographs has been increased by a factor 3,000! This gain in sensitivity allows for the discovery of planets with smaller masses.

Figure 23. Francesco Pepe, the principal investigator of the ESPRESSO spectrograph. This instrument, installed at Cerro Paranal (ESO Chile), can feed either one or four 8.2-meter unit telescopes of the VLT, thereby achieving a collecting power equivalent to a 16-meter telescope! ESPRESSO represents the latest generation of the series of our cross-correlation spectrographs and was designed to achieve a precision of 0.1 m/s. (Picture: Charly Rappo/arkive.ch.)

One of our current projects in Geneva is to build a small catalogue of bright stars with rocky planets in the habitable zone. We need such an input catalogue for the next generation of instruments to explore planets like Earth. If we were ever to build an ambitious space mission, we would want to have a list of likely stars to look at.

More than 2,000 years ago, Greek philosophers were already discussing of the plurality of worlds in the universe and speculating on the possibility that some of these worlds could have living species. Today, exobiology emerges as a new multidisciplinary domain of science.

Do we have living organisms outside the solar systems?

I do not know how many years will be necessary to answer that fundamental question. However, I am certain that it will remain on the agenda of all scientific agencies. Today we are close to having the technology to detect biomarkers in the atmospheric spectra of exoplanets. For more than 2,000 years, humanity has been waiting for an answer to the possible existence of life on other worlds, so we can afford to wait a few decades.

"Do there exist many worlds, or is but a single world? This is one of the most noble and exalted questions in the study of Nature."

Albertus Magnus (1200–1280)

REFERENCES

Baranne, A., M. Mayor and J.L. Poncet. 1979. Coravel: a new tool for radial velocity measurements. *Vistas in Astronomy* **23**, 279–316.

Baranne, A., D. Queloz, M. Mayor, G. Adrianzyk, G. Knispel, D. Kohler, D. Lacroix, J.P. Meunier, G. Rimbaud and A. Vin. 1996. Elodie: A spectrograph for accurate radial velocity measurements. *Astronomy and Astrophysics Supplement Series* **119**, 373–390.

Belorizky, D. 1938, Le Soleil, étoile variable, *l'Astronomie* **52**, 359–361.

Benz, W., Mayor, M. 1981, A new method for determining the rotation of late spectral type stars, *Astronomy and Astrophysics* **93**, 235–240.

Benz, W., Ida, S., Alibert, Y., Lin, D., Mordasini, C., Planet population synthesis. In *Protostars and Planets VI*. pp 691–713, eds. H. Beuther *et al.*, Univesity of Arizona press.

Boss, A.P. 1995, Proximity of Jupiter-like planets to low mass stars. *Science* **267**, Issue 5196, pp 360–362.

Butler, R.P. and G.W. Marcy. 1996. A planet orbiting 47 UMa. *Astrophysical Journal Letters* **464**, L153–156.

Butler, R.P *et al.* 1997, Three new "51 Pegasi-type" planets, *The Astronomical Journal* **474**, L115–118.

Campbell, B. and G.A.H. Walker. 1979. Precision radial velocities with an absorption cell. *Publications of the Astronomical Society of the Pacific* **91**, 540–545.

Campbell, B., G.A.H. Walker and S. Yang. 1988. A search for substellar companions to solar-type stars. *Astrophysical Journal* **331**, 902921.

Charbonneau, D., Brown, T., Latham, D.W., Mayor, M. 2000, Detection of planetary transits accross a sun-like star. *The Astrophysical Journal* **529**, L4548.
Dick, S.J. 1991, in *Bioastronomy: The Search for Extraterrestrial Life*. Eds J. Heidmann & M.J. Klein, 356363.
Duquennoy, A and M. Mayor. 1991. Multiplicity among solar type stars in the solar neighbourhood. II. Distribution of the orbital elements in an unbiased sample. *Astronomy and Astrophysics* **248**, 485524.
Fellgett, P. 1953, A proposal for a radial velocity photometer. *Optica Acta* **2**, 915
Fischer, D.A., G.W. Marcy, R.P. Butler, G. Laughlin and S.S. Vogt. 2002. A second planet orbiting 47 Ursae Majoris. *Astrophysical Journal* **564**, 10281034
Frustagli, G. *et al.*, 2020, An ultra-short period rocky super-Earth orbiting the G2-star HD 80653. *Astronomy and Astrophysics* **633**, 133148.
Fischer, N.C., Valenti, J. 2005, The Planet-metallicity correlation, *The Astrophysical Journal* **622**, 11021117.
Goldreich, P. and S. Tremaine. 1980. Disk-satellite interactions. *Astrophysical Journal* **241**, 425441.
Griffin, R.F. 1967. A photoelectric radial-velocity spectrometer. *Astrophysical Journal* **148**, 465476.
Henry, G.W., Marcy, G.W., Butler, R.P., Vogt, S. 2000, A transiting "51Peg-like planet," The Astrophysical Journal 529, 4144.
Kiefer, F. 2019, Determining the mass of the planetary candidate HD 114762b using GAIA, *Astronomy & Astrophysics* **632**, L9.
Latham, D.W., T. Mazeh, R.P. Stefanik, M. Mayor and G. Burki. 1989b. The unseen companion of HD114762: a probable brown dwarf. *Nature* **339**, 3840.
Lin, D.N.C., P. Bodenheimer and D.C. Richardson. 1996. Orbital migration of the planetary companion of 51 Pegasi to its present location. *Nature* **380**, 606607.
Lin, D.N.C. and J. Papaloizou. 1986. On the tidal interaction between protoplanets and the protoplanetary disk. III – Orbital migration of protoplanets. *Astrophysical Journal* **309**, 846857.
Marcy, G.W. and R.P. Butler. 1992. Precision radial velocities with an iodine absorption cell. *Publications of the Astronomical Society of the Pacific* **104**, 270277.
Marcy, G.W. and R.P. Butler. 1996. A planetary companion to 70 Vir. *Astrophysical Journal Letters* **464**, L147151.
Martins, H.C., N.C. Santos, P. Figueira, J.P. Faria, M. Montalto, I. Boisse, D. Ehrenreich, C. Lovis, M. Mayor, C. Melo, F. Pepe, S.G. Sousa, S. Udry and D. Cunha. 2015. Evidence for a spectroscopic direct detection of reflected light from 51 Pegasi b. *Astronomy and Astrophysics* **576**, A134.
Mayor, M. 1980, Metal abundances of F and G dwarfs determined by the radial velocity scanner CORAVEL, *Astronomy and Astrophysics* **87**, L12
Mayor, M., C. Lovis and N.C. Santos. 2014. Doppler spectroscopy as a path to the detection of Earth-like planets *Nature* **513**, 328335.
Mayor, M. and D. Queloz. 1995. A Jupiter-mass companion to a solar-type star. *Nature* **378**, 355359.
Mayor, M., D. Queloz, S. Udry and J.L. Halbwachs. 1997. From brown dwarfs to planets. Proceedings of the 5th International Conference on Bioastronomy, IAU colloquium n. 161. (Capri, Italy, 1–5 July 1996), on *Astronomical and biochemical origins and the search for life in the universe*, edited by B. Cosmovici, S.C. Bowyer, D. Werthimer, Bologna, Editrice Compositori, pp. 313330.

Mayor, M.; Pepe, F.; Queloz, et al. 2003, Setting new standards with HARPS, *The Messenger,* **114**, 20.
Mayor, M., Udry, S., 2008, The Quest for very low-mass planets. In Phys. Scripta T, 130, 014010.
Mc Caughrean, M.J., O'dell, C.R. 1996, Direct imaging of circumstellar disks in the Orion Nebula, *The Astronomical Journal* **111**, 19771986.
Queloz,D., *et al.* 2001, No planet for HD 166435, *Astronomy and Astrophysics* **379**, 279, 287.
Queloz, D., Eggenberger, A., Mayor, M., Perrier, C., Beuzit, J.L., Naef, D., Sivan, J.P., Udry, S. 2000, Detection of a spectroscopic transit by the planet orbiting the star HD209458, *Astronomy and Astrophysics* **359**, L13L17.
Papaloizou, J., Lin, D.N.C 1984, On the tidal interaction between protoplanets and the primordial solar nebula. I – Linear calculation of the role of angular exchange. *The Astrophysical Journal* **285**, 818834.
Santos, N.C., Israelian, G., Mayor, M. 2001, The metal-rich nature of stars with planets. *Astronomy and Astrophysics* **373**,1019 1031.
Santos, N.C., Israelian, G., Mayor, M. 2004, Spectroscopic (Fe/H) for 98 extrasolar planet-host stars. *Astronomy and Astrophysics* **415**, 1153 .
Sousa, S.G. et al. 2011, Spectroscopic stellar parameters for 582 FGK stars in the HARPS volume-limited sample. Revisiting the metallicity-planet correlation. *Astronomy and Astrophysics* **533**,141149.
Struve, O. 1952, Proposal for a project of high-precision stellar radial velocity work, Observatory 72, 199200.
Udry, S., Santos, N.C. 2007, Statistical properties of exoplanets. *Ann. Rev. of Astonomy and Astrophysics* **45**, 397–439.
Walker, G.A.H., Walker, A.R., Irwin, A.W., Larson, A M. Yang, S.L.S., Richardson, D.C. 1995, *Icarus* **116**, 359.
Ward, W.R. 1986, Density waves in the solar nebula: Differential Lindblad torque, *Icarus*, **67**, 164180.
Wolszczan, A. 1994. Confirmation of Earth-mass planets orbiting the millisecond pulsar PSR 1257+12. *Science* **264**, 538–342.
Wolszczan, A. and D.A. Frail. 1992. A planetary system around the millisecond pulsar PSR 1257+12. *Nature* **355**, 145–147.
Zeng, L., Jacobsen, S.B., Sasselov, D.D. et al. 2019, *Proceedings of the National Academy of Sciences,* **116**, 9723.

Didier Queloz. Photo: A. Mahmoud

Didier Queloz

Biography

DIDIER QUELOZ IS A PROFESSOR of physics at the University of Cambridge's Cavendish Laboratory and professor of astronomy at the University of Geneva (part time). He is one of the originators of the "exoplanet revolution" in astrophysics. In 1995, as part of his PhD, he and his supervisor announced the first discovery of a giant planet orbiting another star, outside the solar system[1]. The planet was detected by the measurement of small periodic changes in stellar radial velocity produced by the orbiting planet. Detecting this small variability using the Doppler effect was possible thanks to the development of a new type of spectrograph combining stability, high resolution and a creative approach to measure precise stellar radial velocity[2]. For his work he was awarded, with Michel Mayor, the 2011 BBVA Foundation Frontiers of Knowledge Award of Basic Sciences for "developing new astronomical instruments and experimental techniques that led to the first observation of planets outside the Solar System."

This seminal discovery spawned a revolution in astronomy and kick-started the research field of exoplanet systems. Over the next 25 years, Didier Queloz's main scientific contributions have essentially focused on expanding the detection and measurement capabilities of these systems to retrieve information on their physical structure. The goal is to better understand their formation and evolution by comparison with our solar system. In the course of his career, he has developed new astronomical equipment, novel observational approaches and detection algorithms. He has participated in and conducted programs leading to the detection of hundreds of planets, include breakthrough results.

Early in his career, he identified stellar activity as a potential limitation for planet detection. He published a reference paper[6] describing how to disentangle stellar activity from a planetary signal using proxies, including new algorithms that have become standard practice in all planet publications based of precise Doppler spectroscopy data. His team[i] and his

Geneva colleagues[ii] established standards to optimise measurements of stellar radial velocity that are still in use today.[5]

Shortly after the start of the ELODIE planet survey at the Haute-Provence Observatory (OHP), he led the installation of an improved version (CORALIE), on the Swiss Euler telescope in Chile. Very quickly this new facility started to detect exoplanets on stars visible in the southern hemisphere.[4] In 2000 he took responsibility, as project scientist, for the development of HARPS[7], a new type of spectrograph for the European Southern Observatory (ESO) 3.6m telescope. This instrument, commissioned in 2003, was to set a new standard in the business of precise Doppler spectroscopy[8]. HARPS' performance, allied with the development of new analysis software inherited from data gathered by ELODIE and CORALIE, would considerably improve the precision of the Doppler technique. Eventually it would deliver spectacular detections of smaller exoplanets in the realm of Neptune, super-Earth systems[10,11] before NASA's Kepler space telescope would massively detect them and establish their statistical occurrence.

After the announcement of the detection of the first transiting planet (in 1999), Didier Queloz's research interests broadened, with the objective to combine capabilities offered by transiting planets and follow-up Doppler spectroscopy measurements. In 2000 he achieved the first spectroscopic transit detection of an exoplanet using the so-called Rossiter-McLaughlin effect.[3] This type of measurement essentially tells us about the projected angle between the stellar angular momentum vector and the planet orbital angular momentum vector. The pinnacle of this program would be reached 10 years later, after he led a significant upgrade of CORALIE, and established collaboration with the Wide Angle Search for Planets (WASP) consortium in the UK. With his PhD student[iii] he demonstrated that a significant number of planets were surprisingly misaligned or on retrograde orbits, providing new insights about their formation process. In 2017 he received the 2017 Wolf Prize in Physics for his "seminal work on revealing an incredible diversity of exoplanets and his contribution to the discovery of more than 250 additional exoplanets, including several multi-planetary systems and measurement of the first Rossiter-McLaughlin effect for a transiting planet, which allowed the measurement of the projected angle between the stellar spin axis and the planets orbital axis [...]"

The special geometry of transiting planets, combined with precise Doppler spectroscopic observations, allows us to measure the mass and

i. Francois Bouchy (postdoc)
ii. Francesco Pepe (HARPS project manager)
iii. Amaury Triaud (PhD student)

radius of planets and to compute their bulk densities in order to gain insights about their physical structure. In 2003, Didier Queloz – recently appointed to a faculty position – with his team[iv] pioneered and established the combination of these techniques by first measuring bulk density of OGLE transiting planets[9]. They also looked for transit opportunities on known radial velocity planets and they found the first transiting Neptune-sized planet (GJ436b).[13] In the course of this program and in collaboration with his colleague Shay Zucker from Tel-Aviv University, he developed the mathematical foundation to compute residual noise they encountered during the analysis of transit they were trying to model. They established a statistical metric to evaluate "red noise."[12] Today this concept is widely used in the field to estimate systematics in light-curves and transit modeling.

In 2007 Didier Queloz became an associate professor. Over the next 5 years following his appointment, his research program – based on a combination of spectroscopy and transit detection – intensified. He took the lead in the spectroscopic follow-up effort of the WASP consortium and the European Space Administration's Corot space mission.[v] The combination of WASP and Corot data with follow-up observations using EulerCam (a CCD imager he developed), CORALIE, HARPS and other main ESO facilities was amazingly successful. It led to more than 100 publications, some of them breakthroughs providing us with new insights on formation and nature of hot Jupiter-type planets. Further, in the same period, the detection of COROT-7b – combined with intensive follow-up work – established the first detection of a planet with a bulk density similar to a rocky planet.[14]

All the follow-up expertise he developed naturally extended to the Kepler mission era, with the HARPS-N consortium confirming the Earth-like bulk density of Kepler10.[16] In ground-based transit programs, Didier Queloz was deeply involved in the design and installation of a new generation survey telescope: the Next Generation Transit Survey (NGTS) observatory. His role was decisive during system tests in Europe and in establishing the facility at Paranal, Chile.[20]

At the time Didier Queloz moved to Cambridge, he essentially focused on setting up comprehensive research activities directed to the detection of Earth-like planets and life in the universe and to the further development of the exoplanet community in the UK. When he left Switzerland, he was co-directing a major national initiative[vi] which eventually got funded. At Cambridge, with the help of his colleagues at the Institute of Astronomy (IoA) and the Department of Applied Mathematics and Theo-

---

iv. Francois Bouchy, Michael Gillon, Frédric Pont (Postdocs)
v. COROT CNES webpage

retical Physics (DAMTP), he established the Cambridge Exoplanet Research Centre[vii] to stimulate joint coordinated efforts and collaboration between departments. In the UK he organized the first "exoplanet community meeting" and installed the idea of a regular yearly community workshop. In the European context, he is leading at Geneva (through his joint professorial appointment) the development of the ground segment of the CHaracterizing ExOPlanet Satellite (CHEOPS)[viii] space mission and he chairs the science team.

His most recent research highlights are related to the search for transiting Earth-like planets near low-mass stars and for extraterrestrial life. This program, carried out in collaboration with Michael Gillon of the University of Liège, is the origin of the detection of Trappist-1[18], a planetary system of potential interest to further searches for atmospheres and signs of life. Another successful avenue of research is the characterization of the rocky surface or atmosphere of hot small planets, including the work on 55 Cnc.[17] The recent extension of this program towards extraterrestrial life is being carried out in the context of an international research initiative supported by the Simons Foundation. One major result of this collaboration is the definition – combining chemistry and astrophysical constraints – of minimum conditions for the origins of RNA precursors on exoplanets ("abiogenesis zone").[19]

Discoveries of exoplanets attract a lot of attention from the public and the media. In parallel with his research and teaching activities, Didier Queloz has been involved in numerous documentary films, has written articles and has done TV and radio interviews to share his excitement, explain his findings and promote a general interest in science.

PUBLICATIONS

1. Michel Mayor and Didier Queloz. "A Jupiter-mass companion to a solar-type star." In: *Nature* **378**, 6555 (Nov. 1995), pp. 355–359. doi: 10.1038/ 378355a0.
2. A. Baranne, D. Queloz, M. Mayor, G. Adrianzyk, G. Knispel, D. Kohler, D. Lacroix, J. -P. Meunier, G. Rimbaud, and A. Vin. "ELODIE: A spectrograph for accurate radial velocity measurements." *In: Astronomy & Astrophysics* **119** (Oct. 1996), pp. 373–390.
3. D. Queloz, A. Eggenberger, M. Mayor, C. Perrier, J. L. Beuzit, D. Naef, J. P. Sivan, and S. Udry. "Detection of a spectroscopic transit by the planet orbiting the star HD209458." In: *Astronomy & Astrophysics* **359** (July 2000), pp. L13–L17. arXiv: astro-ph/0006213 [astro-ph].
4. D. Queloz, M. Mayor, L. Weber, A. Blécha, M. Burnet, B. Confino, D. Naef, F. Pepe, N. Santos, and S. Udry. "The CORALIE survey for southern extra-solar

vi. PlanetS webpage: http://nccr-planets.ch/
vii. Cambridge Exoplanet Centre
viii. CHEOPS webpage: http://exoplanets.phy.cam.ac.uk/

planets. I. A planet orbiting the star Gliese 86." In: *Astronomy & Astrophysics* **354** (Feb. 2000), pp. 99–102.
5. F. Bouchy, F. Pepe, and D. Queloz. "Fundamental photon noise limit to radial velocity measurements." In: *Astronomy & Astrophysics* **374** (Aug. 2001), pp. 733–739. doi: 10.1051/0004-6361:20010730.
6. D. Queloz, G. W. Henry, J. P. Sivan, S. L. Baliunas, J. L. Beuzit, R. A. Donahue, M. Mayor, D. Naef, C. Perrier, and S. Udry. "No planet for HD 166435." In: *Astronomy & Astrophysics* **379** (Nov. 2001), pp. 279–287. doi: 10.1051/0004-6361: 20011308. arXiv: astro-ph/0109491 [astro-ph].
7. D. Queloz, M. Mayor, S. Udry, M. Burnet, F. Carrier, A. Eggenberger, D. Naef, N. Santos, F. Pepe, G. Rupprecht, G. Avila, F. Baeza, W. Benz, J. L. Bertaux, F. Bouchy, C. Cavadore, B. Delabre, W. Eckert, J. Fischer, M. Fleury, A. Gilliotte, D. Goyak, J. C. Guzman, D. Kohler, D. Lacroix, J. L. Lizon, D. Megevand, J. -P. Sivan, D. Sosnowska, and U. Weilenmann. "From CORALIE to HARPS. The way towards 1 m s–1 precision Doppler measurements." In: *The Messenger* **105** (Sept. 2001), pp. 1–7.
8. M. Mayor, F. Pepe, D. Queloz, F. Bouchy, G. Rupprecht, G. Lo Curto, G. Avila, W. Benz, J. -L. Bertaux, X. Bonfils, Th. Dall, H. Dekker, B. Delabre, W. Eckert, M. Fleury, A. Gilliotte, D. Gojak, J. C. Guzman, D. Kohler, J. -L. Lizon, A. Longinotti, C. Lovis, D. Megevand, L. Pasquini, J. Reyes, J. -P. Sivan, D. Sosnowska, R. Soto, S. Udry, A. van Kesteren, L. Weber, and U. Weilenmann. "Setting New Standards with HARPS." In: *The Messenger* **114** (Dec. 2003), pp. 20–24.
9. F. Pont, F. Bouchy, D. Queloz, N. C. Santos, C. Melo, M. Mayor, and S. Udry. "The "missing link:" A 4-day period transiting exoplanet around OGLE-TR-111." In: *Astronomy & Astrophysics* **426** (Oct. 2004), pp. L15–L18. doi: 10.1051/ 0004-6361:200400066. arXiv: astro-ph/0408499 [astro-ph].
10. N. C. Santos, F. Bouchy, M. Mayor, F. Pepe, D. Queloz, S. Udry, C. Lovis, M. Bazot, W. Benz, J. -L. Bertaux, G. Lo Curto, X. Delfosse, C. Mordasini, D. Naef, J. -P. Sivan, and S. Vauclair. "The HARPS survey for southern extra-solar planets. II. A 14 Earth-masses exoplanet around μ Arae." In: *Astronomy & Astrophysics* **426** (Oct. 2004), pp. L19–L23. doi: 10.1051/0004-6361: 200400076. arXiv: astro-ph/0408471 [astro-ph].
11. Christophe Lovis, Michel Mayor, Francesco Pepe, Yann Alibert, Willy Benz, Francois Bouchy, Alexandre C. M. Correia, Jacques Laskar, Christoph Mordasini, Didier Queloz, Nuno C. Santos, Stéphane Udry, Jean-Loup Bertaux, and Jean-Pierre Sivan. "An extrasolar planetary system with three Neptune-mass planets." In: *Nature* **441**, 7091 (May 2006), pp. 305– 309. doi: 10.1038/ nature04828. arXiv: astro-ph/0703024 [astro-ph].
12. Frédéric Pont, Shay Zucker, and Didier Queloz. "The effect of red noise on planetary transit detection." In: *Monthly Notices of the Royal Astronomical Society* **373**, 1 (Nov. 2006), pp. 231– 242. doi: 10.1111/j.1365-2966.2006.11012.x. arXiv: astro-ph/0608597 [astro-ph].
13. M. Gillon, F. Pont, B. -O. Demory, F. Mallmann, M. Mayor, T. Mazeh, D. Queloz, A. Shporer, S. Udry, and C. Vuissoz. "Detection of transits of the nearby hot Neptune GJ 436 b." In: *Astronomy & Astrophysics* **472**, 2 (Sept. 2007), pp. L13–L16. doi: 10.1051/0004-6361:20077799. arXiv: 0705.2219 [astro-ph].
14. D. Queloz, F. Bouchy, C. Moutou, A. Hatzes, G. Hébrard, R. Alonso, M. Auvergne, A. Baglin, M. Barbieri, P. Barge, W. Benz, P. Bordé, H. J. Deeg, M. Deleuil, R. Dvorak, A. Erikson, S. Ferraz Mello, M. Fridlund, D. Gandolfi, M. Gillon, E. Guenther, T. Guillot, L. Jorda, M. Hartmann, H. Lammer, A. Léger, A. Llebaria, C. Lovis, P. Magain, M. Mayor, T. Mazeh, M. Ollivier, M. Pätzold, F.

Pepe, H. Rauer, D. Rouan, J. Schneider, D. Segransan, S. Udry, and G. Wuchterl. "The CoRoT-7 planetary system: two orbiting super-Earths." In: *Astronomy & Astrophysics* **506**, 1 (Oct. 2009), pp. 303–319. doi: 10.1051/0004-6361/200913096.

15. A. H. M. J. Triaud, A. Collier Cameron, D. Queloz, D. R. Anderson, M. Gillon, L. Hebb, C. Hellier, B. Loeillet, P. F. L. Maxted, M. Mayor, F. Pepe, D. Pollacco, D. Ségransan, B. Smalley, S. Udry, R. G. West, and P. J. Wheatley. "Spin-orbit angle measurements for six southern transiting planets. New insights into the dynamical origins of hot Jupiters." In: *Astronomy & Astrophysics* **524**, A25 (Dec. 2010), A25. doi: 10.1051/0004-6361/201014525. arXiv: 1008.2353 [astro-ph.EP].
16. Francesco Pepe, Andrew Collier Cameron, David W. Latham, Emilio Molinari, Stéphane Udry, Aldo S. Bonomo, Lars A. Buchhave, David Charbonneau, Rosario Cosentino, Courtney D. Dressing, Xavier Dumusque, Pedro Figueira, Aldo F. M. Fiorenzano, Sara Gettel, Avet Harutyunyan, Raphaëlle D. Haywood, Keith Horne, Mercedes Lopez-Morales, Christophe Lovis, Luca Malavolta, Michel Mayor, Giusi Micela, Fatemeh Motalebi, Valerio Nascimbeni, David Phillips, Giampaolo Piotto, Don Pollacco, Didier Queloz, Ken Rice, Dimitar Sasselov, Damien Ségransan, Alessandro Sozzetti, Andrew Szentgyorgyi, and Christopher A. Watson. "An Earth-sized planet with an Earth-like density." In: *Nature* **503**, 7476 (Nov. 2013), pp. 377–380. doi: 10.1038/nature12768. arXiv: 1310.7987 [astro-ph.EP].
17. Brice-Olivier Demory, Michael Gillon, Nikku Madhusudhan, and Didier Queloz. "Variability in the super-Earth 55 Cnc e." In: *Monthly Notices of the Royal Astronomical Society* **455**, 2 (Jan. 2016), pp. 2018–2027. doi: 10.1093/mnras/stv2239. arXiv: 1505.00269 [astro-ph.EP].
18. Michaël Gillon, Emmanuël Jehin, Susan M. Lederer, Laetitia Delrez, Julien de Wit, Artem Burdanov, Valérie Van Grootel, Adam J. Burgasser, Amaury H. M. J. Triaud, Cyrielle Opitom, Brice-Olivier Demory, Devendra K. Sahu, Daniella Bardalez Gagliuffi, Pierre Magain, and Didier Queloz. "Temperate Earth-sized planets transiting a nearby ultracool dwarf star." In: *Nature* **533**, 7602 (May 2016), pp. 221–224. doi: 10.1038/nature17448. arXiv: 1605.07211 [astro-ph.EP].
19. Paul B. Rimmer, Jianfeng Xu, Samantha J. Thompson, Ed Gillen, John D. Sutherland, and Didier Queloz. "The origin of RNA precursors on exoplanets." In: *Science Advances* **4**, 8 (Aug. 2018), eaar3302. doi: 10. 1126/sciadv.aar3302. arXiv: 1808.02718 [astro-ph.EP].
20. Peter J. Wheatley, Richard G. West, Michael R. Goad, James S. Jenkins, Don L. Pollacco, Didier Queloz, Heike Rauer, Stéphane Udry, Christopher A. Watson, Bruno Chazelas, Philipp Eigmüller, Gregory Lambert, Ludovic Genolet, James McCormac, Simon Walker, David J. Armstrong, Daniel Bayliss, Joao Bento, Francois Bouchy, Matthew R. Burleigh, Juan Cabrera, Sarah L. Casewell, Alexander Chaushev, Paul Chote, Szilárd Csizmadia, Anders Erikson, Francesca Faedi, Emma Foxell, Boris T. Gänsicke, Edward Gillen, Andrew Grange, Maximilian N. Günther, Simon T. Hodgkin, James Jackman, Andrés Jordán, Tom Louden, Lionel Metrailler, Maximiliano Moyano, Louise D. Nielsen, Hugh P. Osborn, Katja Poppenhaeger, Roberto Raddi, Liam Raynard, Alexis M. S. Smith, Maritza Soto, and Ruth Titz-Weider. "The Next Generation Transit Survey (NGTS)." In: *Monthly Notices of the Royal Astronomical Society* **475**, 4 (Apr. 2018), pp. 4476–4493. doi: 10.1093/mnras/stx2836. arXiv: 1710.11100 [astro-ph.EP].

51 Pegasi b, and the Exoplanet Revolution

Nobel Lecture, December 8, 2019 by
Didier Queloz
University of Cambridge, Cavendish Laboratory, UK & University of Geneva, Department of Astronomy, Switzerland.

Αλλὰ μὴν καὶ κόσμοι ἄπειροί εἰσιν, οἵ θ΄ ὅμοιοι τούτῳ καὶ ἀνόμοιοι
'the worlds also are infinite, whether they resemble this one of ours or whether they are different from it'

Epicurus 300BC[1]

I. FOREWORD

Scientific experiments leading to a paradigm shift are rare and unexpected. They are the combined result of hard work, opportunity, technology readiness, and contributions by many people. With a bit of luck, all these elements play together in harmony and converge to create an exceptional moment where knowledge makes a step forward. Eventually, only a small number of key contributors get the chance to be rewarded for results that include the contributions and ideas of many others. I feel indebted to all these people. I would particularly like to express my deep gratitude to all engineers, technicians and collaborators of the Observatoire de Haute Provence (OHP) and Geneva Observatory that contributed to the construction and operations of the ELODIE spectrograph and the 193-cm telescope of the OHP. Without their professionalism and unfailing motivation, the discovery of the first exoplanet would have been different, and my story as well.

This paper is about the story of the discovery of 51Pegasis b, an exo-planet, a planet orbiting another star than our Sun. I will describe methods and challenges faced at that time. I will elaborate on the profound impact this discovery had on our general knowledge and understanding about planet formation and why it has been a seminal moment for the emergence of a new field of research in astrophysics, as well as a formidable incentive to kick-start the exploration of life in the universe.

II. PRECISE DOPPLER SPECTROSCOPY

An orbiting planet can be inferred by the observation of reflex motion of its parent star. The orbital trajectory of the host star around the centre-of-gravity set by the star-planet system may be detected either through its astrometric orbit or periodic radial velocity changes. When by chance the geometry of the planetary orbital plane is such that the line of sight between the observer and the star is crossed by the planet a transit event occurs. Any of these "indirect" methods may be considered to detect a planet as an alternative to "direct" detection by spatially resolving a planet from its star, a formidable technical challenge still today.

In the 20th century, various exoplanet discovery claims by astrometric techniques have been made, to be later dismissed on the basis of new data[2]. For half a century, astrometry was essentially the only technique considered to detect a giant planet in an orbital configuration, similar to Jupiter. Nobody had really considered searching for planets by measuring stellar radial velocities. They had a good reason for that. A giant planet orbiting at a few astronomical units away would produce a change of radial velocity of its parent star in the order of $10 ms^{-1}$. Detecting a variation of that order of magnitude with available technology at that time was a utopian perspective.

In 1952, Struve published a surprising visionary short note[3] mentioning conducting "high -precision radial velocity work" to look for planets "much closer to their parent stars than is the case in the Solar System." This idea was way ahead of its time until a series of innovations would significantly reduce uncertainties on radial velocity measurements. Nobody considered seriously searching for planets using Doppler spectroscopy methods at the time for the next decade.

In 1967, the successful implementation of spectral matching techniques to derive stellar radial velocity[4] by Griffin, followed a few years later by a publication:[5] "On the possibility of determining stellar radial velocities to $0.01 kms^{-1}$," changed the perspective. It opened a realistic prospect to reach the required performance to eventually detect planets by precise Doppler spectroscopy.

Campbell & Walker achieved the first successful implementation of ideas earlier sketched by Griffin & Griffin, a spectroscopic line reference source superimposed on the stellar light optical path, using an absorption cell located at the spectrograph entrance and filled with hydrogen fluoride (HF) gas.[6] Despite the safety and handling challenges to operate this equipment, they conducted, during 12 years, the first survey looking for "substellar companions to solar-type stars" using precise Doppler Spectroscopy measurements[7,8]. The use of a gas cell as a self-reference to obtain precise radial velocities was later perfected by Marcy & Butler by replacing the meter-long lethally corrosive HF cell with a more compact and easy to handle cell fill with iodine (I_2).[9] The ease and flexibility offered by the use of an I_2 cell would open the possibility for almost any existing high-resolution spectrograph to produce precise radial velocity measurements and to be used for a planet search survey. The apparent simplicity of this technical solution would, however, face the arduous challenge of dealing with non-trivial data analysis inherent to the dense and blended forest of molecular line transitions of I_2.[10]

The alternative to the self-calibration method with a gas cell is to operate a stable and precise spectrograph. In 1990, in a comprehensive review, Brown considered design optimisation trade-offs needed to build such an instrument.[11] Use of échelle spectrograph design is essential to produce, with the same exposure, spectra with high resolution and large wavelength range. These two characteristics allowed us to observe enough stellar spectral lines to precisely compute radial velocity from the Doppler effect by cross-correlation with a match filter (correlation numerical mask)[12] and to reach $10\,ms^{-1}$considering realistic observation sequences with telescopes.[13]

In the 90s only a handful of instruments have been successfully developed along these guidelines, reaching their design purpose to deliver high precision radial velocities. The successful ones[14–16] are essentially built with similar concepts. Optics are mounted on a static bench located in stable environment away from all kinds of telescope and dome mechanical, thermal and acoustic perturbations. They use a multi-mode optical fiber to illuminate the spectrograph entrance (slit) with the image obtained by the telescope and another fiber to track instrument and air index variations in the spectrograph. In addition to removing the instrument away from the noisy telescope environment, optical fiber injection of the stellar image has the essential intrinsic property of scrambling the intensity distribution of the telescope image and producing a nearly uniform illuminated disk at the entrance slit almost suppressing guiding and seeing effects.[17]

ELODIE

The ELODIE spectrograph (see Fig. 1) started its scientific operation in 1994 on the 193cm telescope of the Observatoire de Haute Provence (OHP). Its construction began in 1989 as a collaboration between OHP and the astronomy department of the University of Geneva. Its main purpose was to offer a new modern observation capability particularly for "bright time" period (when the Moon is visible) while at the same time, a twin copy (CORALIE) was built in parallel to be later mounted on the 1.2m Swiss telescope at La Silla (European Southern Observatory, ESO) in Chile.[18]

The spectrograph had been designed to achieve precise Doppler spectroscopy measurements. The optical concept was constrained by the requirement to have a compact, stable instrument and to maximize the use of all available area of the E2V 1024x1024 pixels CCD detector to obtain a recorded échelle spectra with the highest possible resolution over the whole visible range, from 390nm to 681nm. This was made possible by using a large and high angle of incidence diffraction échelle grat-

Figure 1. ELODIE spectrograph on display at OHP. On the left we see the échelle grating with the grooves facing us. On the left side, the optical fibers feeding the spectrograph are clearly visible (in orange). On the top sits the cryostat with the CCD detector inside. The "cross-dispersing" optic (not visible) is located in the vertical dark painted holding structure. ©Collection Photothèque OHP/CNRS.

ing recently produced by Milton and Roy manufacture. To improve slit illumination stability, an efficient double scrambler was included in the fiber-feed train.

In addition, ELODIE was uniquely equipped with a data reduction pipeline delivering radial velocity by numerical cross-correlation shortly after observation.[14]

The development of an on-line data reduction pipeline, routinely delivering high precision radial velocities, was at that time a challenging task only made possible by the opportunity offered by the generous RAM and clock speed of the newly available *SPARCstation* minicomputer by Sun Microsystem. The ELODIE spectral information that is recorded on CCD is distributed over 67 curved and overlapping orders. This complex data structure of échelle spectra creates various software algorithmic challenges. For example, the spectroscopy resolution element was only about 10 kms^{-1}, a thousand times bigger than the Doppler precision we were aiming at. Inspired by the work of Griffin Photometric Velocimeter and CORAVEL[19] implementation, algorithms based on match filter (correlation mask) have been developed. These optimally combined together in an optimal way all of the Doppler spectroscopic information

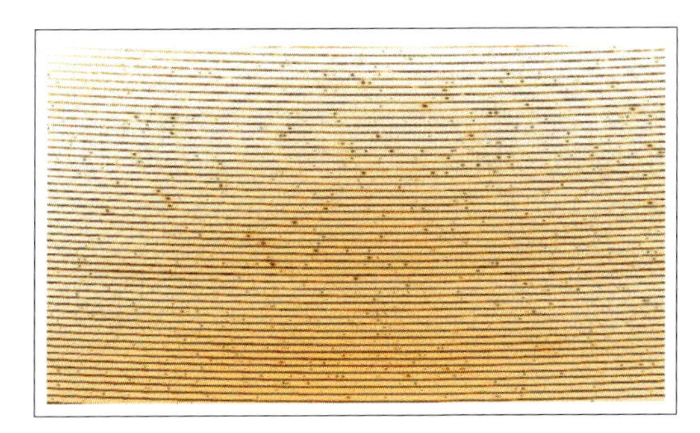

Figure 2. Middle cut of ELODIE image of a stellar spectra observed with simultaneous thorium recorded on the CCD. One clearly distinguishes the curved spectroscopic order of the stellar spectra from the interlaced emission spectrum due to simultaneous thorium lamp illuminating the second fiber.

recorded on the spectra. The use of a reference fiber, fed by a thorium lamp during the exposure, produced a reference spectrum the reduction pipeline was using to correct for mechanical variability and air index changes occurring between the time of wavelength calibrations and actual observations of stars (see Fig. 2). The implementation of "simultaneous referencing" was one of the cleverest tricks at the heart of data analysis to reach high precision in radial velocity measurements. The ELODIE spectrograph design and software development implemented with success a whole set of new concepts that have become standards in succeeding generations of stable spectrograph allowing further improvements in precision performances.[20–22]

III. A PLANET THAT SHOULD NOT EXIST

51 Pegasi

In spring 1994, with ELODIE barely operational, we started our survey. Our goal was to determine the occurrence of sub-stellar companions in the solar neighborhood. Finding giant planets were not the only objective of the survey. It is worth recalling that in the 90s, the search for brown dwarfs was a fashionable theme of research that stretched to the planet regime.[23,24] Moreover our compelling need to make a convincing and realistic case for the Telescope Allocation Committee to obtain access to telescope observations could not be neglected.

The original target sample included 142 F,G,K main sequence stars[25] selected on the basis they were not spectroscopic binaries, located in a 25 pc neighborhood and – to our knowledge – not yet observed by another high precision Doppler survey. Our strategy was to start with a sample size significantly larger than the one previously observed by Campbell *et al.* that didn't succeed in detecting sub-stellar companions.[7,26]

In autumn 1994, Michel Mayor (my PhD advisor), literally left me keys of operation and went to Hawaii on sabbatical leave for a 6 month period. I was delighted and excited to be left in charge of the program, regularly going observing with ELODIE I, which I considered a bit as "my baby", and incidentally gathering more data for my PhD which was due the year after.

In the original survey sample, we had previously identified 24 bright stars equally distributed in the sky. One would observe this subset a bit more frequently than others to serve us as precision validation. The star HD217014 known as 51 Peg was part of this group. We had an observing mission about every two months and they typically lasted one week.

In January 1995, it is fair to say that my first reaction was a moment of panic when I realized that the star HD217014 exhibited radial velocity variations larger than the sole effect of Doppler precision I expected from the spectroscopic information available. I thought something was going wrong in the spectrograph or with the data analysis. After days and nights anxiously spent alone checking any element and software step I could think about and gathering more data, I eventually came to the only conclusion I could think about to explain the variability pattern: A planet of Jupiter's mass is orbiting the star 51 Peg with a 4.25d period corresponding to an orbital distance of 0.05 astronomical units. The planet is literally roasted, and its atmosphere is 1,000K degrees hot. When retrospectively I think about it, I realize how fearless and foolish this idea was, the privilege of an enthusiastic PhD student ...

When later I reported to Michel Mayor that I had found a planet, unsurprisingly he reacted with restrained enthusiasm. I think he couldn't believe it. That was fair enough. When we started the survey, I still remember him telling me I should not expect to find any planets for my PhD, it would take years! He eventually changed his mind when additional radial velocity measurements collected in July 1995 confirmed my initial ephemerides based on previous observations.

We spent summer 1995 writing the paper to report our discovery. We had a fantastic challenge to overcome to convince our peers, considering our planet had no counterpart in the Solar System and no theoretical back-up to explain a hot Jupiter configuration. Moreover, ELODIE was a brand-new challenger without yet any demonstrated results and the field was historically littered with series of misjudgments and mistakes in data analysis. Finally, small changes in Doppler shift may potentially be due to stellar photosphere effects and explain our data as well. It was an impossible job! In the following years we would be confronted with a wave of skepticism. It would take years for the community at large to accept the reality of 51 Peg hot Jupiter and to modify the paradigm about the universality of solar system planetary architecture.

Alternative to planet hypothesis

The strongest resistance we faced about our interpretation was related to the fact that the measurement of radial velocity variation from stellar emerging spectra does not always imply the star is moving due to an orbiting planet. Convective transport of heat in Sun-like stars is carried out by about a million gas cells in motion with typical vertical velocities of kilometers per second.

The resulting visible effect at stellar surface is described as "photosphere granulation". A magnetic field is generated from the sheer motion of the convection mechanism through the alpha dynamo process, producing active regions on the photosphere that may display dark spots at the location of emerging strong magnetic field lines. Magnetic flux tubes form and decay on timescales typically comparable to the stellar rotation period and long-term magnetic cycles modify the convection patterns. The combined result of all these effects is to produce spectral lines of variable shape with underlying periodic and pseudo-periodic patterns. Practically, when measuring radial velocity, it is rather easy to observe variations produced by a combination of all these effects, in particular when the star is young and active.[27]

In the discovery paper[28] we carefully addressed all possible ways to produce the observed changes in radial velocity by stellar atmospheric effects. We looked for records of photometric amplitude changes indicating a young and fast rotating star. We used the property of the correlation function to look for stellar line profile changes. We clearly ruled out all alternative origins by stellar atmosphere features but the idea of "hot Jupiter" planets was so awkward it had hard time being accepted. The main issue was that it didn't fit in the planetary formation paradigm without seriously tweaking this paradigm. Changing a well-established theory is rarely the first idea a physicist is considering out of an unusual experimental result. And yet the foundation of planet formation theory needed to be revised.

Challenging planetary formation

The process of forming a planet is based on core accretion mechanisms in the disk. The underlying principle is a series of steps where a planet grows by stages by accreting material available in the disk. In early stages, proto-planetary disks are dominated by H and He gas. The disk also contains a small fraction of solids. Close to the star one finds refractory dust. In the outer part of the disk, "beyond snow lines,"[29] there are frozen ices originating from the solid phase transition of molecular gas (H_2O, CO, CO_2, CH_4...).

In the disk, solid materials rapidly, dynamically decouple from the gas and settle down on the disk mid-plane where they agglomerate by sticking together. The result is a swarm of planetesimals that grow by collision amongst themselves to eventually form planetary embryos.[30] The formation of giant planets proceeds from these embryos by accreting the gas left in the disk.[31] The outcome depends on two competing processes: on one hand the dispersion of the gas disk, on the other hand the formation of a massive core big enough to efficiently accrete all the gas left around.

The fact that 90% of Jupiter's mass is made of H and He means that the core formed quick enough to accrete a significant amount of gas before it got dispersed. Such favorable timing requires a high solid surface density of planetesimals available when the gas is still around. It is only encountered in the outer part of the disk at a few astronomical units.[32] Therefore, for the formation theory to account for the presence of close-in giant planets one must consider strong and efficient dynamical interaction with the disk (migration) and other massive bodies in the system to change the initial orbital configuration.[33] This element was never seriously considered or looked at by researchers working on planetary formation models despite being explicitly mentioned and computed fifteen years before.[34] The 80s Goldreich & Tremaine paper prediction resurfaced at the time the first migration model was published,[35] shortly after 51 Peg b was announced.

IV. A FEAST OF EXOPLANETS

I concluded my PhD defense with a prophetic statement that the discovery of 51 Peg b exoplanet was just the tip of the iceberg and more planets of that kind would soon be detected. I simply couldn't believe we had, by some extraordinary luck, detected an extremely rare planetary configuration. I didn't have to wait long to be proven right.

Here comes the transit

A few months after the publication of 51 Peg b, two exoplanets detected by the radial velocity technique were announced[36,37]. Three years later, eight exoplanets had been found, all with mass in the range of giant planets and three hot Jupiter planets[38]. Then in late 1999 a new hot Jupiter was found orbiting the star HD209458 and luckily it happened to be transiting. This result, concluding on a similar interpretation from two independent techniques, had the final word and swept any reservations left on the reality of exoplanets discoveries.[39,40]

When the community realized that hot Jupiters truly existed, we saw blossoming dedicated exoplanet transit surveys.[41] A hot Jupiter exoplanet is 10 times smaller than our Sun and has a 10% chance of being seen with an orbital configuration aligned with its host star, making it a good target to look for transit by ground-based differential photometry. The transit method provides us with an alternative to Doppler spectroscopy searches for planets. It allows us to derive the size of a planet instead of its mass.

In 2006 the first exoplanet transit survey from space – the COROT satellite – was launched and rapidly brought us evidence of the first rocky exoplanet COROT-7b.[43,44] The Kepler mission, launched three years later,

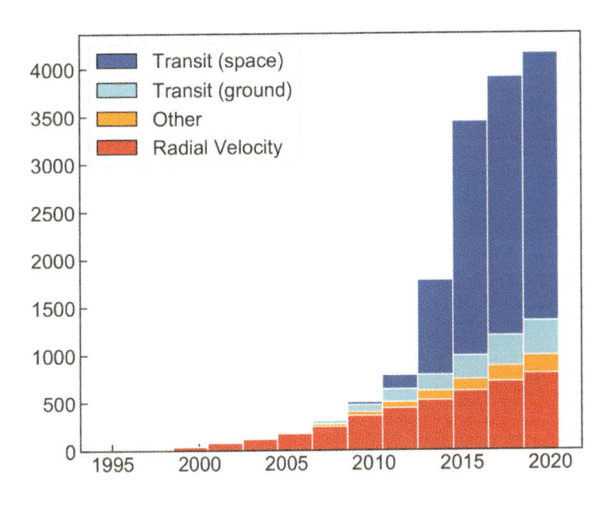

Figure 3. Cumulative histogram of exoplanet discoveries[42] through time by various detection techniques. The spectacular growth of transit detection from space is due to the Kepler mission.

eventually produced a stream of discoveries of small multi-planetary systems.[45] In barely a decade, planet hunting activity went from repeated failures to an exoplanet gold rush, involving big survey and space missions carried out by large international consortiums. As a result of this rapid expansion of survey capabilities, the number of exoplanet detections spectacularly increased (see Fig. 3), lifting the veil on the extraordinary diverse exoplanet realm.

Change of perspectives

The discovery of the exoplanet 51 Peg b kick-started a new field of research of contemporary astrophysics. It acted as a stimulus to develop new instruments and observing facilities. A quarter century later, combined results from precise Doppler spectroscopy surveys, transit search space missions and wide field transit ground-based surveys have completely modified our perspective on the architecture and nature of planetary systems in the universe. We have learned that our solar system architecture is far from being the norm. The wealth of diversity observed in exoplanet structures and orbital configurations (Fig. 4) is oddly contrasting with our Solar System.

Transit and precise Doppler spectroscopic methods favor detection of exoplanets with short orbital period. The significant number of planets orbiting close to their star, so embarrassing for planetary formation theory, ironically turns out to be a fortunate situation from a detection point of view. It is fascinating to think that if the Solar System would be the norm, Fig. 4 would display few measurement points.

The interest and spectacular growth of the detection community would not be the same as we see today.

Exoplanets with characteristics comparable to our solar system's planets are far more challenging to detect than most of the planets so far discovered. It explains the lack of Earth twin ("Goldilocks" planet) in current findings. By comparison to telluric planets a "Jupiter-twin" exoplanet is easier and within reach of Doppler surveys. It still needs long-term series of measurements and extensive telescope time access. In the near future, with the release of the GAIA mission final catalogue, Fig. 4 is likely to display more data points in the mass-period region similar to Jupiter.[46,47]

Measurement by transit method of the planet radius is efficient when the orbital period is short. For long period exoplanets, the probability of getting the right geometrical alignment of line of sight is so negligibly low that it becomes impractical. This limitation is clearly visible on Fig. 4 by the increased scarcity of radius measurements gathered for exoplanets with orbital period typically longer than about 100 days.

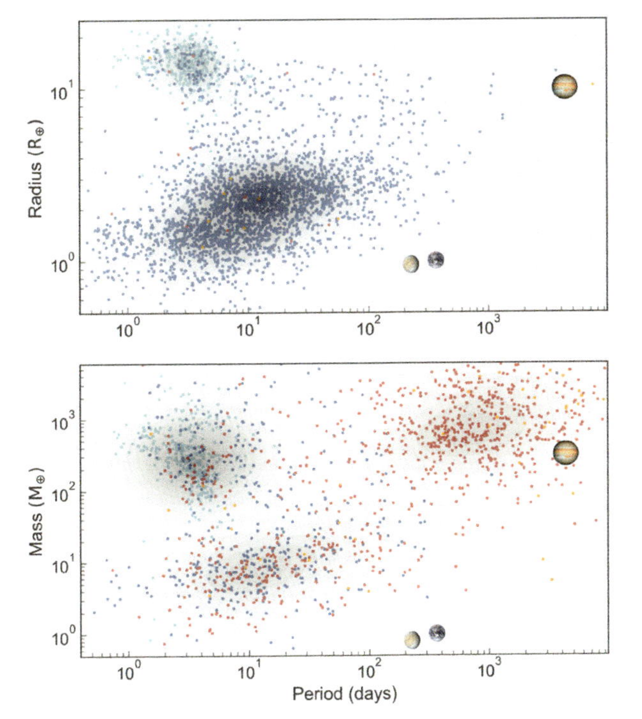

Figure 4. Measured mass, radius and orbital period of all known exoplanets[42]. Color codes indicate techniques used to discover the planet (same as Fig. 3). For mass measured by Doppler spectroscopy *sin* i = 1 is considered. Location of Jupiter, Earth and Venus are indicated for the sake of comparison. A grey scale density map is overlaid to locate "cluster of similar exoplanets" on these diagrams.

On Fig. 4 three distinct groups of exoplanets are visible. The hot Jupiter population is the group of giant planets found on short period (less than 10 days), with 51 Peg b its most emblematic member. On the colder end, further out, one finds "classical" giant planets like our own Jupiter. Then one sees a cluster of smaller exoplanets mostly on short orbit, casually named "Super Earth" or "Mini-Neptune" compact systems. This group of planets is a mixed bag of anything fitting in a range defined on one side by Earth's physical characteristics and on the other side by Neptune.

Detailed statistical analysis of the occurrence of each group of exoplanets is not a trivial task. The apparent number of discoveries can't be simply converted to the occurrence of each type of exoplanet per star. Limitations in the techniques used to detect them and diversity of thresholds of each survey considered need to be carefully taken into account to produce a robust result.[48] Hot Jupiter planets, easy to detect by both techniques, are actually not that frequently found orbiting stars. An average occurrence rate of 1% is derived, with a tendency to be more frequently present when the host star's metallicity is higher.[49] The occurrence of cold Jupiters is about 10% for Jupiter analogs.[48] If a broader definition is considered, including any exoplanet more massive than Neptune and up to $20M_J$ planets, the occurrence rises almost to 50%. Note that this large group of exoplanets clearly distinguishes themselves from outer planets of our Solar System with a wider range of orbital eccentricities. The planetary configuration corresponding to the group of "superearth & mini-neptune" exoplanets seem to be the most commonly found configuration in our galaxy. One derives, on average, an occurrence of about 60% per star with orbital periods less than 100 days.[50] The discovery of such a massive population of planets during short periods is a challenge to planetary formation theory. It is understood as a failure to properly account for dynamic effects occurring during planet formation.[51] It raises the perplexing possibility as well that our Solar System's configuration may be far less common than expected.

Exoplanetary science begins

The exoplanet discovery bonanza not only unveiled the diversity of planetary orbital configurations but also a large range of physical structures. The combination of transit and precise Doppler spectroscopy allows us to measure exoplanet bulk density and to gain insights into the structure of planet interiors. The mass and radius measurement diagram in Fig. 5 displays all exoplanets for which these two physical parameters have been measured as well as a set of superimposed computed bulk density relations for different planet interiors.[52]

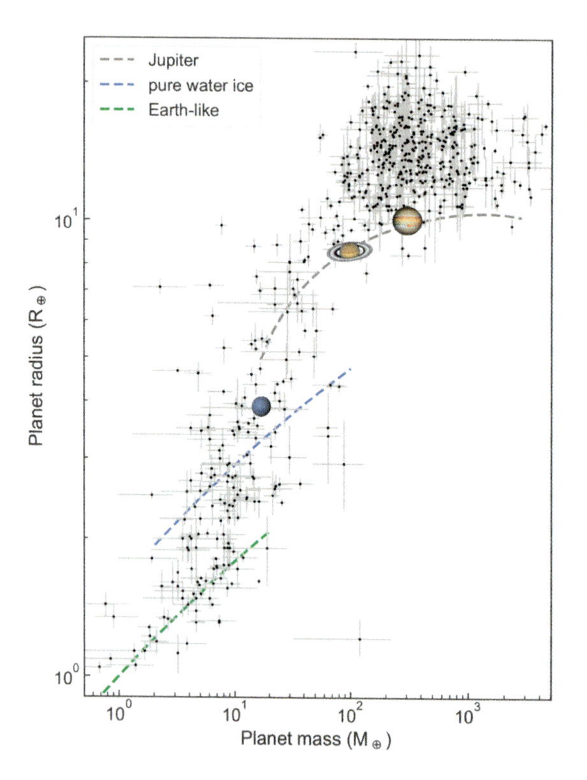

Figure 5. All known exoplanets[42] with a measurement of their mass and radius. Hatched lines indicate model of bulk density for three different compositions. Jupiter, Saturn and Neptune are indicated for the sake of comparison.

The computed bulk density for hydrogen-helium composition dominated planets, applicable to Jupiter, lies on the upper value boundary of observed giant exoplanets densities. The fact that most giant planet measurements displayed on this diagram indicate lower densities than Jupiter's is the consequence of a bias that favors short period exoplanet detections, and the fact that Hot Jupiter planet diameters are observed to be inflated.[53] Some exoplanets have been found with barely 10% Jupiter bulk density.[54] Physical mechanisms at the origins of their bloated nature may be related to the combination of different effects due to their proximity to their host star and their formation process.[22]

In the case of exoplanets with mass smaller than Saturn, for any given range, computed bulk density shows a large dispersion, suggesting a mixture of planet interior structures. Some exoplanets have a bulk density that could be understood as a down-scaling extrapolation of Jupiter's interior. Others with denser values can be modelled by decreasing the value of H and He to 10% and increasing "heavy" elements (such as H_2O, NH_3, CO_2, ...) in planet interiors like in the planet interiors of Uranus and Neptune for example. Going further down in the sub-Saturn mass range one finds exoplanets having bulk density too high to be simply accounted by downscaling Jupiter or Neptune planetary interiors. New structure without H and He should be considered.

The core accretion planet formation scenario produces a composite interior structure with schematically three distinct layers: core (densest component), envelope and atmosphere (visible part). The level of freedom that one can play with by balancing these three components produces naturally a confusing range of bulk density values. Practically, a given bulk density can correspond to different ratios between these components. To simplify, the interpretation on Fig. 5 is displayed a "pure ice" hypothetical planet model. It is revealing to compare it with Earth-like bulk structure extrapolated in the same mass range. The group of exoplanets in the super-earth & mini-neptune range (Fig. 5) exhibits a large dispersion suggesting an underlying diversity of planetary models. For example, some planets with $5M_{\oplus}$ have been found compatible with Earth-like bulk density, while others with Neptune-like structures. This situation is reflected by the fact that for that group of planets, we do not observe a direct relation between mass and radius. A careful inspection of this diagram would demonstrate the statistical significance of two groups of bulk density structures: One more "water-like" and the other more "Earth-like"[50]. This suggests the super-earth & mini-neptune exoplanet category is potentially a group of mixed origins with different interior compositions.[55]

V. PROSPECTS

The fascinating diversity of bulk density encountered among compact super-earth and minineptune exoplanets, and the fact they have no equivalent to be readily compared with solar system planets, is a challenge to modeling their interior as well as tracing their origins. Fortunately, it is likely to change with the launch of the James Webb Space Telescope (JWST) and the availability of large ground-based facilities currently under construction (like for example extremely large telescopes, ELTs). Using transit spectroscopy observations and occultation combined-light techniques, it will be possible to learn far more about these exoplanets.[56] Insights about atmospheric and surface composition[57,58] will offer an exciting opportunity to clarify their nature and their origins.

The imminent prospect of measuring atmospheric features of small transiting exoplanets opens the fascinating possibility of addressing the remote detection of life in these systems. The habitable zone that expresses a range of distances from its host star to maintain liquid water[59] is largely considered as a minimum condition for an exoplanet orbit to be of potential interest for the purpose of searching for biomarkers.[60]

Practically, the concept of habitable zones is a guideline for planning future observations. The habitable zone assumes an ad-hoc atmosphere and planetary surface conditions[61] and scaled the illumination S_{eff} received by the planet to maintain liquid water (assuming water is present ...). For small and cooler stars (Mdwarfs) the inner boundary of the habitable zone gets close enough to overlap the range of short period small exoplanets discovered. Among them, the recent confirmation of rocky planets with a bulk density similar to the Earth,[62,63] located in the habitable zone, reasonably questions the possibility of life on these systems. The prospect of eventually getting insights into the atmosphere and geochemical conditions in these systems is drawing attention beyond the usual astronomy community.

Answering the big question about life on exoplanets will require a combined effort between astrophysics, planetary scientists, geophysicists, biochemists and molecular biologists. Recent developments on the origin of life on Earth[64] as a planetary phenomenon and its relevance to the search for life on another planet is steering us on a new exciting research route. Current efforts to identify true Earth-twin planetary systems on nearby stars[65] will eventually lead to the development of a series of research programs and future facilities to look for bio-signatures and to address the origin, nature, and prevalence of life in the universe. Near us, Mars, Venus and satellites of giant planets in our Solar System are obvious locations to look closer for life signatures.

The discovery of the exoplanet realm is an extraordinary moment in mankind's pursuit of knowledge and natural inclination to be curious. It follows the steps of the Copernican revolution, extending it further out by placing our Solar System among countless planetary systems and by addressing the physical conditions conducive to the emergence of life. The large diversity and high occurrence of exoplanets orbiting stars in our galaxy offers so many opportunities for the chemistry of life to happen, eventually if we shall detect it on another planet. It is just a matter of time ...

Acknowledgements

I thank the Nobel Foundation and the Royal Swedish Academy for the great honor and privilege of receiving and sharing this award with Prof. Peebles and my mentor and colleague Prof. Michel Mayor. I am grateful to the University of Geneva and University of Cambridge and all my collaborators for their trust and support in my endeavor to develop a comprehensive research program on exoplanets and the search for life in the universe. I am delighted and feel fortunate to enjoy and share this exceptional moment with my family and particularly Tina my wonderful wife. Their unfailing support and love is a priceless gift. Thank you!

REFERENCES

1. Epicurus. *The Letter of Epicurus to Herodotus in Diogenes Laertius: The Letter of Epicurus to Herodotus, Book 10, Sections 45,* translation is by C.D. Yonge (1895).
2. *Boss, A. Looking for Earths: The race to find new solar systems* (1998).
3. Struve, O. Proposal for a project of high-precision stellar radial velocity work. *The Observatory* **72**, 199–200 (1952).
4. Griffin, R. F. A Photoelectric Radial Velocity Spectrometer. *Astrophysical Journal* (1967).
5. Griffin, R. F. & Griffin, R. E. M. On the possibility of determining stellar radial velocities to 0.01 km s^{-1}. *Monthly Notices of the Royal Astronomical Society* **162**, 243–253 (1973).
6. Campbell, B. & Walker, G. A. H. Precision radial velocities with an absorption cell. *Astronomical Society of the Pacific* (1979).
7. Campbell, B., Walker, G. A. H. & Yang, S. A search for substellar companions to solar-type stars. *Astrophysical Journal* (1988).
8. Walker, G. A. H. *et al.* A search for Jupiter-mass companions to nearby stars. *Icarus* (1995).
9. Marcy, G. W. & Butler, R. P. Precision radial velocities with an iodine absorption cell. *Astronomical Society of the Pacific* (1992).

10. Butler, R. P. *et al.* Attaining Doppler Precision of 3 M s-1. *Publications of the Astronomical Society of the Pacific* (1996).
11. Brown, T. M. *High precision Doppler measurements via echelle spectroscopy* in *IN: CCDs in astronomy; Proceedings of the Conference* (1990).
12. Queloz, D. Echelle Spectroscopy with a CCD at Low Signal-To-Noise Ratio. *IAU Symposium No. 167* (1995).
13. Bouchy, F., Pepe, F. & Queloz, D. Fundamental photon noise limit to radial velocity measurements. *Astronomy & Astrophysics* (2001).
14. Baranne, A. *et al.* ELODIE: A spectrograph for accurate radial velocity measurements. *Astronomy and Astrophysics Supplement* (1996).
15. Kaufer, A., Wolf, B., Andersen, J. & Pasquini, L. FEROS, the fiber-fed extended range optical spectrograph for the ESO 1.52-m telescope. *The Messenger* (1997).
16. Brown, T. M., Noyes, R. W., Nisenson, P., Korzennik, S. G. & Horner, S. The AFOE: A spectrograph for precise Doppler studies. *Astronomical Society of the Pacific* (1994).
17. Heacox, W. D. *Wavelength-precise slit spectroscopy with optical fiber image scramblers in: Fiber optics in astronomy; Proceedings of the Conference* (1988).
18. Queloz, D. *et al.* The CORALIE survey for southern extra-solar planets. I. A planet orbiting the star Gliese 86. *Astronomy & Astrophysics* (2000).
19. Baranne, A., Mayor, M. & Poncet, J. L. CORAVEL – A new tool for radial velocity measurements. *Vistas in Astronomy* (1979).
20. Queloz, D. *et al.* From CORALIE to HARPS. The way towards 1 m s-1 precision Doppler measurements. *The Messenger* (2001).
21. Plavchan, P. *et al.* Radial Velocity Prospects Current and Future: A White Paper Report prepared by the Study Analysis Group 8 for the Exoplanet Program Analysis Group (ExoPAG). *arXiv.org*. arXiv: 1503.01770 (2015).
22. Pepe, F., Bouchy, F., Mayor, M. & Udry, S. High-Precision Spectrographs for Exoplanet Research: CORAVEL, ELODIE, CORALIE, SOPHIE and HARPS. *Handbook of Exoplanets* (2018).
23. Marcy, G. W. & Butler, R. P. A Search for Brown Dwarfs using Doppler Shifts. *Cool Stars; Stellar Systems; and the Sun; Eighth Cambridge Workshop. Astronomical Society of the Pacific Conference Series* (1994).
24. Latham, D. W., Mazeh, T., Stefanik, R. P., Mayor, M. & Burki, G. The unseen companion of HD114762 - A probable brown dwarf. *Nature (ISSN 0028-0836)* (1989).
25. Queloz, D. et al. The Observatoire de Haute-Provence Search for Extrasolar Planets with ELODIE. *Brown dwarfs and extrasolar planets* (1998).
26. Campbell, B., Yang, S., Irwin, A. W. & Walker, G. A. H. *Towards an Estimate of the Fraction of Stars with Planets from Velocities of High Precision* in *Bioastronomy The Search for Extraterrestial Life — The Exploration Broadens: Proceedings of the Third International Symposium on Bioastronomy Held at Val Cenis* (1991).
27. Queloz, D. *et al.* No planet for HD 166435. *Astronomy & Astrophysics* (2001).
28. Mayor, M. & Queloz, D. A Jupiter-mass companion to a solar-type star. *Nature* (1995).
29. Williams, J. P. & Cieza, L. A. Protoplanetary Disks and Their Evolution. *Annual Review of Astronomy and Astrophysics* (2011).
30. Safronov, V. S. & Zvjagina, E. V. Relative Sizes of the Largest Bodies during the Accumulation of Planets. *Icarus* (1969).
31. Pollack, J. B. et al. Formation of the Giant Planets by Concurrent Accretion of Solids and Gas. *Icarus* (1996).

32. Lecar, M., Podolak, M., Sasselov, D. & Chiang, E. On the Location of the Snow Line in a Protoplanetary Disk. *The Astrophysical Journal* (2006).
33. Dawson, R. I. & Johnson, J. A. Origins of Hot Jupiters. *arXiv.org*. arXiv: 1801.06117 (2018).
34. Goldreich, P. & Tremaine, S. Disk-satellite interactions. *Astrophysical Journal (1980).*
35. Lin, D. N. C., Bodenheimer, P. & Richardson, D. C. Orbital migration of the planetary companion of 51 Pegasi to its present location. *Nature* (1996).
36. Marcy, G. W. & Butler, R. P. A Planetary Companion to 70 Virginis. *Astrophysical Journal Letters v.464* (1996).
37. Butler, R. P. & Marcy, G. W. A Planet Orbiting 47 Ursae Majoris. *Astrophysical Journal Letters v.464* (1996).
38. Marcy, G. W. & Butler, R. P. Detection of Extrasolar Giant Planets. *Annual Review of Astronomy and Astrophysics* (1998).
39. Charbonneau, D., Brown, T. M., Latham, D. W. & Mayor, M. Detection of Planetary Transits Across a Sun-like Star. *The Astrophysical Journal* (2000).
40. Mazeh, T. *et al.* The Spectroscopic Orbit of the Planetary Companion Transiting HD 209458. *The Astrophysical Journal* (2000).
41. Pollacco, D. L. et al. The WASP Project and the SuperWASP Cameras. *The Publications of the Astronomical Society of the Pacific* (2006).
42. Akeson, R. L. et al. The NASA Exoplanet Archive: Data and Tools for Exoplanet Research. *Publications of the Astronomical Society of the Pacific* (2013).
43. Leger, A., Rouan, D., Schneider, J., Barge, P. & Fridlund. Transiting exoplanets from the CoRoT space mission. VIII. CoRoT7b: the first super-Earth with measured radius. *Astronomy & Astrophysics* (2009).
44. Queloz, D. et al. The CoRoT-7 planetary system: two orbiting super-Earths. *Astronomy & Astrophysics* (2009).
45. Lissauer, J. J., Dawson, R. I. & Tremaine, S. Advances in exoplanet science from Kepler. arXiv.org. arXiv: 1409.1595v1 (2014).
46. Lattanzi, M. G. & Sozzetti, A. Gaia and the Astrometry of Giant Planets. *arXiv.org*. arXiv: 1003.3921v1 (2010).
47. Perryman, M., Hartman, J., Bakos, G. & Lindegren, L. Astrometric exoplanet detection with Gaia. *arXiv.org*. arXiv: 1411.1173v1 (2014).
48. Winn, J. N. Planet Occurrence: Doppler and Transit Surveys. *arXiv.org*. arXiv: 1801.08543v5 (2018).
49. Santos, N. C., Israelian, G. & Mayor, M. The metal-rich nature of stars with planets. *Astronomy & Astrophysics* (2001).
50. Fulton, B. J. et al. The California-Kepler Survey. III. A Gap in the Radius Distribution of Small Planets. *The Astronomical Journal* (2017).
51. Winn, J. N. & Fabrycky, D. C. The Occurrence and Architecture of Exoplanetary Systems. *arXiv.org*. arXiv: 1410.4199v4 (2014).
52. Fortney, J. J., Marley, M. S. & Barnes, J. W. Planetary Radii across Five Orders of Magnitude in Mass and Stellar Insolation: Application to Transits. *The Astrophysical Journal* (2007).
53. Guillot, T. & Gautier, D. Giant Planets. *arXiv.org*. arXiv: 0912.2019 (2009).
54. Anderson, D. R. et al. WASP-17b: An Ultra-Low Density Planet in a Probable Retrograde Orbit. *The Astrophysical Journal* (2010).
55. Neil, A. R. & Rogers, L. A. A Joint MassRadius-Period Distribution of Exoplanets. *arXiv.org*. arXiv: 1911.03582v1 (2019).
56. Winn, J. N. Exoplanet Transits and Occultations. *Exoplanets* (2010).

57. Demory, B.-O. *et* al. A map of the large day-night temperature gradient of a superEarth exoplanet. *Nature* (2016).
58. Kreidberg, L. *et al.* Absence of a thick atmosphere on the terrestrial exoplanet LHS 3844b. *arXiv.org*. arXiv: 1908.06834v1 (2019).
59. Hart, M. H. The evolution of the atmosphere of the earth. *Icarus* (1978).
60. Kaltenegger, L. How to Characterize Habitable Worlds and Signs of Life. *arXiv.org*. arXiv: 1911.05597v1 (2019).
61. Kopparapu, R. K. et al. Habitable Zones Around Main-Sequence Stars: New Estimates. *arXiv.org*. arXiv: 1301.6674v2(2013).
62. Gillon, M. *et al.* Seven temperate terrestrial planets around the nearby ultra-cool dwarf star TRAPPIST-1. *Nature* (2017).
63. Grimm, S. L. *et al.* The nature of the TRAPPIST-1 exoplanets. *arXiv.org*. arXiv: 1802.01377v1 (2018).
64. Sasselov, D. D., Grotzinger, J. P. & Sutherland, J. D. The origin of life as a planetary phenomenon. *Science Advances* **6**, eaax3419 (2020).
65. Hall, R. D., Thompson, S. J., Handley, W. & Queloz, D. On the Feasibility of Intense Radial Velocity Surveys for Earth-Twin Discoveries. *Monthly Notices of the Royal Astronomical Society* (2018).

Physics 2020

one half to

Roger Penrose

"for the discovery that black hole formation is a robust prediction of the general theory of relativity"

and the other half jointly to

Reinhard Genzel and Andrea Ghez

"for the discovery of a supermassive compact object at the centre of our galaxy"

The Nobel Prize in Physics, 2020

Presentation speech by Professor Ulf Danielsson, of the Royal Swedish Academy of Sciences.

Your Majesties, Your Royal Highnesses, Esteemed Nobel Laureates, Ladies and Gentlemen,

This year's Laureates have made groundbreaking discoveries concerning the most mythical and strange objects in physics: black holes.

Black holes are bodies with gravity so strong that not even light can escape. To create a black hole, the earth would have to be squeezed down to the size of a pea and the sun compressed into a sphere comparable in size to the central parts of Stockholm.

The first speculation that such a thing could exist in our universe dates back to the end of the 18th century. But it was not until Einstein presented his general theory of relativity in 1915 that one had a theory powerful enough to describe them.

According to Einstein's theory of gravity, time and gravity are intimately connected. Time goes by a trillionth of a second per hour slower at my feet than at my head. Although the difference is small, it gives rise to what we call gravity. It is strange that only by lifting a glass of water with my own hand can I feel this small difference in the passage of time.

At the horizon of a black hole, on the other hand, gravity is so strong that time seems to stand still. And within the horizon, time curves so that it points inward toward the center of the black hole. Anyone who is unlucky enough to fall into a black hole therefore has as much difficulty getting out again as traveling back in time. And in the middle of the black hole, in the unfortunate traveler's near future, lurks the singularity where the known laws of nature cease to apply. Black holes are truly the most extreme objects in physics.

But the mathematics of the general theory of relativity is difficult, and for a long time black holes were mere speculation. And many physicists, for good reasons, doubted their existence. Not even Einstein thought they could exist. It was only in 1965, half a century after Einstein formulated his theory, that Roger Penrose was able to master the mathematics and show that black holes are an inevitable consequence of general relativity. If matter accumulates densely enough, nothing can prevent a collapse into a black hole.

But, where are they? For a long time, it was suspected that black holes could hide in the center of many galaxies, where they could explain eruptions of extreme energy. Could there perhaps also be a black hole in the center of our own galaxy, the Milky Way?

Reinhard Genzel and Andrea Ghez used powerful telescopes to study the mysterious interior of the Milky Way. The heart of our galaxy is hidden inside a cloud of dust and it is only by using infrared light that one can discern what is going on.

With the help of the instruments they developed, they were able to follow the motion of individual stars and discover how these stars revolved around something they could not see. Calculations showed that the invisible object must have a mass about four million times that of the sun.

Within the framework of current theories, there is no other explanation than a black hole. A more than two hundred year old riddle has been solved. But black holes also mark the limit of our knowledge of the physical world. As we look into the dark abysses of black holes, there may be new secrets to reveal.

Professors Penrose, Genzel and Ghez:

You have been awarded the 2020 Nobel Prize in Physics for your outstanding contributions to our understanding of black holes. It is an honor and a privilege to convey to you, on behalf of the Royal Swedish Academy of Sciences, our warmest congratulations.

Roger Penrose. © Nobel Prize Outreach AB. Photo: Fergus Kennedy

Roger Penrose

Biography

ROGER PENROSE was born on August 8th, 1931 in Colchester, Essex, England. His father Lionel Sharples Penrose FRS was a distinguished medically trained scientist, who worked primarily on human genetics, studying mainly the genetic origins of numerous mental conditions, particularly the issue of Downs Syndrome. He had earlier, in around 1920, studied psychoanalysis, travelling to Vienna to attend lectures by Sigmund Freud, then returning to England to obtain a medical degree in Cambridge. Lionel had considerable talents also as an artist and as a chess problemist. He enjoyed playing the piano and a spinet (small harpsichord). He frequently occupied himself making things out of wood such as puzzles of various kinds and, in later life, gadgets that would reproduce themselves if shaken together appropriately. Lionel's father, James Doyle Penrose was a Quaker and very distinguished painter of portraits and religious topics, and Lionel's mother, Elizabeth Josephine Peckover, came from a very well-to-do religious Quaker pacifist banking family. Lionel eventually donated his entire inheritance to various good causes.

Lionel was one of four brothers, one of whom, Sir Roland Algernon Penrose CBE, became one of Britain's leading surrealists, and was a good friend of Pablo Picasso and many others of the surrealist community. Roland's second marriage was to an American, Lee Miller, a well-known 2nd World War photographer. Both Lionel and Roland served in the 1st World War in the Friends' Ambulance Unit.

Roger's mother, Margaret (Leathes) had studied medicine at Newnham College Cambridge. Her father, John Beresford Leathes FRS, was a professor of physiology at the University of Sheffield. Margaret's mother, Sonia Marie Natanson (though named "Sara Mara Natansohn" on her birth certificate) was Jewish – although completely secretive about her heritage, even keeping her maiden name a secret. She had lived with her strict Jewish family in in St Petersburg, Russia, though born in Latvia. She was a concert-level pianist, having known various Russian composers and classical music performers. She had apparently left her family to study in Switzerland when she met John Leathes, and returned to England with him, whom she subsequently married.

Margaret first met Lionel on a Swiss mountain trip (the Grossglockner). She was a highly intelligent, talented, and attractive young woman with a medical degree and a talent for amusing writing. She had been Head Girl at her school, Bedales, in southern England. However, after her marriage to Lionel, Margaret found that her circumstances with him made it particularly difficult for her to develop these talents, except in relation to her children.

After Lionel died in 1972, Margaret gained some freedom to express herself. She married Maxwell Herman Alexander Newman FRS, a prominent mathematician, who had been a close friend of both Lionel and Margaret in earlier days (Max having accompanied Lionel on his early trip to Vienna, Max's purpose there to meet the Viennese mathematical logicians). Max Newman was born in 1897 and died in 1984. Margaret was born in 1901 and died in 1989.

Max Newman had done particularly important work at Bletchley Park, during World War 2, being in charge of the "Newmanry" which was responsible for decoding Hitler's especially secret *Lorenz* code that he used for communication with his generals. It should be mentioned also, that there was a crucial component to their being able to crack the Lorenz code, coming from a brilliant contribution from the mathematician W.T. Tutte. This decoding is not so well known publicly as that, associated with the name of Alan Turing, who cracked the Nazi Enigma code using electronic devices referred to as "*Bombes*". Newman concluded that cracking the Lorenz code required the construction of a more flexible computer assembly, termed "*Colossus*". This was designed and constructed by the research telephone engineer Thomas H. Flowers, under Newman's guidance. It first worked on 1 June 1944, just in time for the Normandy landings on D-Day, 6 June 1944. The existence of the Colossus was kept secret until the mid-1970s, Colossus is now being commonly regarded as being the world's first programmable electronic digital computer.

Margaret and Lionel had four children, Roger being the second-born of their three sons, the fourth being a girl, Shirley, born in Canada much later. The eldest, Oliver, was born on June 6th, 1929 and the youngest of the sons, Jonathan, was born on October 7th, 1933, and in early 1939 the five of them sailed to the USA on the Aquitania, in view of the impending danger of the 2nd World War, Lionel taking up a job in Philadelphia, Pennsylvania. There, the three boys attended a Quaker school, the main thing that Roger could remember about it being the forced nap that pupils had to have in the afternoon during which he never once succeeded in sleeping. In the summer of 1939, the family travelled to London, Ontario, Canada, where Lionel had obtained a job at the Ontario Hospital, and the family lived at 1000 Wellington Street, where they stayed until the end of the war with Germany, in 1945, when the family had also been able to celebrate, with their 4th child – a lovely baby daughter, Shirley Victoria, born on February 22nd 1945.

As for the boys' schooling, while at London, Ontario, Oliver demonstrated his considerable precocity by moving rapidly up, owing to his extremely high rating on an IQ test and consequent attendance at what was called an "advancement class", thereby skipping elementary school altogether and transferring to *Central Collegiate*, where he was two years younger than anyone else in the same class, and nevertheless being first or second in each subject. Jonathan was precocious in a different way, showing natural skills in most games, but particularly in *chess*. Roger did not excel at school while in Canada, having a "bumpy ride" at best, though the main problem was spotted by one of his teachers, Mr Stenett, who, with considerable insight, realized that Roger was basically just *slow*, and when subsequently given as long as needed in mathematics tests, could consistently get marks in the 90s where without the extra time would obtain perhaps 40% to 50%. This helped Roger's confidence in later years! At home, Roger liked making things with moving parts that he had designed, like a perpetual calendar and a moon clock, and also some "pop-up" books that described simple stories that he wrote. He also found a close relation with Lionel constructing various regular polyhedra. Roger also gained much from his older brother Oliver, who read science-related books to him and, on one occasion, showing him what could be achieved with the manipulations of simple algebra. The family returned to England by boat in the summer of 1949, after the war with Germany had ended.

On returning to England, Lionel took up a professorship at University College, London, named the *Galton Chair of Eugenics*. However, Lionel, being strongly against the notion of "eugenics", regarding this as a distinctly unsavoury political movement, rather than a scientifically motivated pursuit, was determined to change the name of his chair, which he finally succeeded in doing, changing it to its current title "Galton Chair of Human Genetics", though this change took many more years than Lionel had anticipated. As far as the schooling of the three of their boys was concerned, Oliver was already at the stage of going to University, so he embarked on a 3-year degree course in Physics at University College. However, it was perhaps the jolt of moving directly to University in a different country, at the young age of 16, had been a bit more than had been anticipated, and he ended up getting what he referred to as "a miserable 2nd". However, it was good enough for him to get a place to do research in statistical mechanics at Cambridge in 1948, at the age of 19.

In contrast, Roger started at University College School (UCS), in Hampstead, north London at the age of 14 in Grade 3, in which he was one year *older* than the normal age of 13 for that grade, apparently due to his grossly insufficient knowledge of Latin, for which he had obtained a little private tuition in Canada.

Fortunately, he was able to be top of the class that year and was promoted up to Grade 5 for the next year. After that, he was able to impress his teachers with his natural grasp of mathematics – geometry in particular – once finding a nice geometrical result unknown to his teacher.

At one point, Roger told Lionel that his maths teacher had informed his class that on the following day he would explain about the ideas of *calculus*. Lionel, with an element of desperation, immediately took Roger to a table in the corner of the room and gave him a rapid account of the ideas and the beauty of calculus! This made a big impression on Roger, not necessarily that he understood everything that Lionel told him, but more that Lionel regarded calculus as being such a beautiful and powerful set of ideas that he couldn't bear to have someone else have the privilege of being the first to reveal this magic to Roger! Later, Roger learnt from Oliver that Lionel had done the same thing with him, several years earlier!

One day in Roger's second year at UCS, there was an event when each pupil in the class, in turn, had to go up to a table in the room where the headmaster (Mr Walton) would discuss which specialist subjects would be appropriate for that pupil, for his final two years of study. When Roger's time came, as he walked up to the table, he was of the clear opinion that he would be the one to carry on the family tradition to study medicine. Both Lionel and Margaret were keen on this. For many years, they had regarded Roger as being the obvious one to carry on the family tradition. Clearly Oliver would not do, as he was dedicated to the study of physics. Moreover, Jonathan's primary interest was chess, having never expressed any interest in science at all. As Roger walked up to the headmaster's table, he also believed that he would become a doctor – or perhaps a *brain surgeon* as even at that time he had hopes that he might find something out about how that strange and wonderful organ actually worked! Accordingly, Roger sat down, facing the headmaster, firmly believing that he would be a doctor. When the headmaster asked Roger what subjects did he want to study in his final two years, Roger asserted: "biology, chemistry, and mathematics". However, the headmaster immediately responded: "No, that combination is not possible. If you want to do mathematics you can't do biology; if you want to do biology, you can't do mathematics. Make your choice." Without hesitation, Roger said "Mathematics, chemistry and physics". At that moment his medical career evaporated! His love for mathematics had become too strong for him to leave that subject behind.

When Roger returned home, and explained to his parents what had happened, they were furious, thinking that Roger had been too influenced by a schoolfriend who wanted to study nuclear physics, which they regarded as a taboo subject because of the nuclear bomb. How could he give up his medical career so easily? In the end, however, they won their case. Not only did Roger's sister Shirley become a doctor, but she married one: Humphrey Hodgson – two for the price of one!

The next such conflict came when Roger wanted to study for a BSc Mathematics degree at University College London. Again, Lionel was unhappy about this, arguing that just studying mathematics alone was too limiting, and a broader perspective on scientific life would be much preferable. After having some difficulties in persuading Roger, Lionel consulted one of his mathematical colleagues at UCL, Hyman Kestelman, who very generously constructed a collection of around 12 different somewhat unusual mathematical problems, giving Roger the rest of the day to see whether he could answer perhaps two or three of them. By the end of the day, Roger had answered all of them, apparently almost all correctly. This impressed Kestelman enough to persuade Lionel to allow Roger to study for his mathematics degree, which he completed, in 1952, after three years, obtaining 1st Class honours. It may be mentioned that he did not just concentrate on his degree work during this period, but also spent time developing other ideas with colleagues, particularly Ian Percival (subsequently FRS) and Peter Ungar. During his 2nd year, Roger gave a seminar at UCL (which Lionel attended), on a geometrical theorem that Roger had found, concerning 8 conics, with numerous remarkable specializations – currently still unpublished!

It is undoubtedly true that from his particular family background – especially from Lionel, but definitely also, on occasion, from Oliver – Roger had grown up with a deep appreciation for science, mathematics, games, puzzles, and geometrical patterns. His siblings all became distinguished intellectuals. His older brother Oliver became a highly respected professor of statistical mechanics and FRS, having done important work on liquid helium and Bose-Einstein condensates, partly collaborating with Lars Onsager (1968 Nobel Prize in Chemistry). Roger's younger brother Jonathan was a chess prodigy, winning the British Chess Championship a record 10 times (7 consecutively), and once beating the reigning world champion Mikhail Tal in a chess game, becoming a grandmaster and world leader at correspondence chess. Moreover, Roger's sister Shirley became a distinguished geneticist.

In such an intellectual environment Roger's entry into a research career was hardly unexpected, and officially started when he was accepted for research at Cambridge, in pure mathematics, specifically in *algebraic geometry* under the distinguished Cambridge mathematician

William V.D. Hodge. Roger had perhaps felt a little uneasy about his choice of specific topic to do research on, having been given a list of possible topics to choose from, only one of which he could really understand. This concerned what are called "Cayley forms" (or "Chow forms"), a Cayley form being an intriguing but unusual way of representing an "algebraic variety" of any particular dimension, where an *algebraic variety* is, in essence, a curved space defined by algebraic equations. The problem that Hodge had suggested was to find a formula for the Cayley form of the *intersection* of two algebraic varieties, in terms of the Cayley form of each of them individually. Although this seemed like a complicated problem, its nature was clear, not require an understanding of the more sophisticated abstract conceptual ideas that Roger had not yet come to terms with.

Perhaps he had suffered somewhat from the fact that practically all of the graduate students starting mathematical research at Cambridge would have done an extra year, following their undergraduate degrees, probably doing Cambridge "Part 3", or possibly, for a student coming from outside Cambridge, some other qualification judged as effectively equivalent. But Roger had had no such prior preparation. After a few weeks, Hodge, perhaps sensing an unease in Roger's particular choice of research topic, suggested that he might sit in on a supervision session of one of his other beginning graduate students, working on a different topic, to see whether that might be more to Roger's liking.

At this point, it should be mentioned that Roger was one of four graduate students taken on by Hodge that year. As it eventually turned out, one of the students gave up, after several weeks, while another (Michael Hoskin) did his three years of research, and wrote a very decent thesis to obtain his PhD, but then gave up mathematics to become a philosopher and historian of science, becoming a pre-eminent historian of astronomy. It was the third student in the group whose research session Hodge suggested that Roger might sit in on. Roger did this, but felt totally bewildered, finding that he could not really understand a word of what was going on! He came away thinking "If they are all like this, what am I doing here?"

What Roger didn't know was that this student, a "Mr Atiyah" was no ordinary student. He would subsequently become Sir Michael Francis Atiyah OM FRS, President of the Royal Society, First Director of the New Isaac Newton Institute in Cambridge, winner of the highly distinguished 1966 Fields Medal in mathematics and a very early winner of the new Abel Prize for 2004 (considered the effective equivalent of a Nobel Prize, in Mathematics) and, indeed, Britain's leading mathematician. So, they were not "all like this"! Roger stuck with his choice of topic, and after various twists and turns taking him in strange directions, he did eventually write a thesis which grew out of this "Cayley form" topic, finally obtaining his PhD somewhat belatedly in 1957.

As a way of overcoming his uncertainty and unease, with such high-power mathematical colleagues, Roger brought with him to Cambridge a 6-piece assembly puzzle – which he had designed and constructed from Perspex acrylic (using a hacksaw and a broken file) some 6 months before coming to Cambridge. The pieces would have to be put together by means of a complicated locking mechanism involving confusing-looking angular parts, to make a regular tetrahedron when assembled. It went the rounds among the various mathematics graduate students (including Atiyah) taking each of them about 5 hours or more of puzzling time to find the solution. Roger found this to be a good way to gain respect from his mathematical colleagues, despite his lack of confidence with the high-power mathematical activity going on, which he felt unable to keep up with.

The research problem that Roger chose to work on was not really "mainstream", and it led him in some unorthodox directions, such as devising an unusual diagrammatic notation for the algebra of complicated systems of tensors, these being regarded by Roger as abstract algebraic entities. Certain such "abstract tensors" would have to be taken to be *negative dimensional* and these provided a formulation of Roger's concept of "spin-networks" which much later were picked up by others in attempts to provide a combinatorial basis for a "quantized space", such as in the theory of *loop quantum gravity* (developed by Ashtekar, Rovelli and Smolin).

One of the more orthodox concepts arising from this work on Cayley forms was the concept of a "generalized inverse", which exists uniquely for any matrix with complex elements, and was the topic of one of Roger's very earliest papers. It turned out that this idea was not actually new, having been initially found by E.H. Moore, who was primarily a philosopher, and Moore's announcement of this discovery apparently lay deep within one of his philosophical treatises. It seems that this "Moore-Penrose pseudo-inverse" finds application in certain statistical problems, arising from Roger's follow-up paper on this topic.

Roger's interest in this "generalized inverse" notion arose from its use in the Cayley form problem, in the special case when both the algebraic varieties involved are simply collections of hyperplanes. The required algebraic expression is then a generalized inverse. However, this expression would need to work also when the algebraic varieties are not just collections of hyperplanes, but then it turns out that the expression does not always work and consequently there cannot be a polynomial solution to this general Cayley form problem.

But why should we expect a polynomial solution in any case? This expectation arose from an observation that Roger made very early on, that if this Cayley form problem were expressed in terms of "dual variables", rather than the original coordinates, then the problem looked much neater than in its original form, and when Roger explained this to Hodge, he was very encouraged, and it looked as though a polynomial solution to the whole problem was very likely. But when Roger subsequently informed Hodge of his negative conclusion for an overall polynomial solution, it appears that Hodge didn't believe him, but was too polite to say so directly! Instead, at the end of Roger's first year Hodge decided that John A. Todd would be better as a supervisor for Roger, apparently because he thought that Todd would be better than himself at dealing with the very complicated expressions that Roger had become involved with.

However, there were two misconceptions involved in this decision. The first was that Roger was happy dealing with complicated algebraic equations! This misconception arose because Roger was able to use his diagrammatic notation, whereby certain equations can look pretty simple, although when written out in conventional notation can appear extremely complicated. The other misconception seems to be that Hodge thought Roger must have made a mistake in his calculations, and that Todd should be able to sort him out. Roger eventually found out that Hodge held this view only when, later in Roger's third year, Todd had repeated (in conventional notation) a particular critical case of Roger's calculations, finding that Roger was correct about the failure of polynomial solutions, and suggested to Hodge that he, also, might repeat that calculation. Roger was very struck by the *delight* in Hodge's expression when coming up to Roger to tell him he had been correct all along!

A couple of years later, in 1957, Roger wrote a much more general document for his application for a Research Fellowship at St John's College Cambridge, in which he provided an argument to show that whereas polynomial solutions do not always exist for algebraic/geometrical problems of this general kind, nevertheless, there is always a solution in terms of *quotients* of polynomials or as *factors* of polynomial outer products. There was no indication, however, of what this could look like, in the case of the Cayley form problem.

When he had been a graduate student in Cambridge, still working for his PhD in pure mathematics, Roger developed a strong friendship with Dennis Sciama, who had been a graduate student of the great physicist Paul Dirac. Dennis had been a colleague of Roger's brother Oliver and had first met Roger in the Kingswood Restaurant in Cambridge, when Roger had come up from London and was visiting Oliver. Roger posed a query concerning Fred Hoyle's very stimulating radio talks on cosmology at that time, but Oliver referred Roger to Dennis, who was sitting at another table. Dennis had no immediate answer but was impressed by Roger's genuine interest in cosmology, so that when Roger later came to Cambridge as a graduate student, Dennis felt that it was worth developing Roger's cosmological interests further, particularly in relation to the "steady state" model of cosmology, of which both Hoyle and Sciama were strong proponents. As it turned out, this friendship proved very valuable to Roger, as he learned a great deal of physics from Dennis. Not only did Dennis have a considerable knowledge of physics over a broad range, but he was an excellent expositor and had friends who were experts in several areas of physics, and often made efforts to bring such people together if he felt that it could be valuable in promoting research. Their friendship continued at a high level until Dennis died in 1999.

Despite Dennis's important influence on Roger in opening his eyes to the wonders of physics, and how he might divert Roger's talents in that direction, it should be mentioned that there were also other influences on Roger in that direction. In his early years as a graduate student, he attended three courses of lectures that could be said to have had greater influences on his future research than the pure-mathematical courses that were of direct importance to his official research project. These were an impressive course by Herman Bondi on general relativity (clearly of great relevance to Roger's later work in that area), a course by Dirac on basic quantum mechanics (clearly also of later relevance), and a course by S.W.P. Steen on mathematical logic.

The importance of Steen's course to Roger was that he described the notion of *computability* (Turing machines, etc.) and Gödel's theorem(s), the latter being a revelation to Roger, providing the case that the quality of human *understanding* cannot be a computational process. Many years later, this insight led to two of Roger's semi-popular books *"The Emperor's New Mind"* and *"Shadows of the Mind"* (Oxford University Press 1989 and 1994), where Roger presented his case that the phenomenon of consciousness (specifically "conscious understanding") could not arise from classical-physics processes, nor even from Schrödinger's quantum evolution of the wave-function, but had to be an effect of the other part of quantum mechanics, namely the *"collapse of the wave-function"* (denoted **OR** = *objective reduction*) which Roger later provided arguments for it being an objective *gravitational* effect – a "gravitisation of quantum mechanics," rather the more usual reverse endeavour of "quantized gravity". When Stuart Hameroff, (University of Arizona) read *"The Emperor's New Mind"*, he contacted Roger to suggest that neuronal microtubules (then unknown to Roger) might be promising locations for preserving quantum coherence long enough for this **OR** effect to be appropriately "orchestrated" thereby providing the macroscopic effects of consciously controlled actions. Though undoubtedly speculative in various respects, this "Orch-**OR**" proposal is now regarded, after some 20 years, as a serious contender amongst current theories of consciousness.

There were also other topics that Roger worked on from time to time. He constructed self-contradictory pictures, stimulated by the works of M.C Escher, called "impossible objects". He also produced quasi-symmetric tiling patterns, which have close relations to the *quasicrystals* discovered by Dan Shechtman (2011 Chemistry Nobel Prize).

Yet, Roger had a particular respect for Dennis Sciama, not only for his broad understanding and promotion of physical science, but also for his *scientific integrity.* For over a decade, Dennis had been a strong promoter of steady-state cosmology, according to which the universe had no beginning, its eternal expansion being sustained by the continual creation of hydrogen to compensate for the depletion of material due to the expansion. However, when in 1964 Penzias and Wilson (1978 Physics Nobel Prize) provided a convincing refutation of the steady-state model, with their discovery of the *microwave background* (CMB), Dennis made gave numerous powerful lectures refuting his earlier viewpoint, and now firmly supporting the idea of a "Big-Bang" origin for our universe and strongly encouraged research into the physical nature of this initial state.

This momentous shift in viewpoint strongly influenced Roger's own thinking. How is one to deal with the physics of this initial state, where space-time curvatures appear to have to diverge to infinity, resulting in what is referred to as a "singularity" in the classical space-time structure. Large curvatures mean small radii of curvature, so one appears to be forced into considering the nature of physics at the ridiculously tiny "Planck-scale" lengths and times ($\sim 10^{-33}$ cm and $\sim 10^{-43}$ s) which are enormously smaller than those encountered in ordinary particle physics. This is indeed taken to be the realm of *quantum gravity*, where it is supposed that the very nature of space-time must itself be treated in some kind of quantum-theory terms, whether or not the above-mentioned "gravitization" of quantum mechanics also plays a role in unifying these two great theories of 20th century physics.

Roger's own concern with such "singular" space-time states occurred several years earlier, being basically initiated by a lecture given by David Finkelstein, in early 1959, in the second year of Roger's Research Fellowship at St Johns College Cambridge. This talk was given at Kings College London, and Dennis drove he two of them there, having persuaded Roger that the talk would be of interest to him – as indeed it turned out to be! Finkelstein's talk was to show how, using an appropriate choice of coordinates, one can eliminate what had been commonly referred to as the "Schwarzschild singularity" in Karl Schwarzschild's famous solution of Einstein's equations for the curved vacuum space-time for the gravitational field of a static spherically symmetrically symmetrical body. This "singularity" occurs at the radius $r = 2m$, in units where Newton's gravitational constant G and the speed of light c are both taken to be unity: G = 1, c = 1.

At this radius, a term in Schwarzschild's expression for the space-time metric becomes *infinite*, and this is why this radius was referred to as a "singularity". However, this feature arises from a demand that the solution is *time-symmetrical*, and Finkelstein, in his talk, showed how an elegant time-*asymmetrical* change of coordinates can remove this singularity, providing a picture that we currently refer to as a "black hole". This was the first time that Roger had seen this extension to within the $r = 2m$ radius (though he later learned of other such ways of extending the Schwarzschild solution to within this radius, though not done with the insight or elegance that Finkelstein had demonstrated in his lecture).

After the talk, Roger told Finkelstein about spin-networks, which intrigued him. He later told Roger that on that day they had "swapped topics" since Roger subsequently worked on general relativity and he on combinatorial space-time, this being his approach to the issue of *quantum gravity.*

Roger began to wonder about the singularity at $r = 0$, that persists despite the coordinate change. It had seemed to him that this central singularity was much more robust than that at $r = 2m$, especially since the space-time curvature becomes *infinite* at $r = 0$. He began to wonder whether there might be some theorem which showed that even if we perturb the solution away from the spherical symmetry assumed by Schwarzschild, a space-time singularity would persist. He had not heard of such a theorem and began to wonder whether some different slant on the equations of general relativity might be helpful for this. His thoughts turned to the theory of *2-spinors* that provides a distinctive way of analysing space-time. This had been greatly clarified for him by Dirac's 2nd-term lectures in early 1958. In Roger's diagrammatic notation, the tensor lines in the diagram become double stranded 2-spinor lines, the formalism allowing manipulation of strands independently, thereby providing a way of examining general relativity in ways that are not so immediately addressed in the standard tensor formalism. Many years later, Roger collaborated with Wolfgang Rindler to provide a 2-volume account explaining these procedures in detail (including the "spin-coefficient techniques introduced by Roer and his close colleague Ezra T. Newman): "*Spinors and Space-Time*" (Cambridge University Press, 1984 and 1986). Volume 2 also gave an extended account of the theory of *twistors*, that Roger had introduced in 1967, and which became a major part of Roger's later research, involving many of his students.

The 2-spinor formalism has an advantage over the standard tensor formalism which makes it much clearer that the free gravitational degrees of freedom are described by the *Weyl conformal tensor,* closely analogously to the way that the electromagnetic degrees of freedom are described by the Maxwell field tensor. Here "conformal" refers to the structure given by the metric only up to proportionality. The conformal structure of space-time is effectively its *causal structure*, i.e., it determines which points can be connected by time-like or null (light-like) curves, this being central to the analysis that Roger eventually used to show that singularities cannot be avoided in gravitational collapse, irrespective of the spherical symmetry of Schwarzschild's space-time, thereby providing the *theorem* that Roger had wondered about in 1958, following Finkelstein's lecture. Roger's 1965 paper on this, earned his share of the 2020 Physics Nobel Prize. It had been motivated by the discovery of the violently energetic *quasars*, indicating the presence of very distant enormous gravitational collapse events – *black holes*!

The 2-spinor formalism had earlier been found useful by Roger in relation to work being carried out mainly in the 1960s, concerning gravitationally radiating systems and the mass-energy carried away by the gravitational waves. Following initial contributions by Andrzej Trautman and Bondi, with collaborators, it had indeed become clear that in an asymptotically flat space-time, there was a clear-cut contribution to the *mass* in the waves, carried away from the sources.

In this work, an intriguing an effect, known as the "peeling-off" property of the *Weyl* tensor, was pointed out by Rainer Sachs. This was best understood in terms of 2-spinors, and Roger showed, in a 1965 Royal Society paper, that Sachs's peeling property can be understood as the spin-2 massless field described by the Weyl tensor being *finite* at a conformally defined *boundary* $\mathscr{I}$, attached smoothly to the space-time, $\mathscr{I}$ being referred to as the space -time's "conformal infinity".

Roger also studied cosmological models, using this same conformal technique for "bringing infinity in" to provide a smooth conformal boundary $\mathscr{I}$. He obtained a clear role for Einstein's cosmological constant Λ, finding that $\Lambda>0$ (the observed "dark energy" value) corresponds to $\mathscr{I}$ being *spacelike*, where in the case of asymptotically flat space-times, $\mathscr{I}$ is null.

This conformal "squashing down" of the future can be used also in the reverse sense of conformally "stretching out" the big-bang singularity of a cosmological model to obtain a (normally spacelike) initial boundary $\mathscr{B}$. However, this situation is very different from the *time-reverse* of a realistic collapsing universe, which would involve the congealing of numerous black-hole singularities, very probably involving wildly complicated oscillating and diverging Weyl curvature, as suggested by the BKLM picture (Misner in 1969 and, separately, Beinskii, Khalatnikov and Lifshitz in 1970). One may regard BKLM as presenting an enormous entropy in the gravitational degrees of freedom, hugely exceeding the already enormous

Bekenstein-Hawking entropy in the individual congealing black holes, the latter already vastly swamping all other forms of entropy in our observed universe. All this is consistent with the 2nd Law of Thermodynamics and with direct observations, where the *earliest* direct observations reveal a very uniform CMB, indicating that this enormous reservoir of gravitational degrees of freedom had still hardly been touched, 380,000 years after our Big Bang!

No conventional form of quantum gravity could provide such a *vast* time-asymmetry in the past/future singularity structure. Nor could the mere introduction of an "inflation field", the assumed source of an inflationary very early stage of the universe, so we appear to be presented with a fundamental conundrum.

Accordingly, in 2006 – as described in his books *Cycles of Time* (Knopf 2010), and *Fashion, Faith, and Fantasy in the New Physics of the Universe* (Princeton University Press 2016) – Roger put forward his exotic "conformal cyclic cosmology" proposal (CCC), proposing that the Big Bang was in fact the *conformal continuation* of an earlier "aeon" whose conformally squashed remote future joins smoothly to our conformally stretched Big Bang of our own "aeon", this continuing in both directions with an unending succession of such aeons, joined successively in this way. This incorporates a proposal due to Paul Tod that the conformal smoothness of our $\mathscr{B}$ suffices to characterize its extraordinary specialness. Though highly unconventional, this model not only resolves the aforementioned conundrum, but has also obtained some remarkable observational support, the strongest, published in 2020 in the *Monthly Notices of the Royal Astronomical* Society by Daniel An, Krzysztof Meissner, Pawel Nurowski, and Roger, confirming a predicted effect of CCC at a 99.98% confidence level.

Roger Penrose had two marriages: in 1958 to Joan Isabel Wedge (divorced 1980; died 2019) and in 1988 to Vanessa Dee Thomas. There were four sons altogether: 1963 Christopher Shaun, 1964 Toby Nicholas, 1966 Eric Alexander; and then in 2000 Maxwell Sebastian.

Black Holes, Cosmology and Space-Time Singularities

Nobel Lecture, December 8, 2020 by
Roger Penrose
University of Oxford, Oxford, United Kingdom.

IN 1908 HERMANN MINKOWSKI introduced the idea of space-time, which is a 4-dimensional space that encapsulates pretty well all of Einstein's 1905 theory of special relativity. At first, Einstein didn't like the idea very much. He initially thought it was just mathematical sophistry or something like that, but subsequently he picked up on it, realizing the power of the 4-dimensional geometrical perspective, and it became central to its generalization to Einstein's *general* theory of relativity. In Fig. 1, I have indicated three coordinate axes for ordinary 3-dimensional space, but in Fig. 2, we move on to introduce a time axis to describe 4-dimensional space-time.

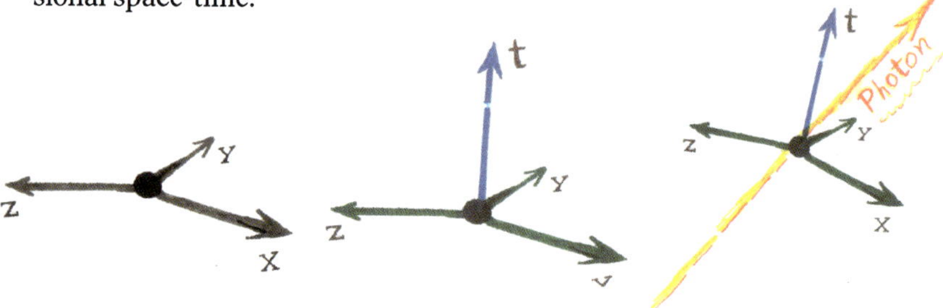

Figure 1. *Figure 2.* *Figure 3.*

Now, the most important thing about this description is to represent the speed of light, so in Fig. 3 we have a light ray (or photon history) where, in order that it does not simply almost lie along the spatial "floor", we need to choose space and time units so that the light ray can be represented as tilted at some reasonable angle to this "floor", such as at ~45°. In Fig. 4 we can depict the *null cone*, which represents all the space-time directions of light rays through our chosen origin point. These cones are very important for the structure of space-time, and we shall frequently be concerned with the null cones themselves rather than with any particular light ray (Fig. 5). Moreover, we need not be concerned with the choice of axes either (Fig. 6). What is physically important is this null cone itself, at each point of the space-time.

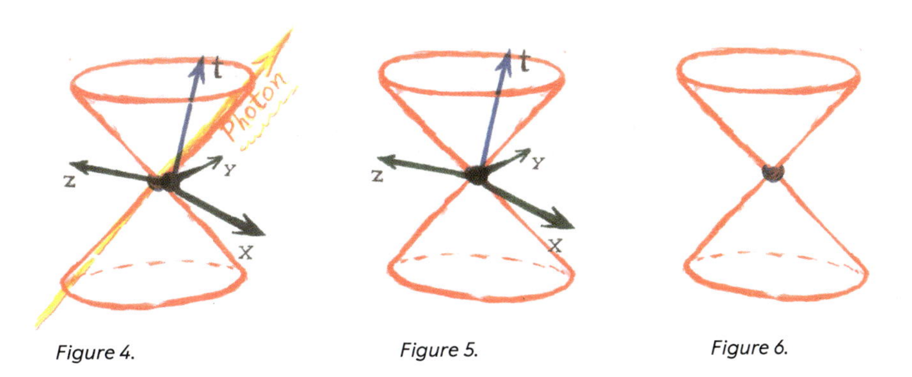

Figure 4. Figure 5. Figure 6.

In general relativity we likewise have a null cone at each point of a (now curved) space-time, representing the local speed of light at each space-time point, but the cones can be more or less all over the place (Fig. 7). We can imagine a space-time point p with all the light rays coming out of *p*. In Einstein's general theory of relativity, these light rays are geometrically what are called null geodesics. In what follows, we shall refer to these *null geodesics* simply as rays. When these rays are extended outwards into the future, away from p in the space-time, we get what is called the future *light cone* of *p* (Fig. 8). The null cones would be *tangent* to this light cone wherever it goes. But, as you can see at the back, at the top right-hand side of Fig. 8, the light rays may start to cross each other, and this sort of thing can make light cones complicated. However, it is important for what I am going to discuss shortly that you appreciate how to deal with such things. In general situations, you certainly are likely to get such crossover regions, caustics and things like that, and how to deal with them is a central feature of the following discussion.

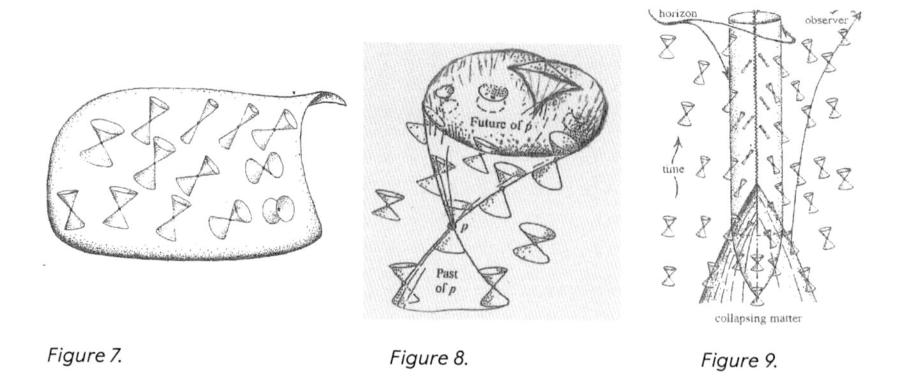

Figure 7. Figure 8. Figure 9.

Now, let us consider the next picture (Fig. 9), where we see a space-time depiction of what is, in effect, the Oppenheimer-Snyder (O-S) evolution of the collapse of a dust cloud to what we now refer to as a *black hole* – although that particular aspect of Fig. 9 was not something that was properly appreciated in 1939, when J. Robert Oppenheimer and his student Hartland Snyder published their paper [1]. What they were concerned with was the behaviour of the material body, indicated in the central lower part of Fig. 9, depicted as falling inwards (as we move up the picture to express the passage of time). This gravitationally collapsing material was to be what is referred to as "dust", which simply means a fluid material with no pressure. These authors had been considering how a very massive individual star might behave, according to Einstein's general relativity, in the star's late stages, when nuclear energy and other relevant resources would have become exhausted so that the material's pressure would become too small to be able to prevent the star's collapse. Whereas such an idealization of the gravitational collapse of an individual star need not be regarded as particularly realistic, such a picture might also be applied to much larger collections of material where, in its outer regions at least, the "dust" approximation is not at all unreasonable.

It should be mentioned that this collapse phenomenon had been encountered much earlier, initially by Subrahmanyan Chandrasekhar [2], in his 1931 discussion of white dwarfs, where he had calculated that such a star, when it had cooled off, would not be able to sustain itself (by its electron degeneracy pressure) if its mass were greater than around 1½ times the mass of the sun, and his considerations are now, not unreasonably, considered to constitute the "birth" of the idea of the plausible existence of what we currently refer to as "black holes". Nevertheless, in my considerations here, I shall refer only to the O-S picture, simply because with the very much larger collections of material that will be our main concern, gravity will be the dominant physical interaction, and such things as nuclear forces and degeneracy pressure become unimportant, so that the simpler O-S picture suffices for our purposes.

The collapse is taken to be exactly spherically symmetrical, in the O-S situation, so that all the matter falls directly inwards towards the central point, where the material's density becomes *infinite*, in this idealization, so the space-time *curvature* must also become infinite, in accordance with Einstein's general relativity. According to the picture of Fig. 9, depicted above where the infalling material is represented, we see this singularity to be extending further upwards, i.e. maintaining its existence with respect to an external time measure, so even in the vacuum region of the picture, we appear to have to maintain the singularity, long after the material has seemed to be absorbed into the singularity. It does not appear that Oppenheimer and Snyder were particularly concerned with such an aspect of their collapse picture – i.e., with the apparent future of the vacuum region following the collapse – but it is, indeed, a matter that we shall shortly need to address in a serious way.

Although the O-S picture of such a gravitational collapse might well not be considered to be altogether realistic under general circumstances, the space-time picture of Fig. 9 can nevertheless be regarded as providing us with a reasonable first impression of the kind of situation that might well arise, at least in the early stages of a large-scale gravitational collapse, before the effects of departure from spherical symmetry and the presence of pressure may indeed begin to have significant implications. It is thus important to know what aspects of this picture might or might not be expected to be maintained under more general circumstances.

The O-S idealized description of a gravitational collapse was fairly well appreciated at the much later time when quasars were discovered in the early 1960s, these bring extra-ordinarily powerful very distant sources of radio signals, and theorists then started to speculate as to whether there might be something like an O-S collapse involved, yet without the understandably gross simplifications of spherical symmetry and pressure-free material, assumed in the O-S situation.

My own acquaintance with the kind of space-time geometry that is required for this O-S collapse picture had occurred somewhat earlier than this, namely in 1959, when I was a mathematics Research Fellow at St John's College Cambridge. In January 1959, I had not yet become aware of the O-S paper, but I went to a lecture by David Finkelstein, given at King's College London, accompanied by my good friend and mentor Dennis Sciama, from whom I had learnt a good deal of relevant physics. He drove me there from Cambridge, having encouraged me that the talk would be interesting to me. Finkelstein's seminar described how you can smoothly pass through what had then been referred to as the "*Schwarzschild singularity*", a feature occurring at a certain radius out from the centre, according to the well-known spherically symmetrical *Schwarzschild solution* of Einstein's vacuum equations, for the external gravitational field of a spherically symmetrical body. This radius is:

$$r = 2m$$

in units where the speed of light c and the gravitational constant G are both chosen to have the value unity:

$$c = 1, G = 1.$$

(Without this choice of units, we have $r = 2Gm/c^2$.) In the coordinate description that Schwarzschild had used, we find that a metric component indeed becomes infinite, and this cannot be avoided in the seemingly very reasonable time-symmetrical description adopted by Schwarzschild.

Karl Schwarzschild had found this very basic solution of Einstein's vacuum equations in 1916, not long after Einstein introduced his general theory of relativity in 1915. Schwarzschild had solved the equations of Einstein's general relativity for the vacuum gravitational field outside a spherically symmetrical body. He also provided a solution for the material of the body itself, but that was not an altogether realistic material, and is not important for our discussion here.

In fact, *the top* part of Fig. 9, depicting the situation arising after all the actual matter has disappeared, apparently all having been absorbed into the central space-time singularity, depicts essentially what Finkelstein described, where you see a portion of the Schwarzschild vacuum space-time, but in a coordinate description that allows a smooth extension to within the $r = 2m$ Schwarzschild radius. Such a situation was described by Finkelstein in a coordinate description that allows for a non-singular extension of the space-time to within the Schwarzschild radius of $r = 2m$, thereby allowing a description for which the infalling material can be seen as actually falling smoothly through the $r = 2m$ Schwarzschild radius that had appeared to be a singularity in Schwarzschild's original description.

Indeed, the description shown in Fig. 9 was not the one presented by Schwarzschild's choice of coordinates, his reasonable-seeming choice providing a *static, time-symmetric* picture, which would not allow the inward-tilting null cones that we see at this radius in the upper part of Fig. 9. Instead, as remarked previously, the expression that Schwarzschild obtained becomes *infinite* at the Schwarzschild radius $r = 2m$, so that the term "Schwarzschild singularity" is now seen to be inappropriate. Finkelstein, in his talk, provided a coordinate change to obtain an elegant time-asymmetric form [3] that extended the solution inwards to values of the radius r that lie in the full range

$$0 < r < \infty,$$

the metric form remaining perfectly smooth across $r = 2m$. This "Schwarzschild radius" now describes the upper cylinder of Fig. 9, where the inward tilting null cones become tangential to this $r = 2m$ cylinder. Clearly, the tilting of the cones presents a time-asymmetric description at that radius, which explains why Schwarzschild's time-symmetric assumption leads to what appears to be a singularity, rather than the *horizon* that was made evident by Finkelstein's choice of coordinates, and which describes the situation of the upper part of Fig. 9.

It should be pointed out that in the early days of general relativity, various other theorists – going back to Painlevé in 1921 [4] – had also found coordinate changes that could eliminate this Schwarzschild singularity, but most did not fully appreciate the physical implications of the "horizon" character of this Schwarzschild radius. Most noteworthy among those who did properly appreciate this physical situation was Abbé Georges Lemaître [5], who understood that infalling material could cross $r = 2m$ into an interior region, consistently with the situation depicted in Fig.9.

I came away from Finkelstein's lecture thinking that, whereas by a suitable coordinate choice you can – somewhat remarkably – get rid of this "Schwarzschild singularity", located at a certain distance $2m$ out from the centre, you nevertheless still have a genuine singularity in the middle (at $r = 0$) where space-time curvatures become infinite, so no coordinate change can help. Accordingly, I began to wonder whether there might perhaps be a general theorem, or something like that, which showed that whatever you might do to complicate the space-time metric, to describe a similar but very irregular collapse situation, you might still have to get a genuine singularity. I had not heard of any such theorem, nor did I have any idea how one might prove such a thing, so I started to think to myself: "what might I know, plausibly relevant to general relativity, that maybe other people in the field don't generally know much about – so possibly this could be helpful to me to achieve something along these lines, that theorists working in the field seem not be familiar with?"

The area that I thought might be helpful was the theory of *2-component spinors*. I should explain that I finally understood properly about 2-spinors from lectures by the great quantum physicist Paul Dirac, earlier in the spring of 1958, in the academic year before Finkelstein's lecture. Dirac's course was basically on quantum field theory, but he appeared to have deviated from his normal course when he talked about 2-spinors. He was famous for discovering the dynamical equation for the electron, but this originally involved the introduction of 4-spinors. However, you can break each of them down into a pair of 2-spinors and Dirac had become well acquainted with this fact, having written an important paper [6] describing higher-spin fields in these terms. Yet, these techniques were not very familiar to most physicists at that time. I had myself heard of 2-spinors and had been intrigued by them, but I did not really understand them until Dirac's lectures had made them abundantly clear to me!

So, soon after Finkelstein's lecture, I began to wonder whether 2-spinor theory might be something I could apply to general relativity, which possibly might supply me with certain significant insights into general relativity that were, perhaps, unfamiliar to most theorists working in the field. But for this I needed to have a clear geometrical way of thinking about a 2-spinor. How was this to be achieved?

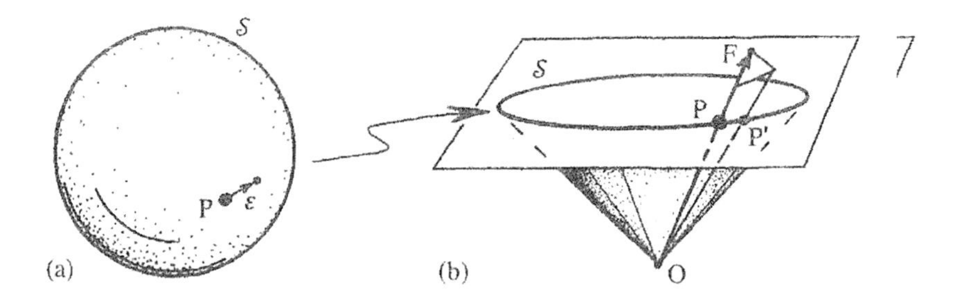

Figure 10.

In Fig. 10 we have a geometrical picture representing this remarkable concept. On the right-hand side of the picture (Fig. 10b) I have, in effect, depicted a 2-spinor in space-time terms, but this needs some clarification. On the left-hand side of Fig. 10 (10a), we see the celestial sphere S – although it is better to think of this as the *future* celestial sphere, rather than the past one (i.e. where light rays go to, rather than where they come from), and Fig. 10b depicts the different directions in which the 2-spinor's "flagpole" can point, this flagpole being a future-pointing null vector. We must bear in mind that there needs to be an additional spatial dimension in the picture of Fig. 10b, so that the bounding "ring" at the top is actually a *sphere S*, as in Fig. 10a. The 2-spinor also has a little flag attached to this null vector, whose plane touches this sphere S and determines the 2-spinor's *phase* (Payne [7], Penrose and Rindler [8]). We have to bear in mind that the upper ring in Fig. 10b is actually a sphere, so there is a freedom for the flag plane to rotate while still touching the sphere, because of the extra dimension, and it is better to refer back to Fig. 10a, where we note that the direction of the flag plane corresponds to a tangent vector ε on the sphere S of Fig. 10a, corresponding to the direction towards P', out from P in Fig. 10b. The "spinorial" nature of this flag corresponds to the fact that a rotation of the flag through 360° changes the 2-spinor's sign, while a 720° rotation returns it to its original value.

The important thing about using the 2-spinor formalism in general relativity, for our purposes, is that it provides us with some insights about space-time curvature that are not so evident in the standard tensor formalism. Most particularly, there is a particular part of the space-time curvature, referred to as the *conformal curvature,* that is determined by a tensor quantity referred to as the *Weyl curvature tensor,* named after the highly esteemed mathematician Hermann Weyl. The Weyl tensor is defined from the full Riemann curvature tensor by a somewhat complicated-looking formula, but the 2-spinor formalism brings out an essential simplicity of the Weyl curvature that is not at all evident in the conventional tensor formalism.

It is not my purpose, here, to provide explicit expressions for all the tensor or spinor expressions for the quantities that are involved, these making use of the delicate interplay between tensor and spinor indices, often involving the metric tensor in its very elementary spinor form. The details of these algebraic manipulations are, of course, important, and are ultimately based on fundamentally simple rules [9], [8], but it is not necessary for us to go into this in detail here.

The Riemann curvature tensor, when written in its 2-spinor form, splits into three parts, one of which is the spinor form of the trace-free Ricci tensor, another being the trace of the Ricci tensor (namely the scalar curvature), and the third part being the spinor form of the Weyl tensor. Einstein's field equations, when expressed in 2-spinor form, effectively tell us that the spinor form of the trace-free Ricci part plays a role as a *source* of the gravitational field, the latter being described by the spinor form of the Weyl tensor, this "Weyl spinor" describing the free gravitational field. There is a strong analogy with electromagnetism revealed here, where the Ricci tensor (really in trace-reversed form) is analogous to the charge-current vector of electromagnetism, the Weyl spinor being analogous to a "Maxwell spinor", which is the spinor form of the Maxwell field tensor, which describes the free electromagnetic field. This analogy is made much more evident in the 2-spinor form than in the tensor form.

Indeed, in the 2-spinor formalism, we find a particularly simple expression for the Weyl curvature, which is not at all evident in the tensor formulation. We find that this "Weyl curvature spinor" is a 4-index 2-spinor which is *totally symmetric* in all its four 2-spinor indices, and it satisfies a very simple free-field equation in vacuum, where there are no gravitational sources (vanishing Ricci tensor). This is strikingly analogous to

Maxwell's electromagnetic theory in 2-spinor form, where the spinor form of the Maxwell field tensor is a 2-index 2-spinor which is *symmetric* in its two 2-spinor indices, and it satisfies the completely analogous free-field equations when the charge-current vector vanishes [9], [8]. All this fits in closely with a general study, made by Dirac in 1936, of field equations for fields of arbitrary spin, written in 2-spinor form. (Curiously, I once had a private discussion with Dirac, when I explained to him how Einstein's gravitational theory fits in with the other spin fields in Dirac's own 2-spinor analysis of higher-spin fields [6]. He had not been previously familiar with this aspect of Einstein's theory, and he found it interesting.)

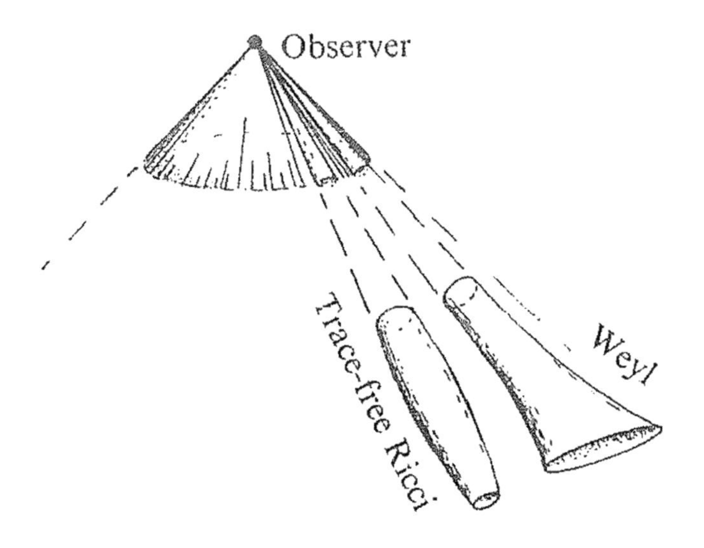

Figure 11.

In order to appreciate the geometrical effect of the Weyl and Ricci curvatures, we can have a look at Fig. 11. This depicts an observer, located at the top of the picture, looking back into the past and sees the inward focusing effect due to the trace-free Ricci curvature (behaving like an ordinary positive convex lens), and also sees the *distortion* due to the Weyl curvature W (like a purely astigmatic lens). In this way, we see the direct roles of the (trace-free) Ricci curvature and the Weyl curvature in their effects on light rays. The trace-free Ricci tensor acts as a lens which is positively focusing (like an ordinary magnifying lens) when the energy flux across the light ray is positive. (The trace part of the Ricci tensor, being proportional to the metric tensor, does not affect light rays.)

A few years after my time as a Cambridge Research Fellow, the quasars ("quasi-stellar objects") were being discovered, this being around 1962–1964. These objects were producing enormous quantities of energy and yet they seemed to be remarkably small, considering this output. They appeared to have an energy output of perhaps 100 to 1000 times an entire galaxy's emission, but yet they seemed to vary substantially in a few hours or days, which meant that they had to be very small compared with a galaxy, probably not larger than the solar system, but would have to be extremely massive to emit so much energy. How could all that mass-energy be squashed into that small volume? Astrophysicists, such as Fred Hoyle, started to speculate upon whether something gravitationally concentrated as in the O-S collapse might be relevant. But when you think just of a collapse that falls radially inwards, it doesn't give you any scope for signals coming out. If you were to have any involvement of gravitational waves, these would need to have at least a quadrupole structure. Moreover, the highly varying nature of the radio signals suggests that something very complicated must be involved. Certainly, the possibility of a very irregular gravitational collapse must be considered, which could be very different from the O-S picture of matter falling radially inwards. Perhaps material falling inwards in a complicated way might swirl round an come out again. Such possibilities had been suggested to me by John Wheeler and others, so I began to think again about such matters as a general gravitational collapse, in a serious way.

At around that time, in 1963, there had been a paper published by two Russians, Lifshitz and Khalatnikov [10], which seemed to have established that in general you would *not* get singularities, these occurring only in very special circumstances such as in the particular O-S collapse picture. Accordingly, in a physically realistic situation you might indeed consider that in a generic collapse the material would indeed just swish around and come swirling out again. As part of my worries about this problem, I had a good look at the Russians' paper. In fact, there was actually a serious mistake in the paper, but I didn't examine it carefully enough to notice that. What I did feel, however, was that the methods they were using were not altogether convincing to me, and that it was worth trying to think independently about whether or not you would get singularities in a generic gravitational collapse.

I remember walking in the woods near where I lived at the time, in Stanmore at the north of London, and trying to imagine that I was in the midst of such a gravitationally collapsing situation, I came to the conclusion that it could not just be a local curvature blow-up, but it had to involve a build-up of curvature due to some overall excessiveness of the material concentration. Some kind of non-local criterion would be needed to tell you that a "point of no return" had, in an appropriate sense, been passed.

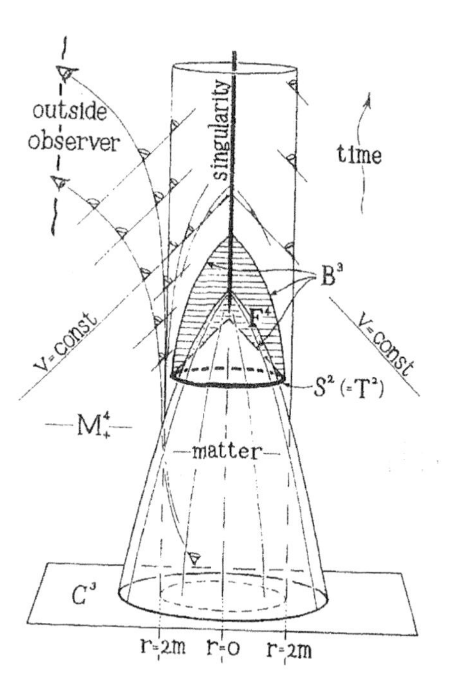

Figure 12.

A few weeks later, I hit upon the idea of a "*trapped surface*" which seemed to supply the kind of non-local criterion that I was looking for. The picture you see in Fig. 12 is a diagram that subsequently appeared in my 1965 paper [11], though a little earlier, in late 1964, I gave a talk at King's College London about it. The argument was that if you have a collapse which is generic, but in which you happen to have a trapped surface, then you will still have problems with singularities even though the infalling material would not be all aimed at a central point. In Fig. 12, you see what is basically the O-S collapse of Fig. 9, but you can also see that there is this little ring in the middle of the picture, marked S^2 ($=T^2$) surrounding the infalling matter. This is a trapped surface, and you have to realize that it is not actually a "ring" because I am only depicting two spatial dimensions, together with the time dimension and (as with Fig. 10b) the whole thing should actually be a 4-dimensional space-time, so that the "ring" is really a 2-dimensional surface, topologically like an ordinary sphere. In a general collapse, it might be distorted, not necessarily being a precise geometrical sphere.

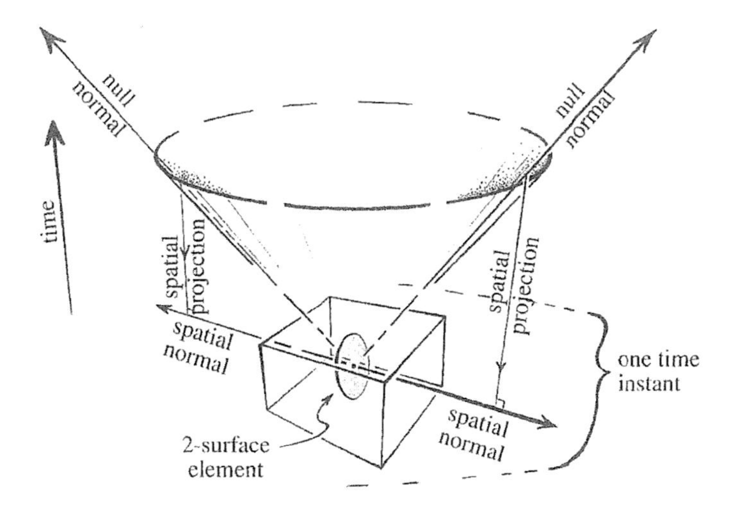

Figure 13.

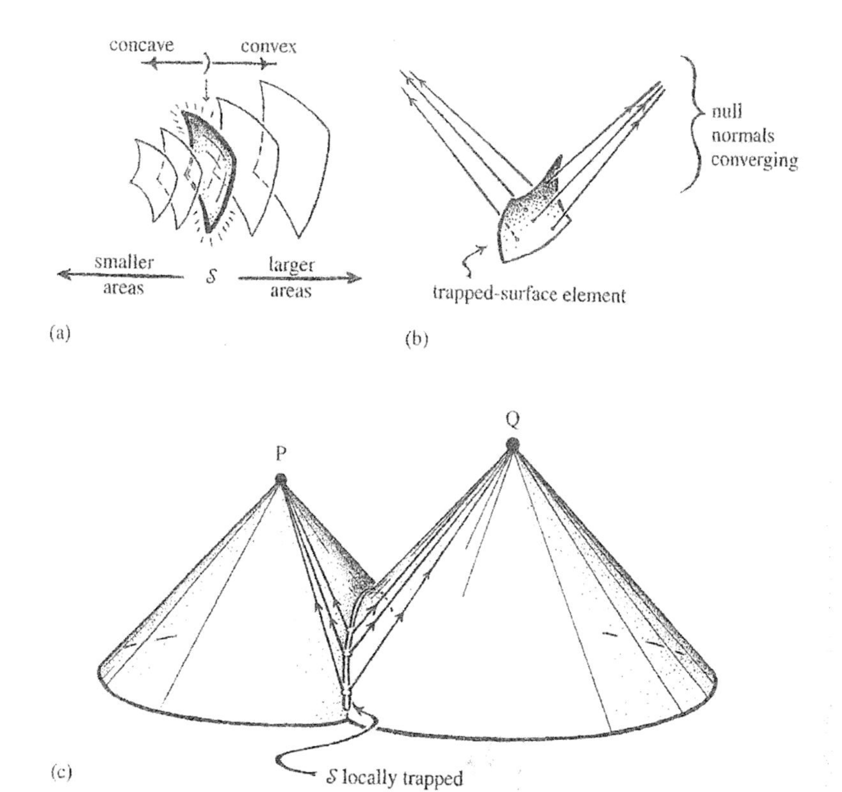

Figure 14.

But what exactly is a trapped surface? You have to imagine that there is a flash of light emitted simultaneously all over that surface and that this flash has a certain characteristic property. In the lower middle part of Fig. 13, you see a little 2-dimensional spatial surface element and you are to imagine that the surface emits a flash of light, which moves directly away from it in opposite directions, represented by light rays in the 4-space-time-dimensionally indicated top part of the picture. In Fig. 14, at the top left (Fig. 14a), we have a *curved* surface element which is concave on one side, where the light-flash rays from it are converging, and convex on the other side, with light-flash rays which diverge. That would be a normal thing for a curved surface element. Now, on the top right (Fig. 14b) you have something that's harder to imagine, namely that light-flash rays can converge on *both* sides. This is what we need for a trapped surface. In fact, in space-time terms, this is not really a problem for a curved surface element. At the bottom (Fig. 14c) you see a situation in ordinary flat (Minkowski) space-time, where the surface is the intersection of two past light cones (with vertices P and Q) and, indeed, the light flashes coming from *both* sides of such a surface element would indeed be converging. The definition of a *trapped surface* S^2, however, is that it is a *compact* spacelike 2-surface which has this *both-way* converging property occurs all over the surface S^2. The essential compactness condition means that it is closed up without any boundary (like an ordinary sphere, but which could be distorted in various ways without disturbing its topology). Thus, each point of this S^2 is *locally* like the intersection surface of Fig. 14c (i.e. like Fig. 14b) which, in itself, would present no problem, but what makes it the problematic "trapped" surface is that it is also compact, a feature that would certainly be the case for rotationally symmetrical 2-surfaces within the Schwarzschild horizon of Fig. 9.

A key aspect of this definition is that it is stable under small deformations. Because of the compactness of the surface S^2, if we were to vary the situation very slightly, the degree of the convergence of light rays moving out orthogonally away from S^2 must still be bounded away from zero, for a small-enough change in the geometry of the situation. Accordingly, a small-enough generic perturbation away from the spherically symmetrical case will not disturb the "trappedness" of the 2-surface S^2. We can certainly have many generic collapse situations in which a trapped surface indeed arises.

The next question for us is: why does the presence of a trapped surface cause us a problem? We need to consider that the trapped surface S^2 lies in a space-time region M^+ that is the future time-evolution away from a *non-compact* spacelike 3-surface C^3. (We are thinking of a reasonably local phenomenon, not a cosmological one for which we might, on the other hand, actually want C^3 to be some compact spacelike 3-surface.) We shall be concerned with the region F^4 (shown shaded in Fig. 12), within the future evolution of C^3, which is the *chronological future* of S^2 [12], [13] That is to say, F^4 is the region swept out by (i.e. the union of) all the time-like curves with past end-points in S^2. What we are particularly concerned with is the *boundary* B^3 of F^4.

It follows from general results (see [12], [13]) that any point p of B^3, not on S^2, is the future end-point of a ray (i.e. null geodesic) lying on B^3 which is either past endless or else has a past end-point on the initial region S^2. The past-endless case is anomalous, because it implies that the region M^+ is not a standard time-evolution away from C^3, since it contains rays that wind endlessly into the past without ever reaching the initial surface C^3. It would not be unreasonable to regard such a departure from normal causality as "singular" behaviour, and no less anomalous than regions where the space-time curvature diverges to infinity. Accordingly, I shall here refer to such anomalous space-times as being "singular" whether or not their "incomplete" status actually arises from a curvature divergence.

We have just seen that the boundary B^3 of F^4 consists entirely of rays, and if M^+ is to be non-singular in the above sense, then all these rays have past end-points on S^2, but what do these rays do in their futures? For this, we need the "trapped" nature of S^2, which tells us that the divergence of these rays starts off as *negative* as they leave S^2. Now there is a result known as the "Raichaudhury effect" for rays [12], [13], that tells us that if we have a null hypersurface (such as B^3) for which the divergence of its generating rays is initially negative (which is here the trapped-surface condition) and for which the curvature along the rays is non-negative (i.e., according to Einstein's equations, that the energy flux across these ray s is non-negative, as indicated in Fig. 11), then after a finite affine distance along the ray its separation from some of its neighbouring rays becomes zero, which is what happens at a caustic point, as illustrated at the top right of Fig. 8. This tells us that – either at that caustic point, or, more usually earlier than that, at a crossing region of B^3 – the ray ceases to lie on the boundary B^3 of F^4 and enters into the interior of F^4. Either way, it is only a finite segment of the ray which lies on B^3, telling us that B^3 must, in fact, be *compact*, consisting entirely of a union of finite segments of rays, each of which has an initial end-point on the compact surface S^2.

It is this compactness of B^3 that actually leads us to a contradiction that drives us to the conclusion that there must be a *singularity* somewhere within F^4 (or perhaps at its boundary). However, in my original paper [11] I used a rather clumsy argument to demonstrate this, and afterwards Charles W. Misner pointed out to me that I could have used a much simpler argument, using the fact that there is a general theorem concerning Lorentzian manifolds (i.e. manifolds with the standard metric signature, with one time-like dimension and the rest space-like) that there exists a smooth time-like vector field all over the whole manifold. We can follow along such a vector field to map B^3 homeomorphically (i.e. preserving its topology) into C^3. Since B^3 is compact without boundary, the image of this map must also be compact without boundary. Since this 3-dimensional image indeed has no boundary, it must be the whole of C^3. But C^3 is non-compact, which immediately provides us with the required contradiction. I had been aware of the result that Misner was using in his subsequent contribution to the argument, but for some reason I had not thought of using it. In all my later accounts of this result I took advantage of Misner's simplification.

In the autumn of 1965 there was a conference at Imperial College, London about progress in general relativity, and Igor Novikov from the Russian school tried to present the aforementioned 1963 result, but Misner then pointed out the conflict with my result. Subsequently Belinskii joined with Lifschitz and Khalatnikov to provide a corrected paper with a conclusion opposite from what they had done before [14], and Misner provided his own version [15], these papers demonstrating how immensely complicated the singularities in a general gravitational collapse can be, with the Weyl tensor itself diverging in an extremely complicated way. I shall refer to these as *BKLM-type* singularities.

At this point, I should draw attention to the fact that what I showed was that the occurrence of *singularities* is a robust prediction of Einstein's general theory of relativity, not necessarily that *black holes* must necessarily be the consequence of a realistic gravitational collapse. Another possibility might be that "naked singularities" could arise, these being, in effect, space-time singularities that are not hidden behind event horizons, and so might actually be directly visible from a distance away. Unlike the case of a black hole, however, there is no reason to expect that actual naked singularities should exist in nature (although there are many exact solutions of the Einstein equations which do possess naked singularities, such as the Schwarzschild solution for a negative mass). The normal view – probably correct, in my opinion – is that naked singularities cannot occur in ordinary gravitational collapse situations but, as far as I am aware, no mathematical theorem has yet been provided to demonstrate what I have referred to as the "Cosmic censorship hypothesis", which might, in effect, prove that naked singularities are unstable, or something like that (though I did once write a paper exploring some mathematical implications of the production of naked singularities [16]).

After I gave my talk at King's College London, Dennis Sciama asked me to give a repeat in Cambridge, which I did in early 1965 and at that occasion Stephen Hawking was present (though he had not been present at my London talk). Immediately following my Cambridge lecture, I had a private session with Stephen, accompanied also by George Ellis, who had been collaborating with him on the use of certain ideas that they had hoped might be useful in addressing the necessity of the Big Bang singularity. I explained much more about the details of my arguments. Stephen Hawking very quickly picked up on the ideas and applied a version of my own result to cosmology in an original way. Subsequently, Hawking developed these techniques very considerably, often with the benefit of critical corrections from Brandon Carter. Hawking's developments were published in a series of three papers in the *Proceedings of the Royal Society* [17], [18], [19], [20]. Eventually I came back to collaborate with Hawking to provide a very general result which encompassed practically all that had gone before [21] (although with a slightly stronger energy condition than the one used in my original argument [11]).

Hawking's considerations were directly concerned with the problem of the Big Bang. The issue under consideration was whether the "singular" description of that event is necessary, and might it have been the case that there had been a previously collapsing phase of the universe which, through some extreme complication of the earlier collapse, the universe might have "bounced" into the expansion that we now perceive in our "Big Bang".

In Fig. 15 I have indicated our current picture of the overall history of the universe, from its Big Bang origin to the currently observed exponential expansion. The frilly part at the back is just to accommodate the currently popular view that the universe may well not be spatially closed, but may continue indefinitely in spatial directions. The exponential expansion is a conclusion arrived at by the 2011 Physics Nobel Prize laureates Saul Perlmutter, Brian Schmidt and Adam Riess, and it can be most directly explained by the presence of a positive value for the cosmological constant Λ in Einstein's modified gravitational field equations. Einstein introduced his Λ-term for the wrong reason in 1917. At that time, Einstein was hoping to be able to accommodate a spatially closed static universe into his equations. However, he later rejected that term in his equations after being convinced by Hubble and others that the universe is actually expanding. Nevertheless, a positive value for Λ is the most economical explanation for what is now rather misleadingly referred to as "dark energy", namely the cause of the exponential expansion that I have schematically depicted in Fig. 15, this having been convincingly observed at around the turn of the 21st century by Perlmutter, Schmidt and Riess.

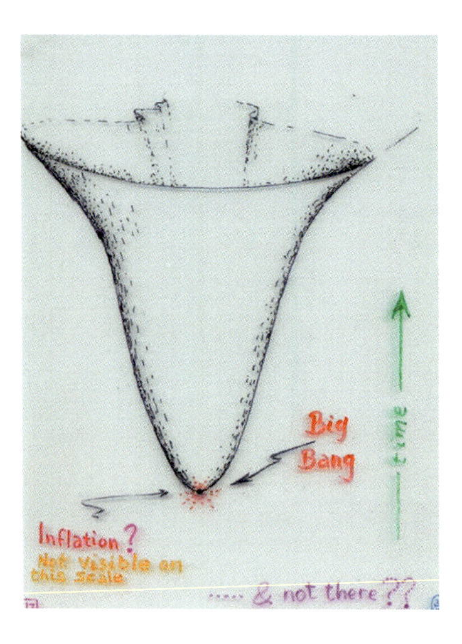

Figure 15.

The "singularity theorems" that Stephen Hawking and I and some others had developed, all have the character of being completely insensitive to the direction of time. I remember being puzzled by the fact that whereas one might well expect that some very complicated solutions of the Einstein equations might apply in the future, perhaps being realized through the interplay by systems of black holes, while, on the other hand, cosmologists seemed to restrict their attention to the very simplest of possibilities. I recall being very puzzled by why cosmologists did not study any of the many other kinds of possible singular origins for the universe. I remember an occasion when I was in Princeton and was about to go to one of the frequent conferences at Stevens Institute in Hoboken, New Jersey. We used to drive up in several cars, and I noticed in the back of one of the cars was James Peebles (later to become the 2019 Nobel Prize laureate in physics), so I asked him why serious cosmologists never seemed to consider any of these complicated possible alternative singularities that you might have for the description of a Big Bang, rather than just this simple highly symmetrical special case. Why, I asked, do cosmologists never consider any of these more complicated alternatives? He just looked at me and said, "because the universe is not like that". So, I thought to myself: "my gosh, it isn't like that is it – but why?"

I presumed that he was partly thinking about the uniformity of the *cosmic microwave background* radiation (CMB) which is indeed very uniform over the whole sky, having been discovered in 1963 by Arno Penzias and Robert Wilson, the 1978 Physics Nobel Prize laureates (shared also with

Pyotr Kapitsa), and this remarkable uniformity tells you that the universe really is indeed very spatially uniform. It struck me that there is something very strange about all these various singularities, namely that the big bang singularity is utterly different from the kinds that you might see in the future, namely in the collapses in black holes, with the distinct likelihood of things like the BKLM type of singularity arising! I was very puzzled by this, particularly since everybody seemed to think that the solution to the singularity problem would lie in combining general relativity with quantum mechanics. Accordingly, you need to find a *quantum gravity* theory to resolve the singularities – so everyone had supposed. But it seemed to me that it must be a very peculiar quantum gravity theory which is grossly asymmetrical in time, in order to give you a theory which makes the singularities quite different in structure in the past from in the future!

I held this view for quite a long time, but then I began to worry more about this problem in relation to the entropy in the universe. This is a key issue, which I shall come to shortly, but let us first consider what is currently a very popular view about what the very early universe was actually like. In Fig. 15, I have indeed sketched the history of our universe according to observation (and some theory), with time going up the picture. At the bottom is the big bang and at the top we see the beginnings of the exponential expansion which became an observationally established feature of current cosmology, through the work of Reiss, Schmidt and Perlmutter and which can be taken as an implication for a *positive* cosmological constant Λ in Einstein's 1917 version of his theory.

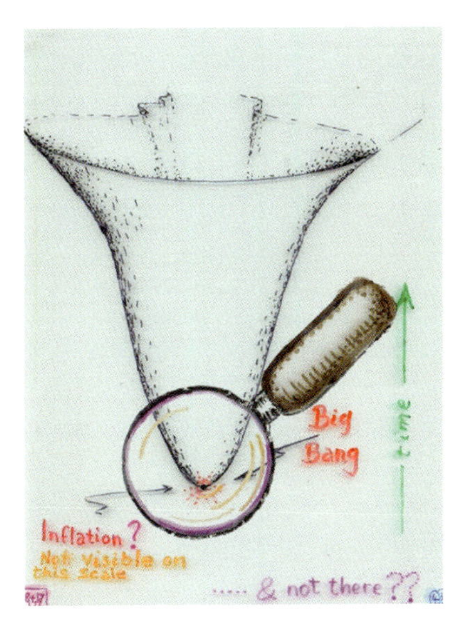

Figure 16.

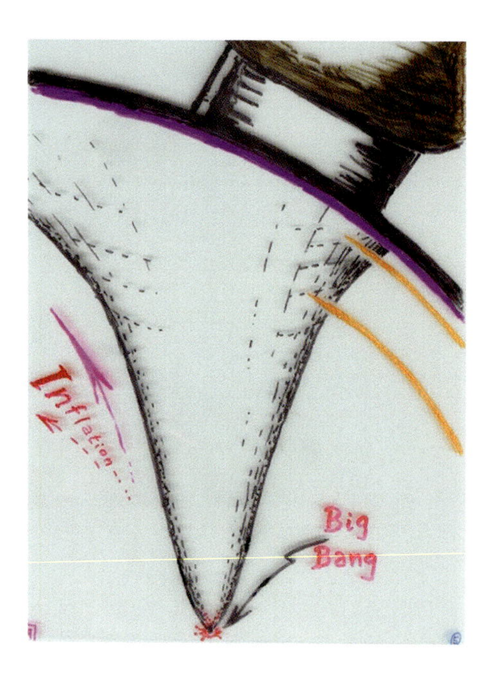

Figure 17.

However, if you want to get a picture of what seems to be currently believed by most cosmologists concerning the extremely early universe, you need a very powerful magnifying glass (Fig. 16) in order have a good look at it in the picture. What you would see (Fig. 17), according to current "inflationary cosmology" would be a much earlier exponential expansion, supposedly all taking place within the absurdly tiny initial 10^{-32} seconds, or so, of the universe's existence, this being referred to as the *inflationary* phase of the universe's expansion. Inflation was initially introduced by Starobinski and Andrei Linde [22], and by Alan Guth [23], [24] in the early 1980s for various reasons, but a particularly important one in the present context was in order to explain an observed striking feature of the tiny variations in the temperature of the CMB radiation over the sky, namely that these variations are extremely closely *scale invariant,* which suggests some sort of exponentially expanding origin for these disturbances.

Another claim often made for the existence of this early inflationary phase was that it would iron out the very early universe immediately following the big bang so as to provide us the very uniform universe that James Peebles had pointed out to me.

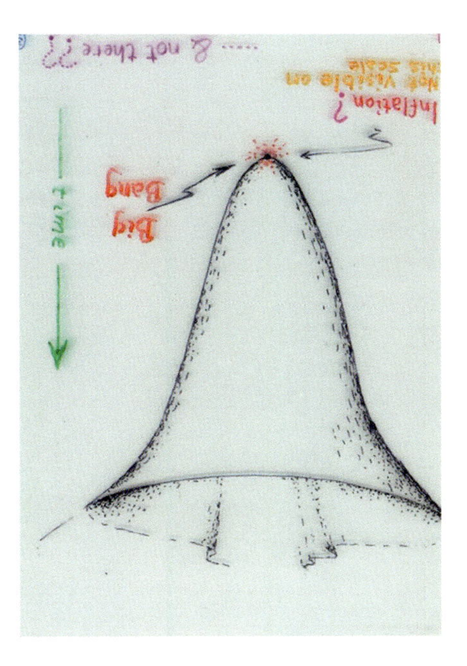

Figure 18.

However, I could never believe that it would actually do that, for the following reason. Imagine that the universe was contracting, rather than expanding, as shown in Fig. 18, this being simply Fig. 15 turned upside down, so the progression of time is still represented as going up the picture. We should take note of the fact that all the relevant dynamical equations – including those of the "inflation field", introduced solely for the purpose of making inflation work – are all unchanged if we reverse the direction of time. Accordingly, Fig. 18 represents a possible universe evolution. However, if we introduce some perturbations into the mass distribution, we must expect black holes to arise in this collapsing situation, which will start to merge with one another and finally produce a horrendous non-uniform mess of a "generic" singularity at the end, as indicated in Fig. 19, the presence of an "inflation field" making no essential difference at all to the picture. Now, we reverse the time-direction back again, as shown in Fig. 20, and we ask: why did the universe not have this "far more probable" type of big bang singularity, as opposed to the uniform one illustrated in indicated in Fig. 15? Inflation provides no answer to this fundamental problem.

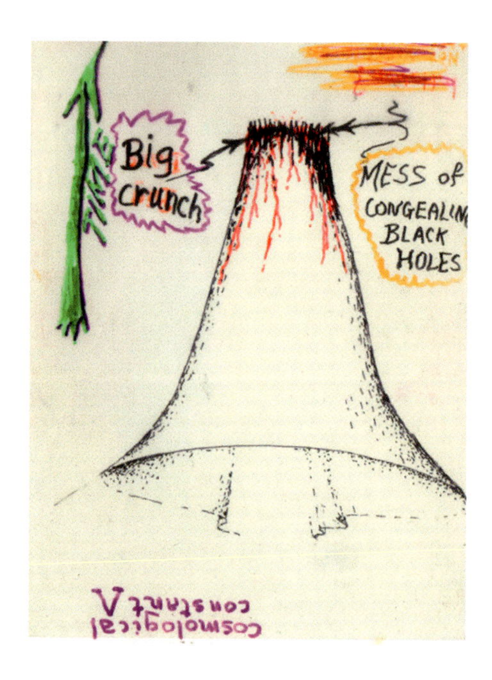

Figure 19.

In order to quantify this issue a little better, we must turn to thermodynamics, and most particularly to the Bekenstein-Hawking formula for the entropy of a black hole. We find, for the totality of the black holes currently within our observable universe, that the entropy is utterly dominated by that in the black holes. Moreover, in a collapse situation like that in Fig. 19, where we take into consideration only the amount of material (including dark matter) within our current observable universe, we find a value for the entropy in a collapse like that of Fig. 19 would be something like 10^{124}. Then, taking into consideration that entropies are really logarithms of probabilities, we come to the conclusion that if our big bang had come about as a singularity chosen somehow "by chance", then the odds against the uniform situation in the big bang that we actually appear to see (i.e. like Fig. 20, rather than the observed Fig. 15, now with time proceeding upwards in the picture), would be something like the utterly absurd figure of around $\exp(10^{124})$:1 (i.e. the probability of our observed universe having the uniformity that we see, this having been a chance occurrence – inflation or no inflation – would be the reciprokal of a number with around 10^{124} digits) which can hardly be the right answer! We need another explanation for the extraordinary specialness of our big bang.

Figure 20.

In fact, the issue is even more curious than this. Let us consider Fig. 21. The top three pictures represent a gas in a box where in the top left-hand picture we see the gas initially constrained to be in a smaller box in its lower right-hand corner. Then you open this smaller box and the gas spreads out within the large box, so that as we move from the left-hand to the right-hand of the top three pictures, the gas gets more and more uniform. This illustrates the action of the 2nd law of thermodynamics, where the entropy (or the randomness) increases with time – time being represented as going from left to right in Fig. 21. Now let us consider the bottom three pictures in Fig. 21. These represent an imagined galactic-scale box containing a large number of stars, initially (bottom left picture) taken to be pretty uniformly distributed. Then, because of the universally attractive nature of gravity, the star distribution gets more and more clumpy, as we move from the lower left picture to the lower right one with, perhaps, the formation of a black hole, in the lower right-hand picture. Again, the time increases from left to right, and so also does the entropy.

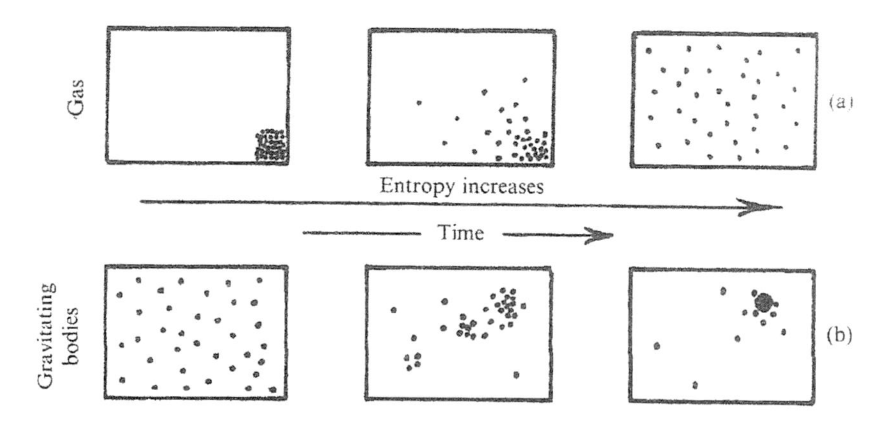

Figure 21.

Now, what is it that we see in the very early stages of our actual *universe*? It is uniformity which, as described in Fig. 21 by a combination of the upper right and lower left pictures. Moreover, one of the most striking features of the CMB observations is the extraordinarily precise Planck curve in temperature distribution for different frequencies, telling us that the matter – by which term I include both the electromagnetic field together with the actual material particles – is indeed in a very closely maximally entropic state. Where the initial very low entropy (a necessity for the 2nd law) resided, was in *gravity* and *only* in the gravity! The puzzle about the origin of the 2nd law lies in the very curious nature of the big-bang singularity, where the gravitational degrees of freedom appear to have been completely suppressed i.e. like Fig. 15 and not like Fig. 20.

Why is there such an extraordinary difference between the past-type and future-type singularities? As mentioned earlier, I used to think that there must be a very peculiar, blatantly time-asymmetric theory of quantum gravity governing this past/future distinction. Later, I simply *postulated* that past-type singularities must have vanishing Weyl curvature – i.e. vanishing gravitational degrees of freedom (what I called "the Weyl curvature hypothesis"). But this does not provide us with any kind of "physical reason" that, whereas the singularities of gravitational collapse must almost always have wildly diverging Weyl curvature, perhaps like the very exotic BKLM situation, the structure of the actual big bang appears to be quite the opposite.

As an approach to studying both the singularities and the asymptotic features of space-times, and also the massless fiends within them, I had come to realize the enormous value of looking at a space-time from the *conformal* point of view, rather than just its more restricting metric structure. The Weyl curvature is, after all, distinguished as describing the *conformal* curvature of a space-time. Moreover, massless fields, most notably Maxwell's electromagnetic field, all exhibit conformal invariance. Even more importantly, the conformal structure of any physically reasonable space-time is effectively the same as its *causal structure*, as we shall come to see very shortly.

Figure 22.

For the understanding of the asymptotic structure of a space-time, we need to understand what its "infinity" might be like. To this end, it is helpful to turn to Fig. 22, which shows (in his print "Circle limit I") how the Dutch artist M.C. Escher exquisitely illustrates how the infinity of the hyperbolic plane can be represented as a smooth circular boundary (Beltrami-Poincaré disc). Conformal maps preserve angles, rather than distances, so small shapes are accurately represented, but not necessarily their sizes. Note that, in this picture, the eyes of the fish-creatures remain exact circles, no matter how closely we approach the boundary.

Figure 23.

We can use similar representations for space-times, and to understand what is involved, it is useful to look first at Fig. 23 where, in addition to the null cone of Fig. 6, I have added some hill-shaped and bowl-shaped surfaces, marked in brown. These brown surfaces (in full 4-dimensional space-time being 3-dimensional surfaces) enable us to represent the full metric stricture at a point of space-time. They represent the ticks of identical clocks travelling at different velocities through the vertex point, as indicated in Fig. 24. At the bottom of this picture, I have written the two most famous formulae of 20th century physics, one of these being, of course, Einstein's well-known formula $E = mc^2$ (fundamental to relativity theory), and the other being Planck's earlier formula $E = h\nu$, (where ν is a frequency, fundamental to quantum mechanics). Einstein tells us that energy and mass are equivalent, and Planck tells us that energy and frequency are equivalent, so putting the two together we see that mass and frequency are equivalent (c and h being just conversion constants). This tells us that any stable massive particle is, in effect, a perfect clock! (The frequencies would be extremely high for individual fundamental particles, but appropriately scaled down, this gives us what is, in effect, the basis for atomic and nuclear clocks.) In fact, it is in the amount of "crowding" of these surfaces that the metric of space-time is defined.

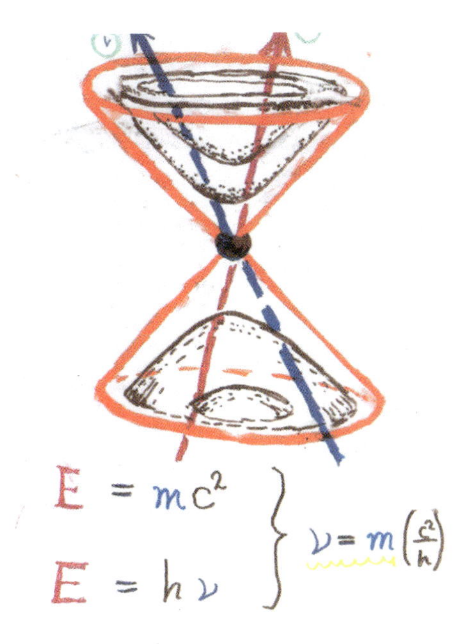

Figure 24.

Perhaps one usually thinks of a space-time's metric to be defining *distances* on an infinitesimal scale. However, it is much more physically direct to think of *times* as defining the metric structure. Spatial distances then arise as a secondary concept, determined in terms of times of transit, the time measures along world lines of particles being primary.

Figure 25.

But what about massless particles, where we now think of a photon in free space? We see from Fig. 25. that the photon does not even notice these scale-determining brown surfaces and does not "experience" the passage of time at all. To a photon, there is no elapse of time from one end of its trajectory to the other, so these scaling surfaces now play no role whatever, and we may as well remove them altogether, so we are left simply with the null cones themselves. Without the scaling, we simply have the metric up to proportionality, i.e. the space-time's *conformal* structure, i.e. that defined by the null cones themselves (Fig. 26, as in Fig. 6 and Fig. 7). It may be mentioned that, since causal signals are transmitted by effects on or within the null cones, the conformal stricture of space-time also defines its causal structure.

Figure 26.

Moreover, it is not just the *classical* picture of a photon that is concerned with only the conformal structure, rather than the full metric structure of space-time. James Clerk Maxwell's famous equations for the electromagnetic field can also be seen (when phrased appropriately) to be insensitive to the metric scaling, needing only the null-cone structure and not the full metric, i.e. Maxwell's equations are *conformally* invariant. Moreover, the Schrödinger equation for a photon's wave-function is, in effect, just Maxwell's equations and is therefore also conformally invariant. This conformal invariance would extend also to other massless particles and, in an appropriate sense, to gravitational wave propagation.

Now, as with Escher's representation of the infinity of the hyperbolic plane shown in Fig. 22, we can also use conformal re-scalings of the metric to get a good picture of the future infinity of cosmological models.

When there is a positive value for the cosmological constant Λ, we find that this future conformal infinity is space-like [25], so it represents this infinity as a temporal "moment", albeit a moment that would be at time "infinity" according to the normal space-time metric. This conformal "squashing down" of temporal infinity is represented at the top part of Fig. 27. This procedure is very general for space-times with positive Λ, as has been demonstrated by Helmut Friedrich [26], who showed that such a conformal future boundary is generic for space-times with $\Lambda > 0$ and massless field sources.

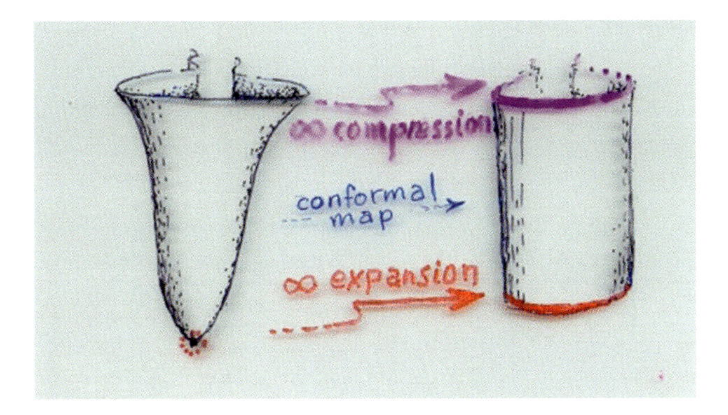

Figure 27.

This conformal squashing of the future can be accompanied by a conformal stretching out of the big bang to obtain a smooth spacelike past boundary, as illustrated at the bottom part of Fig. 27. This is a standard procedure that can be applied to most conventional cosmological models [25], [27]. However, these models assume isotropy and homogeneity, and therefore do not address the issue, raised earlier, that singularities of the type illustrated in Fig. 20 would have been vastly more probable, and we need some form of "Weyl curvature hypothesis", asserting that "past-type" space-time singularities must have highly restricted (perhaps zero) Weyl curvature, while no such restriction would be appropriate for future-type space-time singularities.

Yet, we definitely need such a hypothesis to restrict the possibilities for the *past* if we are to describe space-time models which have any real hope of describing the actual universe, in which there is a 2nd Law of Thermodynamics in accordance with what is observed, for which the low entropy in the early universe arises from the initial suppression of gravitational degrees of freedom. As was remarked earlier, the presence of a very early phase of the universe in which there was an inflationary expansion does not in itself resolve this issue (and it is my own opinion that there was actually no such inflationary phase – an issue that we shall need to return to shortly). In any case, inflation or no inflation a huge constraint on the Big Bang is required, which is indeed of the nature of some kind of "Weyl curvature hypothesis", as appears to be a feature of the actual universe in which we find ourselves. Accordingly, the ontological status of adopting each of the two conformal boundaries, depicted in Fig. 27, could hardly be more different, the one in the future being a generic procedure, very broadly applicable, and imposing no significant constraint on the applicability of the procedure being presented, whereas the one in the past provides an enormous restriction on the type of universe model under serious consideration.

Nevertheless, we can regard such a restriction as indeed formulating a version of Weyl curvature hypothesis, where we may regard such a hypothesis as an essential feature of any universe model having a chance of representing the actual world that we see around us. In fact, it was Paul Tod (of the university of Oxford, and a former graduate student of mine) who first formally proposed, and then studied in detail, this form of Weyl curvature hypothesis [28] – namely that for our actual universe, the stretching procedure, illustrated at the bottom of Fig. 27 should result in a smooth initial spacelike hypersurface boundary, this being a far more attractive and mathematically tractable procedure than my original rather vague form of this hypothesis. Tod's procedure allows detailed calculations to be performed, and this enables the implications of the hypothesis to be studied in some considerable detail.

With regard to the future conformal infinity, it had been a useful "mathematical trick" in the study of gravitational radiation, etc., to imagine that the future infinity could be conformally extended smoothly to a fictional space-time continuation beyond this future infinity [29], [30]. But now we can imagine that a kind of "time-reverse" of this trick is applied to the Big Bang, where we contemplate a fictional pre-Big-Bang extension of our universe. Thus, not only can we imagine a fictional extension of the universe's future to some kind of world beyond, but we can also contemplate a fictional world that extends our universe in a conformally smooth way to some world prior to our Big Bang.

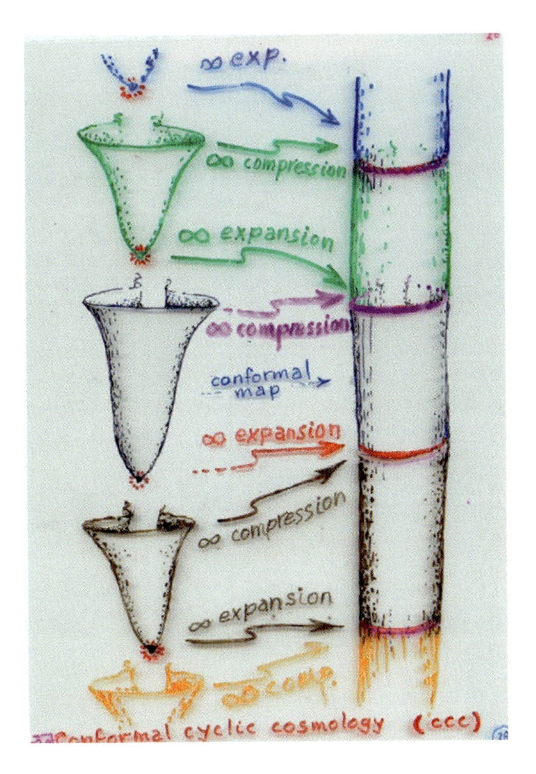

Figure 28.

This may not be quite the usual way that cosmologists have thought to picture the universe, but there is nothing outrageous about it. What may actually be regarded as outrageous, on the other hand, is the picture presented in Fig. 28, where the view is taken that neither of these conformal extensions is taken to be "fictional", but *both* are regarded as being actually *real*. The new feature about these extensions is that we do not continue the Big Bang singularity to another future-like singularity or perhaps a remote future boundary to a remote past-type boundary, but we preserve the time directions, to continue our non-singular remote future boundary to a succeeding big bang singularity, and precede our Big Bang singularity by a non-singular remote future boundary, thereby automatically *forcing* our Big Bang to satisfy a Weyl curvature hypothesis, as appears to be required.

Such a picture makes geometrical sense, but we must ask whether it can possibly make *physical* sense. On first consideration one might well take the view that this is unreasonable because the remote future is extremely cold, and the density very low, whereas at the Big Bang, things were very much the opposite, with an extraordinarily high temperature and density. However, when you conformally rescale things, the conjugate variables go the opposite way. Time scales oppositely to energy and space scales oppositely to momentum. Thus, the very cold and rarefied remote future re-scales to the very hot and dense next big bang, this being consistent with the very large measures of space and time in the remote future rescaling to very tiny measures of space and time in the next big bang, all this being consistent with the model that I have been proposing [31].

There is also an issue about the cosmological constant Λ. This needs to be positive for the scheme to work. The stretched-out Big Bang appears to be space-like in all serious models, but for the scheme to work we are restricted to those for which the conformal infinity is also space-like, this corresponding to $\Lambda > 0$ which, fortunately for the scheme, appears to be the case [25]!

I refer to this scheme as *conformal cyclic cosmology* or CCC for short. The portion of the sequence from a big bang moment to its following remote future I call an aeon. I adopt the conceptually simplest version of CCC that there is an infinite succession of aeons, infinite in both directions, though other possibilities might also be considered. I also adopt the view that the aeons are qualitatively similar to one another so that, the constants of nature do not vary from aeon to aeon, but other possibilities are certainly open to consideration. I tend to use the capitalised "Big Bang", when this refers to the specific moment that initiated our current aeon, and "big bang" otherwise.

A comment needs to be made in relation to inflation. Certainly, CCC is not compatible with the version of inflation that is currently favoured by many cosmologists, because this would provide a causal gap between aeons that would ruin the observational issues that will be described below, though a very small inflationary phase could be considered. A more satisfactory CCC picture would be to eliminate inflation altogether, the hope being that the things that inflation is useful for in cosmology can be taken over by the final exponential behaviour of the previous aeon, which plays any role that in conventional cosmology an exponentially expanding phase seems to be required. In short, CCC does provide an "inflationary phase" in effect, but it occurred prior to the Big Bang rather than following it!

Our final issue has to do with possible observational tests of CCC. In fact, there are several of these, especially if one takes the view that the various cosmic aeons are necessarily qualitatively similar to each other. Then there are certainly issues about physical parameters having to match on two sides of the crossover 3-surface joining one aeon to the next. Another possibility might be signals, such as electromagnetic ones getting through from one aeon to the next, as indicated in Fig. 29. These might be important magnetic fields getting across, for example, but that has not yet been looked at.

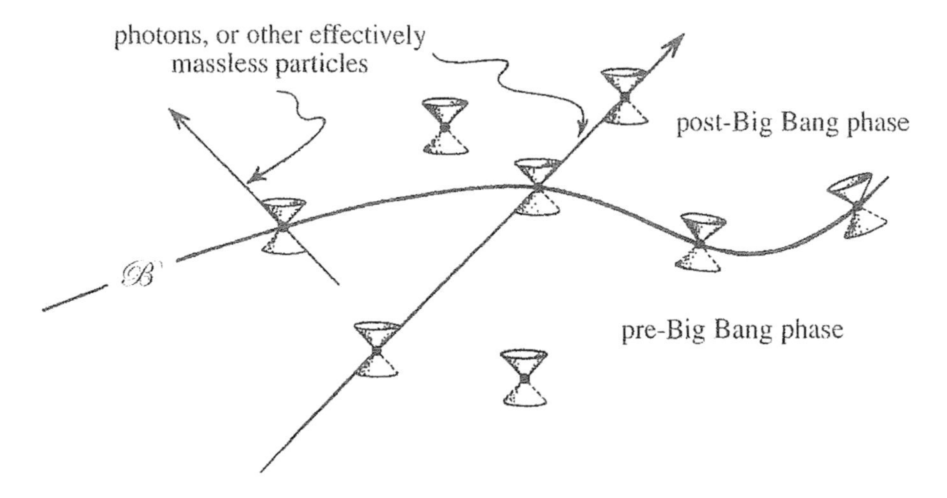

Figure 29.

An alternative, which has indeed been examined is the possibility of gravitational wave signals getting across. These could certainly get through, in the CCC picture. We can imagine that encounters between supermassive black holes within a previous-aeon galactic cluster should encounter one another from time to time, and in doing so, should emit enormous amounts of energy in the form of gravitational waves. Such waves ought certainly to be able to get through into our aeon after smoothly propagating through the crossover between the two. When in our aeon, such waves would transfer some of their energy into electromagnetic form and slightly affect the temperature of our microwave background, each wave often making a large ring of such slightly increased temperature. Such a ring can be understood as being the intersection of the sphere W^2 here the gravitational wave encounters our last scattering 3-surface L^3 win the sphere M^2 which is where our past light cone meets L^3, so that M^2 is, is fact, our own microwave background sky. The circle in the microwave sky that we are looking for is the intersection W2∩M2 of these two spheres. Rather remarkably, two groups appear to have independently observed rings consistently with such expectations. One was a Polish group, [32], [33], who regard their observations to support our CCC expectations, with around 99.5% confidence level.

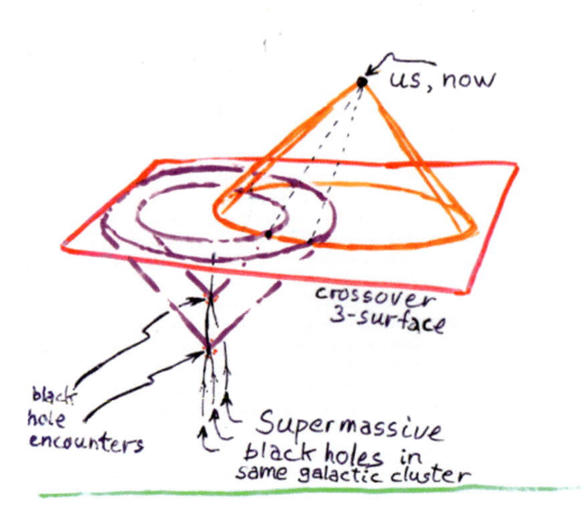

Figure 30.

The other is a collaboration between myself and my Armenian colleague Vahe Gurzadyan. In order to get a significant enough signal, we took advantage of the fact that within any particular large-enough galactic cluster in the previous aeon there ought normally to be several such encounters between supermassive black holes. Each such black-hole encounter should provide one of these rings in our CMB, but since there should be several of these events within the same cluster, these should provide rings with the same centre (see Fig. 30). In our papers [34], [35] we considered that the temperature disturbance in the CMB should be able to be seen as a reduced temperature variance around the ring, and to get a strong enough signal, we looked for occurrences where there are at least 3 different low-variance rings with the same centre, being from the same cluster. This would be plotted as a single point in the map of our CMB sky. The results are illustrated in Fig. 31, and we find what we regard as a rather remarkable effect, First of all, we notice that the points, indicating the ring centres and therefore, according to the theory, the locations of galactic clusters in the previous aeon, are extremely clustered, and certainly not uniformly distributed across the sky, as would have been expected according to conventional ideas about the uniformity of the universe, that should reveal itself on a large-enough scale. Moreover, the points are also distinctively clustered with regard to colour, this colour coding referring to the temperature assigned to the individual ring whose centre is being marked in the picture. We must also recall that the criterion for selecting the points was not the overall energy (temperature) of the ring, but the fact that the *variance* in the temperature was low. The actual temperature is an independent parameter, and (according to the theory) should be a signal of the distance of the source of the signal from us. The theory says that the red points in the picture are extremely distant and the blue ones somewhat less distant. Accordingly, according to the theory, the galactic clusters that we seem to be seeing are very non-uniformly distributed, not only in their angular separation across the sky, but also in the distance away from us!

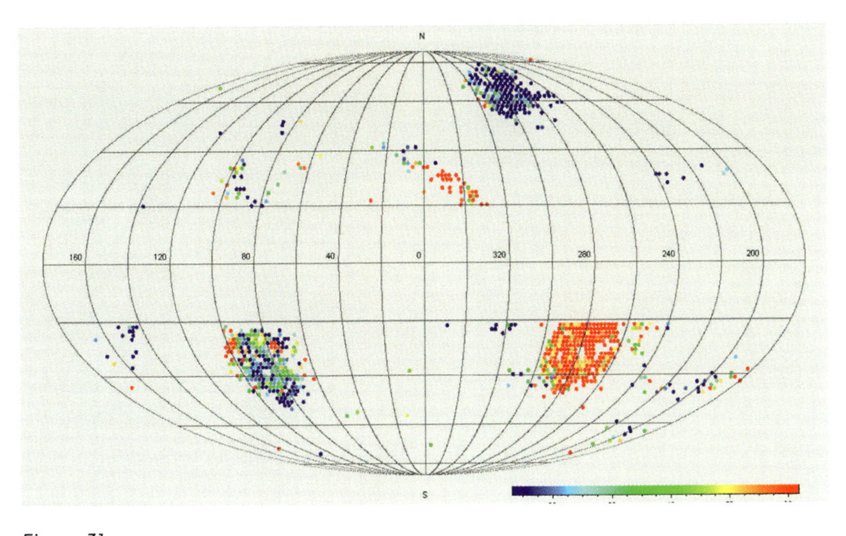

Figure 31.

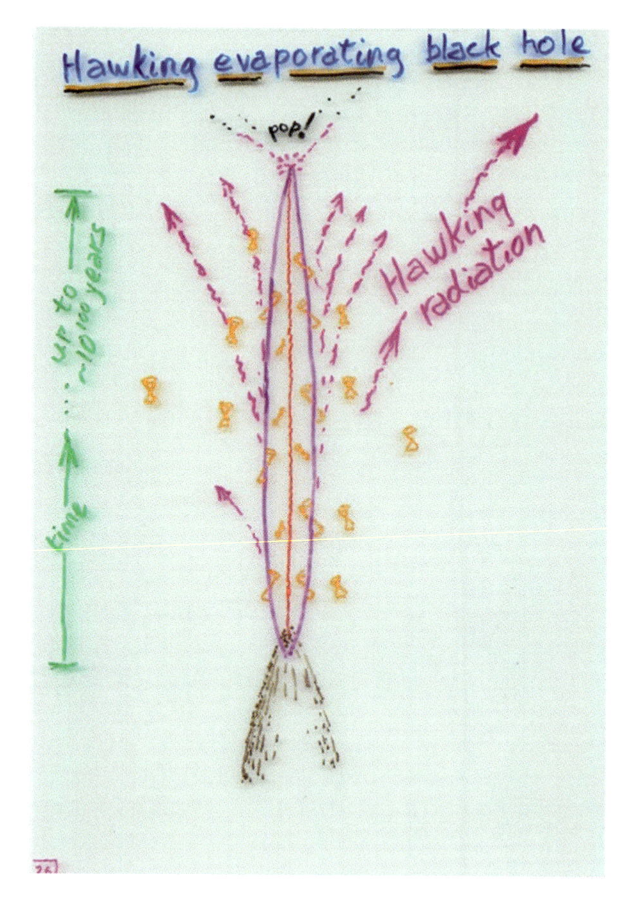

Figure 32.

Finally, we should consider the completely different effect illustrated in Fig. 32. This concerns the ultimate phenomenon that we presume to have taken place in the previous aeon. According to the effect predicted by Hawking, a black hole ought to have a very tiny temperature, referred to as the Hawking temperature, which for a very large black hole would be exceptionally tiny. Nevertheless, as the universe expands, the temperature of the universe gets lower and lower until it becomes smaller even than the Hawking temperature of the supermassive black hole, at which point the hole itself begins to evaporate away, all of its enormous mass being eventually radiated away into this Hawking radiation. However, because this occurs so extremely late in that aeon's existence (perhaps at least 10^{100} years), this entire activity occupies what is effectively a single point on the crossover surface, as exhibited in Fig. 33. However, the mass-energy cannot be lost, and it bursts through into the subsequent aeon (our aeon) at a single point that we refer to as a *Hawking point*. The energy bursting through at that point would disperse itself through that early material reaching a certain diameter until revealing itself as a heated spot of a certain diameter on the last scattering surface. See Fig. 34. It turns out that we actually see such spots in the CMB sky with an angular diameter of about 4° radians (about 8 times the diameter of the full moon), which is close to what one should expect on theoretical grounds, with a confidence level of about 99.98%. [36]

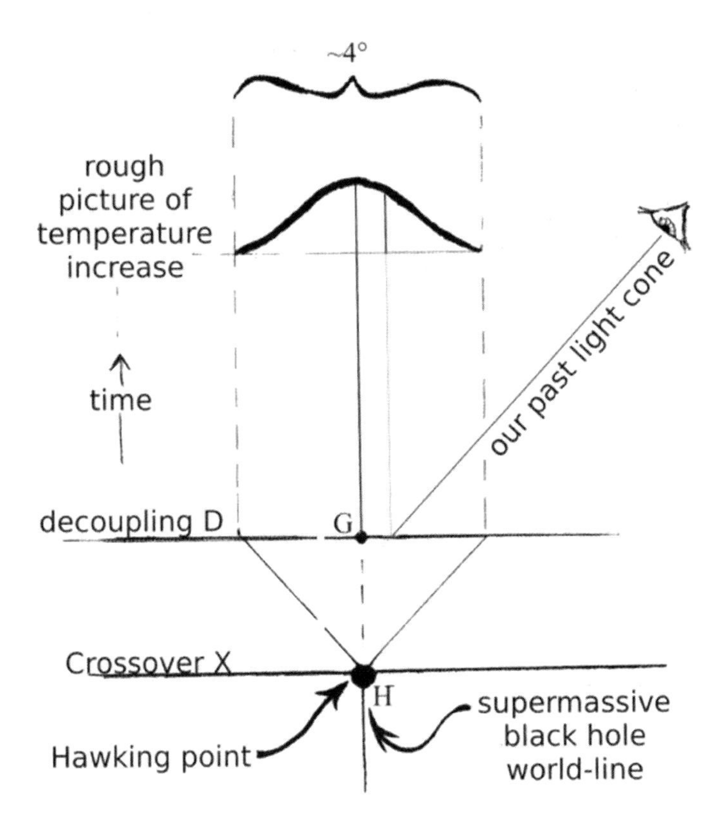

Figure 33.

We have been provided with two different kinds of signal that seem to exhibit rather strong evidence for a cosmic aeon prior tours, in accordance with CCC. A natural question to ask is whether there might, in addition, be some interconnection between the two. This ought to be the case, but with the data currently available the connection remains a somewhat weak one so far. The gravitational waves accessible to us from a galactic cluster that ends up as a Hawking point that we can directly see would have to b "edge-on"' to our line of sight and would therefore be marked as "green" (middle-range temperature) in Fig. 31. Of the 5 points that are clearly seen in both the Planck and WMAP data at exactly the same places in each [36], it is noteworthy that 3 of then are remarkably close to such green points in [35]. Various factors could contribute to any slight discrepancies, such as imprecision in locating the exact centres of the rings in question or even proper motions building up over the vast intervening timescales involved.

Warm thanks go to Dennis Lehmkuhl for several historical clarifications.

REFERENCES

It should be noted that the "phi" coordinates in [35] and [36] differ by a right/left reflection.

1. Oppenheimer, J.R. and Snyder, H. (1939) On continued gravitational contraction. *Phys. Rev.* **56**, 455–9.
2. Chandrasekhar, S. (1931) The maximum mass of ideal white dwarfs. *Astrophys. J.*, **74**, 81–2.
3. Finkelstein, D. (1958) Past-future asymmetry of the gravitational field of a point particle, *Phys. Rev.* **110**, 965–7.
4. Painleve, P (1921) La mécanique Classique et la théorie de la relativitité. *R. Acad. Sci.* (Paris) **173**, 677–80
5. Lemaître, G. (1933) L'universe en expansion *Ann. Soc. Sci. Bruxelles* I **A53**, 51–85 (cf. p. 82).
6. Dirac, P.A.M. (1936) Relativistic wave equations. *Proc. Roy. Soc.* (Lond.) **A155**, 447–59.
7. Payne, W.T. (1952) Elementary spinor theory *Am. J. Phys.* **20**, 253–62.
8. Penrose, R. and Rindler, W. (1984) *Spinors and Space-Time, Vol. 1: Two-Spinor Calculus and Relativistic Fields* (Cambridge University Press, Cambridge).
9. Bade, W.L. and Jehle, H. (1953). An introduction to spinors. *Rev. Mod. Phys.* **25**, 714–28.
10. Lifshitz, E.M. and Khalatnikov, I.M. (1963) Investigations in relativistic cosmology, in *Advances in Physics* **12**,185–249.
11. Penrose, R. (1965) Gravitational collapse and space–time singularities, *Phys. Rev. Lett.* **14**, 57–59.
12. Penrose, R. (1972) Techniques of Differential Topology in Relativity, *CBMS Regional Conf. Ser. in Appl. Math.*, No. 7 (S.I.A.M., Philadelphia).
13. Hawking, S.W. and Ellis, G.F.R. (1973) *The Large-Scale Structure of Space-Time* (Cambridge University Press, Cambridge).
14. Belinskii, V.A., Khalatnikov, I.M. and Lifshitz, E.M. (1970) Oscilliatory

approach to a singular point in the relativistic cosmology, *Usp. Fiz. Nauk* 102, 463–500. Engl. transl. in *Adv. in Phys.* **19**, 525–73.
15. Misner, C.W. (1969) Mixmaster universe. *Phys. Rev. Lett.*, **22**, 1071–4.
16. Penrose, R. (1973) Naked singularities, *Ann. N.Y. Acad. Sci.*, **224**, 125–34.
17. Hawking, S.W. (1965) Occurrence of singularities in open universes. *Phys. Rev. Lett.* **15**, 689–90.
18. Hawking, S.W. (1966) The occurrence of singularities in cosmology. *Proc. Roy. Soc.* (London) **A294**, 511–21.
19. Hawking, S.W. (1966) The occurrence of singularities in cosmology II. ..... . *Proc. Roy. Soc.* (London) **A295**, 490–3. [Adams Prize Essay: Singularities in the geometry of space-time.]
20. Hawking, S.W. (1967) The occurrence of singularities in cosmology III. Causality and singularities. *Proc. Roy. Soc.* (London) **A300**, 187–201.
21. Hawking, S.W. and Penrose, R. (1970) The singularities of gravitational collapse and cosmology *Proc. Roy. Soc.* (London) **A314** (1519) 529–48.
22. Linde, A.D. (1982) A new inflationary universe scenario: a possible solution of the horizon, flatness, homogeneity, isotropy and primordial monopole problems. *Physics Letters B* **108**:389–93
23. Guth, A.H. (1981) Inflationary universe: A possible solution to the horizon and flatness problems. *Phys. Rev.* **D23**, 347–56.
24. Guth, A.H. (1997) *The Inflationary Universe* (Jonathan Cape, London).
25. Penrose, R. (1964) Conformal approach to infinity, in Relativity, Groups and Topology: *The 1963 Les Houches Lectures*, eds. B.S. DeWitt and C.M. DeWitt (Gordon and Breach, New York).
26. Friedrich, H. (1998) Einstein's equation and conformal structure. In T*he Geometric Universe; Science, Geometry, and the Work of Roger Penrose*, eds. S.A. Huggett, L.J. Mason, K.P. Tod, S.T. Tsou, and N.M.J. Woodhouse (Oxford Univ. Press, Oxford).
27. Penrose, R. (2004) *The Road to Reality: A Complete Guide to the Laws of the Universe* (Jonathan Cape, London). Vintage IBN: 9780-679-77631-4.
28. Tod, K.P. (2003) Isotropic cosmological singularities: other matter models. *Class. Quantum Grav.* **20**, 521–534 doi:10.1088/0264-9381/20/3/309.
29. Penrose, R. (1963) Asymptotic properties of fields and space-times, *Phys. Rev. Lett.* **10**, 66–8.
30. Penrose, R. (1965) Zero rest-mass fields including gravitation: asymptotic behaviour, *Proc. Roy. Soc. London*, **A284**, 159–203.
31. Penrose, R. (2010) *Cycles of Time: An Extraordinary New View of the Universe.* (Bodley Head, London). ISBN 978-0-224-08036-1; 9780099505945.
32. Meissner, K. A., Nurowski, P. and Ruszczycki, B. (2013) Structures in the microwave background radiation, *Proc. R. Soc.* A469:2155, 20130116, arXiv:1207.2498[astro-ph.CO].
33. D. An, Meissner, K. A. Nurowski, P, *Mon. Not. Roy. Astron. Soc.* **473** (2018) no. 3, 3251.
34. Gurzadyan, V.G. and Penrose R. (2013) On CCC-predicted concentric low-variance circles in the CMB sky. *Eur. Phys. J. Plus* 128: 22 DOI 10.1140/epjp/i2013-13022-4.
35. Gurzadyan, V.G. and Penrose R. (2016) CCC and the Fermi paradox *Eur Phys. J. Plus* (2016) 131:11.
36. An, D, Meissner, A, Nurowski,P, Penrose, R, (2020) Apparent evidence for Hawking Points, *Mon. Not. Roy. Astron. Soc.* **495** no. 3 3403.

Reinhard Genzel. © Nobel Prize Outreach AB. Photo: Bernhard Ludewig

Reinhard Genzel

Biography

I WAS BORN IN BAD HOMBURG, near Frankfurt, Germany in 1952. My parents and I lived in an apartment right above the physics laboratories of the University, where my father was an instructor. I also started elementary school there. In 1960 my parents moved to southwestern Germany, to the idyllic Black Forest town of Freiburg, bordering France (Alsace) and Switzerland. There I went to high school at a 'humanistic gymnasium', featuring 9 years of Latin and 6 years of Greek. Perhaps as a result I have enjoyed a lifetime interest in history and archeology.

My father was a well-known experimental solid state physicist and a gifted university teacher. I learned most of my early physics from him. I will always treasure how he showed me (aged 16) how to build a decent optical spectrometer from its basic optical components, which could resolve the sodium D lines. My mother had studied economics and then worked, together with her father, in managing a leather factory near Frankfurt. Once we had moved to Freiburg, she stopped her professional career and devoted herself to her husband and only child, as was typical of most German women of her generation. Only at the end of her life did she tell me that she regretted having given up her career.

During my high school years, I spent a lot of my time doing intense sports. To this day I am proud of having been one of Germany's best young javelin throwers, as well as being on my school's handball team. I even made it into the national German junior track and field team training for the 1972 Munich Olympics. An elbow injury and the increasing incompatibility with university studies brought my track and field career to a rapid end in the early 1970s. I still feel compelled to do some daily workout activity and until recently went frequently on mountain hikes.

Figure 1. Family and youth. Top (from left to right): My father Ludwig, my mother Eva-Maria, and I as boy during the Frankfurt years. Bottom: With my parents on a walk (left), and on the Handball team of the Berthold-Gymnasium in Freiburg (around 1968, I am on the left in the top row).

After completing my undergraduate physics education at the Albert Ludwig University in Freiburg I moved to Bonn for my graduate education (1970–1974). At that time my father had also moved from Freiburg, to become one of the founding Directors of the new Max-Planck Institute for solid state research in Stuttgart. In a discussion about my future in physics, he advised me against nuclear and particle physics (and in any case his own field, solid state physics). He mentioned that the Max Planck Society (MPG) had just founded a new Institute for Radio Astronomy in Bonn (MPIfR), with a 100m single dish telescope in the Eifel mountains as its key new research instrument. Would I perhaps be interested in doing astronomy? And so, I followed this advice and began my astronomy career initially as a master (Diplom) student, and then as a PhD student (Doktorand) student at the MPIfR, under Peter Mezger, one of the Institute's Directors.

Working with the then largest radio telescope in the world was a fantastic opportunity for a young student. I was particularly fascinated by

Figure 2. Stations of my life. From top left, counter-clockwise to top right: The Rathaus in Frankfurt, called the 'Römer' (1952–1960). The idyllic town of Freiburg im Breisgau, with its medieval old core, and its famous Gothic Münster (1960–1972). Statue of Ludwig van Beethoven in the center of his birth-town, Bonn (1972–1978). Harvard University and the Charles River near Boston, Massachusetts (1978–1980). Looking west from the Campanile of the University of California across the Bay, with San Francisco, Alcatraz Island and the Golden Gate Bridge in the background (1980–1986, and then part-time 1999–2021). Looking south across the center of Munich, on a windy day with 'Föhn', where the Alps 100 km south of the City seem be just outside of town (1985–2021).

the emerging field of molecular spectroscopy. Working in close collaboration with Dennis Downes, my PhD thesis work (1976–78) was on the phenomenon of interstellar water vapor masers, which had been discovered ten years previously by the group of Charles Townes at the University of California, Berkeley (UCB). Townes, Nobel Laureate for the invention of the maser and laser, had switched to experimental astrophysics research in the late 1960s, and was at the forefront of the fledgling field of molecular astrophysics. My thesis work showed that the H2O masers originated in dense, dusty clouds in the process of forming massive stars. Working together with colleagues in the USA and the Soviet Union, we used the

100m telescope as part of a network of intercontinental "Very Long Baseline" Interferometry (VLBI), generated by wave interference milli-arcsecond angular resolution. We demonstrated that the masers were extremely dense little cloudlets, probably compressed and excited by supersonic shocks and the intense radiation of the newly formed stars.

My thesis work was completed in 1978 and led seamlessly to postdoctoral work with James Moran at the Harvard-Smithsonian Center for Astrophysics (CfA). In the next two years, during my time as a CfA Postdoc, we expanded further on the maser VLBI studies and could show that the maser phenomenon was triggered by rapid gas outflows during the protostellar stages.

In 1980 I was offered the unique opportunity to change fields and join the group of Charles Townes in Berkeley, California, initially as a Miller Postdoctoral Fellow and then one year later as Associate Professor in the Physics Department. My luck of becoming tenured faculty at Berkeley at

Figure 3. My mentors. Top row (left): My father (left) who taught me physics and the love of experiments and remained a key advisor until his death in 2003. Bottom row (left): Peter Mezger, my official 'Doktorvater' during the Bonn period. Top row (right): James Moran, from whom I learned about interferometry and masers during my postdoc period at the CfA, who has remained a good friend up to this day. Bottom (middle/right): Charles Townes, whom I regard as my second father, and from whom I learned about infrared experimental astrophysics, about styles of research, including when to move on to newer pastures, about decency in scientific work, and about US political developments as seen by a prominent Nobel Laureate and government science adviser.

age 29 was mainly the result of Charlie's support and invaluable mentorship, but perhaps also because I had in the meantime received a competing offer for a professorship at Caltech. The scientific life with Charlie and his large group was an incredible experience. I finally did experimental physics in the lab, as in my youth, was polishing my own mirrors, and learned to work with cryogenic far-infrared detectors. The Townes group was carrying out ground-breaking experimental astrophysics work across the mid-infrared to submillimeter wave-bands, which fascinated me enormously. Life for the next six years at Berkeley was intense and exciting. We were flying an innovative high-resolution Fabry-Perot spectrometer on NASA's Kuiper Airborne Observatory, to explore for the first time Galactic star formation regions, the Galactic Center, and nearby galaxies, in far-infrared fine-structure and molecular rotation lines. Interestingly the Fabry-Perot reflectors and other filters in the instrument were based on a stretched metal-mesh technology which my father had invented some years prior. I was trying to combine teaching physics, with intense research in infrared and submillimeter spectroscopy – developing new instruments, working with my first graduate students, and finding time for the family. During that time, I was very much helped by Dan Jaffe, an old friend from the CfA, who had joined my fledgling group as a Senior Fellow at the UCB Space Sciences Lab, and taking on many tasks I really disliked and was not good at, in particular hunting for grants. Another key person was my first student, Andy Harris, who has also remained a lifelong friend.

At that time, my family was also developing. In 1976, I had married Orsolya Boroviczény, whom I had got to know in my late high school days. She had studied medicine at the University in Bonn and at the Harvard Medical School in Boston, and then carried out her pediatrics residency at Children's Hospital in Oakland. Our older daughter, Daria, was born in Boston in 1979 and the younger one, Lisa, in Berkeley in 1983. California and Berkeley became more than a temporary residence. Even after returning to Germany in 1987, the family considered Berkeley as our 'Heimat', that sentiment being an important factor in my taking on a commitment as part-time professor in the UCB Physics department again in 1999. The current status of the family, including two grandchildren is summarized in Figure 4.

In 1986 I was offered and accepted a directorship at the Max Planck Institute for Extraterrestrial Physics (MPE) in Garching, near Munich, in significant part due to the support of Gerhard Haerendel and Joachim Trümper. At the time, I was quite unsure whether returning to Germany was the best choice. Berkeley was wonderful, Europe looked pretty un-dynamic. However, I never regretted it. The older I get the more I am grateful to the Max Planck Society (MPG) for providing such an abso-

Figure 4. Family. On the left side is my wife, Orsolya, who was born in Budapest, Hungary and then fled in 1956 with her parents from the communist regime to settle in Germany. Orsolya is a now retired neonatologist at the Ludwig Maximilian University in Munich. Top: a flying baby in 1985 (my younger daughter Lisa, who is also seen in her current incarnation at the bottom right). Lisa is a neuroscientist at Radboud University in Nijmegen (Netherlands). Lisa's son Nelu is at bottom, center, and is presently 4 months old. Center: My older daughter Daria with her husband Marco and their daughter Pika, presently 1.5 of age. The same young lady is seen in the upper right in my arms, when she was a few days old in 2019. Daria is a biologist/neuroscientist now working at Spike Gadgets[1] in San Francisco, and Marco is a Lab Coordinator and Adjunct Physics Faculty at Skyline College, San Bruno.

lutely wonderful opportunity to pursue basic research at the top level with so few strings attached.

Building up a substantial infrared/submillimeter experimental astrophysics group at MPE has been a fantastic and all-consuming opportunity for me. We started from a small core of a dozen people around senior scientist Siegfried Drapatz, and grew to more than sixty over two decades, when we were engaged in two large space experiments with the European Space Agency (ESA). Our style was similar to what I had learned from Charlie in Berkeley. The most important aspects driving our research are the fundamental large science questions in Galactic and extragalactic star formation, physics of the interstellar medium, galaxy

1. https://spikegadgets.com/

evolution, massive black holes (the Galactic Center being the most important goal) and their activity, and co-evolution over cosmic time of the activity of these massive black holes with their host galaxies (active galactic nuclei: AGN). As I have described in more detail in my Nobel Lecture about our Galactic Center research, we next identified how we could tackle these questions with novel infrared to millimeter experiments. These longer wavelengths were not yet explored, but are of great interest astronomically for studying cold, dusty regions, or regions with large cosmological redshifts. Novel technologies (adaptive optics, integral field spectroscopy, spatial interferometry, photonics, new semi-conductor detectors etc.) promised very substantial progress. Typically, each project has a senior scientist project leader, and 2 to 7 additional scientists and postdocs, a few masters and PhD students, plus additional technicians and engineers 'on delegation' from the MPE central divisions. Our team culture calls for sharing as best as possible in the various challenges, from hardware/software design, building and testing, to observing, data analysis and interpretation to, finally, publications and conference presentations. As an umbrella of these individual projects groups, we have a circle of senior scientists (chaired for many years by Linda Tacconi and Dieter Lutz) who bring the individual parts back to an overall strategy.

Having obtained first science results with one of these new instruments, new ideas and questions came up and the cycle repeated. A key difference from Charlie's approach has been the scale of the experiments, and of our teams, driven by the possibilities available at the MPG and at the MPE. Because of the long space research tradition of MPE since its inception in 1963, we dared to do bigger, more ambitious and sometimes riskier experiments, not only relying on students and postdocs, but also on professional physics, astronomy, engineering and IT staff in the group, and in the central divisions at MPE.

Initially we carried out projects without immediate connection to a specific telescope. Rather we developed an instrument based on our science questions and then made a proposal to a suitable observatory to bring the instrument to their telescope for a campaign typically lasting 1–2 months, sometimes several times. To conclude such deals typically meant that in return we made our instrument available to other scientists in the community of the observatory. The SHARP speckle camera for the European Southern Observatory's New Technology Telescope (ESO NTT) was one example of such an instrument, '3D', the world's first infrared integral field spectrometer, a second, and 'Receiver G', a heterodyne spectrometer for submillimeter observations on the James Clerk Maxwell Telescope (JCMT) on Hawaii's Mauna Kea, a third.

This style changed with the advent of the Very Large Telescope (VLT). ESO was unlikely to allow visitor instruments on its brand-new facility any time soon, so we changed our style to competing for instrumentation developments ESO wanted to have on their new 8m class telescopes, and bidding for a building contract. The adaptive optics camera CONICA (with the Max Planck Institute for Astronomy, MPIA Heidelberg) as part of the NACO adaptive optics system (with the Observatoire de Paris), the SPIFFI-SINFONI integral field spectrometer (with the Observatory of Munich (LMU), the PARSEC laser guide star facility, the KMOS multiplexed integral field spectrometer (with the UK Universities Durham and Oxford, with the UK ATC in Edinburgh, and the Observatory of Munich (LMU) and finally, largest of all, the GRAVITY interferometric beam combiner, were all examples of this approach. In this way we earned privileged (guaranteed) telescope time (GTO), which we could use for our top science. The same happened for our involvement in the European Space Agency (ESA) infrared space telescopes ISO (as co-PI Institute on the Short Wavelength Spectrometer (SWS), with Leo Haser as co-PI) and on the Herschel space telescope (as lead institute for the PACS instrument, with Albrecht Poglitsch as PI). In retrospect I am sure that most of our most important science results in the Galactic Center, in active galactic nucleus (AGN) studies and in galaxy evolution would not have been possible without this GTO mechanism, which permitted devoting significant telescope time over sustained periods to a single or a few core science projects.

In the millimeter range, we gave up our own developments fairly early on, and instead relied on and strongly supported the Institut de Radioastronomie Millimetrique IRAM), a German-French-Spanish joint institution in Grenoble, France. IRAM operates a 30m diameter single dish telescope on Pico Veleta (Spain), and an increasingly powerful interferometer on the Plateau de Bure in southern France and develops all necessary focal plane instrumentation for these world-class facilities. Our IRAM involvement was and continues to be a great success. With the recent upgrade to the Northern Extended Millimeter Array (NOEMA), the interferometer is by far the most capable millimeter interferometer in the northern hemisphere, and in several aspects comparable and competitive with the larger international Atacama Large Millimeter Array (ALMA) facility in Chile.

At the current count, we have built 25 instruments, most in a leadership or co-leadership role. Over my past 35 years as a Director here at MPE, the IR-submm group has had 70 PhD students, 85 postdocs and scientists, 15 senior scientists and 25 technical staff. I am particularly proud about the diversity of our group, both in terms of geographical distribution (>25 different countries) and gender (our female contingent has lately been varying between 30 and 36%, as compared to 12% in 1986).

I have to admit that my intense and continuing engagement in many

Figure 5. The Mountain Man. The opportunities for many, sometimes extensive mountain hikes and climbs, in the Bavarian Alps, in the Sierra Nevada of California and in Alaska, together with good friends and family, have been precious to me. Our greatest trip was a 200+ mile hike along the John Muir trail from north to south, along the ridge of the Sierra Nevada in 2002 (with Linda Tacconi, and Ali and Ric Davies).

aspects and areas of astrophysical research – trying to keep our group at the top level, and anticipating the next steps (e.g. with the remarkable breakthroughs our VLTI infrared interferometer activities led by Frank Eisenhauer, and our leadership of the first-light camera MICADO, P.I. Ric Davies, for the next-generation, 39m diameter giant extremely large telescope of ESO, the ESO-ELT), leaves little time for other things. My regular visits to Berkeley in winter and summer have become welcome interludes, with the possibility of reading books, writing papers in peace, hiking along the beautiful California coast and thinking about the next steps.

It has been a privilege for me to be able to have had such a wonderful, supportive family and a group of many outstanding and capable colleagues and friends, who were able to join and tolerate a sometimes grumpy or "single channel" guy. It has been (and continues to be) a lot of fun! Thank you, Linda, Natascha, Adriane, Susanne, Albrecht, Amiel, Dieter, Eckhard, Frank, Helmut, Sebastian, Stefan, Ric, Taro and Thomas.

A Forty-Year Journey

Nobel Lecture, December 8, 2020 by
Reinhard Genzel
Max Planck Institute for Extraterrestrial Physics, Garching, Germany
Departments of Physics & Astronomy, University of California, Berkeley, USA.

1. Prologue

A *'black hole'* (e.g. Wheeler 1968) conceptually is a region of space-time where gravity is so strong that within its event horizon neither particles with mass, nor even electromagnetic radiation (massless photons), can escape from it. Based on Newton's theory of gravity Rev. John Michell (in 1784) and Pierre-Simon Laplace (in 1795) were the first to note that a sufficiently compact, massive star may have a surface escape velocity exceeding the speed of light. Such an object would thus be 'dark' or invisible. A proper mathematical treatment of this remarkable proposition had to await Albert Einstein's theory of General Relativity in 1915/1916 (henceforth GR). Karl Schwarzschild's (1916) solution of the vacuum field equations in spherical symmetry demonstrated the existence of a characteristic *event horizon* of a mass M, the *Schwarzschild radius* $R_s=2GM/c^2$, within which no communication is possible with external observers. It is a 'one way door'. Roy Kerr (1963) generalized this solution to spinning black holes. However, these solutions refer to configurations with sufficiently high symmetry, so that Einstein's equations can be solved analytically, and there was doubt about whether such cases were typical. Roger Penrose, one of the other recipients of this year's Nobel Prize, dropped the assumption of spherical symmetry, and analyzed the problem topologically (Penrose 1963, 1965). Using the key concept of 'trapped surfaces' he showed that any arbitrarily shaped surface with a radius less than the Schwarzschild radius is a trapped surface, and the radial direction becomes time-like as one passes through the horizon. Any observer is then inexorably pulled towards the center where time ends. All the matter that forms the black hole resides at this single moment in time, the singularity.

From considerations of the information content of black holes, there is significant tension between the predictions of GR and general concepts of quantum theory (e.g. Susskind 1995, Maldacena 1998, Bousso 2002). It is likely that a proper quantum theory of gravity will modify the concepts of GR on scales comparable to or smaller than the Planck length, $l_{Pl} \sim 1.6 \times 10^{-33}$ cm, remove the concept of the central singularity, and potentially challenge the interpretation of the GR event horizon (Almheiri et al. 2013).

But are these bizarre objects of GR actually realized in Nature?

2. Overture: X-ray Binaries and Quasars

Astronomical evidence for the existence of black holes started to emerge sixty years ago with the discovery of variable X-ray emitting binaries in the Milky Way (Giacconi et al. 1962, Giacconi 2003 (Nobel Lecture 2002)) on the one hand, and of distant luminous 'quasi-stellar-radio-sources/objects' (QSOs, Schmidt 1963) on the other. For about two dozen X-ray binaries, dynamical mass determinations from Doppler spectroscopy of the visible primary star established that the mass of the X-ray emitting secondary is significantly larger than the maximum stable neutron star mass, ~2.3 solar masses (McClintock & Remillard 2004, Remillard & McClintock 2006, Özel et al. 2010, Rezzolla et al. 2018). The binary X-ray sources thus are excellent candidates for stellar black holes (SBH). They are probably formed when a massive star explodes as a supernova at the end of its fusion lifetime and the compact remnant collapses to an SBH. The measurements of gravitational waves from in-spiraling binaries with LIGO (Abbott et al. 2016a, b, Nobel Prize 2017) have recently provided very strong and arguably conclusive evidence for the existence of SBHs.

The luminosities of QSOs often exceed by 3 to 4 orders of magnitude the entire energy output of the Milky Way Galaxy. Furthermore, their strong high energy emission in the UV-, X-ray and γ-ray bands, as well as their spectacular relativistic jets, can most plausibly be explained by accretion of matter onto massive black holes (henceforth MBHs, e.g. Lynden-Bell 1969, Shakura & Sunyaev 1973, Blandford 1999, Yuan & Narayan 2014, Blandford, Meier & Readhead 2019). Between 7% (for a non-rotating Schwarzschild hole) and 40% (for a maximally rotating Kerr hole) of the rest energy of an infalling particle can, in principle, be converted to radiation outside the event horizon, one to two orders of magnitude greater than nuclear fusion in stars. To explain powerful QSOs by this mechanism, black hole masses of 10^8 to 10^9 solar masses and accretion flows between 0.1 to 10 solar masses per year are required. QSOs are located (without exception) in the nuclei of large, massive galaxies (e.g. Osmer 2004). QSOs represent the most extreme and spectacular among the general nuclear activity of most galaxies.

A conclusive experimental proof of the existence of a SBH or MBH, as defined by GR, requires the *determination of the gravitational potential on the scale of the event horizon*. This gravitational potential can be inferred from spatially resolved measurements of the motions of test particles (interstellar gas, stars, other black holes, or photons) in close orbit around the black hole (Lynden-Bell & Rees 1971). Until very recently this ambitious test was not feasible. A more modest goal then is to show that the gravitational potential of a galaxy nucleus is dominated by a compact non-stellar mass and that this central mass concentration cannot be anything but a black hole, because all other conceivable configurations either are more extended, are not stable, or produce more light (e.g. Maoz 1995, 1998). Even this test cannot be conducted (yet) in distant QSOs. Lynden-Bell (1969) and Lynden-Bell & Rees (1971) proposed that MBHs might be common in most galaxies (although in a low state of accretion). If so, dynamical tests are feasible in nearby galaxy nuclei, including the center of our Milky Way.

Over the past fifty years, since these seminal papers, increasingly solid *evidence for central 'dark' (i.e. non-stellar) mass concentrations* has emerged for about one hundred galaxies (e.g. Kormendy 2004, Gültekin et al. 2009, Kormendy & Ho 2013, McConnell & Ma 2013, Saglia et al. 2016, Greene et al. 2016), from optical/infrared imaging and spectroscopy on the Hubble Space Telescope (HST) and large ground-based telescopes, as well as from Very Long Baseline radio Interferometry (VLBI). Further evidence comes from relativistically broadened, redshifted iron Kα line emission in nearby Seyfert galaxies (e.g. Tanaka et al. 1995, Nandra et al. 1997, Fabian et al. 2000). In external galaxies the most compelling case that such a dark mass concentration cannot just be a dense nuclear cluster of white dwarfs, neutron stars and perhaps stellar black holes emerged in the mid-1990s from spectacular VLBI observations of the nucleus of NGC 4258, a mildly active galaxy at a distance of 7 Mpc (Miyoshi et al. 1995, Moran 2008). The VLBI observations show that the galaxy nucleus contains a thin, slightly warped disk of H_2O masers (viewed almost edge on) in Keplerian rotation around an unresolved mass of 40 million solar masses. The inferred density of this mass exceeds a few 10^9 solar masses pc^{-3} and thus cannot be a long-lived cluster of 'dark' astrophysical objects of the type mentioned above (Maoz 1995). As we will discuss below, the Galactic Center provides a yet more compelling case.

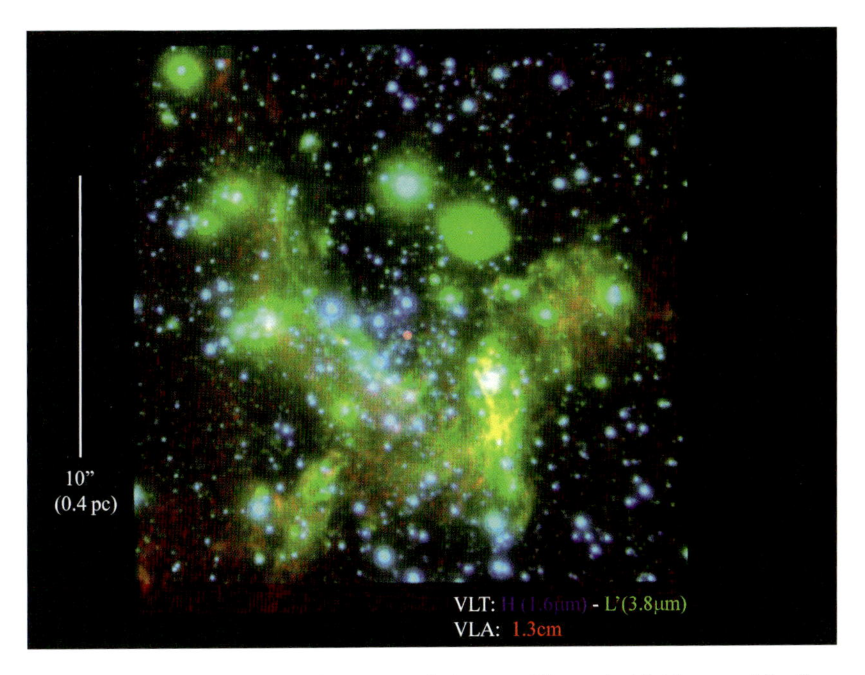

Figure 1. Near-infrared/radio, color-composite image of the central light years of the Galactic Center. The blue and green colors represent the 1.6 and 3.8μm broad-band near-infrared emission, at the diffraction limit (~0.05") of the 8m Very Large Telescope (VLT) of the European Southern Observatory (ESO), and taken with the 'NACO' AO camera and an infrared wave-front sensor (adapted from Genzel et al. 2003a). Similar work has been carried out at the 10 m Keck telescope (Ghez et al. 2003, 2005). The red color image is the 1.3cm radio continuum emission taken with the Very Large Array (VLA) of the US National Radio Astronomy Observatory (NRAO). The red dot in the center of the image is the compact, non-thermal radio source Sgr A*. Many of the bright blue stars are young, massive O/B- and Wolf-Rayet (WR) stars that have formed recently. Other bright stars are giants and asymptotic giant branch stars in the old nuclear star cluster. The extended streamers/wisps of 3.8μm emission and radio emission are dusty filaments of ionized gas orbiting in the central light years (adapted from Genzel, Eisenhauer & Gillessen 2010).

3. Scherzo: Sgr A* and Gas Motions

The central light years of our Galaxy contain a dense and luminous star cluster, as well as several components of neutral, ionized and extremely hot gas (Figure 1, Genzel & Townes 1987, Genzel, Hollenbach & Townes 1994, Melia & Falcke 2001, Genzel, Eisenhauer & Gillessen 2010, Morris, Meyer & Ghez 2012, Reid 2013). Compared to the distant QSOs, the Galactic Center is 'just around the corner' (R_0=8.25 kilo-parsecs or kpc, 27,000 light years). High resolution observations of the Milky Way nucleus thus offer the unique opportunity of carrying out a stringent test of the MBH-paradigm deep within its gravitational 'sphere of influence' where gravity is dominated by the central mass (R<1-3 pc). Since the center of the Milky Way is highly obscured by interstellar dust particles in the plane of the Galactic disk, observations in the visible part of the electromagnetic spectrum are not possible. The veil of dust, however, becomes transparent at longer wavelengths (the infrared, microwave and radio bands), as well as at shorter wavelengths (hard X-ray and γ-ray bands), where observations of the Galactic Center thus become feasible (Oort 1977).

The stellar density in the nuclear cluster increases inward from a scale of tens of parsecs to within the central 0.04 parsec (Becklin & Neugebauer 1968, Genzel et al. 2003a). At its center is a *very compact radio source, Sgr A** (Fig. 1, Balick & Brown, 1974, Lo et al. 1985, Backer et al. 1993). Millimeter inter-continental Very Long Baseline Interferometry (VLBI) observations have established that its intrinsic radius is a mere 20-50 micro-arcseconds (μas Figure 2), 2-5 R_S for a 4x10^6 $M_\odot$ (solar masses) MBH (Krichbaum et al. 1993, Bower et al. 2004, Shen et al. 2005, Doeleman et al. 2008, Johnson et al. 2015, Lu et al. 2014, 2018, Issaoun et al. 2019). Sgr A* thus is the prime candidate for the location and immediate environment of a possible MBH.

VLBI observations also have set an upper limit of about 0.6 km/s and 1 km/s to the motion of Sgr A* itself, along and perpendicular to the plane of the Milky Way, respectively (Reid & Brunthaler 2004, 2020). When compared to the two orders of magnitude greater velocities of the stars in the immediate vicinity of Sgr A* (see below), this demonstrates that the radio source must indeed be massive, with simulations giving a lower limit to the mass of Sgr A* of ~10^5 $M_\odot$ (Chatterjee, Hernquist & Loeb 2002, but see Tremaine, Kocsis & Loeb 2021).

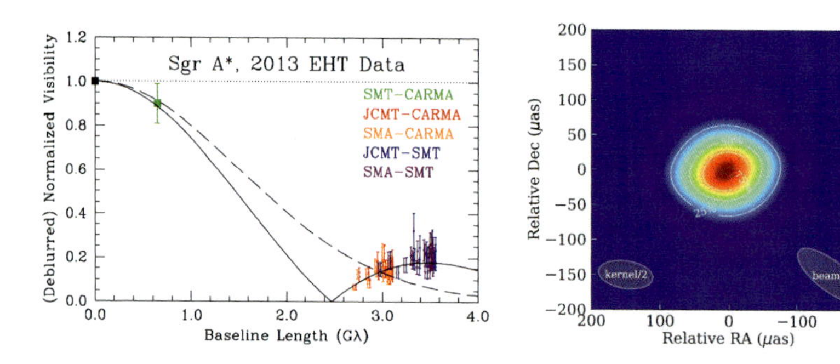

Figure 2. Total-intensity mm-VLBI of Sgr A*. Left: Normalized, de-blurred visibilities at 1.3mm taken with the Event Horizon Telescope are shown as a function of baseline length; errors are ±1σ. The dashed line shows the best fit circular Gaussian (FWHM: 52 μas). An annulus of uniform intensity (inner diameter: 21 μas, outer diameter: 97 μas), shown with a solid line, is perhaps the most plausible model that is consistent with the data (adapted from Figure S5 in Johnson et al. 2015, Supplement). Right: 3mm Global mm-VLBI image of Sgr A*, after removal of the scattering screen. The reconstructed image has an intrinsic Gaussian source diameter of θ_{maj} = 120 ± 34 μas and θ_{min} = 100 ± 18 μas. The ellipses at the bottom indicate half the size of the scatter-broadening kernel (θ_{maj} = 159.9 μas, θ_{min} = 79.5 μas, PA = 81.9°) and of the observing beam (adapted from Figure 5 in Issaoun et al. 2019).

The first *dynamical evidence for the presence of a non-stellar mass concentration of 2-4 million times the mass of the Sun ($M_\odot$)* and plausibly centered on or near Sgr A* came from infrared imaging spectroscopy of interstellar gas clouds, carried out by Charles Townes' group in Berkeley[1] (Wollman et al. 1977, Lacy et al. 1980, Serabyn & Lacy 1985, Crawford et al. 1985). In their 1985 Nature paper Crawford et al. summarized the then available evidence on the mass distribution obtained from the infrared and submillimeter spectroscopy that traced the ionized and neutral gas components. They concluded that "...the measurements fit a point mass of $\sim 4 \times 10^6\ M_\odot$ but are also consistent with a cluster where stellar density decreases with radius *(R)* at least as fast as $R^{-2.7}$, or a combination of a point mass and a stellar cluster....." However, many considered this dynamical evidence not compelling because of the possibility that the ionized gas is affected by non-gravitational forces (shocks, winds, magnetic fields).

4. Escursione: Ever sharper, Ever deeper

The most critical aspect in testing the MBH paradigm obviously lies in the ability of sensitive, very high angular resolution observations. The Schwarzschild radius of a 4 million solar mass black hole at the Galactic Center subtends a mere 10^{-5} arc-seconds, or 10μas[2].

In the *radio and millimeter* bands, such high resolution can be obtained from VLBI. Starting in the 1980s, ever higher resolution VLBI measurements showed that the radio size of Sgr A* decreases with decreasing wavelength, due to scattering by intervening electrons between Sgr A* and the Earth (Shen et al. 2005, Bower et al. 2006). Measuring the intrinsic size of the source and imaging its two dimensional distribution requires short millimeter VLBI observations, which are technically very challenging (Event Horizon Telescope Project in the USA: Doeleman 2010, Black Hole Cam Project in Europe: Goddi et al. 2017)[3].

For high resolution *infrared* imaging from the ground an important technical hurdle is the correction of the distortions of an incoming electromagnetic wave by the turbulent, refractive Earth atmosphere. In the optical/near-infrared wave-band the atmosphere distorts the incoming electromagnetic waves on time scales of milli-seconds and smears out long-exposure images to a diameter of more than an order of magnitude greater than the diffraction limited resolution of large ground-based tele-

---

1. I had joined Townes' group in 1980 as a Miller Postdoctoral Fellow, and then became Associate Professor in the Physics Department in 1981.
2. 10 μas correspond to about 2 cm at the distance of the Moon
3. https://eventhorizontelescope.org/ , https://blackholecam.org/

scopes. The enormous progress in testing the MBH paradigm in the Galactic Center carried out by our group at MPE (at the telescopes of the European Southern Observatory in Chile), and by Andrea Ghez and her collaborators (at the Keck telescopes in Hawaii), described in the following sections, largely rests on substantial, continuous improvements in the angular resolution, astrometric precision and sensitivity of near-IR imaging and spectroscopy (by factors between one hundred to one hundred thousand over three decades).

From the early 1990s onward, short exposure imaging with new infrared imaging detectors was made possible with 'speckle imaging', resulting in diffraction-limited resolution (0.05–0.1") near-infrared stellar images (Sibille, Chelli & Léna 1979, Christou 1991, Hofmann et al. 1993, Matthews & Soifer 1994). Because of the short exposures and detector noise, speckle imaging is not able to go very deep. In the early 1990s 'adaptive optics' techniques (AO: correcting the wave distortions on-line) became available (Rousset et al. 1990, Lena 1991, Tyson & Wizinowich 1992), with upgraded imaging cameras (Lenzen et al. 2003, Lenzen & Hofmann 1995), which have since allowed increasingly precise high resolution near-infrared observations with the currently largest (10 m diameter) ground-based telescopes. If bright natural guide stars near the science target are not available, laser guide star beacons can be employed for AO corrections (Max et al. 1997, Rabien et al. 1998, Bonaccini-Calia et al. 2006, Ghez et al. 2005). Increasingly powerful integral field spectrometers (IFUs) coupled with AO have opened up deep imaging spectroscopy near the diffraction limit (Weitzel et al. 1994, Eisenhauer et al. 2003b, Larkin et al. 2006). The most recent step forward in the capability of the impressive record of instrumental innovation brought to bear on Galactic Center MBH studies is spatial interferometry, which I discuss separately below (Glindemann et al. 2003, Eisenhauer et al. 2008, 2011, GRAVITY collaboration et al. 2017).

5. Menuetto: Stellar Motions and Orbits

A more reliable probe of the gravitational field is stellar motions, which started to become available from Doppler spectroscopy of stellar absorption and emission lines in the late 1980s. They broadly confirmed the results obtained in the first phase from gas motions (Rieke & Rieke 1988, McGinn et al. 1989, Sellgren et al. 1990, Krabbe et al. 1991, 1995, Haller et al. 1996, Genzel et al. 1996). As described in the last section, the ultimate breakthrough came from the combination of AO techniques with IFU imaging spectroscopy (Eisenhauer et al. 2003b), opening deep near-infrared spectroscopy of thousands of O/B and WR stars and GKM giants (e.g. Trippe et al. 2008, Do et al. 2013, 2018, Feldmeier et al. 2014, Fritz et al. 2016, Habibi et al. 2019).

With diffraction-limited 'speckle' imagery starting in 1991/1992 on the 3.5m New Technology Telescope (NTT) of the European Southern Observatory (ESO) in La Silla, Chile our group at MPE was able to determine proper motions of stars as close as ~0.1" from Sgr A* (Eckart & Genzel 1996, 1997, Genzel et al. 1997). In 1995 Andrea Ghez's group at the University of California, Los Angeles started a similar program with the 10m diameter Keck telescope in Hawaii (Ghez et al. 1998). Both groups independently found that the stellar velocities follow a 'Kepler' law ($v \sim R^{-1/2}$) as a function of distance from Sgr A* and reach $\geq 10^3$ km/s within the central light month. Assuming that the mass in the center is the sum of a point mass and an isothermal star cluster, the central mass inferred from projected mass estimators (Bahcall & Tremaine 1981) is ~2.5 million solar masses, for an isotropic velocity distribution (Figure 3), in excellent agreement between the two groups. For more elliptical orbits the inferred mass increases (Bahcall & Tremaine 1981). We now know that the velocity distribution of the innermost stars favors highly elliptical orbits (Schödel et al. 2003, Gillessen et al. 2017), so that the appropriately corrected estimate of M(0) would be 3.5-4.7 $\times 10^6$ $M_\odot$, for R(GC)=8.25 kpc.

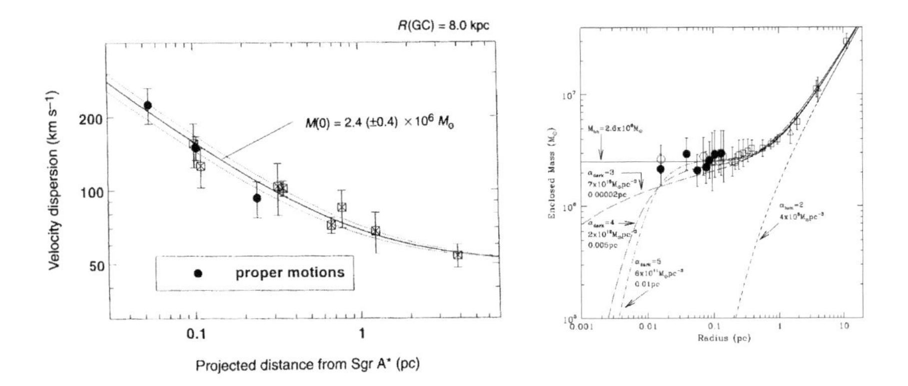

Figure 3. Mass distribution in the central parsec of the Galactic Center after the second phase (1996/1998). The left graph shows the projected 1D stellar velocity dispersion as a function of projected distance from Sgr A*, obtained from proper motions (filled circles) and Doppler velocities (crossed squares) (adapted from Figure 2 of Eckart & Genzel 1996). Each point is derived from averaging the motions of 9-20 stars. The solid curve is a model assuming that the stars move with an isotropic velocity distribution in the potential of a point mass (M(0)) plus an isothermal star cluster of velocity dispersion 50 km/s. The distance of the Galactic Center is assumed to be 8.0 kpc (from Eckart & Genzel 1996). The right graph shows the mass distribution derived from stellar proper motions published by the Keck group in 1998 (Ghez et al. 1998, filled black circles), and compared to the Eckart & Genzel (1996, 1997) proper motions (open circles), the Genzel et al. (1996) stellar radial velocities (squares), and the Guesten et al. (1987) measurement of the rotating gas disk (triangles). From 0.1 to 0.015 pc the enclosed mass appears to be constant with a value of 2.6×10^6 $M_\odot$. For comparison there are several power law distributions (adapted from Figure 7 of Ghez et al. 1998). *The agreement between the results of the MPE and UCLA groups is excellent.*

In the next phase, the MPE group moved in 2002 onto ESO's 8.2m Very Large Telescope (VLT) at the Paranal Observatory in Chile, and both groups improved their imagery with adaptive optics and upgraded cameras, improving the astrometry to a few hundred μas in the next decade (Schödel et al. 2002, 2003, 2005, Ghez et al. 2003, 2008, Gillessen et al. 2009a,b, 2017, Meyer et al. 2012, Boehle et al. 2016, Jia et al. 2019). Ghez et al. (2000) detected accelerations for three of the innermost 'S'-stars (subsequently confirmed by Eckart et al. 2002), opening the prospect of much more precise mass determinations from individual orbits, instead of the statistical evaluation through mass estimators.

In 2001/2002 the star S2 (S02) approached Sgr A* to 15 mas and made a sharp turn around the radio source during 2002 (Schödel et al. 2002, Ghez et al. 2003). S2/S02 is on a highly elliptical orbit (e=0.88), with a peri-distance of 14 mas (17 light hours or 1400 R_S, for M(0) = 4.26x10^6 $M_\odot$ Figure 4) and an orbital period of a mere 16 years. Ghez et al. (2003, 2005) and Eisenhauer et al. (2003a, 2005) also obtained Doppler velocities and accelerations of S2/S02 and several other orbiting stars, allowing precision measurement of the three-dimensional structure of the orbits, as well as the distance to the Galactic Center. Figure 4 shows the data and best fitting GR orbit for S2/S02 in its most recent version (from GRAVITY collaboration et al. 2020a, see below). At the time of writing, the two groups have determined individual orbits for more than 40 stars in the central light month. These orbits show that *the gravitational potential indeed is dominated by a point mass, whose position is identical within a mas uncertainty with that of the radio source Sgr A** (Plewa et al. 2015, Sakai et al. 2019).

At the end of the third phase (~2017), it is clear that > 98% of the 4 million solar mass central mass concentration identified in the previous phases is indeed confined to a region < 17 light hours around the compact radio source (in a volume a million times smaller than inferred in 1985). The intrinsic size in turn is only a few times the event horizon of that mass. This evidence *eliminates all astrophysically plausible alternatives to a massive black hole.* These include astrophysical clusters of neutron stars, stellar black holes, brown dwarfs and stellar remnants (e.g., Maoz 1995, 1998; Genzel et al. 1997, 2000; Ghez et al. 1998, 2005), and even fermion balls (Viollier, Trautmann & Tupper 1993, Munyaneza, Tsiklauri & Viollier 1998, Ghez et al. 2005; Genzel, Eisenhauer & Gillessen 2010). Clusters of a very large number of mini-black holes and boson balls (Torres, Capozziello & Lambiase 2000; Schunck & Mielke 2003; Liebling & Palenzuela 2012) are harder to exclude. The former have a large relaxation and collapse time, the latter have no hard surfaces that could exclude them from luminosity arguments (Broderick, Loeb & Narayan 2009), and they are consistent with the dynamical mass

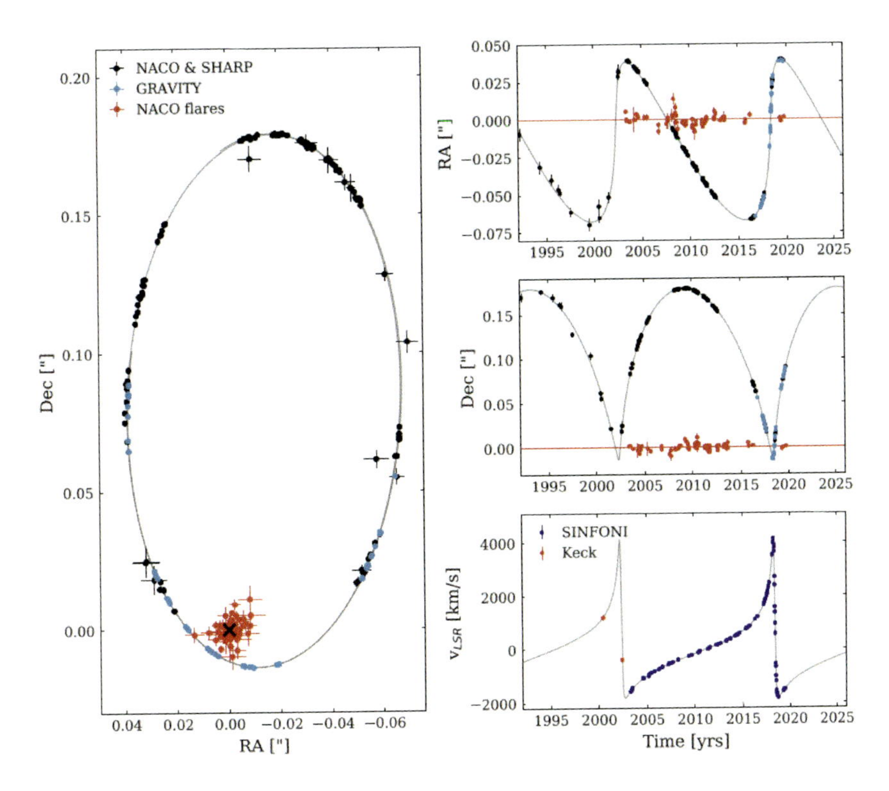

Figure 4. Summary of the MPE-ESO observational results of monitoring the S2 - Sgr A* orbit from 1992 to the end of 2019. Left: SHARP (black points with large error bars), NACO (black points), and GRAVITY (blue points) astrometric positions of the star S2, along with the best-fitting GR orbit (grey line). The orbit does not close as a result of the Schwarzschild precession (see text). The mass center is at (0,0), marked by the black cross. All NACO and SHARP points were corrected for a zero-point onset and drift of the reference frame in RA and Dec. The red data points mark the positions of the infrared emission from Sgr A* during bright states, where the separation of S2 and Sgr A* can be directly inferred from differential imaging. Right: RA (top) and Dec (middle) onset of S2 and of the infrared emission from Sgr A* relative to the position of Sgr A* (assumed to be identical with the mass center) (same symbols as in the left panel). Grey is the best-fitting GR-orbit including the Rømer effect (finite speed of light), special relativity, and GR to 'parameterized post-Newtonian' approximation PPN1 (Will 2008). Bottom right: same for the line-of-sight velocity of the star. Position on the sky as a function of time (left) and Doppler velocity (relative to the Local Standard of Rest) as a function of time (right) of the star S2 orbiting the compact radio source Sgr A*. Blue filled circles denote data taken with the SINFONI red open circles denote data taken with the Keck telescope as part of the UCLA monitoring project (Do et al. 2019, adapted from Figure 1 of GRAVITY collaboration 2020a).

and size constraints. However, such a boson 'star' would be unstable to collapse to a MBH when continuously accreting baryons (as in the Galactic Center), and it is very unclear how it could have formed. *Under the assumption of the validity of General Relativity the Galactic Center thus provides the best quantitative evidence that MBHs do indeed exist.*

6. Rondo Allegretto: Testing General Relativity with Sgr A*

At peri-passage S2 moves at v ~ 7650 km/s and β = v/c ~ 0.026 so that the first order Post-Newtonian effects of GR (PPN1: ~ β^2 ~ 6.5 x 10^{-4}, Will 2008), namely the gravitational redshift and the Schwarzschild in-plane orbital precession, can be realistically detected in the spectra and the astrometry of the star near peri-center. Knowing that S2 would return in 2018 for its next peri-passage, we proposed in 2005 to ESO to build a novel near-infrared beam combiner instrument (GRAVITY) combining the light of all four 8m telescopes of the VLT (Eisenhauer et al. 2008, Paumard et al. 2008). GRAVITY would improve the angular resolution and astrometry by more than an order of magnitude and thus reach the required precision to detect the GR effects (Eisenhauer et al. 2011). GRAVITY was designed and built in the next decade by a French-German-Portuguese Consortium of six institutes (plus ESO), under the PI-ship of Frank Eisenhauer at MPE[4], and installed on Cerro Paranal in July 2015. A detailed discussion of this complex and challenging instrument is given in GRAVITY collaboration et al. (2017), and several other publications.

Our goal and hope were that the combination of SINFONI, NACO and GRAVITY data would allow us to turn the problem around and use Sgr A* as a '*laboratory*' to test General Relativity and the MBH paradigm in a hitherto unexplored regime (e.g. Johannsen 2016). As already mentioned, the peri-passage of S2 in May 2018 is a unique opportunity to test GR to PPN1 (e.g. Zucker et al. 2006). Waisberg et al. (2018) showed that a star with a peri-passage 3-5 times smaller than that of S2 may be used to measure the MBH spin through the Lense-Thirring precession of its orbit. Finally, Sgr A* itself exhibits continuous variability (Baganoff et al. 2001, Genzel et al. 2003b, Dodds-Eden et al. 2011, Witzel et al. 2018), and in some cases the fluxes of these 'flares' approach the flux of S2 (K~14), such that 20μas-astrometry on time scales of a few minutes becomes feasible. Several authors had previously speculated that such flares might come from strongly magnetized 'hot spots' of accelerated electrons whose orbital motions might be detectable and used for exploring the innermost accretion zone on the scale of the innermost stable circular orbit, ISCO (R_{ISCO}< 6 R_S, Broderick & Loeb 2006, Genzel, Eisenhauer & Gillessen 2010, GRAVITY collaboration et al. 2020c,2021).

4. https://www.mpe.mpg.de/938240/Overview, https://www.eso.org/public/teles-instr/paranal-observatory/vlt/vlt-instr/gravity/, https://www.eso.org/sci/facilities/paranal/instruments/gravity.html

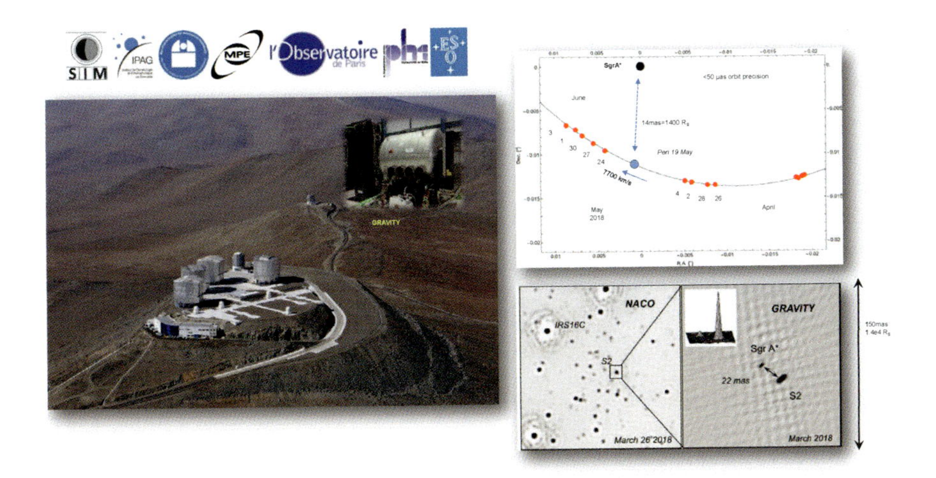

Figure 5. Left: The ESO-Very Large Telescope (VLT) on Cerro Paranal (Chile), where most of the observations by our group were obtained. The Observatory in the Atacama desert is at 2635 m altitude and -24.7° latitude. It hosts four 8.2m telescopes (large silvered structures), as well as four 1.8m Auxiliary Telescopes (white round domes). Both arrays can be combined optically as a spatial interferometer (VLTI) through mirror trains, where the relative geometric path lengths to a given celestial source can be compensated by movable delay line mirrors in the linear white structure underneath the platform (Glindemann et al. 2003). The final combined set of four beams finally arrives at the beam combiner facility structure underneath the rectangular building in the center of the array. Here the light beams are brought together in the cryogenic beam combiner instrument GRAVITY (built by a French-German-Portuguese consortium of six Institutions plus ESO itself (logos above the VLT image). In GRAVITY we calibrate and optimize the data and extract the visibilities and relative phases of the science object, as well as that of a nearby, fringe tracking reference object, as a function of wavelength, guiding and manipulating the infrared light in single-mode fibers and combining the six two-telescope combinations in a micro-chip (GRAVITY collaboration et al. 2017). Bottom right: After calibration of the phases using laser metrology, images with 2×4 mas FWHM resolution are reconstructed by Fourier transformation. In the case shown the VLTI science fibers were placed on the star S2/Sgr A* in the left image, while the interferometer phases were tracked on the bright star IRS16C 1″ NE of Sgr A*, in the top left of the AO image. All four telescopes are equipped with infrared adaptive optics, which uses the K=7 bright star IRS7 5″ north of S2/Sgr A* as a natural guide star to flatten the wavefronts. The image in the bottom right was taken in March 2018, about 2 months before the peri-passage of S2, and both S2 and SgrA* can be clearly detected and its ~22 mas separation measured to ~40 - 100 μas precision. Top right: During the peri-passage in 2018, the motion of S2 can be easily detected night for night, then moving at ~7700 km/s at ~ 1400 Schwarzschild radii from Sgr A* (adapted from Figure 2 of Gravity collaboration et al. 2018a).

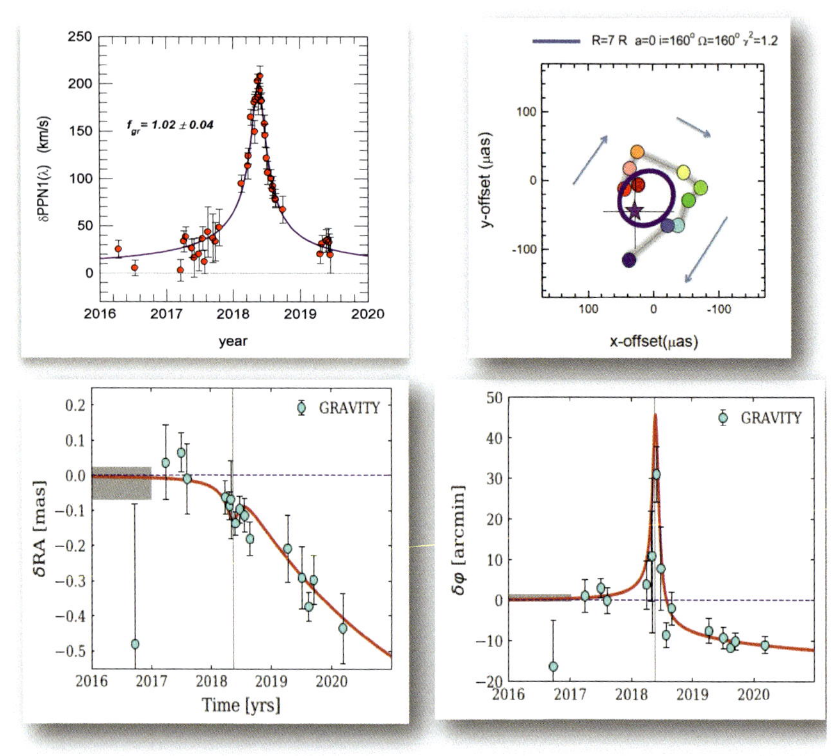

Figure 6. Testing GR and the MBH paradigm with relativistic effects near Sgr A*. Top left: Residuals between the SINFONI HeI/HI Brγ line centroid velocities in the local standard of rest (filled red circles with 1σ uncertainties) and the best fitting Newton/Kepler orbit of all spectroscopic and astrometric data over the past three decades (grey horizontal line at 0). The blue line is the best fitting relativistic orbit including all PPN1 terms (as well as Rømer effect), and fitting a free parameter f_{gr} to the PPN1 wavelength term including gravitational redshift and transverse Doppler effect. GR has $f_{gr} = \pm 1$ and our best fit yields $f_{gr} = 1.02 \pm 0.04$ (GRAVITY collaboration et al. 2018a, 2019a, 2020a). Bottom plots: Residuals in RA (left) and angle on the sky φ (right) between the GRAVITY (filled cyan circles and 1σ uncertainties) and average NACO astrometry before 2017 (grey bar) and the best fitting relativistic orbit without precession ($f_{SP} = 0$, blue dotted horizontal line at 0). The best fitting relativistic orbit including precession has $f_{SP} = 1.1 \pm 0.19$ (adapted from Figures 3 and B2 in GRAVITY collaboration et al 2020a). Top right: Residual motion of the 2μm light centroid of Sgr A* (originating from polarized synchrotron emission from γ>1000 accelerated electrons in the inner accretion zone in a bright 'flare' on July 22nd, 2018, cf. Genzel, Eisenhauer & Gillessen 2010) as a function of time over about 30 minutes, and relative to the location of the mass as estimated from the S2 orbit (dark grey asterisk and 1σ errors). The blue curve denotes a circular particle orbit at 3.5 R_S around a non-spinning MBH of 4.3 million solar masses, inclined at 160° (Figure 1 in GRAVITY collaboration et al. 2018b, 2020b, c, 2021).

It is remarkable to look back in late 2020, two and half years after the peri of S2 on May 19, 2018 and realize that most of these hopes actually turned into reality (Figure 6). The gravitational redshift of S2 has been well determined (5–50σ) by both groups (GRAVITY collaboration et al. 2018a, 2019a, 2020a, Do et al. 2019). The Schwarzschild precession has been detected at ~5σ (GRAVITY collaboration et al. 2020a). Flare motions in three flares of 2018 were consistent with the orbital motions near ISCO around a 4 million solar mass MBH (GRAVITY collaboration et al. 2018b, 2020b). Using the HeI and HI lines as independent 'clocks' GRAVITY collaboration et al. (2019b) have confirmed the local positional invariance of Einstein's equivalence principle to about 5%. Significant upper limits can be placed on the presence of a hypothetical 'fifth force' (Hees et al. 2017, GRAVITY collaboration et al. 2020a). Faint stars close to Sgr A* have also been recently detected (GRAVITY collaboration et al. 2021) but are likely not inside the S2 orbit. Overall these discoveries have strengthened the MBH paradigm and GR yet significantly further (Figure 7).

7. Coda

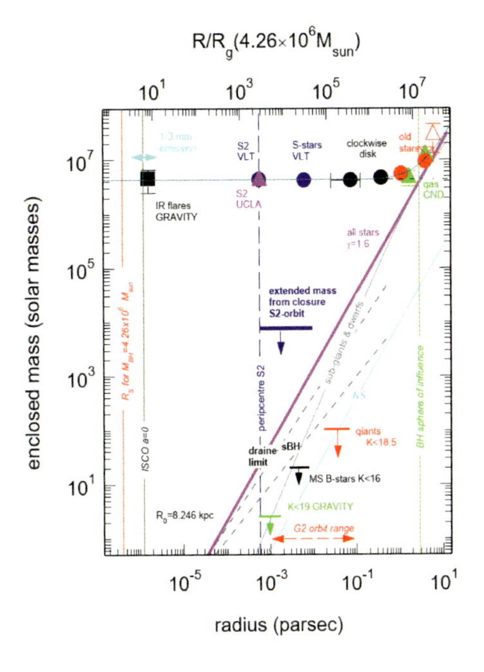

Figure 7. Status of the Galactic Center mass distribution after the fourth phase. Constraints on the enclosed mass in the central 10 pc of the Galaxy. The blue, black and red circles, the pink, green and red triangles are estimates of the enclosed mass at different radii obtained from stellar and gas motions (see Genzel et al. 2010, GRAVITY collaboration 2021a for details). The filled black rectangle comes from the clockwise loop-motions of synchrotron near-infrared flares (Gravity Collaboration et al. 2018b). The cyan double arrow denotes current VLBI estimates of the 3 mm size of Sgr A* (Issaoun et al. 2019). The continuous magenta line shows the total mass model from all stars and stellar remnants (Alexander 2017). The grey line mark the distribution of K < 18.5 sub-giants and dwarfs from Schödel et al. (2018). The grey dashed lines indicate the distribution of stellar black holes and neutron stars from theoretical simulations of Alexander (2017) and Baumgardt et al. (2018), which span a range of roughly a factor 5. Red, black and green upper limits denote upper limits on giants, main-sequence B stars and K < 19 GRAVITY sources. The Schwarzschild radius of a $4.26 \times 10^6 M_\odot$ black hole and the inner-most stable circular orbit radius for a non-spinning black hole are given by orange and dark green vertical lines. The peri-center radius of S2 is the dashed vertical blue line and the sphere of influence of the black hole is given by the vertical light green line. The blue horizontal line denotes the 2σ upper limit of any extended mass around Sgr A* obtained from the lack of retrograde precession in the S2 orbit (adapted from Figure D1 in Gravity collaboration 2020a).

Besides its role at the center stage of testing the black hole paradigm, the Galactic Center has also provided many important *discoveries and surprises on the astrophysics side,* which I have not described in this paper so far. One is the fact that the central parsec contains ~200 massive, early type stars (O/B and Wolf-Rayet stars), which must have formed in the last few million years (c.f. Sanders 1998, Genzel et al. 1996, 2000, 2010, Paumard et al. 2006, Lu et al. 2009, Bartko et al. 2009). This 'paradox of youth' (Ghez et al. 2003) is completely unexpected, as the MBH should disrupt moderately dense gas clouds tidally, and prevent star formation through local gravitational instabilities and cloud collapse. Perhaps the most likely solution of this riddle is that a large gas cloud fell in a few million years ago, was initially tidally disrupted and shocked, but then cooled and became denser over time, so that gravitational collapse did become possible (cf. Morris & Serabyn 1996, Bonnell & Rice 2008, Hobbs & Nayakshin 2009, Genzel et al. 2010, Alexander 2017).

Possibly connected is the question how the 'S-stars' were captured so close to the MBH, on solar system scales. These B, A, G and K stars could have never migrated in their lifetime to their current position through normal two-body relaxation processes, which take several Gyrs. Instead, rapid stochastic injection of binaries into 'loss-cone' radial orbits from large distances (Hills 1988), and perhaps assisted by massive perturbers (Perets et al. 2007), could have led to a capture of one member of the binary near peri-center, and rapid ejection of the second as a hyper-velocity star (cf. Alexander 2005, 2017, Genzel et al. 2010).

A tidal disruption of a star by the MBH is expected to occur only once every 30,000 years (Alexander 2005, 2017). In 2012 Gillessen et al. (2012, 2019, and references therein) reported the near-radial infall, tidal disruption and eventual slowing down by drag forces near ~2000 R_S of an ionized gas cloud ('G2'). The discussion is ongoing whether this gas cloud is isolated, or whether the gas is the envelope of a central single or binary star.

A third riddle is the lack of a strong cusp of old late type stars around the MBH (Do et al. 2009, Buchholz et al. 2009, Schödel et al. 2018, Figure 7), which is expected in equilibrium ($\rho \sim R^{-1.5...-1.75}$, Bahcall & Wolf 1977, Alexander 2005, 2017). Finally, the lack of any substantial mass close to Sgr A*-MBH greater than a few hundred to one thousand solar masses (Figure 7, GRAVITY collaboration et al. 2020a) is highly exciting and important for other MBH systems and needs to be confirmed by further measurements.

Another aspect I did not cover is the important role MBHs apparently had in the cosmological *co-evolution with their galactic hosts* (e.g. Fabian 2012, Kormendy & Ho 2013, Madau & Dickinson 2014).

I have tried to describe in this paper the stepwise progress in proving that massive black holes do exist in the Universe. As compared to the first phase forty years ago, these measurements have pushed the 'size' of the 4 million solar mass concentration downward by almost 10^6, and its density up by 10^{18}! Looking ahead toward the future the question is probably no longer whether Sgr A* must be an MBH, but rather whether GR is correct on the scales of the event horizon, whether space-time is described by the Kerr metric and whether the 'no hair theorem' holds. Further improvements of GRAVITY (to GRAVITY$^+$) and the next generation 25–40m telescopes (the ESO-ELT, the TMT and the GMT) promise further progress. A test of the no hair theorem in the Galactic Center might come from combining the stellar dynamics with EHT measurements of the photon ring of Sgr A* (Falcke, Melia & Algol 2000, Psaltis & Johannsen 2011, Psaltis, Wex & Kramer 2016, Johannsen 2016). The gravitational waves emanating from the extreme mass ratio in-spiral of a stellar black hole into a massive black hole with the LISA space mission[5] might provide the ultimate culmination of this exciting journey, which Albert Einstein started more than a century ago.

Acknowledgements

I would like to thank Odele Straub, Tim de Zeeuw, Frank Eisenhauer, Stefan Gillessen, Hannelore Hämmerle, Luis Ho, Pierre Léna, Alvio Renzini, Luciano Rezzolla, Linda Tacconi, Scott Tremaine & Hannah Übler for substantial help with, and comments on this manuscript. I have tried to describe in this paper the journey my colleagues and I took for the past 40 years. The road was long, took patience and hard work, but was enormously rewarding. I am deeply grateful to the many outstanding colleagues who have been willing to work with me on this project, and to the Max-Planck Gesellschaft and the European Southern Observatory for supporting us.

5. https://www.elisascience.org/

REFERENCES

Abbott, B.P. et al. 2016 *Ph.Rev.L.* **116**, 1102.
Abbott, B.P. et al. 2016 *Ph.Rev.L.* **116**, 1101.
Alexander, T. 2005, *Ph.Rev.* **419**, 65.
Alexander, T. 2017, *ARA&A* **55**, 17.
Almheiri, A. Marolf, D., Polchinski, J. & Sully, J. 2013. *JHEP* 02. 62.
Argon, A.L., Greenhill, L.J., Reid, M.J., Moran, J.M. & Humphreys, E.M.L. 2007, *ApJ*, **659**, 1040.
Backer, D. C., Zensus, J. A., Kellermann, K. I., Reid, M., Moran, J. M. & Lo, K. Y. 1993, *Science* **262**, 1414.
Baganoff, F. et al.2001, *Nature*, **413**, 45.
Bahcall, J. N. & Wolf, R.A. 1977, *ApJ* **216**, 883.
Balick, B. & Brown, R.L. 1974, *ApJ* **194**, 265.
Bardeen J. M., 1973, in *Black holes* (Les astres occlus), DeWitt B. S., DeWitt C., eds., New York: Gordon and Breach, p. 215.
Bartko, H. et al. 2009, *ApJ* **697**, 1741.
Baumgardt, H., Amaro-Seoane, P. & Schödel ,R. 2018 *A&A* **609**, 28.
Becklin, E.E. & Neugebauer, G. 1968, *ApJ* **151**, 145.
Blandford, R.D. 1999, *ASPC*, **160**, 265.
Blandford, R.D., Meier, D. & Readhead, A. 2019, *ARA&A* **57**, 467.
Bonaccini-Calia, D., et al. 2006, *SPIE* **6272**, 207.
Boehle, A., Ghez, A.M., Schödel, R. et al. 2016, *ApJ* **830**, 17.
Bonnell, I.A. & Rice, W.K.M. 2008, *Science* **321**, 1060.
Bousso, R. 2002, *Rev.Mod.Phys.* **74**, 825.
Bower, G.C. et al. 2004, *Science* **304**, 704.
Bower, G. C., Goss, W. M., Falcke, H., Backer, D. C. & Lithwick, Y. 2006, *ApJ* **648**, L127.
Broderick, A. & Loeb, A. 2006, *MNRAS* **367**, 905.
Broderick, A., Loeb, A. & Narayan, R. 2009, *ApJ*, **701**, 1357.
Buchholz, R.M. et al. 2009, *A&A* **499**, 483.
Chatterjee, P., Hernquist, L. & Loeb, A. 2002, *ApJ*, 572, 371.
Chatzopoulos, S., Fritz, T., Gerhard, O., Gillessen, S., Wegg, C., Genzel, R. & Pfuhl, O. 2015, *MNRAS* **447**, 948.
Christou, J.C. 1991, *PASP* **103**, 1040.
Crawford, M.K., Genzel,R. Harris, A.I., Jaffe, D. T., Lacy, J. H., Lugten, J. B., Serabyn, E. & Townes, C. H. 1985, *Nature* **315**, 467.
Do, T., Ghez, A. M., Morris, M. R., Yelda, S., Meyer, L., Lu, J. R., Hornstein, S. D.& Matthews, K. 2009, *ApJ*, **691**, 1021.
Do, T. et al. 2013, *ApJ* **779**, L6.
Do, T. et al. 2018, *ApJ* **855**, L5.
Do, T. et al. 2019, *Science* **365**, 664.
Dodds-Eden, K. et al. 2011, *ApJ*, **728**, 37.
Doeleman, S.S. et al. 2008, *Nature*, **455**, 78.
Doeleman, S.S. 2010, in *Proceedings of the 10th European VLBI Network Symposium and EVN Users Meeting: VLBI and the new generation of radio arrays.* September 20–24, 2010. Manchester, UK. Published online at http://pos.sissa.it/cgi-bin/reader/conf.cgi?confid=125, id.53
Eckart, A. & Genzel, R. 1996, *Nature* **383**, 415.
Eckart, A. & Genzel, R. 1997, *MNRAS*, *284*, 576.
Eckart, A., Genzel, R., Ott, T. & Schödel, R. 2002, *MNRAS* **331**, 917.
Eckart, A. et al., 2006, *A&A* **450**, 535.

Einstein, A. 1916, *Ann.Phys.* **49**, 50.
Eisenhauer, F. et al. 2003a, *ApJ* **597**, L121.
Eisenhauer, F., Tecza, M., Thatte, N. et al. 2003b, *ESO Messenger* **113**, 17.
Eisenhauer, F. et al. 2005, *ApJ* **628**, 246.
Eisenhauer et al. 2008, in "The Power of Optical/IR Interferometry: Recent Scientific Results and 2nd Generation Instrumentation", ESO Astrophysics Symposia. ISBN 978-3-540-74253-1. Springer, 2008, p. 431.
Eisenhauer, F. et al. 2011, *ESO Msngr.***143**, 16.
Fabian, A.C., Iwasawa, K., Reynolds, C.S. & Young, A.J. *PASP* **112**, 1145.
Fabian, A.C. 2012, *ARA&A* **50**,455.
Falcke, H., Melia, F. & Algol, E. 2000, *ApJ* **528**, L13.
Feldmeier, A. et al. 2014, *A&A* **570**, 2.
Fritz, T.K. et al. 2016, *ApJ* **821**, 44.
Genzel, R. & Townes, C.H. 1987, *ARA&A* **25**, 377.
Genzel, R., Hollenbach, D., & Townes, C. H., 1994, *Rep. Prog. Phys.*, **57**, 417.
Genzel, R., Thatte, N., Krabbe, A., Kroker, H. & Tacconi-Garman, L.E. 1996, *ApJ*, **472**, 153.
Genzel, R., Eckart, A., Ott, T. & Eisenhauer, F. 1997, *MNRAS*, **291**, 219.
Genzel, R., Pichon, C., Eckart, A., Gerhard, O.E. & Ott, T. 2000, *MNRAS*, **317**, 348.
Genzel, R. et al. 2003a, *ApJ* **594**, 633.
Genzel, R. et al. 2003b, *Nature*, **425**, 934.
Genzel, R., Eisenhauer, F. & Gillessen, S. 2010, *Rev.Mod.Phys.* **82**, 3121.
Ghez, A.M., Klein, B.L., Morris, M. & Becklin, E.E. 1998, *ApJ* **509**, 678.
Ghez, A.M. et al. 2000, *Nature*, **407**, 349.
Ghez, A.M. et al. 2003, *ApJ* **586**, L127.
Ghez, A.M. et al. 2005, *ApJ*, **620**, 744.
Ghez, A.M. et al. 2008, *ApJ*, **689**, 1044.
Giacconi, R., Gursky, H., Paolini, F. & Rossi, B.B. 1962, *Phys.Rev.Lett.* **9**, 439.
Giacconi, R. 2003, *Rev.Mod.Phys.* **75**, 995.
Gillessen, S. et al. 2009a, *ApJ*, **692**, 1075.
Gillessen, S. et al. 2009b, *ApJ* **707**, L114.
Gillessen, S. et al. 2012, *Nature* **481**, 51.
Gillessen, S. et al. 2017, *ApJ* **837**, 30.
Glindeman, A. et al. 2003, *Astroph.Sp.Sci.* **286**, 35.
Goddi, C. Falcke, H., Kramer, M. et al. 2017, *IJMPD*, 2630001.
GRAVITY Collaboration et al. 2017, *A&A* **602**, 94.
GRAVITY Collaboration et al. 2018a, *A&A* **615**, L15.
GRAVITY Collaboration et al. 2018b, *A&A* **618**, L10.
GRAVITY Collaboration et al. 2019a, *A&A* **625**, L10.
GRAVITY Collaboration et al. 2019b, *Phys.Rev.Lett.* **122**, 1102.
GRAVITY Collaboration et al. 2020a, *A&A* **636**, L5.
GRAVITY Collaboration et al. 2020b, *A&A* **635**, 143.
GRAVITY Collaboration et al. 2020c, *A&A* **643**, 56.
GRAVITY Collaboration et al. 2021, *A&A* in press (ArXiv201103058)
Greene, J. et al. 2016, *ApJ* **826**, L32.
Gültekin, K. et al. 2009, *ApJ* **698**, 198.
Habibi, M. et al. 2019, *ApJ* **872**, L15.
Haller, J.W., Rieke, M.J., Rieke, G.H., Tamblyn, P., Close, L. & Melia, F. 1996, *ApJ*, **456**, 194.
Hees, A. et al. 2017, *Ph.RevL.* **118**, 1101.
Hills, J.G. 1988, *Nature* **331**, 687.

Hobbs, A. & Nayakshin, S. 2009, *MNRAS* **392**, 191.
Hofmann, K.H. & Eigelt, H. 1993, *A&A* **278**, 328.
Issaoun, S. et al. 2019, *ApJ* **871**, 30.
Jia, S., Lu, J.R., Sakai, S. et al. 2019, *ApJ* **873**, 9.
Johannsen, T. 2016, *CQGra* **33**, 3001.
Johnson, M.D. et al. 2015, *Science* **350**, 1242.
Kerr, R.1963, *Ph.Rev.Lett.,* **11**, 237.
Kormendy, J. 2004, in 'Coevolution of Black Holes and Galaxies', Carnegie Observatories Centennial Symposia. Cambridge University Press, Ed. L.C. Ho, p. 1.
Kormendy, J. & Ho, L. 2013, *ARAA* **51**, 511.
Krabbe, A., Genzel, R., Drapatz, S. & Rotaciuc, V. 1991, *ApJ* **382**, L81.
Krabbe, A. et al. 1995, *ApJ*, **447**, L95.
Krichbaum, T.P. et al. 1993, *A&A* **274**, 37.
Lacy, J.H., Townes, C.H., Geballe, T.R. & Hollenbach, D.J. 1980, *ApJ* **241**, 132.
Laplace, S.P. 1795, Le Systeme de Monde, Vol.II. Paris
Larkin, J., Barczys, M., Krabbe, A. et al. 2006, *New AR* **50**, 362.
Léna, P. 1991, *Science* **251**, 854.
Lenzen, R. & Hofmann, R. 1995, *SPIE* **2475**, 268.
Lenzen, R. et al. 2003, *SPIE* **4841**, 944.
Liebling, S.L. & Palenzuela, C. 2012, *LRR*, **15**, 6.
Lo, K.Y. Backer, D.C., Ekers, R.D. et al. 1985, *Nature* **315**, 124.
Lynden-Bell, D. 1969, *Nature* **223**, 690.
Lynden-Bell, D. & Rees, M. 1971, *MNRAS* **152**, 461.
Lu, J.R. et al. 2009, *ApJ* **690**, 1463.
Lu, R.-S. et al. 2014, *ApJ* **788**, L120.
Lu, R.-S. et al. 2018, *ApJ* **859**, 60.
Madau, P. & Dickinson, M. 2014, *ARA&A* **52**, 415.
Magorrian, J. et al. 1998, *AJ*, **115,** 2285.
Maldacena, J. 1998, *Ad. Th.Math.Phys.* **2**, 231.
Maoz, E. 1995, *ApJ*, **447**, L91.
Maoz, E., 1998, *ApJ* **494**, L181.
Matthews, K. & Soifer, B.T. 1994, *Exp.Astr.* **3**, 77.
Max, C.E. et al. 1997, *Science* **277**, 1649.
Mayer, L., Kazantzidis, S., Escala, A.& Callegari, S. 2010, *Nature*, **466**, 1082.
McClintock, J. & R. Remillard 2004, in *Compact Stellar X-ray sources*, eds. W. Lewin and M. van der Klis, Cambridge Univ. Press (astro-ph/0306123)
McConnell, N. & Ma, C.-P. 2013, *ApJ* **764**, 184.
McGinn, M.T., Sellgren, K., Becklin, E.E. & Hall, D.N.B. 1989, *ApJ*, **338**, 824
Melia, F. & Falcke, H. 2001, ARA&A 39, 309.
Meyer, L., Ghez, A. M., Schödel, R., Yelda, S., Boehle, A., Lu, J. R., Do, T., Morris, M. R., Becklin, E. E.& Matthews, K. 2012, *Sci*, **338**, 84.
Michell, J. 1784, Phil. *Trans.Royal Soc.London*, **74**, 35.
Miyoshi, M. et al. 1995, *Nature* **373**, 127.
Moran, J.M. 2008, *ASPC*, **395**, 87.
Morris, M.R. & Serabyn, E. 1996, *ARA&A* **34**, 645.
Morris, M.R., Meyer, L. & Ghez, A.M.2012, *RAA* **12**, 995.
Munyaneza, F, Tsiklauri, D. & Viollier, R.D. 1998, *ApJ*, **509**, L105.
Nandra, K., George, I.M., Mushotzky, R.F., Turner, T.J. & Yaqoob, T.1997, *ApJ* **477**, 602.
Oort, J. 1977, *ARA&A* **15**, 295.
Osmer, P.S. 2004, in *Coevolution of Black Holes and Galaxies*, from the Carnegie

Observatories Centennial Symposia. Published by Cambridge University Press, as part of the Carnegie Observatories Astrophysics Series. Edited by L. C. Ho, p. 324.
Özel, F., Psaltis, D., Narayan, R. & McClintock, J. E. 2010, *ApJ* **725**, 1918.
Paumard, T. et al. 2006, *ApJ* **643**, 1011.
Paumard, T. et al. 2008, in "The Power of Optical/IR Interferometry: Recent Scientific Results and 2nd Generation Instrumentation", *ESO Astrophysics Symposia.* ISBN 978-3-540-74253-1. Springer, 2008, p. 313.
Penrose, R. 1963, *Ph.Rev.L.* **10**, 66.
Penrose, R. 1965, *Ph.Rev.L.* **14**, 57.
Perets, H. et al. 2007 *ApJ* 656, 709.
Plewa, P.M. et al. 2015, *MNRAS* **453**, 3234.
Psaltis, D. & Johanssen, T. 2011, *JPhCS*, **283**, 2030.
Psaltis, D., Wex, N. & Kramer, M. 2016, *ApJ* **818**,121.
Rabien, S., Ott, T., Hackenberg, W. et al. 1998, *ApJ* **498**, 278.
Reid, M.J. & Brunthaler, A. 2004, *ApJ* **616**, 872.
Reid, M.J. 2009, *IJMPD* **18**, 889.
Reid, M.J. Braatz, J. A., Condon, J. J., Lo, K. Y., Kuo, C. Y., Impellizzeri, C. M. V.& Henkel, C. 2013, *ApJ*, **767**, 154.
Reid, M.J. et al. 2014, *ApJ*, **783**, 130.
Reid, M.J. & Brunthaler, A. 2020, *ApJ* **892**, 39.
Remillard, R.A. & McClintock, J.E. 2006, *ARA&A 44*, **49**.
Rezzolla, L. et al. 2018, *ApJ* **852**, L25.
Rieke, G.H. & Rieke, M.J. 1988, *ApJ*, **330**, L33.
Rousset, G., Fontanella, J.C., Kern, P., Gigan, P., Rigaut, F., Léna, P., Boyer, C., Jagourel, P., Gaffard, J.P. & Merkle, F. 1990, *A&A* **230**, L29.
Saglia, R. et al. 2016, *ApJ* **818**, 47.
Sakai, S. et al. 2019, *ApJ* **873**, 65.
Sanders, R.H. 1998, *MNRAS*, **294**, 35.
Schödel, R. et al. 2002, *Nature* **419**, 694.
Schödel, R. et al. 2003, *ApJ* **596**, 1015.
Schödel, R. et al. 2018 *A&A* **609**, 27.
Schmidt, M. 1963, *Nature* **197**, 1040.
Schunck, F.E. & Mielke, E.W. 2003, *CQW* **20**, R301.
Schwarzschild, K., 1916, *Sitzungsber. Preuss. Akad.Wiss.*, **424**.
Sellgren, K., McGinn, M.T., Becklin, E.E. & Hall, D.N. 1990, *ApJ* **359**, 112.
Serabyn, E. & Lacy, J.H. 1985, *ApJ* **293**, 445.
Shakura, N.I. & Sunyaev, R.A. 1973, *A&A* **24**, 337.
Shen, Z.-Q., Lo, K.Y., Liang, M.C., Ho, P.T.P. & Zhao, J.H. 2005, *Nature* **438**, 62.
Sibille, F. Chelli, A.& Léna, P. 1979 *A&A*, **79**, 315.
Susskind, L. 1995, *JMP* **36**, 6377.
Tanaka, Y., Nandra, K., Fabian, A.C. et al. 1995, *Nature* **375**, 659.
Torres, D.F., Capoziello, S. & Lambiase, G. 2000, *PhRv D*, **62**, 4012.
Townes, C. H.; Lacy, J. H.; Geballe, T. R. & Hollenbach, D. J. 1982, *Nature* **301**, 661.
Tremaine, S., Kocsis, N. & Loeb A. 2021, ArXiv:2012.13273.
Trippe, S. et al. 2008, *A&A* **492**, 419.
Tsiklauri, D. & Viollier, R. 1998, *ApJ* **500**, 591.
Tyson, R.K. & Wizinowich, P. 1992, *Phys.Tod.* **45**, 100.
Yelda, S. et al. 2014, *ApJ* **783**, 131.
Yuan, F. & Narayan, R. 2014, *ARA&A* **52**, 529.
Viollier, R.D, Trautmann, D. & Tupper 1993, *PhLB*, **306**, 79.

Waisberg, I. et al. 2018, *MNRAS* **476**, 3600.
Weinberg, N. N., Milosavljevic, M. & Ghez, A. M. 2005, *ApJ* **622**, 878.
Weitzel, L., Cameron, M., Drapatz, S., Genzel, R. & Krabbe, A. 1994, *Exp.Astr.* **3**,1.
Wheeler, J.A. 1968, *Amer.Scient.* **56**, 1.
Will, C.M. 2008, *ApJ*, **674**, L25.
Witzel, G. et al. 2018, *ApJ* **863**, 15.
Wollman, E. R.; Geballe, T. R.; Lacy, J. H.; Townes, C. H.; Rank, D. M.1977, *ApJ* **218**, L103.
Zucker, S., Alexander, T., Gillessen, S., Eisenhauer, F. & Genzel, R. 2006, *ApJ* **639**, L21.

Andrea Ghez. © Nobel Prize Outreach AB. Photo: Stefan Bladh

Andrea Ghez did not submit her autobiography and lecture. See
https://www.nobelprize.org/prizes/physics/2020/ghez/facts/
https://www.nobelprize.org/prizes/physics/2020/ghez/lecture/